THE MULTIFUNCTIONAL
GUT OF FISH

This is Volume 30 in the

FISH PHYSIOLOGY series
Edited by Anthony P. Farrell and Colin J. Brauner
Honorary Editors: William S. Hoar and David J. Randall

A complete list of books in this series appears at the end of the volume

THE MULTIFUNCTIONAL GUT OF FISH

Edited by

MARTIN GROSELL
Marine Biology and Fisheries Department
University of Miami-RSMAS
Miami, Florida, USA

ANTHONY P. FARRELL
Faculty of Agricultural Sciences
The University of British Columbia
Vancouver, British Columbia
Canada

COLIN J. BRAUNER
Department of Zoology
The University of British Columbia
Vancouver, British Columbia
Canada

AMSTERDAM • BOSTON • HEIDELBERG • LONDON • OXFORD
NEW YORK • PARIS • SAN DIEGO • SAN FRANCISCO
SINGAPORE • SYDNEY • TOKYO

Academic Press is an imprint of Elsevier

Academic Press is an imprint of Elsevier
32 Jamestown Road, London NW1 7BY, UK
30 Corporate Drive, Suite 400, Burlington, MA 01803, USA
525 B Street, Suite 1800, San Diego, CA 92101-4495, USA

First edition 2011

Notice
No responsibility is assumed by the publisher for any injury and/or damage to persons or
property as a matter of products liability, negligence or otherwise, or from any use or operation
of any methods, products, instructions or ideas contained in the material herein. Because of
rapid advances in the medical sciences, in particular, independent verification of diagnoses and
drug dosages should be made

British Library Cataloguing-in-Publication Data
A catalogue record for this book is available from the British Library

Library of Congress Cataloging-in-Publication Data
A catalog record for this book is available from the Library of Congress

ISBN: 978-0-1237-4982-6
ISSN: 1546-5098

For information on all Academic Press publications
visit our web site at www.elsevierdirect.com

Typeset by MPS Limited, a Macmillan Company, Chennai, India
www.macmillansolutions.com

Printed and bound by CPI Group (UK) Ltd, Croydon, CR0 4YY

Transferred to digital print 2012

Working together to grow
libraries in developing countries

www.elsevier.com | www.bookaid.org | www.sabre.org

ELSEVIER   BOOK AID International   Sabre Foundation

CONTENTS

CONTRIBUTORS

The numbers in parentheses indicate the pages on which the authors' contributions begin.

MICHAEL AXELSSON *(351), Department of Zoology, University of Gothenburg, Gothenburg, Sweden*

ANNE MARIE BAKKE *(57), Aquaculture Protein Centre, a Centre of Excellence (Norway), Norwegian School of Veterinary Science, Oslo, Norway*

CAROL BUCKING *(165), McMaster University, Canada and University of British Columbia, Ontario, Canada*

KENNETH CAIN *(111) University of Idaho, Moscow, ID, USA*

L. FILIPE C. CASTRO *(1) Centre for Marine and Environmental Research (CIMAR), Porto, Portugal*

C. A. COOPER *(213) University of Guelph, Ontario, Canada*

A. MICKEY DEHN *(395), Towson University, Baltimore, MD, USA*

ANTHONY P. FARRELL *(351), University of British Columbia, Vancouver, Canada*

CHRIS GLOVER *(57), University of Canterbury, Christchurch, New Zealand*

MARTIN GROSELL *(135), RSMAS University of Miami, Miami, FL, USA*

ÅSHILD KROGDAHL *(57), Aquaculture Protein Centre, a Centre of Excellence (Norway), Norwegian School of Veterinary Science, Oslo, Norway*

CHRISTOPHER A. LORETZ *(261), University of Buffalo, New York, NY, USA*

T. P. MOMMSEN *(213), University of Victoria, British Columbia, Canada*

JAY A. NELSON *(395), Towson University, Baltimore, MD, USA*

CATHARINA OLSSON *(319), University of Gothenburg, Gothenburg, Sweden*

HENRIK SETH *(351)*, *Department of Zoology, University of Gothenburg, Gothenburg, Sweden*

CHRISTINE SWAN *(111)*, *Utah Division of Wildlife Resources, Fisheries Experiment Station, Logan, UT, USA*

YOSHIO TAKEI *(261)*, *University of Tokyo, Tokyo, Japan*

JOSI R. TAYLOR *(213)*, *Monterey Bay Aquarium Research Institute, CA, USA*

JONATHAN M. WILSON *(1)*, *Centre for Marine and Environmental Research (CIMAR), Porto, Portugal*

CHRIS M. WOOD *(165)*, *McMaster University, Canada and University of Miami, Miami, FL, USA*

PREFACE

It is our pleasure to contribute this volume on the multifunctional gut to the prestigious *Fish Physiology* series which has served as the primary reference for fish physiologists for over 40 years. The gastrointestinal tract serves a diverse range of functions in fish from nutrient absorption to ionic and osmotic regulation and even air breathing. Although these main functions have been considered separately in the past, recent years have seen a shift toward considering interactions between functions in different environments. Furthermore, fish are a highly diverse and the most abundant vertebrate group. As such, they display an incredible diversity of morphology, anatomy and histology of the gastrointestinal tract in association with its numerous specialized functions. The rich diversity of form reflect to some extent these functions, including respiration via specialized gastrointestinal respiratory surfaces and nitrogen excretion. In addition, gut function in general has pronounced implications for whole animal integrative physiology and compensatory demands for non-gastrointestinal organs. Despite a longstanding knowledge concerning the multifunctional role of the gut in fish, this volume represents the first comprehensive discussion of the many vital functions conducted by this central organ in the context of integrative and comparative physiology.

We very much hope you will enjoy the outcome, which we see as a celebration of the myriad of exquisite adaptations and the multifunctional role displayed by the fish gastrointestinal tract.

This volume is the product of firm commitment and hard work by some of the world's leading authors in areas of fish gastrointestinal physiology and we hope it will serve the teaching and research programs of current and future fish physiologists. We are extremely grateful for the immense contribution of our authors and for the guidance provided by the Elsevier staff, Pat Gonzalez, Kristi Gomez and Caroline Jones, who were invaluable in keeping this project on track for a timely completion. Last, but by no means least, this volume would not have seen the day without the very generous investment of time by the more than 20 anonymous experts acting as external reviewers. We are deeply indebted to these colleagues, who returned insightful and constructive reviews of draft chapters, often under extreme time constraints and in many cases over the 2009/2010 holidays.

Martin Grosell
Anthony P. Farrell
Colin J. Brauner

1

MORPHOLOGICAL DIVERSITY OF THE GASTROINTESTINAL TRACT IN FISHES

J.M. WILSON

L.F.C. CASTRO

The anatomy of the gastrointestinal tract of fishes follows the same basic plan as in other vertebrates with a degree of variation reflected in phylogeny and ontogeny, diet, and environment. Morphological studies provide us with a context for understanding the spatial organization and relationship of physiological and biochemical data, and the molecular machinery that is rapidly being elucidated through molecular techniques directed at the genome, transcriptome, and proteome. Morphological data are also key to understanding fish nutrition in ecology and aquaculture, and during development as well as mechanisms for physiological adaptations to a changing environment.

1

The Multifunctional Gut of Fish: Volume 30
FISH PHYSIOLOGY

A number of the multifunctional roles of the fish gut that are discussed in the following chapters (e.g. respiration and ion regulation) of this volume incorporate distinctive morphological features that will be highlighted in this chapter. The stomach represents a significant vertebrate innovation and is also the most highly diversified region of the gut, yet it has undergone a number of independent secondary losses with stomachless fishes accounting for approximately 20% of species. The diversity of fishes makes the identification of the causes of stomach loss elusive. An updated survey is given.

1. INTRODUCTION

With fishes the properties of the gut and stomach are similar (to other vertebrates); that is they have a stomach single and simple but variable in shape according to species...And the whole length of the gut is simple, and if it have a reduplication or kink it loosens out again into a simple form.

(Aristotle 345 BC[1])

From this unassuming first description of the gross anatomy of the fish gut by Aristotle over 2000 years ago in his work *Historium Animalia*, the story of the investigation of fish gut morphology begins. However, it was not until the 19th century with the advent of comparative anatomy that significant effort was put toward determining the anatomical diversity and understanding of its significance in the fishes and vertebrates in general, starting with the works of Cuvier (1805), Rathke (1820, 1837), Edinger (1877), Oppel (1896), and Hopkins (1895). With advances in technology over the past 200 years, from light to electron microscopy and staining techniques from histochemistry, to enzyme histochemistry, immunohisto-chemistry and *in situ* hybridization of RNA, the level of detail has progressed from organ, to tissue, to cell and finally to the molecular level.

This review will focus on the "back-end" of the gut where digestion and absorption take place as opposed to the "front-end" which is associated with the headgut and food capture and initial mechanical processing (Horn 1997; Clements and Raubenheimer 2005; Schwenk and Rubega 2005). Also since the literature is extensive, spanning over 200 years, a number of reviews of the gastrointestinal tract of fishes are referred to (e.g. Jacobshagen 1937; Pernkopf and Lehner 1937; Barrington 1957; Bertin 1958; Adam 1963; Kapoor et al. 1975; Harder 1975; Fange and Grove 1979; Youson 1981). Descriptions of the gastrointestinal tract (GIT) of the different taxonomic groups are given,

---

[1]English translation (Wentworth-Thompson 1910).

highlighting functional significance with a focus on the mucosal epithelium in adult fish as well as during development. Special attention is paid to morphological adaptations specifically related to the multifunctional aspects of the fish gut that will be described in greater detail in the following chapters of this volume. The chapter will end with a re-examination of stomach loss, which is surprisingly common in fishes.

2. GUT ORGANIZATION

The gastrointestinal tract of fishes can be subdivided into four topographical regions: the headgut, foregut, midgut and hindgut (Harder 1975). Further morphofunctional subdivisions can be superimposed on this basic plan. The headgut is composed of the mouth and pharynx, and its function is to acquire food and mechanically process it (Horn 1997; Clements and Raubenheimer 2005). The foregut follows and is comprised of the esophagus and stomach, where chemical digestion of food begins. In some fishes, the mechanical breakdown of food may also occur partially or fully in the stomach. The midgut or intestine accounts for the greatest proportion of the gut length and is where chemical digestion is continued and absorption mainly occurs. The hindgut is the final section of the gut, which includes the rectum; although in some cases there is no clear morphological distinction between midgut and hindgut. The foregut epithelium is of ectodermal and the midgut of endodermal origin.

Radially the gut wall from foregut to hindgut consists of four concentric layers. (1) The *tunica mucosa* consisting of the mucosal epithelium and *lamina propria*, vascularized connective tissue containing nerves and leukocytes. (2) The submucosa[2] is an additional connective tissue layer. (3) The *tunica muscularis* consists of circular and longitudinal layers of either striated or smooth muscle. However, the lampreys only have a layer of oblique muscle. (4) The *tunica serosa* is only present within the coelomic cavity and corresponds to mesothelial cells and loose connective tissue containing blood vessels.

Unlike the rest of the GIT, the *tunica muscularis* of the esophagus is composed of skeletal muscle instead of smooth muscle. The circular muscle layer dominates and is found exterior to the longitudinal muscle layer when present; a relationship that is reversed in the rest of the digestive tract (Harder 1975).

---

[2]Term commonly used in all fishes, although technically a submucosa is only present in the groups with a mucosal smooth muscle layer, which excludes the teleosts and agnathans.

2.1. Esophagus

2.1.1. Esophageal Morphology

The esophagus of fishes is generally a short and straight thick-walled tube, connecting the pharynx to the stomach, or intestine in agastric fishes. Morphologically the esophagus is designed primarily for the passage of food. The mucosa is arranged in longitudinal folds or papillae that allow distension of the lumen for the passage of large food items during swallowing. In chondrichthians (Holmgren and Nilsson 1999) and chondrosteins (Hopkins 1895; Buddington and Christofferson 1985), conical backward-facing papillae are more common while in osteicthians longitudinal folds predominate and can range in number from 6 to 50 (Suyehiro 1942) and also include additional secondary and tertiary folding of the mucosa. The esophagus is lined with a mucous layer secreted by epithelial mucous cells that protect it from chemical and mechanical damage, acting as a lubricant to aid the passage of food, as well as having a role in osmoregulation (Humbert et al. 1984; Shephard 1994; Abaurrea-Esquisoaín and Ostos-Garrido 1996a; Figs 1.1A, 1.2A, D, 1.3A). The mucosal epithelium in teleost and elasmobranch fishes is stratified (= multilayered squamous); however, towards the posterior region the epithelium may gradually become more columnar, depending on species and environmental salinity (Sullivan 1907; Yamamoto and Hirano 1978; Meister et al. 1983). The stratified epithelium is generally dominated by large saccular mucous cells at its surface, which are absent in the areas of columnar epithelium. In lampreys, the esophagus is lined with columnar mucus-secreting cells, and lacks saccular and goblet-type mucous cells (Langille and Youson 1985). Ciliated cells are found in the esophagus of most fishes with the exception of adult teleost fishes. Enteroendocrine cells and lymphocytes are also found scattered throughout the esophageal mucosal epithelium.

The mucosal epithelium rests on a vascularized connective tissue layer with no clear distinction between the *lamina propria* and the submucosa. This layer can be subdivided into a densely packed layer (*stratum compactum*) composed of collagen fibers that provide mechanical strength, followed by a looser layer of connective tissue containing eosinophilic granular cells arranged in a definitive layer, the *stratum granulosum* (Ezeasor and Stokoe 1980b; Reite 2005). Stem cells are found throughout the esophageal epithelium rather than being concentrated in a specific region as in other parts of the GIT (Fig. 1.2D). In some elasmobranchs, masses of lymphomyeloid tissue (organ of Leydig) are found (Fange and Grove 1979; Chapter 3).

Generally glands are not associated with the esophagus, although complex racemose, alveolar and tubular glands have all been described (Kapoor et al. 1975). However, given the gradual transition from esophagus

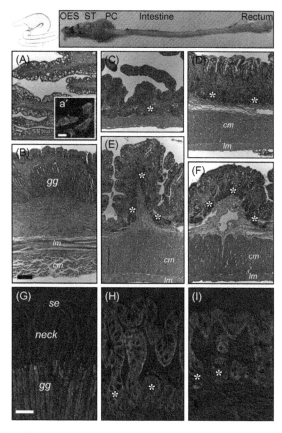

Fig. 1.1. Gastrointestinal tract of the cod *Gadus morhua*, a carnivorous teleost fish. Schematic and view of the laid-out excised GIT (top). Longitudinal sections of the esophagus (A), cardiac stomach (B), pyloric ceca (C), anterior intestine (D), posterior intestine (E), and (F) rectum illustrating glandular tissues and different tissue layers. (a') Immunofluorescent localization of Na$^+$/K$^+$-ATPase (green) and cystic fibrosis transmembrane regulator (CFTR) chloride channel (red) in esophagus columnar epithelium indicating a role in ion regulation. Stem cells are identified by PCNA (red) immunoreactivity in (G) stomach, (H) intestine and (I) rectum sections that are double labeled with gastric (G) H$^+$/K$^+$-ATPase or (H, I) Na$^+$/K$^+$-ATPase which are gastric gland and enterocyte markers, respectively. Abbreviations: *gg* gastric glands, *neck* neck region, *se* surface epithelium, * crypts of Lieberkuhn, *cm* circular muscle, *lm* longitudinal muscle. Sections are either stained with (A–F) hematoxylin and eosin or (a', G–I) for immunofluorescence DAPI (4',6-diamidino-2-phenylindole) nuclear stain with the differential interference contrast (DIC) image overlaid. Scale bar (A–F) 100 μm and (a', G–I) 25 μm. (J. M. Wilson, O. Gonçalves, L. F. Castro unpublished micrographs). See color plate section.

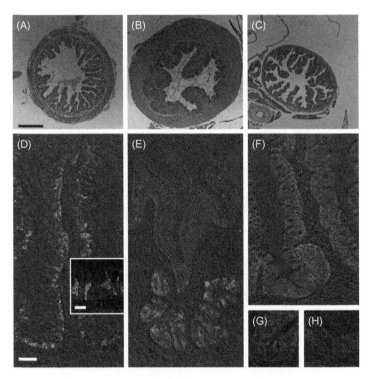

Fig. 1.2. Gastrointestinal tract of the eel pout *Zoarces viviparous* a carnivorous teleost fish with acinar gastric glands. Cross-sections through the (A) esophagus, (B) stomach and (C) intestine. Double immunofluorescent labeling of proliferating cells (PCNA, red) and either (D, F) Na^+/K^+-ATPase or (E) H^+/K^+-ATPase (green). Sections of intestine were also double labeled with Na^+/K^+-ATPase (green) and either (G) NKCC or (H) CFTR (red). Immunolabeled sections (D–H) were counter-stained with DAPI for nuclei and DIC images overlaid for tissue orientation. Histological sections were stained with hematoxylin and eosin (A–C). Scale bar (A–C) 1 mm, (D, E) 100 μm and (inset D) 25 μm. See color plate section.

to stomach, it seems likely that in some of these examples the glands in question are gastric glands of the esogastric transition zone or cardiac stomach. However, in some examples the glands are clearly distinct from gastric glands such as in the case of the milkfish *Chanos chanos* (Chandy 1956; Ferraris and de la Cruz 1987) and pike (Reifel and Travill 1977).

2.1.2. ESOPHAGEAL CELL TYPES

Stratified and columnar epithelia have been described in the esophagus, with the latter only observed in the distal and esogastric regions in euryhaline and marine teleost fishes (Yamamoto and Hirano 1978; Meister et al. 1983; Figs 1.1A, 1.2A, 1.3A). When the epithelial topography is viewed by scanning electron microscopy (SEM), the stratified epithelial cells are revealed to be covered with fingerprint-like microridges, which dominate the surface with mucus exuding from small crypts between epithelial cells where the mucous cells surface (Yamamoto and Hirano 1978; Ezeasor and Stokoe 1980a; Meister et al. 1983). The microridges are suspected to improve mucus adherence and spread, and to minimize impact damage to the epithelium (reviewed by Abaurrea-Esquisoaín and Ostos-Garrido 1996a). Below the surface, the microridged cells are distorted in shape by being squeezed between the large saccular mucous cells, which dominate. The microridged cells are characterized by an abundance of tonofilaments that are likely associated with the desmosomes that attach to neighboring filament-rich cells conferring mechanical strength to the epithelium. Plasma membrane interdigitations rather than desmosomes attach neighboring filament-rich cells to mucous cells. The filament-rich cells are found throughout the epithelium, although they are smaller in the basal region. The mucous cells have a round to flattened basal nucleus beneath a large vacuole containing mucin granules, which vary in their electron density (Yamamoto and Hirano 1978; Meister et al. 1983). Smaller mucous cells, ribosome-rich cells and undifferentiated cells with a high nucleus to cytoplasm ratio are found in the basal part of the epithelium. The histochemical staining of the mucous cell hyaluronic acid sulfomucins and sialomucins indicate subpopulations of different types of mucous cells in some fishes (e.g. Reifel and Travill 1977; Abaurrea-Esquisoaín and Ostos-Garrido 1996a; Diaz et al. 2005; Manjaksay et al. 2009) with further characterization by lectin histochemistry (e.g. Domeneghini et al. 2005; Marchetti et al. 2006).

The columnar epithelium in contrast is lined with microvilli and tends to be limited to the crest of the folds in distal regions of the esophagus (Yamamoto and Hirano 1978; Meister et al. 1983). Mucous cells are also notably absent within this epithelium. The columnar microvillous cells are characterized by an abundance of subapical mitochondria, and prominent lateral connection with neighboring cells through plasma membrane interdigitations and desmosomes. The columnar epithelium has high Na^+/K^+-ATPase immunoreactivity (Fig. 1.1a′) and is associated with osmo-regulation in marine teleosts (see Section 5.1; Chapter 4).

In chondrichthians, there are few studies of the esophageal epithelium (Sullivan 1907; Jacobshagen 1937; Pernkopf and Lehner 1937; Holmgren and Nilsson 1999) and no studies examining the subcellular details of these cells. The epithelium is ciliated and described as stratified and containing mucous cells similar to teleosts (Sullivan 1907). In the chondrosteans, the epithelium has scattered mucous cells but arranged along the apical plasma membrane of the columnar epithelial cells are electron dense granules presumably containing secretory material (Cataldi et al. 2002). In the lamprey, the simple columnar epithelium is lined with a uniform cell type characterized by the presence of apical mucous globules and strong PAS staining indicating the mucus-secreting function of this epithelium (Barrington 1972; Youson 1981; Langille and Youson 1984a, 1985). The apical surface of the esophagus is covered by sparse microvilli. These cells have a prominent Golgi apparatus and rough endoplasmic reticulum to support mucus production. The intercellular spaces are well dilated. Lipoprotein-like particles, which increase during feeding, are also present in these cells indicating an absorptive function (Langille and Youson 1984a).

2.2. Stomach

2.2.1. GASTRIC MORPHOLOGY

The junction of the esophagus and stomach is generally not clearly demarcated anatomically with the exception of a few elasmobranchs, which have an esogastric valve (Harder 1975; Kapoor et al. 1975; Fange and Grove 1979). However, histologically and topographically, the beginning of the stomach is obvious with the abrupt change in the epithelium to the columnar mucous cells of the stomach and the appearance of gastric glands. Since this transition can be gradual the term esogaster is sometimes applied to the transition zone between these two organs (Kapoor et al. 1975). The striated muscle fibers of the esophagus are gradually replaced by smooth muscle in the cardiac stomach and the circular muscle layer will now be internal to the new longitudinal muscle layer (Pernkopf and Lehner 1937; Harder 1975).

In terms of gross morphology the stomach of fishes can be classified according to shape as either straight (I), siphonal (U or J) or cecal (Y) (Pernkopf and Lehner 1937; Suyehiro 1942; Harder 1975). The straight stomach is relatively rare (*Esox*) and in some cases is actually more indicative of the absence of the stomach (e.g. Tetraodontiformes; Suyehiro 1942), while the siphonal shape is the most common among the osteichthyes and elasmobranchii. The cecal stomach allows the storage of larger quantities of food. The overall size of the stomach can also be quite variable from very large in gluttonous fishes such as *Gadus macrocephalus* and the Lophiidae to very small as in the Salangidae and Oplegnathidae (Pernkopf and Lehner 1937; Suyehiro 1942).

The mucosa of the stomach is the most highly differentiated of any gut region (Harder 1975). Histologically the stomach can be divided into two main regions, the anterior cardiac or fundic region and the posterior pyloric region.

Mucosal differences distinguish these regions with the cardiac[3] region containing gastric or chief glands, which are absent from the pyloric region. In the pyloric region mucous glands are found in only some fishes (e.g. *Salmo*, *Esox*, *Anguilla* and *Perca*) (Harder 1975).

The stomach mucosa is lined with a columnar epithelium, which is interspersed with gastric pits (*foveola*) that lead into the tubular or alveolar gastric glands. Goblet-type mucous cells are rarely observed in this epithelium (Harder 1975; Kapoor et al. 1975), since mucus is secreted by the columnar epithelial cells. Endocrine cells are also found in both epithelia (Noaillac-Depeyre and Gas 1982; Chapter 7). The tubular glands are more or less linear and may branch with some sharing of the same gastric pit (Figs 1.1B, G, 1.4). The tubules are separated by the *lamina propria* and generally packed closely together. The tubules are lined by a single type of cuboidal cell, containing acidophilic granules indicative of zymogen granules, the oxynticopeptic cells (Barrington 1942, 1957; Kapoor et al. 1975). Fish with acinar or alveolar gastric glands tend to be much less densely packed than tubular glands and, depending on the distension of the stomach and the species in question, may appear either close to the surface (catfish *Hypostomus*, Fig. 1.3B, D, E) or deeper within the mucosa (eel pout *Zoarces*, Fig. 1.2B, E; loach *Botia*, Fig. 1.9). Acinar glands are generally associated with less acid secretion (higher gastric pH) compared to tubular glands (see Section 6).

The *lamina propria* contains connective tissue fibers, blood vessels, nerve plexus and mast cells. The circular smooth muscle layer in the pyloric region tends to be more developed than in the cardiac region; however, extreme thickening assumes a globular or spindle-like shape giving the appearance of the gizzard found in birds, as was noted by Aristotle (345 BC) in the mullet (Mugiloidei) and subsequently in a number of other fishes to include members of Clupeodei, Chanoidei, and Characinoidei (Kapoor et al. 1975) and chondrosteans (Buddington and Christofferson 1985) as well as the Gillaroo trout (*Salmo stomachicus*; Bridge 1904). The function of the gizzard in fishes is similar to birds; to grind or triturate food, although in fishes the mucous membrane is lined with a thick layer of mucopolysaccharide rather than keratin (Castro et al. 1961).

The stomach ends at the pylorus, which may be present as a muscular sphincter created from the thickening of circular smooth muscle and/or a

---

[3]It should be noted that the gastric glands in the anterior cardiac region of the fish stomach are not analogous to the cardiac gland-type in mammals which are mucous glands. For this reason some authors prefer to use the term fundic to describe this region (Barrington 1957, Fange and Grove 1979), although the term cardiac is widely used and still valid because it is the region proximal to the heart which lies anteriorly (Kapoor et al. 1975).

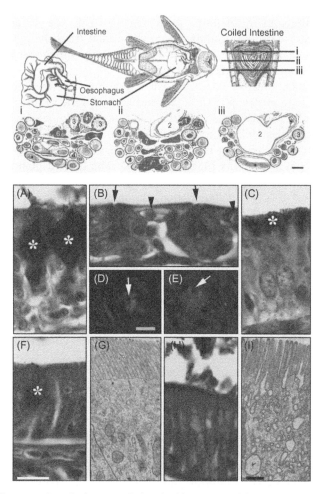

Fig. 1.3. The gastrointestinal tract of the detritivorous catfish *Hypostomus plecostomus*. *H. plecostomus* has a long coiled intestine and its stomach serves the additional function of gas exchange. The esophagus (1) is packed with saccular-type mucous cells that stain with both (A) PAS and Alcian blue. The air-filled cecal stomach (2) is lined by a stratified epithelium (B) with intra-epithelial capillaries (arrowhead) and simple acinar gastric glands (arrows). (D) Eosinophilic (zymogen) granules and (E) apical H^+/K^+-ATPase immunoreactivity are found associated with gastric gland cells. In the bile duct and (C) pyloric region of the stomach, columnar mucous cells (*) are found as well as gastric glands in the latter. Enterocytes are columnar with a prominent brush border with cell morphology showing regional variability (F–I). Sections (i–iii, A–C, F–I) are stained with PAS+Alcian blue (pH 2.5) and hematoxylin. Mucocytes indicated by asterisks (*). 1 Esophagus, 2 stomach, 3–5 anterior to posterior intestine, 6 liver, 7 spleen, 8 pancreas. Scale bars (i, ii, iii) 1 mm, (A–C, F, H) 10 μm, (G, I) 2 μm and (D, E) 10 μm. Drawings modified from Carter (1935). Micrographs (J. M. Wilson, O. Gonçalves, L. F. Castro unpublished). See color plate section.

mucous membrane fold that serves as a valve-like structure. A pylorus-like structure is also found in a number of stomachless fishes such as *Tinca* and *Aspius* (Suyehiro 1942; Al-Hussaini 1947) although since the stomach is absent it may be referred to as the esophageal-intestinal valve. An extra sphincter has been reported in *Leporinus taeniofaciatus* within the stomach (Albrecht et al. 2001).

2.2.2. GASTRIC CELL TYPES

2.2.2.1. Columnar Cells. The columnar mucous cells that line the stomach surface are tall with oval nuclei in the basal third of the cell and characteristic mucous granules in the apical region. Associated with mucous secretion is the presence of a distinct Golgi apparatus, along with numerous mitochondria and apical mucous granules. Transmission electron microscopy (TEM) shows the mucous granules are rod-ovoid in shape, membrane bound and contain highly electron dense material that can be frequently seen to exocytose across the apical plasma membrane (Noaillac-Depeyre and Gas 1978; Ezeasor 1981). The neutral, PAS (Periodic Acid Schiff) reactive mucus secreted by these cells protects the epithelium and underlying tissue from gastric juices (Allen and Flemström 2005). The carbohydrate chemistry of the stomach mucins has been studied by histochemistry and lectin binding in a number of species (e.g. Reifel and Travill 1978; Murray et al. 1994; Domeneghini et al. 2005; Marchetti et al. 2006). Oddly in goboid fishes, mucus is not detected in the surface columnar epithelium (Milward 1974; Kobegenova and Dzhumaliev 1991; Jaroszewska et al. 2008) suggesting that acid-peptic digestion is not significant since a protective mucous coat is not required. The columnar cells in chondrosteans are ciliated (Radaelli et al. 2000), while in other fishes the surface may be smooth or lined with microvilli. Microvilli indicate an absorptive function; however, the importance of the stomach to nutrient uptake has been questioned (Buddington and Christofferson 1985; Jobling 1995; Chapter 2).

In the neck region of the gastric pits, the columnar mucous cells are compact and specialized mucus-secreting neck cells seen in mammals are generally not present (Kapoor et al. 1975; Ezeasor 1981). However, perch (Noaillac-Depeyre and Gas 1978), tilapia (Gargiulo et al. 1997), and mullet (Al-Hussaini 1946a) do have distinct neck cells containing large supra-nuclear mucous granules that are moderately electron dense although mucin PAS histochemistry remains similar (Reifel and Travill 1978; Murray et al. 1994). In fish with tubular gastric glands (Fig. 1.4C), the neck region also contains multipotent stem cells, which migrate to the surface epithelium and gastric glands. However, in fish with acinar gastric glands progenitor cells

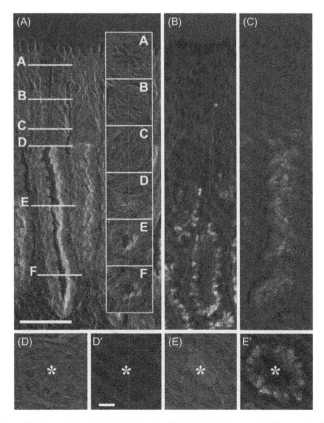

Fig. 1.4. The tubular gastric gland of the dogfish *Scyliorhinus canicula* seen in longitudinal section with (A) immunofluorescent localization of H^+/K^+-ATPase (green) apically and Na^+: K^+:Cl^- cotransporter (red) basolaterally with representative cross-sections (A–F) through a tubule. Nuclei are counterstained with DAPI (blue) and differential interference contrast (DIC) images are overlaid for tissue orientation. Sections stained with eosin (B, D and E) reveal granular (zymogen) staining in the lower tubule in sagittal section (B) and cross-section (E, E′); however, not in the neck of the gland (D, D′). (C) Proliferating cells are identified in the neck region through PCNA (proliferating cell nuclear antigen) immunofluorescence (red) in a section double labeled with H^+/K^+-ATPase (green). The asterisk indicates tubule lumen. Scale bars (A, B) 50 μm and (C, D) 25 μm. (J. M. Wilson, O. Gonçalves, L. F. Castro unpublished). See color plate section.

are found dispersed within the surface epithelium and gastric glands themselves as well (Fig. 1.2E).

2.2.2.2. Oxynticopeptic Cells. Generally only one secretory cell type is observed within fish gastric glands without the differentiation into separate pepsinogen-secreting chief (peptic) cells and acid-secreting parietal (oxyntic) cells observed in mammals (Barrington 1957; Hirschowitz 1957; Smit 1968). These oxynticopeptic or oxyntopeptic cells thus contain the cellular machinery for both functions. Morphologically, acid secretion is associated with a well-developed intracytoplasmic membrane system consisting of a tubulovesicular network of smooth membranes (Noaillac-Depeyre and Gas 1978; Ezeasor 1981). Activation of these cells by distending the stomach prior to fixation results in the fusion of these tubulovesicles with the apical plasma membrane creating an apical labyrinth (Ezeasor 1981). The gastric proton pump has been identified as an H^+/K^+-ATPase, a heterodimer consisting of an α and β subunit belonging to the same P-type ATPase family as Na^+/K^+-ATPase (Yao and Forte 2003). H^+/K^+-ATPase has been immunolocalized to the apical region (apical plasma membrane and tubulovesicular system) of the Atlantic stingray (Smolka et al. 1994) using a porcine α subunit antibody. This same antibody shows apical immunoreactivity with a broad range of fishes (Figs 1.2B, 1.3D, 1.4A, C, 1.9B). The expression of the gastric proton pump has also been confirmed by *in situ* hybridization in winter flounder (Gawlicka et al. 2001) and porgy (Darias et al. 2007).

Pepsinogen is found in round electron-dense zymogen granules in the supranuclear region of the cell (Hirschowitz 1957). Zymogen granules are most often identified based on their eosinophilic staining (Figs 1.3E, 1.4B, E′; Barrington 1957; Kapoor et al. 1975); however, Orange G staining (Weisel 1979), and Bowie's zymogen method (Michelangeli et al. 1988) have also been used. Pepsinogen has also been localized to oxynticopeptic cells in fish using *in situ* hybridization (*Pseudopleuronectes americanus*, Gawlicka et al. 2001; *Pagrus pagrus*, Darias et al. 2007) and immunohistochemistry (osteichthyes: e.g. *Sebastes inemis*, *Gimnothorax kikado*, and *O. mykiss*, and chondrichthyes *Narke japonica* and *Triakis Scylla*, Yasugi 1987; Yasugi et al. 1988). As a corollary immunoreactivity was absent in stomachless fishes (e.g. *Fugu crysopus*, *Pseudolabrus japonicas*, and *Eptatretus burger*, Yasugi 1987; Yasugi et al. 1988). Pepsinogen production and secretion are associated with the Golgi apparatus and rough endoplasmic reticulum (RER) in the basal region of the cell (Ezeasor 1981; Michelangeli et al. 1988).

The only clear exceptions to date of the single oxynticopeptic cell type in fishes come from some elasmobranchs. In the shark *Hexanchus griseus* (Michelangeli et al. 1988) and Atlantic stingray *Dasyatis sabina* (Smolka et al. 1994) two distinct secretory cell types on the basis of morphological features of acid secretion, H^+/K^+-ATPase immunoreactivity and zymogen production have been observed. However, the elasmobranch parietal cell has the more primitive tubulovesicular network typical of non-mammalian vertebrates rather than a system of canniculi seen in mammals. Reports in the teleost literature of separate cell types are contradictory. In tilapia

(*O. mossambicus, O. Niloticus, O. zilii*) Gargiulo et al. (1997) observed oxyntic cells with only the characteristics of an acid-secreting cell (well-developed apical tubuloviscular system, mitochondria) and not pepsinogen production (absence of zymogen granules, poorly developed RER and Golgi). However, these results contrast with the finding of eosinophilic granules by Osman and Caceti (1991) in wild *O. niloticus*, and activity measurements in stomach extracts and mRNA expression and partial cloning of a pepsinogen (A2) by Lo and Weng (2006) in *O. mossambicus* (AY513876.1). It is unclear if nutritional status or tissue processing may have resulted in these conflicting results, but they are nonetheless significant for the discussion in Section 6.

2.3. Intestine

2.3.1. INTESTINAL MORPHOLOGY

The intestine follows the pylorus or esophagus in gastric and agastric fishes, respectively. In some stomachless fishes, the anterior intestine may bulge to form an intestinal bulb or pseudogaster (*Petromyzon*, Cyprinidae) and functions in temporary food storage; however, gastric glands and a pylorus are lacking. Some parrotfish have a spherically shaped cecal chamber extending from the anterior intestine just after the esophagus to serve a similar function (Al-Hussaini 1946b).

The primary function of the intestine is the completion of the digestive processes started in the stomach and the absorption of nutrients. Central to this is optimizing intestinal surface area within the constraints of the coelomic cavity, which has a marked impact on intestinal morphology. Intestinal surface area is increased in all fish by folding (primary, secondary, and/or tertiary) of the mucosa and by apical plasma membrane amplification through brush border microvilli. In the lampreys, chondrosteans, chondrichtheans and dipnoids which all have short intestines, the mucosa forms a spiral valve whereas in the teleost fishes, surface area is increased by lengthening the intestine through convolution (loops). A wide range of looping and coiled arrangements can be seen within the teleosts with long intestines (e.g. Cyprinidae, Loricadae). However, some fish just have short, straight intestines with no looping or spiral valve (e.g. Cobitidae; Salmonidae, Agnathans). The intestine in some Scarides is constricted at short intervals giving a sacculated structure forming irregular internal pouches (Al-Hussaini 1946b). Blind-ended appendages (pyloric ceca) are found in ostichthys (teleosts and chondrosteans) and elasmobranchs, although rarely in the latter. Notably, intestinal villi (finger-like projections), characteristic of the mammalian intestine, are rare in fishes (Harder 1975; Kapoor et al. 1975). The central villus lacteal (lymphatic vessel) is also absent in fishes. Mucosal glands are lacking in the intestine of almost all fishes with the exception of the Gadidae (cods), which have glands that are similar in appearance to the crypts of Lieberkuhn (Jacobshagen 1937; Bishop and Odense 1966; Fig. 1.1). In mammals, the crypts are the site of Paneth (immune) cells, stem cells and intestinal fluid secretion. However, functionally, very little is known about the Gadidae intestinal glands, such as if they are secretory. Although, in common with their mammalian counterpart the cod crypts are associated with stem cells as indicated by the expression of proliferating cell nuclear antigen (Fig. 1.1H, I). In other fishes, stem cells occur at the base of the primary (Fig. 1.5C′) and/or secondary intestinal folds (*Anarhichas lupus*; Hellberg and Bjerkås 2000; *Zoarces viviparous*, Fig. 1.2F). Interestingly, in the eel pout *Z. viviparous* apical CFTR (cystic fibrosis transmembrane regulator chloride channel) immunoreactivity is found associated with these "crypts", suggesting a fluid-secretory capacity (Fig. 1.2E). Brunner's glands, which secrete an alkaline, mucus-rich fluid in the duodenum, are a mammalian innovation and are not found in fishes (Jacobshagen 1937; Barrington 1957; Harder 1975; Kapoor et al. 1975).

Intestine length is used as a morphological indicator of trophic level in nutritional ecology (Horn 1997). However, intestinal length is influenced by a number of other factors apart from diet, which include fish size (mass and length) and body shape, recent feeding history (starved versus fed), ontogeny, and phylogeny (Horn 1989, 1997; Clements and Raubenheimer 2005). The length and relative mass of the intestine is generally greater in herbivores relative to carnivores, which is thought to allow for additional processing of relatively difficult-to-digest items (Horn 1997; Clements and Raubenheimer 2005); although within omnivorous fishes there is no clear relationship with degree of herbivory or carnivory (Kramer and Bryant 1995). Kramer and Bryant (1995) caution that gut length as a reflection of diet should be applied only to identifying broad catagories. Harder (1975) goes further and argues that there are no clear relationships between intestinal morphology and feeding type and it is not possible to draw conclusions on one from the other. Evolutionary history and phylogeny have been shown to be important when interpreting gut morphometric data (coral reef fishes, Elliott and Bellwood 2003; prickleback (Stichaeidae), German and Horn 2006; minnows, German et al. 2010a). For example, in the prickleback, although *Xiphister atropurpureus* is a carnivore it belongs to an herbivorous taxon suggesting an herbivorous ancestor and consequently its relative intestinal length is more similar to its sister taxon and greater than prickleback from a carnivorous taxon (German and Horn 2006). Similar relationships have been reported in herbivorous taxa of a carnivorous clade of minnows (German et al. 2010a).

Al-Hussaini (1949) estimated the entire surface area of the intestine (however, excluding the contribution of microvilli) in three species having different feeding habits: *Rutilis rutilis* (omnivore), *Gobio gobio* (carnivore) and *Cyprinus carpio* (herbivore). In a comparison of size-matched individuals he found no differences in total surface area even though intestinal lengths between species varied greatly indicating compensation through increased folding. Buddington and Diamond (1987) found similar results including fish with pyloric ceca.

2.3.1.1. Spiral Valve.   The spiral valve not only increases the surface area of the intestine without increasing intestinal length, but also is associated with a slow food passage rate (Parker 1880; Jacobshagen 1937; Wetherbee and Gruber 1993; Figs 1.5, 1.6A, B). The spiral valve is considered a primitive feature and is present in the lamprey and non-teleost gnathostome fishes although not in the Myxini (Jacobshagen 1937; Adam 1963). The walls of the valve are formed from the infolding of the intestinal mucosa and submucosa although in Holocephali and ammocoete larvae the muscularis is also involved. Surface area is increased further by the presence of mucosal folds on the spiral folds (Figs 1.5B, 1.6B). The intestine is twisted along its longitudinal axis giving the spiral or screw-like appearance with the number of turns and height of the fold being variable between species (Parker 1880). A counter spiral is observed in the lamprey but is lost in most other fishes.

2.3.1.2. Pyloric Ceca.   The pyloric ceca are blind-ended sphincterless ducts associated with the anterior intestine in osteichthys and infrequently in chondrichthys (Jacobshagen 1937; Buddington and Diamond 1987). Of the holosteans, only the *Amia* lack pyloric ceca (Hopkins 1895). Hossain and Dutta (1996) estimated that 60% of known fish species possess pyloric ceca, which vary greatly in number (0–1000s), length, and diameter. For example, in the flounder (*Platichthys*) the ceca only appear as a few bumps on the intestinal wall, whereas in the Salmonidae the ceca are long and numerous (Harder 1975). In species with high numbers of ceca, greater individual variation is seen and the numbers of openings to the intestine are limited and thus the ceca appear as tufts (Suyehiro 1942; Harder 1975). In sturgeon and tunas the ceca appear as a fused compact mass surrounded by connective tissue (Jacobshagen 1937; Suyehiro 1942; Buddington and Christofferson 1985) and in *Polypterus* as a pair of racemose ceca (Abdel Magid 1975).

   The mucosa of the pyloric ceca is similar to the intestine and no special cell types or glands are associated with this organ (Kapoor et al. 1975; Buddington and Diamond 1987) although there are fewer goblet cells in the ceca (Hossain and Dutta 1996). Buddington and Diamond (1987) demonstrated that the pyloric ceca increase the surface area for digestion and absorption but do not have a role in fermentation or storage. Harder (1975) and Hossain and Dutta (1996) suggest there is no clear correlation

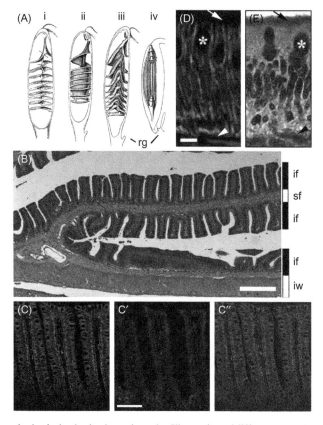

Fig. 1.5. Intestinal spiral valve in elasmobranchs. Illustration of different ventral cut-away views of spiral valves in *Raia* spp. (i, ii, iii) and roll valve in *Sphyrna* (iv) modified from Parker (1880). Light micrograph of longitudinal section through the spiral intestine of *S. caniculae* fasted for 1 week. The spiral fold (sf) originating from the intestinal wall (iw) is pointing cranially. Additional tertiary intestinal folds (if) originate at regular intervals on the spiral folds. The epithelium consists of columnar cells interspersed with goblet-type mucous cells identified by PAS staining (B, E). Na^+/K^+-ATPase, a basolateral membrane marker (green) and important ion pump for secondary transport of nutrients and ions, is present in the enterocytes of the intestinal folds (C, D). (C') Proliferating epithelial cells are identified by PCNA (proliferating cell nuclear antigen; red) immunoreactivity found at the base of the intestinal folds. In panels C', C'' and D, DAPI (blue) is used as a nuclear marker for orientation. (C''). The composite image of C and C' with DIC image overlaid. Scale bars: (B) 500, (C) 200, and (D, E) 10 µm. Abbreviation: rg rectal gland. Micrographs (J. M. Wilson, O. Gonçalves, L. F. Castro unpublished). See color plate section.

between pyloric ceca number or size and intestine length or feeding type in contrast to Buddington and Diamond (1987). The latter authors found a trade-off between ceca and intestine length and thickness. The only undisputed relationship with feeding type is the greater likelihood of the

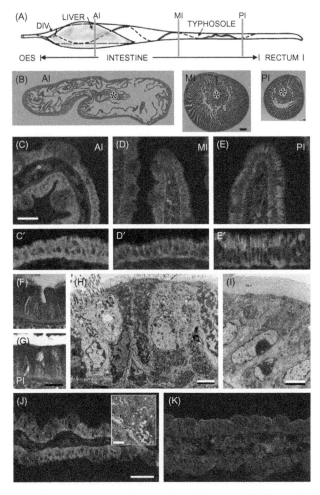

Fig. 1.6. Lamprey (*Petromyzon marinus*) gastrointestinal tract. (A) Illustration depicting the different regions of the GIT (esophagus (OES), intestine and rectum) and the liver modified from Langille and Youson (1984a). At the junction of the esophagus and intestine is situated the diverticulum (DIV). The position of the typhosole or spiral valve is indicated by the solid and dashed lines. (B) Cross-sections of the anterior (AI), mid (MI) and posterior (PI) intestine of an adult lamprey caught on its spawning migration (fasting). The typhosole is indicated by an asterisk (*). (C–E) Immunofluorescent staining of Na$^+$/K$^+$-ATPase (green), a basolateral membrane marker, and H$^+$-ATPase (red), an apical membrane marker of ciliated cells in lamprey with additional 2× magnification insets (C′, D′, E′, respectively). Note the more homogeneous cellular distribution of Na$^+$/K$^+$-ATPase in the anterior intestine A-type enterocytes (C′), due to the presence of a basolateral tubular system (see inset in J) compared to the more distal regions of the intestine (D′, E′) where the staining is more peripheral. In the posterior intestine (E) strong H$^+$-ATPase immunoreactivity is also associated with apical tubulovesicles and vacuoles of B-type enterocytes in this region. (J, K) Type-A enterocytes can

absence of pyloric ceca in herbivores than carnivores (and omnivores) (Buddington and Diamond 1987; Hossain and Dutta 1996).

There is also a strong positive correlation between the presence of pyloric ceca and the presence of a stomach (Barrington 1957; Kapoor et al. 1975). However, the functional relationship has not been firmly established and conclusions based on diet are unclear, although during development the pyloric ceca and stomach gastric glands are the last organs to complete differentiation (Section 3).

2.3.2. Intestinal Cell Types

The intestinal epithelium consists of a single layer of columnar absorptive cells with a distinctive apical brush border. Goblet-type mucous cells, lymphocytes and enteroendocrine cells are scattered throughout the epithelium and rodlet cells are also found in some species of teleost fishes (Kapoor et al. 1975). The columnar epithelial cells that dominate the intestinal epithelium are relatively homogeneous in appearance but show regional differentiation (Yamamoto 1966; Ezeasor and Stokoe 1981; Figs 1.1–1.3, 1.5, 1.6).

2.3.2.1. Enterocytes. Enterocytes are generally tall and narrow, with elongated nuclei located just below the middle of the cell, mitochondria located in both apical and basal regions, a well-developed brush border and lamellar structures running parallel to the lateral plasma membrane (Yamamoto 1966; Ezeasor and Stokoe 1981). The lamellar structures have anastomoses with the basolateral membrane and have been shown by Ruiter et al. (1985) using freeze fracture SEM to form a basal labyrinth that increases the basolateral surface area. It should be noted that fish generally lack the lateral interdigitations characteristic of mammalian enterocytes. In the posterior intestine absorptive cells with large vacuoles in the apical cytoplasm are present (Yamamoto 1966; Ezeasor and Stokoe 1981; Langille

---

also be differentiated from ciliated cells from their strong apical $Na^+:K^+:2Cl^-$ cotransporter staining (red) when double labeled with either (J) Na^+/K^+-ATPase (green) or (K) H^+-ATPase (green). TEM micrographs of (H) A-type (A) cells, ciliated cells (C) enteroendocrine cells (E, arrow) and (I) B-type enterocytes with pinocytotic figures (arrow) and dark body vacuoles (b) or secondary lysosomes in fasted adults. Neutral mucin staining of (F) anterior and (G) posterior intestine determined using PAS method is associated with ciliated cells and B-type enterocytes, respectively. Additional abbreviations: mv microvilli, m mitochondria. Scale bars: (B) 1 mm, (C–E) 100 µm or (C′, D′, E′) 50 µm, (F, G) 25 µm, (H, I) 5 µm, (J, K) 100 µm (inset in J) 1 µm. Drawing and TEM micrographs from Langille and Youson (1984b). LM micrographs (J. M. Wilson, O. Gonçalves, L. F. Castro unpublished). See color plate section.

and Youson 1984b). In lamprey, in addition to the absorptive cell described in teleost fishes, the anterior intestine has a unique columnar cell type with a system of smooth reticulum continuous with the basolateral membrane similar to the tubular system of branchial chloride cells (Youson 1981; Langille and Youson 1984a). Based on this similarity, they have been suggested to be involved in ion regulation as well; however, this relationship has not been established (Youson 1981). In elasmobranchs, interdigitations of the lateral plasma membrane of neighboring enterocytes increase surface area (Teshima and Hara 1983). Enterocytes show strong basolateral expression of Na^+/K^+-ATPase which is essential in driving a number of transepithelial transport processes important for nutrient uptake and ion regulation (Figs 1.1H, 1.2F–H, 1.5C, D, 1.6C–E, J; Chapter 2 and 4). In the lamprey, the differences in the organization of the basolateral membrane are reflected in the staining patterns, with tubular system cells having a whole cell distribution of Na^+/K^+-ATPase immunostaining (Fig. 1.6c'), while in cells with a basal labyrinth, as in all other fishes, staining is limited to the cell periphery (Figs 1.5C, D, 1.6D', E').

The enterocyte apical membrane is characterized by the presence of microvilli that form a brush border that is evident with light microscopy. The brush border contributes greater than 90% to total intestinal surface area (*Tilapia aurea* and *T. zilli*; Frierson and Foltz 1992) and forms the critical digestive/absorptive interface, a functional microenvironment where enzymes involved in further food breakdown are located and where absorption and transport will occur (Crane 1968; Kuz'mina and Gelman 1997). The activities of a number of enzymes have been localized to the brush border membrane in vertebrates including fishes. This list includes alkaline phosphatase, disaccharidases, leucine-aminopeptidase, tri- and di-peptidases which are produced by the epithelial cells while pancreatic lipase and esterase, α-amylase, and carboxypeptidase are adsorbed to the microvilli (Kuz'mina and Gelman 1997; Chapter 2). The presence of these enzymes in the brush border has been demonstrated by enzyme histochemistry and immunohistochemistry. Glucose (SGLT1) and di-/tri-peptide transporters (pepT1) have also been localized to the brush border of enterocytes (Sala-Rabanal et al. 2004; Gonçalves et al. 2007; Yuen et al. 2007). The tight packing of the microvilli may also have a sieving effect, excluding large particles and aggregates from entering the microvillar space (Kuz'mina and Gelman 1997). The density and height of microvilli may vary with species and intestinal region (Frierson and Foltz 1992; German et al. 2010b). The density of microvilli is generally less in the posterior region of the intestine (e.g. Noaillac-Depeyre and Gas 1976; Stroband 1977; German et al. 2010b). Prolonged starvation markedly reduced microvilli density to the point at which they may completely disappear, although in the short term only microvilli height decreases (Gas and Noailliac-Depeyre 1976; German et al. 2010b).

The general appearance of the enterocyte indicates an absorptive function (Yamamoto 1966; Iwai 1968; Rombout et al. 1985; Noaillac-Depeyre and Gas 1976; Fig. 1.3F–I). Notably, the absorption pathways of lipids and macromolecular proteins were initially identified using morphological techniques. In the trout, lipid absorption in the anterior intestine and pyloric ceca have been demonstrated by the presence of 60–100 nm very low density lipoproteins (VLDL) (Sire et al. 1981), and 250–1200 nm lipid droplets, and 400 nm chylomicrons by TEM (Bauermeister et al. 1979). Feeding has a marked effect on the abundance of these features, which disappear with fasting (Sire et al. 1981). Enzymes localized by histochemistry include esterases and lipases secreted by the pancreas that hydrolyze wax esters and triacylglycerols forming free fatty acids and monoacylglycerols which are absorbed into enterocytes. Re-esterification of free fatty acids into triacylglycerols corresponds to the appearance of VLDL in the endoplasmic reticulum, Golgi apparatus, lamellar structures, interstitial spaces, and secondary circulation (Sire et al. 1981). In trout fed zooplankton, which is rich in wax esters, larger osmophillic lipid droplets are observed in these spaces and particles similar in size to chylomicrons released (Bauermeister et al. 1979). Lymph channels equivalent to the lacteals of mammals are lacking in fishes. Ultrastructural studies have also revealed VLDL in RER and Golgi apparatus of tench (Noaillac-Depeyre and Gas 1976), grasscarp (Stroband and Debets 1978) and lamprey (Langille and Youson 1984a, 1985). It is also interesting to note that lampreys are able to efficiently digest lipids without the need for bile salts, since the bile duct is not present in adults (Langille and Youson 1985).

In the distal intestine, columnar epithelial cells with large vacuoles have been identified as the site of macromolecular protein uptake (Fig. 1.3H, I). These cells are characterized by numerous microvilli, large numbers of apical pits or pinocytotic vesicles, a well-developed tubulovesicular network, and numerous vacuoles and lysosomes (Iwai 1968; Ezeasor and Stokoe 1981; Stroband and Kroon 1981; Rombout et al. 1985). Similar cells were also described in the posterior intestine of the lamprey (caveolated cells, Langille and Youson 1984b, 1985). Marker studies with horse radish peroxidise (HRP) and ferritin have demonstrated that intact proteins can be taken up via different routes for transcytosis to the intercellular space or for intracellular degradation in lysosomes, respectively (Noaillac-Depeyre and Gas 1976; Rombout et al. 1985; Langille and Youson 1985). H^+-ATPase can be seen to localize to the apical cytoplasmic area of these cells in lamprey where vacuoles and lysosomes are present (Fig. 1.6E, E′, G, I), with a similar pattern seen in other fishes (J. M. Wilson unpublished).

However, it is important to keep in mind that the majority of protein uptake (80%) takes place in the anterior intestine (Fange and Grove 1979; Stroband and van der Veen 1981; Chapter 2), and that the quantitative significance of pinocytotic protein uptake for digestion is questionable. However, it has been proposed that this pathway may provide a reserve capacity for protein uptake or is of possible immunological significance (Rombout et al. 1985; Chapter 3).

2.3.2.2. Mucous Cells. Goblet cells are the dominant mucous cell type in the intestine of fishes, with the exception of the lampreys. They derive their name from their goblet or challis-like shape. The nucleus is located in the tapered stem, which widens and then constricts to form an apical pore through which mucus is discharged. In lamprey, goblet cells are absent and instead columnar cells of the epithelium secrete mucus (Fig. 1.6F, G) similar to the columnar mucous cells found in the stomach lining in other groups (Figs 1.1B, 1.2C, 1.9C). The mucin granules that dominate the cell can be of variable electron density when viewed by TEM. There is little variation in carbohydrate localization from species to species, unlike in the foregut (Reifel and Travill 1979). The majority of goblet cells contain the acidic mucosubstance sialomucin although with smaller amounts of sulfomucins.

2.3.2.3. Enteroendocrine   Cells. Enteroendocrine   cells   are   present throughout the epithelium of the GIT in all fishes (Holmgren and Olsson 2009; Chapter 7) and with the pancreas constitute the gastroenteropancreatic (GEP) endocrine system. Endocrine cells are readily identifiable by the presence in the cytoplasm of very characteristic secretory vesicles, the so-called dense core vesicles (DCV), which have a halo between the electron-dense core and the surrounding membrane (Holmgren and Olsson 2009; Fig. 1.6H). Subtypes of enteroendocrine cells are classified by their relative position within the epithelium with the open type extending to the apical membrane while the closed type does not, as well as on the basis of differences in secretory granule morphology (size, shape and electron density) and the expression of neuroendocrine   substances   by   immunohistochemistry   (e.g.   gastrin/ cholecystokinin, ghrelin, somatostatin, serotonin; Holmgren and Olsson 2009). The immunohistochemical characterization of EEC is quite extensive and beyond this review (see Nilsson and Holmgren 1994; Holmgren and Olsson 2009 for more detailed descriptions). Enterochromaffin (serotonin) cells are present in species without strong serotinergic neurons (Anderson and Campbell 1988).

2.3.2.4. Ciliated Cells. Ciliated cells are found in the intestinal epithelia of lampreys (Barrington 1972; Youson 1981), chondrosteans (Abdel Magid

1975; Buddington and Christofferson 1985) and dipnoids (Purkerson et al. 1975; Rafn and Wingstrand 1981) as well as early life history stages in some teleosts (Govoni et al. 1986) and are considered a primitive condition although absent in myxines (Adam 1963). Earlier reports of the more widespread occurrence of ciliated cells in the intestine of teleost fishes have proven incorrect (Ishida 1935, see Odense and Bishop 1966). In lampreys, ciliated cells are found in clusters of 3–5 cells and apart from the presence of cilia resemble neighboring absorptive cells (Youson 1981; Fig. 1.6). The presence of alkaline phosphatase activity along the apical microvilli of ciliated cells, combined with the occasional observation of lipid matrix droplets and the apical mucous PAS-positive vesicles, suggests that the ciliated cells of the lamprey intestine are multifunctional and are related in many ways to the absorptive cell (Langille and Youson 1984a,b; Fig. 1.6F, G). Cilated cells also preferentially express apical H^+-ATPase, which may be significant for driving H^+ coupled transport processes (Fig. 1.6C, D, K). However, the distribution of these transporters (e.g. pepT1) is unknown in lampreys.

2.3.2.5. Zymogen Cells. In the lampreys, a discrete exocrine pancreas is non-existent and the pancreatic zymogen-secreting cells are instead present within the intestinal epithelium (Barrington 1972; Youson 1981). Intestinal zymogen cells are also present in myxines although discrete exocrine pancreatic tissue is present (Adam 1963). The intestinal zymogen cells are characterized by electron-dense "zymogen" granules in the apical region, with elongated mitochondria and well-developed endoplasmic reticulum and Gogi apparatus in the supranuclear region. In the Southern hemisphere lampreys, the zymogen cells are concentrated in the diverticulum of the intestine, which is found near the junction with the esophagus forming a protopancreas (Bartels and Potter 1995; Fig. 1.7). In Northern hemisphere lampreys, the zymogen cells are scattered throughout the anterior intestine in what is considered a primitive condition (Langille and Youson 1984a) similar to myxines (Adam 1963).

2.3.2.6. Rodlet Cells. Rodlet cells are an enigmatic cell type found in the epithelia of the GIT and other organs in a wide range of teleost fishes, although expression is not always consistent between individuals of a given species (Manera and Dezfuli 2004; Reite 2005). The mature cells are ovoid, have a basally located nucleus and are characterized by a wide fibrous layer beneath the plasma membrane and the presence of large rod-shaped cytoplasmic granules from which they get their name. They have typical cellular organelles including mitochondria, endoplasmic reticulum, and Golgi, and desmosomes are found between mature rodlet cells and neighboring enterocytes. These cells have been hypothesized to be of

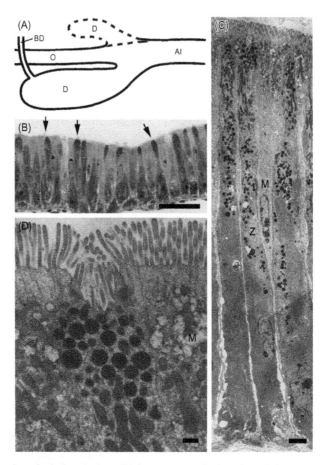

Fig. 1.7. An intestinal diverticulum (D) is present at the junction of the esophagus (O) and anterior intestine (AI) in lamprey. (A) In the Southern hemisphere lampreys *Mordacia* and *Geotria* a prominent diverticulum is present, and is paired in the latter genus. In larvae, the bile duct (BD) empties into the diverticulum. The exocrine pancreas is not present in lamprey and thus zymogen-secreting cells (Z) are present within the intestinal epithelium. In *Mordacia* and *Geotria* they are concentrated in the diverticulum in constrast to Northern hemisphere lamprey where they are scattered throughout the anterior intestine. Mucus-secreting enterocytes (M) are found scattered among the zymogen cells as shown by (B) light and (C, D) electron microscopy. Arrows indicate zymogen cells stained with Alcian blue. Scale bars (B) 50, (C) 5, (D) 1 μm. (Modified from Bartels and Potter 1995.)

either an exogenous (parasitic) or endogenous origin, with recent data supporting the latter (Manera and Dezfuli 2004). Little is known about rodlet cell function although they have been associated with secretory and immune functions (Manera and Dezfuli 2004; Reite 2005).

2.4. Hindgut (Rectum)

The start of the hindgut is usually indicated by the presence of an "ileorectal" valve or a thickening of the circular muscle layer, a sudden change in gut diameter, and/or the mucosal folding pattern (Harder 1975; Kapoor et al. 1975). In the lampreys, the start of the hindgut is marked by the end of the spiral valve (Fig. 1.6A; Youson 1981). In the elasmobranchs there is a reduction in gut diameter, a thickening of the muscular wall and a stratified epithelium in contrast to the columnar epithelium of the spiral intestine (Sullivan 1907; Holmgren and Nilsson 1999). However, these features are not always present so the hindgut can be difficult to distinguish from the midgut such as in cyprinids (Harder 1975). The ileorectal valve is also only found in some teleost fishes (Barrington 1957). The hindgut is generally short and subdivisions are usually not noted in fishes (Harder 1975) although cecal chamber(s) are present in some fishes (e.g. kyphosids, Rimmer and Wiebe 1987). Bacterial fermentation is associated with, although not limited to, these cecal chambers (Rimmer and Wiebe 1987). The rectum ends in a muscular sphincter that empties into either the cloaca or vent. Histologically the hindgut is quite similar to the midgut in the teleosts. In trout, the rectal epithelial cells have vacuolated cells similar to the posterior intestine with their characteristic apical tubulovesicular system representing a differentiation of the inter-microvillous plasma membrane and large irregular vacuoles (Ezeasor and Stokoe 1981). There is also a tendency for greater numbers of goblet cells to be present, lower mucosal folds and a thickening of the musculature of the rectum associated with defecation in some species (e.g. Reifel and Travill 1979; Clarke and Witcomb 1980).

3. GUT DEVELOPMENT

The development of the GIT in teleost fishes accompanies a significant number of changes at morphological and functional (molecular) levels. The initial stage of gut development in fishes is impacted by the level of endogenous food up to the hatching moment. While few fishes develop directly into juveniles with a complete functional GIT at hatch (wolffish *Anarhichas lupus*, Falk-Peterson and Hansen 2001), the vast majority undergo a larval period before entering into the juvenile stage. In these, the alimentary canal at the hatching moment is either a simple tube, dorsal to the yolk sac, with an undifferentiated epithelium (e.g. marine fish with small pelagic eggs) or in the process of differentiation (e.g. fish with large demersal eggs such as salmonids) (Govoni et al. 1986). Unlike other organ systems, the development of the gut does not occur gradually but in spurts.

The sea bass (*Dicentrarchus labrax*) is a well-studied example of gut development in a fish with a true stomach (see Section 6). Newly hatched larvae display a straight gut tube with smooth muscle (Garcia Hernandez et al. 2001). Post-hatch development can be divided into four phases, with the first phase ending when the mouth and anus are open at 6/7 days post-hatch (dph). The epithelium is largely undifferentiated, although some differences can be seen at the ultrastructural level (Garcia Hernandez et al. 2001). During Phase II the esophagus, gastric region, intestine and rectum become distinguishable. The start of Phase III (13/15 dph) corresponds to the total reabsorption of the yolk sac, and the emergence of gastric gland cells and the pyloric ceca. In the final phase (IV from 55 dph onward), gut development is completed, the stomach is differentiated with its definitive morphology and structure and the intestine forms two loops (Garcia Hernandez et al. 2001).

In the Japanese flounder, *Paralichthys olivaceus*, the stomach primordium differentiates during the larval stage at 20/30 dph, although pepsinogen secretion starts only at metamorphosis (45 dph; Kurokawa and Suzuki 1995). In contrast, digestive pancreatic enzymes (e.g. trypsinogen-1) start their expression at 1 dph, which is followed by first exogenous feeding at 3 dph (Srivastava et al. 2002). In the red porgy, *Pagrus pagrus*, pepsinogen and proton pump genes are first detected around 30 dph in the newly formed gastric glands (Darias et al. 2007), which correlate with the start of acid-peptic digestion. Overall, in stomach-bearing fish, the development of gastric glands is the final major morphological transformation in the gastrointestinal canal, which occurs well after first feeding.

In the stomachless zebrafish, *Danio rerio*, the digestive system forms at mid-somite stages (Wallace and Pack 2003). Intestinal morphological formation undergoes three phases covering a window between 26 and 126 h post-fertilization (hpf) (Ng et al. 2005). Stage I encompasses the period between hatching and the opening of the mouth. A simple tubular structure is derived from the endoderm layer, with cuboid cells morphologically indistinguishable. The enteroendocrine cells first appear in the posterior intestine. The second stage is characterized by the opening of the mouth until the opening of the anus. Most importantly, the intestinal epithelium proliferates and cells are columnar in shape. The final stage involves remodeling, compartmentalization and differentiation of the epithelium. Folding of the epithelium appears in the intestinal bulb (an expansion in the anterior intestine), with cell proliferation restricted to the base of the folds. In the mid-intestine cell proliferation is rare at this stage, and the goblet cells containing acid mucins are differentiated. In the posterior intestine a monolayer of enterocytes emerges without any goblet cells. In the stomachless medaka, *Oryzes latipes*, the overall gut development is similar to the zebrafish, except that in the latter the initiation of the tube starts in both anterior and posterior portions (Iwamatsu 2004; Kobayashi et al. 2006). The endoderm provides the major contribution to the embryonic development of the gut. In mammals, where the process of gut development has been extensively studied, the primitive endodermal tube forms in the early somite stages. The digestive epithelium originates from gut endoderm, while the mesenchyme gives rise to muscle and connective tissue. In contrast, in the two model fish species, medaka and zebrafish, the pharynx and the esophagus develop independently of the rostral gut morphogenesis (Wallace and Pack 2003; Kobayashi et al. 2006).

The mechanisms of gut organ specification are well studied in a range of organisms including mouse, xenopus and chicken (Smith et al. 2000; Yasugi and Mizuno 2008). An interaction of molecular signaling takes place between the mesenchyme and associated epithelium, which will later give origin to specific and functionally different sections of the gastrointestinal tract. Specification of the stomach requires the transient signal of the homeodomain protein Barx1 (Kim et al. 2005; Fig. 1.8A). In mice, the expression is localized to the presumptive stomach mesenchyme and leads to the expression of two Wingless (Wnt) antagonists, sFRP1 and sFRP2 (Kim et al. 2005; Fig. 1.8A). Thus, Barx1-mediated inhibition of Wnt signaling in the endoderm is the central mechanism for the development of the gastric epithelium (Kim et al. 2005, 2007; Fig. 1.8A). As for the epithelium of the intestine, the main role is played by the homeodomain protein Cdx2 (Stringer et al. 2008; Fig. 1.8A). The current model argues that the initiation of gut histo-differentiation lies in the endodermal expression of Cdx2 (Stringer et al. 2008). The Wnt signaling is active in the prospective mesoderm, but when suppressed by mesenchyme Barx1-mediated expression gives rise to the stomach epithelium identity (Kim et al. 2005). The cross-talk between Cdx2 and Barx1 results in the final gut phenotype (Stringer et al. 2008; Fig. 1.8A). Supporting this model, in mice the overexpression of sFRPs overrides Barx1 deficiency leading to the inhibition of Wnt signaling (Kim et al. 2005). Furthermore, Cdx2 null mutation leads to the appearance of heterotopic lesions with gastric phenotype, while transgenic expression in the stomach leads to endoderm differentiation of intestinal-type mucosa (Stringer et al. 2008). The final stages of morphological specification, namely that involving the appearance of the gastric glands as well as progenitor cell maintenance, are directed by the FGF10 signaling pathway (Spencer-Dene et al. 2006; Nyeng et al. 2007). FGF10 knock-out mice have an aberrant gland formation, as well as endocrine and parietal cell differentiation attenuated (Nyeng et al. 2007). In fish, the available data are scarce regarding the molecular cascade leading to gut differentiation, in particular that dealing with species, which have a functional stomach. Curiously, in the stomachless zebrafish Barx1 and Sox2 (SRY-related HMG-box transcription factor) orthologues are expressed in

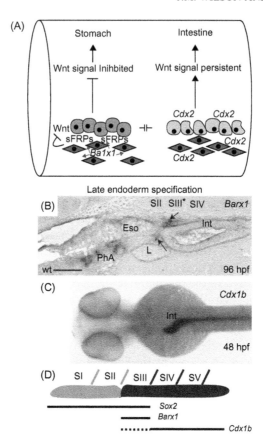

Fig. 1.8. Development of the gastrointestinal tract. (A) Epithelial–mesenchymal signaling in the gut tube, which will originate in the stomach and the intestine. Antagonism to local Wnt signaling is achieved by transient expression of Barx1, which regulates the expression of sFRPs by stomach mesenchymal cells. sFRPs in turn inhibit Wnt canonical expression. In the posterior part of gut, endoderm expression of Cdx2 (probably the default condition during development) promotes intestinal differentiation. Cdx2 maintains its expression in epithelial cells in the adult. (B) Expression of barx1 and (C) cdx1b through *in situ* hybridization in the teleost fish *Danio rerio*. (D) Schematic of expression domains of sox2, barx1, and cdx1b throughout the gastrointestinal tract. Note that due to the absence of gastric glands in zebrafish, an anterior intestinal domain of Barx1 expression is observed in this species (arrow). The potential temporal and spatial relationship of cdx1b and barx1 gene expression in zebrafish has not yet been demonstrated (dashed line). Abbreviations: Int intestine, Eso esophagus, PhA pharynx, somites SII, SIII, and SIV. (A) Adapted from Kim et al. (1995) and Zorn and Whells (2009) using data derived mostly from mouse. Adapted from (B, D) Muncan et al. (2007) and (C) Flores et al. (2008). See color plate section.

the mesenchyme and epithelium, respectively, of the proximal intestine as is observed in mammals (Muncan et al. 2007; Fig. 1.8B, D). This has led the authors to suggest that although the zebrafish is stomachless, a genetic footprint for stomach ontogeny has been retained (Muncan et al. 2007). As for the marker of gastric gland morphogenesis, FGF10, no data are available regarding gut expression. Regarding intestinal differentiation, in fish no Cdx2 gene exists. However, a different Cdx paralogue, Cdx1b, has replaced the function of Cdx2 in gut development, being responsible for the intestinal epithelium identity (Flores et al. 2008). Cdx1b has a distinct expression in the primitive gut endoderm and is first detected at around 24 hpf (Flores et al. 2008). The expression is maintained throughout development, becoming stronger at 36 hpf in embryos, and developing an asymmetric expansion in the rostral endoderm at 48 hpf (Fig. 1.8C), which will form the intestinal bulb in the adult. The molecular setting in *D. rerio* suggests that despite the absence of the gastric phenotype in the adult (see section on stomach loss), there is an apparent conservation in the signaling pathways controlling gut histo-differentiation (Fig. 1.8D). However, with current data it is difficult to discern whether Barx1 and Cdx1b colocalize in time and cell types (Fig. 1.8D dashed line). Finally, Cdx1b expression is maintained through adulthood in all the regions of the gut epithelium as in mammals (Flores et al. 2008).

An interesting challenge for future research involves the analysis of the developmental programs in fish species that have a functional stomach.

4. DIGESTIVE FUNCTION

The main function of the GIT is digestion and it follows that many, if not all, of the morphological features observed serve this end. The jaw was a major vertebrate innovation, which allowed not only an increase in the size of the food items that could be captured (micro- to macrophagous feeding; Mallatt 1984), but to a wider variety of capture techniques (Schwenk and Rubega 2005). It is widely believed that macrophagous feeding led to the enlargement of the foregut for storage and then the innovation of the stomach into the site for the chemical breakdown of food by acid-peptic digestion (Koelz 1992). However, there is recent fossil evidence (sediment infilling of the gut) suggesting that the enlargement of the gut occurred in thelodonts, an extinct agnathan group (Wilson and Caldwell 1993).

Fish can be classified according to trophic level feeding as (1) herbivores, (2) carnivores, (3) omnivores and (4) detritivores although the designation of these catagories is problematic (e.g. classification of marine fish grazing on complex assemblages of algae, bacteria and detritus; Clements and

Raubenheimer 2005). The components of the digestive system that have been described do not work in isolation but in combination with each other and accordingly different digestive mechanisms have been categorized into four main simplified types (Lobel 1981; Horn 1997) although with underlying variation and exceptions (Clements et al. 2009). Low gastric pH characterizes Type I digestion whereby acid lysis has been shown to be as effective as trituration in rupturing plant cell walls (Lobel 1981). Acid lysis leaves the plant cell walls more or less intact but the cellular contents are released for digestion and absorption. Low pH also leads to protein denaturation and the activation of the endopeptidase pepsin (pH < 4) initiating enzymatic digestion of food. The stomach is characterized as thin-walled (relative to the gizzard below) although the musculature is sufficient for trituration and mixing of the food allowing for a degree of mechanical breakdown. Type II digestion involves trituration by a gizzard-like stomach, which includes the families Mugilidae (mullets), Acanthuridae (surgeonfish), Kyphosidae (subfamily Girellinae, nibblers), Clupeidae (subfamily Dorosomatinae, gizzard shad) and Acipenseridae (sturgeon). The thick muscular pyloric region of the stomach in combination with ingested sand grains and a thick mucous coat is effective at triturating foods consisting of bacteria, cyanobacteria, algae, and detritus. Although these fish have gastric glands, the gastric pH is relatively high suggesting acid-peptic digestion is less important (Lobel 1981). In Type III digestion, trituration of food occurs in the pharyngeal apparatus. In some fishes with a well-developed pharyngeal mill, the stomach is completely absent and the finely ground food is passed directly to the intestine, such as in the cyprinids and labrids. Microbial fermentation (Type IV) is used by some herbivorous fishes to produce short-chain fatty acids (SCFA), which provide an energy source for enterocytes (Horn 1997; Clements and Raubenheimer 2005). Although hindgut chambers are present in Pomacanthidae (angelfishes) and Kyphosidae (sea chubs), microbial fermentation also occurs in species without such morphological specializations. The pyloric ceca of fish, in contrast, are not involved in microbial fermentation (Buddington and Diamond 1986).

5. NON-DIGESTIVE ADAPTATIONS

5.1. Osmoregulation

The importance of the GIT in osmoregulation in marine teleost fishes is reviewed by Grosell (Chapter 4). The imbibed seawater is desalted in the esophagus making it isoosmotic (with plasma), which is significant for the digestive processes (Holstein 1979) and the mechanism of water uptake in the intestine. Morphological changes associated with the osmoregulatory role of the esophagus have been observed by Yamamoto and Hirano (1978), Meister et al. (1983) and Cataldi et al. (1988). They have demonstrated that freshwater fish generally have a poorly developed columnar epithelium and that in marine and euryhaline fishes acclimated to seawater there is a significant increase in columnar epithelium in addition to a general increase in the esophageal surface area through increased folding, increased capillary diameter, and change in the columnar epithelial cells. Notably, there are dilations of the intercellular spaces and a high degree of apical interdigitations with neighboring columnar cells. In cod, high Na^+/K^+-ATPase immunoreactivity is found in these cells (Fig. 1.1a′). In the Nile tilapia, which does not osmoregulate well in seawater, Cataldi et al. (1988) did not observe similar modifications to its esophageal epithelium in response to seawater acclimation. The morphological changes in seawater-adapted fish are consistent with a paracellular ion flux route, which is supported by ion flux measurements (Chapter 4). Another interesting observation in *A. lupus* and *Dicentrarchus labrax* is the presence of a loose tubular system in cells similar to branchial chloride cells (Meister et al. 1983; Garcia Hernandez et al. 2001). In the marine eel pout (*Zoarces viviparous*) cells with an immunohistochemical signature identical to branchial chloride cells (basolateral Na^+/K^+-ATPase, NKCC and apical CFTR) were identified (Fig. 1.2D inset). However, the functional significance of these cells in the esophagus requires further work.

In response to seawater acclimation, trout have been shown to increase intestinal and rectal surface area through an increase in intestinal fold height (MacLeod 1977). A significant negative correlation between salinity and mucous cell density in both the intestine and rectum were also observed (MacLeod 1977; Nonnotte et al. 1986). At the subcellular level, greater numbers of mitochondria in the apical region of enterocytes are observed although few other differences in acclimated fish have been observed (Nonnotte et al. 1986). Two morphological types of enterocytes could be identified on the basis of the electron density of the cytoplasm (Abaurrea-Esquisoaín and Ostos-Garrido 1996b) but specifically relating them to ion regulation seems tenuous. The distribution of $Na^+:K^+:Cl^-$ cotransporter/ $Na^+:Cl^-$ cotransporter (NKCC/NCC) in the brush border of fish enterocytes, which facilitates NaCl uptake to osmotically draw water, is quite extensive, suggesting an overlap with nutrient absorptive functions (e.g. Chew et al. 2010; Chapter 4; Figs 1.2G, 1.9D).

Anadromous lamprey, which also drink seawater for marine osmo-regulation, have been found not to alter the morphology of their esophagus

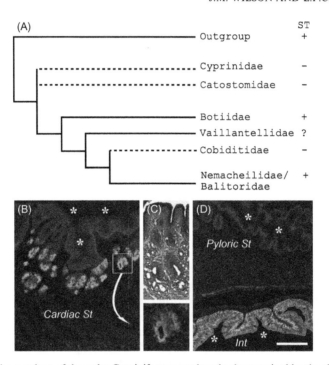

Fig. 1.9. The members of the order Cypriniformes are largely characterized by the absence of a stomach in Cyprinoidae, Catostomidae, Cobitidae; however, the stomach has been retained in the family Nemacheilidae/Balitoridae. (A) A phylogenetic tree redrawn from Slechtova et al. (2007) illustrates the relationship between the different families in this order. The families Nemacheilidae/Balitoridae are known to be gastric and data presented here indicate the same of their recent ancestor Botiidae. Stomach loss thus appears to have arisen independently in the Cobitidae and Cyprinidae/Catostomidae. (B and 2.5 × magnified inset) Presence of acinar-type gastric glands expressing apical H^+/K^+-ATPase (green) as determined by immunofluorescence microscopy. In the gastric glands $Na^+:K^+:2Cl^-$ cotransporter (red) has a basolateral distribution. (C) PAS staining (magenta) indicates the presence of a protective mucous-secreting epithelium in the stomach. (D) The intestinal epithelium (bottom) of Botiidae shows the presence of typical intestinal enterocytes with basolateral Na^+/K^+-ATPase (green) and brush border $Na^+:K^+:2Cl^-$ cotransporter (red) staining, while the pyloric region of the stomach shows no such staining. Asterisks indicate the gut lumen. Scale bar (B–D) 100 μm (inset) 40 μm. Micrographs (J. M. Wilson, O. Gonçalves, L. F. Castro unpublished). See color plate section.

or intestine in response to seawater acclimation (Youson 1981). Dilation of the intercellular spaces is observed in both freshwater- and seawater-adapted animals. More work is needed to address the question of whether seawater is processed in a similar manner as in teleost fishes. Although the largely marine chondrichthys do not drink seawater for osmoregulation, they have special rectal salt-secreting glands for ion regulation (Kirschner 1980; Hyodo et al. 2007).

5.2. Respiration

Accessory air-breathing organs of the GIT have evolved in a number of fishes as a means of surviving periods of aquatic hypoxia (Graham 1997; Chapter 10). Modifications of the intestine in Callichthyidae catfishes and Cobitidae loaches, and the stomach in some Loricariidae and Trichomycteridae catfishes includes the presence of intra-epithelial capillaries, and the loss of the columnar epithelium, the latter of which is otherwise a characteristic of these regions of the GIT (Fig. 1.3B). In the stomach a reduction of the overall thickness of the stomach wall is accompanied by a reduction of the gastric glands to shallow pits. In the catfish *Hypostomus plecostomus*, the cells lining the gastric glands were not even observed to have zymogen granules or the cellular machinery for its production (developed Golgi apparatus and RER) and the tubulovesicular system, characteristic of HCl secretion, was poorly developed and infrequently observed (Podkowa and Goniakowska-Witalinska 2003). The mucosal epithelium also lacked columnar mucous cells, which typically line the stomach in gastric species as protection from gastric juices. However, in the same species we have found weak apical H^+/K^+-ATPase immunoreactivity as well as granular eosinophilic (zymogen) staining associated with the acinar gastric glands (Fig. 1.3D, E). Also columnar mucous cells are present although with a distribution limited to the pyloric region. Taken together, these data would indicate that the stomach air-breathing catfish continues to have acid-peptic digestion although at a greatly diminished capacity.

In intestinal air-breathers as well, the typical intestinal morphology of columnar enterocytes is lost in place of a stratified epithelium with intra-epithelial capillaries (Graham 1997). Despite this apparent loss of absorptive capacity, nutrient transporters (SGLT1 and pepT1) are still expressed in this region of the gut indicating a dual function (Gonçalves et al. 2007).

The esophagus is another region of the GIT that is used as an accessory air-breathing organ. The synbranchoid fish *Monopterus albus* is known to be very hypoxia tolerant and better known for its use of its buccal cavity for gas exchange given that its gills are highly degenerated (Graham 1997). However, intraepithelial capillaries have also been reported in the esophageal epithelium (Liem 1967), although not consistently (Dai et al. 2007). On the other hand, the Alaska blackfish, *Dallia pectoralis* (Crawford 1974), and blenny, *Lipophrys pholis* (Laming et al. 1982), clearly have a well-vascularized esophagus with intraepithelial capillaries and a sphincter preventing air from entering the stomach.

5.3. Defense

The pufferfishes (Tetraodontiformes: Tetradontidae and Diodontidae) are able to defend against predation by inflating their bodies three-fold by swallowing water (Rosen 1912; Brainerd 1994). Rhythmical buccal pumping is used to force water into the "stomach," which will increase 50–100-fold in volume (Brainerd 1994). The "stomach" is a highly extensible sac-like diverticulum of the gut, lined by a thick, folded, stratified epithelium, similar in appearance to the extensible transitional epithelium of the mammalian urinary bladder (Brainerd 1994). When stretched the epithelium thins as the columnar-cuboidal cells become discoid. A strong esophageal sphincter and pylorus hold the water in. The stomach lacks digestive function (gastric glands) as in other Tetraodontiformes (Section 6). However, the stomach still serves for food storage, as pufferfish can be gluttonous feeders. The epithelium also contains mucous cells similar to the rest of the gut, which may aid food passage. The inflation behavior of pufferfishes is an example of a major functional innovation whose origin may lie in the evolutionary transition from coughing to water blowing to inflation in the Tetraodontiformes (Wainwright and Turingan 1997).

6. STOMACH LOSS IN FISHES

The stomach is a highly successful vertebrate innovation that presumably evolved in an early gnathostome ancestor 450 MYA with its success inferred from its conservation (Koelz 1992). The vertebrate stomach is defined by the secretion of gastric juice, which contains HCl and pepsinogen (Smit 1968).[4] The selective advantages of stomach acid secretion include improved digestion, through acid-peptic digestion (Barrington 1942), acid lysis of plant cell walls and invertebrate exoskeletons (Lobel 1981), enhanced dietary calcium (Koelz 1992) and phosphorus uptake (Sugiura et al. 2006), and immune system defense by providing a chemical barrier to pathogens entering the lower digestive tract (Koelz 1992; Chapter 3). Surprisingly, secondary loss of the stomach has occurred independently multiple times within fish lineages. A fact that has attracted the attention of comparative anatomists since the time of Cuvier (1805) and has since appeared in every major review (e.g. Jacobshagen 1937; Pernkopf and Lehner 1937; Barrington 1942, 1957; Bertin 1958; Harder 1975; Kapoor et al. 1975; Fange and Grove 1979). The absence

---

[4]"The stomach is a storage place where food is subjected to the initial stage of digestion. A true stomach always has gastric juice glands and can be closed by a sphincter at its caudal end" (Smit 1968). Incorrectly cited by Koelz (1992).

of the stomach in the extant an agnathan fishes the Myxini (Hagfish) and Cephalaspidomorphi (Lamprey) is assumed to indicate the basal agastric condition of the ancestor invertebrate chordate. However, the fossil record does suggest that an agnathan thelodont had an enlargement of the gut (Wilson and Caldwell 1993) presenting the possibility that the extant agnatha may also represent the condition of secondary loss. The presence of the stomach in the Elasmobranchii indicates that loss in the sister group Holocephali (Chimera), Lipidosireniformes (Lungfish), and in members of several Actinopterygii Orders was clearly secondary and not a transmitted trait from a common ancestor. There are no examples of stomach loss in the elasmobranchii or chondrosteans, and the only stomachless higher vertebrates are the monotremes (Koelz 1992; Stevens and Hume 2004).

It should also be emphasized that the stomach is more than merely an enlargement of the GIT for the temporary storage of food. The intestine bulb or swelling of agastric fishes likely serves a similar function but is not a stomach (Barrington 1942, 1957; Harder 1975; Kapoor et al. 1975). The measurement of stomach pH can be used to indicate the presence of functional gastric acid secretion (Western and Jennings 1970; Lobel 1981) although it can also lead to false-negative predictions for the presence of gastric glands (e.g. as in the mullet, Lobel (1981) and Payne (1978), and eel pout, MacKay (1929), Barrington (1957)) although high gastric pH levels call into question the physiological significance of acid-peptic digestion. Histology has been essential in identifying the presence of gastric glands (Section 2.2.2) although this too is not without controversy. In addition the expression (mRNA, protein) of pepsinogen and the gastric proton pump α and β subunits (ATP4A and ATP4B, respectively) can be used to assess the presence or absence of a stomach (Smolka et al. 1994; Douglas et al. 1999; Gawlicka et al. 2001; Darias et al. 2007; Figs 1.1G, 1.2E, 1.3D, 1.4A, 1.9B).

The analysis of genomic data for four agastric species (zebrafish, *Danio rerio*; medaka, *Oryzias latipes* and pufferfishes, *Tetraodon nigroviridis* and *Fugu rubripes*; www.ensembl.org) representing three lineages (Cypriniformes, Beloniformes, and Tetradontiformes) indicates that loss of the stomach phenotype is accompanied by loss of pepsinogen and gastric proton pump genes (ATP4A and ATP4B) (Castro, Gonçalves, Wilson, unpublished). According to Dollo's law[5] once lost, it is unlikely that the stomach (a complex organ) will re-emerge in a lineage given the almost exclusive gastric expression of pepsinogen, ATP4A and ATP4B genes although there are examples of the reversal of Dollo's law with other traits (Collin and Miglietta 2008).

---

[5]Dollo's law states that, following loss, a complex trait cannot re-evolve in an identical manner.

The identification of stomach loss events in the Actinopterygii is a challenge given the sparsity and temporal spread of the data, contradictory reports, and erroneous use of the name stomach for the foregut, compounded by the sheer size and diversity of this class of 44 orders, 453 families, 4,289 genera and 26,891 species (Nelson 2006). Also the most recent large-scale surveys of gut anatomy by Jacobshagen (1913) on approximately 100 European species, Suyehiro (1942) on 150 species from Japanese waters and Al-Hussaini (1947) on 60 Red Sea species for the most part make use of gross anatomical indicators of stomach presence or absence, and do not apply to the more rigorous definition of the true vertebrate stomach as the site of acid-peptic digestion and thus gastric glands. While in the majority of cases, the categorization of gastric and agastric fishes using these criteria are correct, a number of exceptions do arise.

Jacobshagen (1937) estimated stomach loss in 5% of families and 15% of fish species, while in studies by Pilliet (1885) and Suyehiro (1942) we can calculate corrected values of 20%. We estimate, based on revised phylogeny (Nelson 2006) and a thorough survey of the literature, that 7% of families and 20–27% of species are agastric. The conservative number of loss events (non-transmitted loss) is at least 15. The Cypriniformes and Perciforms contribute to the bulk of the numbers, while uncertainty about the gobies results in variation and also potentially the largest contribution to loss events.

6.1. Review of Stomach Loss in the Teleosts

Harder (1975) lists the orders Cyprinidae (carp) and Cobitidae (loaches), Cypridontidae (killifish), Gobiesocidae (clingfish), Gobioidei (Perciform suborder; gobies), Blennidae (blennies), Scaridae (parrotfish) as having the stomach absent. A careful review of the literature follows.

The cypriniformes comprise a large order (6 families: 321 genera: 3,268 species) that is predominantly agastric. The family Cyprinidae is probably entirely agastric (220 genera and 2,420 species), and in the superfamily Cobitoidea only one gastric family was identified, the Balitoridae ((59:590); Nelson 2006; Slechtova et al. 2007; Chen et al. 2008) with the presence of a stomach observed in *Nemacheilus angorae* (Suicmez and Ulus 2005) and *Barbatula barbatula* (Western 1971). The presence of this most recently diverged gastric family in an otherwise agastric clade indicates that multiple (2–3) loss events occurred (Fig. 1.9A) with the Cyprinidae and Catostomidae, and Cobitidae. Botiidae, which is ancestral to both Cobitidae and Balitoridae, has gastric glands and the attributes of acid-peptic digestion (Fig. 1.9B), thus providing an uninterrupted gastric lineage and making it highly improbable that the stomach was reinvented in the Balitoridae consistent with Dollo's law. It also indicates that the gastric trait should appear elsewhere in the Cobitoidea.

The higher teleost series Atherinomorpha is composed of the Orders of the Atheriniformes (silversides (6:48:312)), Beloniformes (needlefishes (5:36:227)), and Cyprinodontiformes (killifishes (10:109:1,013)) (Nelson 2006; Setiamarga et al. 2008). The Atherinomorpha show diverse phenotypes, yet from the numerous published studies on gut morphology they appear to be a completely agastric clade suggesting a common agastric ancestor (e.g. *Atherina forskali*, Al Hussaini 1947; *Tylosurus gavialoides*, *Strongylura leiura*, Manjaksay et al. 2009; *Fundulus heteroclitus*, Babkin and Bowie 1928). Within this series the Belonidae are an example of a highly carnivorous family that lack a stomach and have a straight, short gut (Manjaksay et al. 2009).

The plecognaths or Tetraodontiformes (9:101:357) include a bizarre collection of body shapes and the well-recognized pufferfishes (families Diodontidae and Tetraodontidae), which inflate their body by holding water in their stomach when they feel threatened (Rosen 1912; Brainerd 1994; Wainwright et al. 1995; see Section 5.3). This defensive feature of the sac-like stomach of the pufferfish likely arose after loss of the gastric glands since they are absent in the more basal members of this group, which appears entirely agastric (Edinger 1877; Rosen 1912).

The Gasterosteiformes (11:71:278) include the seahorses and pipefish (Syngnathidae), which are agastric (Pilliet 1885).[6] The absence of a stomach correlates with the microphagous feeding habit characteristic of this group (Barrington 1957; Bertin 1958). Within the stickleback family Gasterostei-dae, gastric (*Gasterosteus aculeatus, G. spinachia*) and agastric (*Pungitius pungitius*) species have been identified (Rathke 1837; Edinger 1877; Oppel 1896); however, Hale (1965) has reported contradictory evidence for the presence of tubular gastric glands in *P. pungitius*.

The Siluriformes and Pleuronectiformes are two orders typically not associated with the agastric phenotype. The Siluriformes, the catfishes (35:446:2,867), are generally considered to have a true stomach; however, one family has been reported to have an agastric member, the eeltail catfish Plotosidae (10:35). The initial observation was made by Suyehiro (1942) based on gross morphological observations of the GIT of *Plotosus lineatus* (= *P. anguillaris*), which has been confirmed by histology and H^+/K^+-ATPase immunohistochemistry (JM Wilson, LF Castro and OM Gonçalves, unpublished observations). The gastric air-breathing catfishes (Section 5.2)

---

[6]Fange and Grove (1979) incorrectly suggest that syngnathids are gastric. Bertin (1958) is misquoted.

have still retained their gastric glands although the modifications of the stomach for air-breathing have greatly diminished their capacity for acid-peptic digestion (Fig. 1.3; Carter 1935, Podkowa and Goniakowska-Witalinska 2003). They may represent a group in the transition to complete loss of the stomach.

In the flatfishes, Pleuronectiformes (14:134:678), Grove and Campbell (1979) reported *Ammotretis rostrata* and *Rhombosolea tapirina* (family Rhombosoleidae (9:19)) as agastric based on gross morphological observations of the absence of a pyloric valve and pyloric ceca, entry of the bile duct just after the esophagus and the absence of Prussian blue staining, which is indicative of HCl secretion. Within the Pleuronectiformes there is a gradient in stomach size that has been related to feeding type (de Groot 1971) and stomach loss in Rhombosoleidae may be an extreme of this gradient. However, in larval *R. tapirina* Hart (1994) reported the development of the stomach at 20 days post hatch based on the presence of the pyloric valve in histological sections. It is possible that stomach is lost following the larval phase; however, given the uncertainties with the two diagnostic methods employed in these studies, further work is warranted to determine if stomach loss has indeed occurred in the flatfishes.

The Perciformes compose the largest and also the most diversified order of fishes (160:1539:10,033) (Nelson 2006). Three of the eight largest families include agastric members: Gobiidae (gobies) (210:1,950), Labridae (wrasses) (68:453), and Blenniidae (combtooth blennies) (56:360). The families Callionymidae, Gobiosocidae, and within the suborder Labroidei, the Scaridae, Odacidae, and Embiotocidae also contain agastric members.

There are a number of independent reports that the wrasses are agastric (Pilliet 1885; Suyehiro 1942; Chao 1973; Verigina and Kobegenova 1987). In *Pseudolabrus japonicas*, Yasugi (1987) later confirmed the absence of pepsinogen in this species. The Scaridae or parrotfishes (10:88) are marine herbivores, grazing on the algae living on dead coral. They rarely feed on live coral or sea grasses and the absence of the stomach has been well documented (Suyehiro 1942; Al-Hussaini 1946). In the Embiotocidae, or surfperches (13:23), absence of a stomach has been reported by Young and Fox (1936) in *Cymatogaster aggregate*, *Embiotoca jacksoni*, and *Microme-trus minima* ($=Abeona\ minima$) and Suyheiro (1942) reported the absence of a stomach in *Ditrema temmincki* using histological and gross morphological techniques, respectively. The absence of a stomach is not a monophyletic characteristic of this suborder since Cichlidae and Pomacentridae both use acid digestion (Fish 1960; Lobel 1981).

In the combtooth blennies (Blenniidae), the absence of gastric glands has been identified by histology in a number of species (*Blennius pholis* (= *Lipophrys pholis*; Pilliet 1885), *Blennius sanguinolentus* (= *Parablennius sanguinolentus*), *Blennius ocellatus* (Edinger 1877; Verigina 1978), *Lipophrys pavo, Parablennius tentacularis, Aidablennius sphinx* (Kobegenova and Verigina 1988)) and on the basis of morphology (*Salarias fasciatus*, Al-Hussaini 1947). Pilliet (1885) reported the absence of a stomach in *Lepadogaster bimaculatus*, which belongs to the family Gobiosocidae, the clingfishes (36:140). The loss event in the Blenniidae and Gobiosocidae can possibly be traced to a common agastric ancestor (Setiamarga et al. 2008). Separate loss events also occurred in the Callionymidae, or dragonets (10:182) and Odacidae or cales (4:12), with at least one agastric species being identified in each; *Callionymus lyra* by Pilliet (1885) and *Odax pullus* by Clements and Bellmont (1988), respectively.

The gobies, family Gobiidae (210:1,950), represent the second most species-rich family and show a high degree of morphological diversity. Consequently they have a difficult-to-elucidate phylogeny because divergence of a number of groups occurred concurrently or over a short time period (Thacker 2003; Nelson 2006; Neilson and Stepien 2009). There is uncertainty about whether the group is agastric or not since as in other groups the data are patchy and conflicting reports exist. In the subfamilies Gobiinae, Benthophilinae and Gobionellinae, acinar gastric glands have been described histologically in a number of species (*Gobius niger*, *G. cruentatus, Mesogobius batrachocephalus, Neogobius ophiocephalus*, *N. ratan, N. melanostomus, N. fluviatilis, N. gymnotrachelus, Gymnogobius urotaenia, Chaenogobius annularis* (Pailiet 1887; Oppel 1896; Kobegenova and Dzhumaliev 1991; Harada et al. 2003a; Jaroszewska et al. 2008); however, examples of stomach loss have also been noted in these groups as well (*Gillichthys mirabilis*, Van Noorden and Patent 1980; *Leucopsarion petersii*, Harada et al. 2003a,b)). Jaroszewska et al. (2008) and Kobegenova and Dzhumaliev (1991) also argue that gastric glands in gobies are not functional, which is supported by the absence of a protective mucus-secreting epithelium lining the stomach, although positive H^+/K^+-ATPase immunoreactivity would suggest otherwise (J.M. Wilson and S. Malakpoor unpublished). Thus, although gastric glands are present in gobies a lower level of gastric function may be present. Measurements of stomach pH and pepsin active in gobies are required to address this possibility.

The observations of Suyheiro (1942) on the presence of a stomach, although simple in structure, in a number of species of the goby subfamilies Oxudercinae (10:35) (*Boleophthalmus pectinirostris, Periophthalmus canto-nensis* (=*P. modestus*), Gobionellinae (*Pterogobius elapoides, Chasmichthys gulosus* (=*Chaenogobius gulosus*) and Amblyopinae (*Trypauchen vagina*)) should be taken with some caution since these observation were based on gross morphology and not histology. However, he generally notes the small size and cylindrical shape of the stomach and the absence of generally overt stomach indicators such as pyloric ceca. The review of mudskipper biology by Clayton (1993) is ambiguous on the presence of a stomach. Takahashi et al. (2006) did not report the presence of a stomach in *Periophthalmus modestus*, and state that mudskipper fishes do not have a stomach. However, there is supporting histological evidence from Milward (1974) and Kobegenova and Dzhumaliev (1991) who found tubular gastric glands in Oxudercinae species studied, although again without a protective mucus-secretory columnar epithelium.

6.2. Pedomorphosis

The clearest example of stomach loss in the gobies comes from the work of Harada et al. (2003a) on the pedomorphic ice goby *L. petersii*. The ice goby has a nektonic lifestyle that is uncharacteristic for gobies, which are typically benthic. Barrington (1957) first proposed partial neoteny as a potential mechanism for stomach loss using the eel leptocephalus larvae, which has a very long larval period during which stomach development is arrested, as an example in fish along with the better-known examples in amphibians. Tadpoles of Urodela have a normally developed stomach; however, in Anura is the development of the stomach is arrested before production of pepsin and HCl. Only after metamorphosis is development completed. *Pedomorphosis* is defined as any retention of juvenile features by adult descendants, *neoteny* as pedomorphosis by somatic retardation, and *progenesis* as pedomorphosis by accelerated reproductive development (Gould 1977). The underlying mechanism includes *heterochrony*, changes in the timing and/or rates of processes underlying ontogenic formation of morphological traits (Kon and Yoshino 2002). The ice goby is considered pedomorphic because it has retained a scaleless transparent body, swimbladder, and small pelvic fins characteristic of the larval phase, into adulthood. Harada et al. (2003a) also demonstrated that the gastric glands do not emerge in the gut in either juveniles or adults in contrast to the common goby (*G. urotaenia*) where gastric glands emerge during the larval period. Brunelli and Atella (1914) also reported that the gut is short and straight in pedomorphic species of gobies from the Mediterranean, suggestive of stomach loss as well. The phylogenetic analysis of three pedomorphic gobies (*Aphia minuta*, *Crystallogobius linearis* and *Pseudaphya ferreri*) from the Mediterranean strongly supports a polyphyletic origin, indicating that the heterochronic change leading to the retention of larval features in these species seems to have occurred independently in the ancestors of these species (Giovannotti et al. 2007).

The Japanese icefish *Salangichthys microdon* of the Osmeriformes (smelt) subfamily Osmerinae (9 genera, 31 species; Nelson 2006) is also pedomorphic retaining many larval features including the absence of gastric glands (Harada et al. 2005). In contrast, Ayu (*Plecoglossus altivelis*) from the sister subfamily Plecoglossinae develops gastric glands and pyloric ceca. Within the clupeiformes, which is presumably a gastric order (Harder 1975), a pedomorphic taxon has also been identified (see Kon and Yoshino 2002; Lavoué et al. 2008), raising the possibility of stomach loss in this order as well. Joss (2006) has proposed that the extant lungfishes, which are also agastric, may also be neotenic providing a potential mechanism for stomach loss although anatomical evidence for neoteny is still not firmly established in this group.

In both the ice goby (*L. petersii*) and icefish (*S. microdon*) the thyroid gland was poorly developed and pituitary thyroid-stimulating hormone (TSH) cells (thyrotrophs) were absent at the predicted metamorphosis-equivalent phase although well-developed thyroid follicles and thyrotrophs were only found in mature adults (Harada et al. 2003a, 2005). A direct effect of thyroid hormone on gastric gland development has been demonstrated in flounder (*Paralichthys dentatus*; Huang et al. 1998) whereby exogenous T4 stimulated the formation of gastric gland and was necessary for pepsingen expression, which both could be blocked with thiourea (T4 synthesis inhibitor). However, in adult ice gobies exogenous treatment with thyroid hormone (T3) did not alter neotenic features including the agastric phenotype, suggesting an obligatory neoteny (Harada et al. 2003b), which would support gastric gene loss. However, in fishes in general there does not appear to be a clear correlation between stomach absence/presence and thyroid axis development at early life history stages (Power et al. 2001; Liu and Chan 2002).

In summary, pedomorphosis in fishes appears to provide a mechanism for the loss of the stomach; however, one that appears indiscriminate. In common for all the examples given, neoteny maintains the larval pelagic lifestyle and since these nektonic fishes also are microphagous planktivores (e.g. La Mesa et al. 2008) the stomach would be unnecessary (such as in syngnathids; Bertin 1958) and thus its absence not selected against. Although no examples have yet been documented in fishes, stomach loss may not be a general condition of neoteny since in the most well-known example, the Axolotl (*Ambystoma mexicanum*), the stomach is present and functional (personal observation).

6.3. Drivers for Stomach Loss

Presumably, the costs of acid-peptic digestion and maintaining the stomach are high considering the need to pump acid and adjust for resulting systemic acid/base disturbances, and then neutralize the acid in the intestine

(Chapter 5 and 6). H^+ secretion itself requires ATP in a 1:1 ratio (Yao and Forte 2003) to acidify well-buffered food to low pH 2.4–4.2 (in fish, Lobel 1981). A barrier to protect gastric epithelium against the effects of acid also requires maintenance (Allen and Flemström 2005). There are, however, no estimates of the energetic costs of gastric digestion in fishes. Even in vertebrates the only example is the Burmese python (Secor 2009); however, the lack of a significant effect of inhibition of gastric acid secretion by the proton pumb inhibitor omeprazole on specific dynamic action (SDA) suggests it is minor (Andrade et al. 2004). In fishes, the peak metabolic rate difference in agastric and gastric fishes was similar (e.g. $1.7\times$ basal metabolic rate (BMR) in *L. pholis* and $1.6\times$ BMR in *Gadus morhua*, respectively; reviewed by Jobling 1981; Secor 2009); however, this is a cross study comparison. Clearly comparative studies are needed to address this question.

Several hypotheses have been proposed as summarized by Kapoor et al. (1975) and Kobegenova (1988). However, as a note of caution the evolutionary history of the agastric lineage would be required for determining the events leading to stomach loss in a given lineage. Examples of extant species with reduced gastric function may provide convincing correlations but remain speculative with regards to the actual driver of stomach loss in a particular lineage. In addition these factors may have acted in combination or independently to give rise to stomach loss in the different fish lineages but are nevertheless worth discussing.

1. Kobegenova (1988) argues that a historically unstable post-larval food supply was the selective pressure leading to neoteny and stomach loss. This seems a plausible explanation but one that is difficult to confirm.
2. Loss in marine fishes due to the alkalinization of the gastric juices by seawater swallowed with the food. This seems unlikely given the difference in buffer capacity of food versus seawater (Chapter 5 and 6; unpublished observations). Even in the tilapia *Alcolapia grahami* of the soda Lake Magadi (pH 10) with three times the buffer capacity of seawater, acid lysis is still used for digestion (Bergman et al. 2003).
3. A diet rich in calcium carbonate (e.g. coral, mollusk shell) would neutralize acid contents of the stomach leading to loss of peptic digestion (Labridae, Scaridae; Lobel 1981). In the example of the Anarhichadidae wolffish, gastric pH is high[7] (pH 7, Bray 1987) which correlates with the relatively poorly developed acinar gastric glands (Verigina 1974; Hellberg and Bjerkås 2000) and the relatively low pepsin activity (Papoutsoglou

---

[7]Kapoor et al. (1975) and Verigina (1974) misquote Szarski (1956) who they contend provides evidence of enhanced acid secretion in Anarhichadidae to neutralize dietary calcium carbonate. See Bray (1987) for details.

and Lyndon 2006). These factors suggest that acid-peptic digestion is not as important as in other species. In addition, this problem can be partially avoid by ejecting large shell fragments from the mouth or through rapid gastric evacuation. The Gillaroo trout (*S. stomachicus*) with its gizzard-like stomach specialized for a mollusk diet (Bridge 1904) would be an interesting species to study to follow up this point since the prediction would be for reduced gastric glands and a higher gastric pH that could be compared with closely related salmonids.

4. Consumption of food containing a large proportion of indigestible ballast material of low nutritional value, such as in mud-feeding microphagous and herbivorous species, would pose problems for gastric acidification. Storage of such material is useless and generally in these feeding types a reduction in the stomach and rapid gut transit times are observed. The bottom-feeding mullet is an example of the physiological loss of gastric digestion although gastric glands are still present (Lobel 1981). The gizzard performs trituration of the food at an alkaline pH.

5. The emergence of the headgut mechanisms for efficient trituration (teeth, pharageal mill; Rice and Lobel 2003; Schwenk and Rubega 2005) made the gastric glands superfluous. In the Holocephali, Cypriniformes, and Scaridae this would appear to be the case (Lobel 1981). However, there are examples of the operation of both headgut trituration and acid digestion (e.g. cichlids).

6. Improved alkaline digestion is difficult to determine since comparisons of closely related gastric and agastric species with similar diets have not been made and other data are highly variable (e.g. Hidalgo et al. 1999). However, since evolved capacity reflects natural loads, interpretation of such data would be problematic in any case (Diamond 1991).

7. There is limited availability of chloride such as in the freshwater environment for gastric HCl secretion (Cyprinidae). Interestingly, agastric cyprinids (carp and tench) have chloride uptake rates $<10\%$ of gastric species (trout, pike, perch; Williams and Eddy 1986), which correlates with fewer branchial chloride cells (Williams and Eddy 1988). However, gastric secretion of Cl^- would presumably be reabsorbed in the intestine minimizing loss in the feces and the need for uptake from the water.

8. In addition, alternative functions for the stomach such as gas exchange could result in gastric gland loss. In the gastric air-breathing catfishes (Section 2.2), the stomach is highly specialized for gas exchange and gastric glands are greatly reduced although gastric digestion still looks possible and the genes are still likely present. They appear to represent a transition group on the path to loss of gastric function.

It seems fitting to end this section with a quote from Reifel and Travill (1978) regarding the comments from Young (1950). "Fishes have followed their own lines of specialization, often with a confusing amount of parallel evolution and therefore have evolved features which are either unknown in higher forms or if known have appeared quite independently." The story of stomach loss makes an apt example.

7. FUTURE PERSPECTIVES

Far from being a dying line of investigation, morphological techniques still stand to illuminate many of the multifunctional aspects of the fish GIT being investigated. This is especially true with approaches such as *in situ* hybridization and immunohistochemistry, which complement *in vitro* expression (transcriptome and proteome) and physiological approaches where spatial information is essential for the interpretation of results and understanding of mechanisms. There are some outstanding questions which morphological techniques are ideally suited to address such as the origin of some digestive enzymes in the brush border (intestinal or pancreatic?), and if there are functional differences in enterocytes (differential expression patterns of ion and nutrient transporters). In addition the identification of chloride cells in the esophagus deserves a second look as well as the function of intestinal glands in the Gadidae.

Important lines of research will be needed to address the developmental genetics of the gut in species with a functional stomach. So far, the only data collected are in *Danio*, which is a stomachless teleost. We expect that *G. aculeatus*, for example, would be a good gastric candidate to perform future comparative genetics on gut development with mammalian counterparts. The comparison of signaling pathways between teleost lineages, which have a gastric epithelium and those that do not, should provide valuable insights into the origin of this phenotype. Finally, more basal vertebrate lineages such as the cartilaginous fishes and agnathans will provide a crucial evolutionary perspective of gut origin and development.

ACKNOWLEDGMENTS

This work was indirectly supported by grants PTDC/MAR/64016 and PTDC/MAR/098035 from the Portuguese Foundation for Science and Technology (FCT) to JMW and LFCC. The authors are indebted to O. Gonçalves and R. Urbatzka for help with translations of French and German texts. Special thanks to J.F. Steffensen (U. Copenhagen, Denmark) and S. Malakpoor (Gorgan U., Iran) for some fish samples and C. Pereira-Wilson and D.J. Randall for comments on draft versions of the chapter.

REFERENCES

Abaurrea-Esquisoaín, M. A., and Ostos-Garrido, M. V. (1996a). Cell types in the esophageal epithelium of *Anguilla anguilla* (Pisces, Teleostei). Cytochemical and ultrastructural characteristics. *Micron* **27**, 419–429.

Abaurrea-Esquisoaín, M. A., and Ostos-Garrido, M. V. (1996b). Enterocytes in the anterior intestine of *Oncorhynchus mykiss*: cytological characteristics related to osmoregulation. *Aquaculture* **139**, 109–116.

Abdel Magid, A. M. (1975). The epithelium of the gastro-intestinal tract of *Polypterus senegalus* (Pisces: Brachiopterygii). *J. Morph.* **146**, 447–456.

Adam, H. (1963). Structure and histochemistry of the alimentary canal. In: *Biology of Myxine* (A. Brodal and R. Fange, eds), pp. 256–288. University of Oslo Press, Oslo.

Albrecht, M. P., Ferreira, M. F. N., and Caramaschi, E. P. (2001). Anatomical features and histology of the digestive tract of two related neotropical omnivorous fishes (Characiformes; Anostomidae). *J. Fish Biol.* **58**, 419–430.

Al-Hussaini, A. H. (1946a). The anatomy and histology of the alimentary tract of the coral feeding fish *Scarus sordidus*. *Bulletin de l'Institut d'Egypte* **27**, 349–377.

Al-Hussaini, A. H. (1946b). The anatomy and histology of the alimentary tract of the bottom-feeder *Mulloides auriflamma* (Forsk.). *J. Morph.* **78**, 121–153.

Al-Hussaini, A. H. (1947). The anatomy and histology of the alimentary tract of the plankton-feeder *Atherina forskali* Rupp. *J. Morph.* **80**, 251–286.

Al-Hussaini, A. H. (1949). On the functional morphology of the alimentary tract of some fish in relation to differences in their feeding habits: anatomy and histology. *Quarterly Journal of Microscopical Science* **90**, 109–139.

Allen, A., and Flemström, G. (2005). Gastroduodenal mucus bicarbonate barrier: protection against acid and pepsin. *Am. J. Physiol. Cell Physiol.* **288**, 1–19.

Anderson, C., and Campbell, G. (1988). Immunohistochemical study of 5-HT-containing neurons in the teleost intestine: relationship to the presence of enterochromaffin cells. *Cell Tissue Res.* **254**, 553–559.

Andrade, D. V., De Toledo, L. F., Abe, A. S., and Wang, T. (2004). Ventilatory compensation of the alkaline tide during digestion in the snake. Boa constrictor. *J. Experiment. Biol.* **207**, 1379–1385.

Babkin, B. P., and Bowie, D. J. (1928). The digestive system and its function in *Fundulus heteroclitus*. *Biol. Bull.* **54**, 254–278.

Barrington, E. J. W. (1942). Gastric digestion in the lower vertebrates. *Biological Reviews* **17**, 1–27.

Barrington, E. J. W. (1957). The alimentary canal and digestion. In: *The Physiology of Fishes* (M. E. Brown, ed.), pp. 109–161. Academic Press, New York.

Barrington, E. J. W. (1972). The pancreas and intestine. In: *The Biology of Lampreys*, pp. 135–169. Academic Press, New York.

Bartels, H., and Potter, I. C. (1995). Structural organization and epithelial cell types of the intestinal diverticula (protopancreas) of ammocoetes of southern hemisphere lampreys: functional and hylogenetic implications. *Cell Tissue Res.* **280**, 313–324.

Bauermeister, A. E. M., Pirie, B. J. S., and Sargent, J. R. (1979). An electron microscopic study of lipid absorption in the pyloric caeca of rainbow trout (*Salmo gairdneri*) fed wax ester-rich zooplankton. *Cell Tissue Res.* **200**, 475–486.

Bergman, A. N., Laurent, P., Otiang'a-Owiti, G., et al. (2003). Physiological adaptations of the gut in the Lake Magadi tilapia, *Alcolapia grahami*, an alkaline- and saline-adapted teleost fish. *Comp. Biochem. Physiol.* **136A**, 701–715.

Bertin, L. (1958). Appareil digestif. In: *Traité de Zoologie* (E. P. Grassé, ed.), pp. 1248–1302. Masson, Paris.

Bishop, C., and Odense, P. H. (1966). Morphology of the digestive tract of the cod. *Gadus morhua. J. Fisheries Res. Bd. Can.* **23**, 1607–1615.

Brainerd, E. L. (1994). Pufferfish inflation: functional morphology of postcranial structures in *Diiodon holocanthus* (Tetraodontiformes). *J. Morph.* **220**, 243–261.

Bray, R. A. (1987). A study of the helminth parasites of *Anarhichas lupus* (Perciformes: Anarhichadidae) in the North Atlantic. *J. Fish Biol.* **31**, 237–264.

Bridge, T. W. (1904). Fishes. In: *The Cambridge Natural History* (S. F. Harmer and A. E. Shipley, eds), Vol. VII, pp. 141–420. Macmillan and Co., Ltd, London.

Brunelli, G., and Atella, E. (1914). Ricerche sugli adattamenti alla vita planctonica (*I Gobidi planctonici*). *Biol. Zent* **34**, 458–466.

Buddington, R. K., and Christofferson, J. P. (1985). Digestive and feeding characteristics of the chondrosteans. *Env. Biol. Fish.* **14**, 31–41.

Buddington, R. K., and Diamond, J. (1987). Pylori ceca of fish: a "new" absorptive organ. *Am. J. Physiol.* **252**, G65–G76.

Carter, G. S. (1935). Respiratory adaptations of the fishes. *J. Lin. Soc. Lon.* 219–233.

Castro, N. M., Sasso, W. S., and Katchburian, E. (1961). A histological and histochemical study of the gizzard of the Mugil sp. *Acta Anatomica* **45**, 155–163.

Cataldi, E., Crosetti, D., Conte, G., D'Ovidio, D., and Cataudella, S. (1988). Morphological changes in the oesophageal epithelium during adaptation to salinities in *Oreochromis mossambicus, O. niloticus* and their hybrid. *J. Fish Biol.* **32**, 191–196.

Cataldi, E., Albano, C., Boglione, C., et al. (2002). *Acipenser naccarii*: fine structure of the alimentary canal with references to its ontogenesis. *J. Appl. Icthyol.* **18**, 329–337.

Chandy, M. (1956). On the oesophagus of the milk-fish *Chanos chanos* (Forskal). *J. Zool. Soc. India* **8**, 79–84.

Chao, L. N. (1973). Digestive system and feeding habits of the cunner, *Tautogolabrus adspersus*, a stomachless fish. *Fishery Bull.* **71**, 565–586.

Chen, W.-J., Miya, M., Saitoh, K., and Mayden, R. L. (2008). Phylogenetic utility of two existing and four novel nuclear gene loci in reconstructing Tree of Life of ray-finned fishes: the order Cypriniformes (Ostariophysi) as a case study. *Gene* **423**, 125–134.

Chew, S. F., Tng, Y. Y. M., Wee, N. L. J., Tok, C. Y., Wilson, J. M., and Ip, Y. K. (2010). Intestinal osmoregulatory acclimation and nitrogen metabolism in juveniles of the freshwater marble goby exposed to seawater. *J. Comp. Physiol. B.*

Clarke, A. J., and Witcomb, D. M. (1980). A study of the histology and morphology of the digestive tract of the common eel (*Anguilla anguilla*). *J. Fish Biol.* **16**, 159–170.

Clayton, D. A. (1993). Mudskippers. *Oceanogr. Mar. Biol. Annu. Rev.* **31**, 507–577.

Clements, K. D., and Raubenheimer, D. (2005). Feeding and nutrition. In: *The Physiology of Fishes* (D. H. Evans and J. B. Claiborne, eds), pp. 47–82. CRC Press, Boca Raton.

Clements, K. D., Raubenheimer, D., and Choat, J. H. (2009). Nutritional ecology of marine herbivorous fishes: ten years on. *Funct. Ecol.* **23**, 79–92.

Collin, R., and Miglietta, M. P. (2008). Reversing opinions on Dollo's law. *TRENDS Ecol. Evo.* **23**, 602–609.

Crane, R. K. (1968). A concept of the digestive-absorptive surface of the small intestine. In: *Handbook of Physiology Section 6 Alimentary Canal Vol. V Bile, digestion, ruminal physiology* (C. F. Code, ed.), pp. 2535–2542. American Physiological Society, Washington DC.

Crawford, R. H. (1974). Structure of an air-breathing organ and the swim bladder in the Alaska blackfish, *Dallia pectoralis* Bean. *Can. J. Zool.* **52**, 1221–1225.

Cuvier, G. (1805). La première partie des organes de la digestion. In: *Leçons d'anatomie comparée*. Baudouin, Paris.

Dai, X., Shu, M., and Fang, W. (2007). Histological and ultrastructural study of the digestive tract of rice field eel, *Monopterus albus*. *J. Appl. Icthyol.* **23**, 177–183.

Darias, M. J., Murray, H. M., Gallant, J. W., Douglas, S. E., Yúfera, M., and Martínez-Rodríguez, G. (2007). Ontogeny of pepsinogen and gastic proton pump expression in red porgy (*Pagrus pagrus*): determination of stomach functionality. *Aquaculture* **270**, 369–378.

de Groot, S. J. (1971). On the interrelationships between morphology of the alimentary tract, food and feeding behaviour in flatfishes (Pisces: Pleuronectiformes). *Neth. J. Sea Res.* **5**, 121–196.

Diamond, J. (1991). Evolutionary design of intestinal nutrient absorption: enough but not too much. *News Physiol. Sci.* **6**, 92–96.

Diaz, A. O., Escalante, A. H., Garcia, A. M., and Goldemberg, A. L. (2005). Histology and histochemistry of the pharyngeal cavity and oesophagus of the silverside *Odeontesthes bonariensis* (Cuvier and Valenciennes). *Anat. Histol. Embryol.* **35**, 42–46.

Domeneghini, C., Arrighi, S., Radaelli, G., Bosi, G., and Veggetti, A. (2005). Histochemical analysis of glycoconjugate secretion in the alimentary canal of *Anguilla anguilla* L. *Acta Histochem* **106**, 477–487.

Douglas, S. E., Gawlicka, A., Mandla, S., and Gallant, J. W. (1999). Ontogeny of the stomach in winter flounder: characterization and expression of the pepsinogen and proton pump genes and deterimination of pepsin activity. *J. Fish Biol.* **55**, 897–915.

Edinger, L. (1877). Ueber die schleimhaut des fischdarmes, nebst bemerkungen zur phylogenese der drusen des darmrohres. *Arch. Mikrosk. Anat.* **13**, 651–692.

Elliott, J. P., and Bellwood, D. R. (2003). Alimentary tract morphology and diet in three coral reef fish families. *J. Fish Biol.* **63**, 1598–1609.

Ezeasor, D. N. (1981). The fine structure of the gastric epithelium of the rainbow trout, *Salmo gairdneri* Richardson. *J. Fish Biol.* **19**, 611–627.

Ezeasor, D. N., and Stokoe, W. M. (1980a). Scanning electron microscopic study of the gut mucosa of the rainbow trout *Salmo gairdneri* Richardson. *J. Fish Biol.* **17**, 529–539.

Ezeasor, D. N., and Stokoe, W. M. (1980b). A cytochemical, light and electron microscopic study of the eosinophilic granule cells in the gut of the rainbow trout, *Salmo gairdneri* Richardson. *J. Fish Biol.* **17**, 619–634.

Ezeasor, D. N., and Stokoe, W. M. (1981). Light and electron microscopic studies on the absorptive cells of the intestine, caeca and rectum of the adult rainbow trout, *Salmo gairdneri*, Rich. *J. Fish Biol.* **18**, 527–544.

Falk-Peterson, I. B., and Hansen, T. K. (2001). Organ differentiation in newly hatched common wolffish. *J. Fish Biol.* **59**, 1465–1482.

Fange, R., and Grove, D. (1979). Digestion. In: *Fish Physiology* (W. S. Hoar, D. J. Randall and R. Brett, eds), pp. 162–260. Academic Press, New York.

Ferraris, R. P., and de la Cruz, M. C. (1987). Development of the digestive tract of milkfish, *Chanos chanos* (Forsskal): histology and histochemistry. *Aquaculture* **61**, 241–257.

Fish, G. R. (1960). The comparative activity of some digestive enzymes in the alimentary canal of tilapia and perch. *Hydrobiol.* 161–178.

Flores, M. V. C., Hall, C. J., Davidson, A. J., et al. (2008). Intestinal differentiation in zebrafish requires Cdx1b, a functional equivalent of mammalian Cdx2. *Gastroenterol.* **135**, 1665–1675.

Frierson, E. W., and Foltz, J. W. (1992). Comparison and estimation of absorptive intestinal surface areas in two species of cichlid fish. *Trans. Am. Fish. Soc.* **121**, 517–523.

Garcia Hernandez, M. P., Lozano, M. T., Elbal, M. T., and Agulleiro, B. (2001). Development of the digestive tract of sea bass (*Dicentrarchus labrax* L). Light and electron microscopic studies. *Anat. Embryol* **204**, 39–57.

Gargiulo, A. M., Ceccarelli, P., Dall'aglio, C., and Pedini, V. (1997). Ultrastructural study on the stomach of *Tilapia* spp (Teleostei). *Anat. Histol. Embryol.* **26**, 331–336.

Gas, N., and Noaillac-Depeyre, J. (1976). Studies on intestinal epithelium involution during prolonged fasting. *J. Ultrastruct. Res.* **56**, 137–151.

Gawlicka, A., Leggiadro, C. T., Gallant, J. W., and Douglas, S. E. (2001). Cellular expression of the pepsinogen and gastric proton pump genes in the stomach of winter flounder as determined by *in situ* hybridization. *J. Fish Biol.* **58**, 529–536.

German, D. P., and Horn, M. H. (2006). Gut length and mass in herbivorous and carnivorous prickleback fishes (Teleostei: Stichaeidae): ontogenetic, dietary, and phylogenetic effects. *Mar. Biol.* **148**, 1123–1134.

German, D. P., Nagle, B. C., Villeda, J. M., et al. (2010a). Evolution of herbivory in a carnivorous clade of minnows (Teleostei: Cyprinidae): effects on gut size and digestive physiology. *Physiol. Biochem. Zool.* **83**, 1–18.

German, D. P., Neuberger, D. T., Callahan, M. N., Lizardo, N. R. and Evans, D. H. (2010b). Feast to famine: the effects of food quality and quantity on the gut structure and function of a detritivorous catfish (Teleostei: Loricariidae). *Comp. Biochem. Physiol.* In press.

Giovannotti, M., Cerioni, P. N., La Mesa, M., and Caputo, V. (2007). Molecular phylogeny of the three paedomorphic Mediterranean gobies (Perciformes: Gobiidae). *J. Exp. Zool.* **308**, 722–729.

Gonçalves, A. F., Castro, L. F. C., Pereira-Wilson, C., Coimbra, J., and Wilson, J. M. (2007). Is there a compromise between nutrient uptake and gas exchange in the gut of *Misgurnus anguillicaudatus*, an intestinal air-breathing fish? *Comp. Biochem. Physiol.* **2D**, 345–355.

Gould, S. J. (1977). *Ontogeny and Phylogeny*, pp. 357. Belknap Press, Cambridge.

Govoni, J. J., Boehlert, G. W., and Watanabe, Y. (1986). The physiology of digestion in fish larvae. *Env. Biol. Fish.* **16**, 59–77.

Graham, J. B. (1997). *Air Breathing Fishes*, pp. 299. Academic Press, San Diego.

Grove, D. J., and Campbell, G. (1979). The role of extrinsic and intrinsic nerves in the co-ordination of gut motility in the stomachless flatfish *Rhombosolea tapirina* and *Ammotretis rostrata* Guenther. *Comp. Biochem. Physiol.* **63C**, 143–159.

Hale, P. A. (1965). The morphology and histology of the digestive systems of two freshwater teleosts, *Poecilia reticulata* and *Gasterosteus aculeatus*. *J. Zool.* **146**, 132–149.

Harada, Y., Harada, S., Kinoshita, I., Tanaka, M., and Tagawa, M. (2003a). Thyroid gland development in a neotenic goby (ice goby, *Leucopsarion petersii*) and a common goby (ukigori, *Gymnogobius urotaenia*) during early life stages. *Zoological Science* **20**, 883–888.

Harada, Y., Kinoshita, I., Kaneko, T., Moriyama, S., Tanaka, M., and Tagawa, M. (2003b). Response of a neotenic goby, ice goby (*Leucopsarion petersii*) to thyroid hormone and thiourea treatments. *Zoological Science* **20**, 877–882.

Harada, Y., Kuwamura, K., Kinoshita, I., Tanaka, M., and Tagawa, M. (2005). Histological observation of the pituitary-thyroid axis of a neotenic fish (the ice fish, *Salangichthys microdon*). *Fisheries Science* **71**, 115–121.

Harder, W. (1975). *Anatomy of Fishes*, pp. Pt 1 612, Pt 2 132. E. Schweizerbart'sche Verlagsbuchhandlung, Stuttgart.

Hart, P. R. (1994). Factors affecting the early life history stages of hatchery-reared greenback flounder (*Rhombosofea tapirha* Ginther 1879), pp 231. University of Tasmania, Launceston.

Hellberg, H., and Bjerkås, I. (2000). The anatomy of the oesophagus, stomach and intestine in common wolffish (*Anarhichas lupus* L.): a basis for diagnostic work and research. *Acta Vet. Scand.* **41**, 283–297.

Hirschowitz, B. I. (1957). Pepsinogen: Its origins, secretion and excretion. *Physiol. Rev.* **37**, 475–511.

Hidalgo, M. C., Urea, E., and Sanz, A. (1999). Comparative study of digestive enzymes in fish with different nutritional habits. Proteolytic and amylase activities. *Aquaculture* **170**, 267–283.

Holmgren, S. and Nilsson, S. (1999). Digestive system. In: *Sharks, Skates and Rays*, pp. 144–173. Johns Hopkins University Press, Baltimore.

Holmgren, S., and Olsson, C. (2009). The neuronal and endocrine regulation of gut function. In: *Fish Neuroendocrinology* (N. J. Bernier, G. Van der Kraak, A. P. Farrell and C. J. Brauner, eds), pp. 467–512. Academic Press, Amsterdam.

Holstein, B. (1979). Gastric acid secretion and drinking in the Atlantic cod (*Gadus morhua*) during acidic or hyperosmotic perfusion of the intestine. *Acta Physiol. Scand.* **106**, 257–265.

Hopkins, G. S. (1895). On the enteron of American ganoids. *J. Morph.* **11**, 411–442.

Horn, M. H. (1989). Biology of marine herbivorous fishes. *Oceanogr. Mar. Biol. Annu. Rev.* **27**, 167–272.

Horn, M. H. (1997). Feeding and digestion. In: *The Physiology of Fishes* (D. H. Evans, ed.), pp. 43–63. CRC Press, Boca Raton.

Hossain, A. M., and Dutta, H. M. (1996). Phylogeny, ontogeny, structure and function of digestive tract appendages (caeca) in teleost fish. In: *Fish Morphology. Horizon of New Research* (J. S. Datta Munshi and H. M. Dutta, eds), pp. 59–76. AA Balkema Pub., Brookfield VT.

Huang, L., Schreiber, A. M., Soffientino, B., Bengtson, D. A., and Specker, J. L. (1998). Metamorphosis of summer flounder (*Paralichthys dentatus*): thyroid status and the timing of gastric gland formation. *J. Exp. Zool.* **280**, 413–420.

Humbert, W., Kirsch, R., and Meister, M. F. (1984). Scanning electron microscopic study of the oesophageal mucous layer in the eel, *Anguilla anguilla* L. *J. Fish Biol.* **25**, 117–122.

Hyodo, S., Bell, J. D., Healy, J. M., et al. (2007). Osmoregulation in elephant fish *Callorhinchus milii* (Holocephali), with special reference to the rectal gland. *J. Exp. Biol.* **210**, 1303–1310.

Ishida, J. (1935). Ciliated intestinal epithelium in teleosts. *Annot. Zool. Japan* **15**, 158–160.

Iwai, T. (1968). Fine structure and absorption patterns of intestinal epithelial cells in rainbow trout alevins. *Zeitschrift für Zellforschung* **91**, 366–379.

Iwamatsu, T. (2004). Stages of normal development in the medaka *Oryzias latipes*. *Mech. Dev.* **121**, 605–618.

Jacobshagen, E. (1913). Untersuchungen uber das Darmsystem der Fische und Dipnoer. *Jena. Zeit. Naturwiss.* **49**, 373–810.

Jacobshagen, E. (1937). Darmsystem. Mittel und Enddarm. Rumpfdarm. In: *Handbuch der Vergleichende Anatomie der Wirbeltiere* (E. Bolk, E. Gorppert, Kallius and W. Lubosch, eds), pp. 563–724. Urban & Schwarzenberg, Berlin.

Jaroszewska, M., Dabrowski, K., Wilczynska, B., and Kakareko, T. (2008). Structure of the gut of the racer goby *Neogobius gymnotrachelus* (Kessler, 1857). *J. Fish Biol.* **72**, 1773–1786.

Jobling, M. (1981). The influences of feeding on the metabolic rate of fishes: a short review. *J. Fish Biol.* **18**, 385–400.

Jobling, M. (1995). *Environmental Biology of Fishes* (Fish and Fisheries Series 16), pp. 455. Chapman & Hall, London.

Joss, J. M. P. (2006). Lungfish evolution and development. *General and Comparative Endocrinology* **148**, 285–289.

Kapoor, B. G., Smit, H., and Verighina, I. A. (1975). The alimentary canal and digestion in teleosts. *Adv. Mar. Biol* **13**, 109–239.

Kim, B.-M., Buchner, G., Milletich, I., Sharpe, P. T., and Shivdasani, R. A. (2005). The stomach mesenchymal transcription factor Barx1 specifies gastric epithelial identity through inhibition of transient Wnt signaling. *Dev. Cell* **8**, 611–622.

Kim, B.-M., Milletich, I., Mao, J., McMahon, A. P., Sharpe, P. A., and Shivdasani, R. A. (2007). Independent functions and mechanisms for homeobox gene Barx1 in patterning mouse stomach and spleen. *Dev.* **134**, 3603–3613.

Kirschner, L. B. (1980). Comparison of vertebrate salt-excreting organs. *Am. J. Physiol.* **238**, R219–R223.

Kobayashi, D., Jindo, T., Naruse, K., and Takeda, H. (2006). Development of the endoderm and gut in medaka, *Oryzias latipes*. *Develop. Growth and Differ.* **48**, 283–295.

Kobegenova, S. S. (1988). On a possible cause of reduction of the stomach in some teleosts. *Vopr. Ikhtiol.* **28**, 992.

Kobegenova, S. S., and Dzhumaliev, M. K. (1991). Morphofunctional features of the digestive tract in some Gobioidei. *Vopr. Ikhtiol.* **31**, 965–973.

Kobegenova, S. S., and Verigina, I. A. (1988). Structure of the digestive tract of some species of Blennioidei. *Vopr. Ikhtiol.* **28**, 543–591.

Koelz, H. R. (1992). Gastric acid in vertebrates. *Scand. J. Gastroenterol.* **27**, 2–6.

Kon, T., and Yoshino, T. (2002). Diversity and evolution of life histories of gobioid fishes from the viewpoint of heterochrony. *Mar. Freshwater Res.* **53**, 377–402.

Kramer, D. L., and Bryant, M. J. (1995). Intestine length in the fishes of a tropical stream: 2 Relationships to diet—the long and short of a convoluted issue. *Env. Biol. Fish.* **42**, 129–141.

Kurokawa, T., and Suzuki, T. (1995). Structure of the exocrine pancreas of flounder (*Paralichthys olivaceus*): immunological localization of zymogen granules in the digestive tract using anti-trypsinogen antibody. *J. Fish Biol.* **46**, 292–301.

Kuz'mina, V. V., and Gelman, I. L. (1997). Membrane linked digestion in fish. *Rev. Fish. Sci.* **5**, 99–129.

La Mesa, M., Borme, D., Tirelli, V., Di Poi, E., Legovini, S., and Umani, S. F. (2008). Feeding ecology of the transparent goby *Aphia minuta* (Pisces, Gobiidae) in the northwestern Adriatic Sea. *Scientia Marina* **71**, 99–108.

Laming, P. R., Funston, C. W., and Armstrong, M. J. (1982). Behavioural, physiological, and morphological adaptations of the shanny (*Blennius pholis*) to the intertidal habitat. *J. Mar. Biol. Assn. UK* **62**, 329–338.

Langille, R. M., and Youson, J. H. (1984a). Morphology of the intestine of prefeeding and feeding adult lampreys, *Petromyzon marinus* L.: the mucosa of the diverticulum, anterior intestine, and transition zone. *J. Morph.* **182**, 39–61.

Langille, R. M., and Youson, J. H. (1984b). Morphology of the intestine of prefeeding and feeding adult lampreys, *Petromyzon marinus* (L): the mucosa of the posterior intestine and hindgut. *J. Morph.* **182**, 137–152.

Langille, R. M., and Youson, J. H. (1985). Protein and lipid absorption in the intestinal mucosa of adult lampreys (*Petromyzon marinus* L.) following induced feeding. *Can. J. Zool.* **63**, 691–702.

Lavoué, S., Miya, M., Kawaguchi, A., Yoshino, T., and Nishida, M. (2008). The phylogenetic position of an undescribed paedomorphic clupeiform taxon: mitogenomic evidence. *Ichthyological Research* **55**, 328–334.

Liem, K. (1967). Functional morphology of the integumentary, respiratory, and digestive systems of the synbranchoid fish. *Monopterus albus. Copeia* **1967**(2), 375–388.

Liu, Y.-W., and Chan, W.-K. (2002). Thyroid hormones are important for embryonic to larval transitory phase in zebrafish. *Differentiation* **70**, 36–45.

Lo, M.-J., and Weng, C. F. (2006). Developmental regulation of gastric pepsin and pancreatic serine protease in larvae of the euryhaline teleost, *Oreochromis mossambicus. Aquaculture* **261**, 1403–1412.

Lobel, P. S. (1981). Trophic biology of herbivorous reef fishes: an alimentary pH and digestive capabilities. *J. Fish Biol.* **19**, 365–397.

MacKay, M. E. (1929). The digestive system of the eel-pout (*Zoarces anguillaris*). *Biol. Bull.* **56**, 8–23.

Mallatt, J. (1984). Feeding ecology of the earliest vertebrates. *Zool. J. Linnean Soc.* **82**, 261–272.

Manera, M., and Dezfuli, B. S. (2004). Rodlet cells in teleosts: a new insight into their nature and functions. *J. Fish Biol.* **65**, 597–619.

Manjaksay, J. M., Day, R. D., Kemp, A., and Tibbetts, I. R. (2009). Functional morphology of digestion in the stomachless, piscivorous needlefishes *Tylosurus gavialoides* and *Stongylura leiura ferox* (Teleostei: Beloniformes). *J. Morph.* **270**, 1155–1165.

Marchetti, L., Capacchietti, M., Sabbieti, M. G., Accili, D., Materazzi, G., and Menghi, G. (2006). Histology and carbohydrate histochemistry of the alimentary canal in the rainbow trout *Oncorhynchus mykiss. J. Fish Biol.* **68**, 1808–1821.

Meister, M. F., Humbert, W., Kirsch, R., and Vivien-Roels, B. (1983). Structure and ultrastructure of the oesophagus in sea-water and fresh-water teleosts (Pisces). *Zoomorph.* **102**, 33–51.

Michelangeli, F., Ruiz, M. C., Dominguez, M. G., and Parthe, V. (1988). Mammalian-like differentiation of gastric cells in the shark *Hexanchus griseus. Cell Tissue Res.* **251**, 225–227.

Milward, N. E. (1974). Studies on the taxonomy, ecology and physiology of Queensland mudskippers. PhD thesis. University of Queensland, 1–276 pp.

Muncan, V., Faro, A., Haramis, A. P. G., et al. (2007). T-cell factor 4 (Tcf712) maintains proliferative compartments in zebrafish intestine. *EMBO Reports* **8**, 966–973.

Murray, H. M., Wright, G. M., and Goff, G. P. (1994). A comparative histological and histochemical study of the stomach from three species of pleuronectid, the Atlantic halibut, *Hippoglossus hippoglossus*, the yellowtail flounder, *Pleuronectes ferruginea*, and the winter flounder, *Pleuronectes americanus. Can. J. Zool.* **72**, 1199–1210.

Neilson, M. E., and Stepien, C. A. (2009). Escape from the Ponto-Caspian: evolution and biogeography of an endemic goby species flock (Benthophilinae: Gobiidae: Teleostei). *Molecular Phylogenetics and Evolution* **52**, 84–102.

Nelson, J. S. (2006). *Fishes of the World*, pp. 601. John Wiley & Sons, Hoboken.

Ng, A. N. Y., de Jong-Curtain, T. A., Mawdsley, D. J., et al. (2005). Formation of the digestive system in zebrafish: III. Intestinal epithelium morphogenesis. *Developmental Biology* **286**, 114–135.

Nilsson, S., and Holmgren, S. (1994). *Comparative Physiology and Evolution of Autonomic Nervous System.* Harwood Academic Publishers, Chur, Switzerland.

Noaillac-Depeyre, J., and Gas, N. (1976). Electron microscopic study on gut epithelium of the tench (*Tinca tinca* L.) with respect to its absoptive functions. *Tissue and Cell* **8**, 511–530.

Noaillac-Depeyre, J., and Gas, N. (1978). Ultrastructural and cytochemical study of the gastric epithelium in a fresh water teleostean fish (*Perca fluviatilis*). *Tissue and Cell* **10**, 23–27.

Noaillac-Depeyre, J., and Gas, N. (1982). Ultrastructure of endocrine cells in the stomach of two teleost fish, *Perca fluviatilis* L. and *Ameiurus nebulosus* L. *Cell Tissue Res.* **221**, 657–678.

Nonnotte, L., Nonnotte, G., and Leray, C. (1986). Morphological changes in the middle intestine of the rainbow trout, *salmo gairdneri*, induced by a hyperosmotic environment. *Cell Tissue Res.* **243**, 619–628.

Nyeng, P., Norgaard, A., Kobberup, S., and Jensen, J. (2007). FGF10 signaling controls stomach morphogenesis. *Developmental Biology* **303**, 295–310.

Odense, P. H., and Bishop, C. M. (1966). The ultrastructure of the epithelial border of the ileum, pyloric caeca, and rectum of the cod, *Gadus morhua. J. Fish. Res. Bd. Can.* **23**, 1841–1843.

Oppel, M. A. (1896). *Lehrbuch der Vergleichenden Mikroskopischen Anatomie der Wirbeltiere I*, pp. 543. Verlag von Gustav Fischer, Jena.

Osman, A. H. K., and Caceti, T. (1991). Histology of the stomach of *Tilapia nilotica* (Linnaeus, 1758) from the River Nile. *J. Fish Biol.* **38**, 211–223.

Papoutsoglou, E. S., and Lyndon, A. R. (2006). Digestive enzymes of *Anarhichas minor* and the effect of diet composition on their performance. *J. Fish Biol.* **69**, 446–460.

Parker, T. S. (1880). On the intestinal spiral valve in the genus Raia. *Trans. Zool. Soc. Lon.* **11**, 49–61.

Payne, A. I. (1978). Gut pH and digestive strategies in estuarine grey mullet (Mugilidae) and tilapia (Cichlidae). *J. Fish Biol.* **13**, 627–629.

Pernkopf, E., and Lehner, J. (1937). Vorderdam. Vergleichende Beschreinbung des Vorderdarm bei den einzelnen Klassen der Cranioten. In: *Handbuch der Vergleichende Anatomie der Wirbeltiere* (E. Bolk, E. Gorppert, Kallius and W. Lubosch, eds), pp. 349–562. Urban & Schwarzenberg, Berlin.

Pilliet, M. A. (1885). La structure du tube digestif de quelques poissons de mer. *Bull. Soc. Zool. France* 283–308.

Podkowa, D., and Goniakowska-Witalinska, L. (2003). Morphology of the air-breathing stomach of the catfish. *Hypostomus plecostomus. J. Morph.* **257**, 147–163.

Power, D. M., Llewellyn, L., Faustino, M., et al. (2001). Thyroid hormones in growth and development of fish. *Comp. Biochem. Physiol.* **130C**, 459.

Purkerson, M. L., Jarvis, J. U. M., Luse, S. A., and Dempsey, E. W. (1975). Electron microscopy of the intestine of the African lungfish, *Protopterus aethiopicus. Anat. Rec.* **182**, 71–90.

Radaelli, G., Domeneghini, C., Arrighi, S., Francolini, M., and Mascarello, F. (2000). Ultrastructural features of the gut in the white sturgeon, *Acipenser transmontanua. Histol. Histopathol.* **15**, 429–439.

Rafn, S., and Wingstrand, G. (1981). Structure of intestine, pancreas, and spleen of the Australian lungfish, *Neoceratodus forsteri* (Krefft). *Zool. Scripta* **10**, 223–239.

Rathke, H. (1820). *Beiträge zur Geschichte der Thierwelt*, pp. 129. Gedruckt des Carl Heinrick Eduard Müller, Danzig.

Rathke, H. (1837). Anatomie der Fische. *Archiv fur Anatomie, Physiologie und Wissenschaftliche Medicin* **1837**, 335–356.

Reifel, C. W., and Travill, A. A. (1977). Structure and carbohydrate histochemistry of the esophagus in ten teleostean species. *J. Morph.* **152**, 303–314.

Reifel, C. W., and Travill, A. A. (1978). Structure and carbohydrate histochemistry of the stomach in eight species of teleosts. *J. Morph.* **158**, 155–168.

Reifel, C. W., and Travill, A. A. (1979). Structure and carbohydrate histochemistry of the intestine in ten teleostean species. *J. Morph.* **162**, 343–360.

Reite, O. B. (2005). The rodlet cells of teleostean fish: their potential role in host defence in relation to the role of mast cells/eosinophilic granule cells. *Fish & Shellfish Immunology* **19**, 253–267.

Rice, A. N., and Lobel, P. S. (2003). The pharyngeal jaw apparatus of the Cichlidae and Pomacentridae: function in feeding and sound production. *Rev. Fish Biol. Fish.* **13**, 433–444.

Rimmer, W. D., and Wiebe, W. J. (1987). Fermentative microbial digestion in herbivorous fishes. *J. Fish Biol.* **31**, 229–236.

Rombout, J. H. W. M., Lamers, C. H. J., Helfrich, M. H., Dekker, A., and Taverne-Thicle, J. J. (1985). Uptake and transport of intact macromolecules in the intestinal epithelium of carp (*Cyprinus carpio* L.) and the possible immunological implications. *Cell Tissue Res.* **239**, 519–530.

Rosen, N. (1912). Studies on the Plectognaths. *Arkiv for Zoologi* **7**, 1–23.

Ruiter, A. J. H., Hoogeveen, Y. L., and Bonga, S. E. W. (1985). Ultrastructure of intestinal and gall-bladder epithelium in the teleost *Gasterosteus aculeatus L.*, as related to their osmoregulatory function. *Cell Tissue Res.* **240**, 191–198.

Sala-Rabanal, M., Gallardo, M. A., Sánchez, J., and Planas, J. M. (2004). Na-dependent D-glucose transport by intestinal brush border membrane vesicles from gilthead sea bream (*Sparus aurata*). *J. Membrane Biol.* **201**, 85–96.

Schwenk, K., and Rubega, M. (2005). Diversity of vertebrate feeding systems. In: *Physiological and Ecological Adaptations to Feeding in Vertebrates* (J. Matthias Starck and T. Wang, eds). Science Publishers. Enfield, USA.

Secor, S. M. (2009). Specific dynamic action: a review of postprandial metabolic response. *J. Comp. Physiol. B* **179**, 1–56.

Setiamarga, D. H. E., Miya, M., Yamanoue, Y., et al. (2008). Interrationships of Atherinomorpha (medakas, flyingfishes, killifishes, silversides and their relatives): the first evidence based on whole mitogenome sequences. *Mol. Phylogenet. Evol.* **49**, 598–605.

Shephard, K. L. (1994). Functions for fish mucus. *Rev. Fish Biol. Fish.* **4**, 401–429.

Sire, M.-F., Lutton, C., and Vernier, J.-M. (1981). New views on intestinal absorption of lipids in teleostean fishes: an ultrastructural and biochemical study in the rainbow trout. *J. Lipid Res.* **22**, 81–94.

Slechtova, V., Bohlen, J., and Tan, H. H. (2007). Families of Cobitoidea (Teleostei; Cypriniformes) as revealed from nuclear genetic data and the position of the mysterious genera *Barbucca, Psilorhynchus, Serpenticobitis* and *Vaillantella. Mol. Phylogenet. Evol.* **44**, 1358–1365.

Smit, H. (1968). Gastric secretion in the lower vertebrates and birds. In: *Handbook of Physiology Section 6 Alimentary Canal Vol. V Bile, digestion, ruminal physiology* (C. F. Code, ed.), pp. 2791–2805. American Physiological Society, Washington DC.

Smith, D. M., Grasty, R. C., Theodosiou, N. A., Tabin, C. J., and Nascone-Yoder, N. M. (2000). Evolutionary relationships between the amphibian, avian, and mammalian stomachs. *Evo. Dev.* **2**, 348–359.

Smolka, A. J., Lacy, E. R., Luciano, L., and Reale, E. (1994). Identification of gastric H,K-ATPase in an early vertebrate, the Atlantic stingray *Dasyatis sabina. J. Histochem. Cytochem.* **42**, 1323–1332.

Spencer-Dene, B., Sala, F. G., Bellusci, S., Gschmeissner, S., Stamp, G., and Dickson, C. (2006). Stomach development is dependent on fibroblast growth factor 10/fibroblast growth factor receptor 2b-mediated signaling. *Gastroenterol.* **130**, 1233–1244.

Srivastava, A. S., Kurokawa, T., and Suzuki, T. (2002). mRNA expression of pancreatic enzyme precursors and estimation of protein digestibility in first feeding larvae of the Japanese flounder, *Paralichthys olivaceus. Comp. Biochem. Physiol.* **132A**, 629–635.

Stevens, E. D., and Hume, I. D. (2004). *Comparative Physiology of the Vertebrate Digestive System*, pp. 400. Cambridge University Press, Cambridge.

Stringer, E., Pritchard, C. A., and Beck, F. (2008). Cdx2 initiates histodifferentiation of the midgut endoderm. *FEBS Letters* **582**, 2555–2560.

Stroband, H. W. J. (1977). Growth and diet dependent structural adaptations of the digestive tract in juvenile grass carp (*Ctenopharyngodon idella*, Val.). *J. Fish Biol.* **11**, 167–174.

Stroband, H. W. J., and Debets, F. M. H. (1978). The ultrastructure and renewal of the intestinal epithelium of the juvenile grasscarp, *Ctenopharyngodon idella* (Val.). *Cell Tissue Res.* **187**, 181–200.

Stroband, H. W. J., and Kroon, A. G. (1981). The development of the stomach in *Clarias lazera* and the intestinal absorption of protein macromolecules. *Cell Tissue Res.* **215**, 397–415.

Stroband, H. W. J., and van der Veen, F. H. (1981). Localization of protein absorption during transport of food in the intestine of the grass carp, *Ctenopharyngodon idella* (Val). *J. exp. Zool.* **218**, 149–156.

Sugiura, S. H., Roy, P. K., and Ferraris, R. P. (2006). Dietary acidification enhances phosphorus digestibility but decreases H^+/K^+-ATPase expression in rainbow trout. *J. Exp. Biol.* **209**, 3719–3728.

Suicmez, M., and Ulus, E. (2005). A study of the anatomy, histology and ultrastructure of the digestive tract of *Orthrias angorae* Steindachner, 1897. *Folia Biologica (Krakow)* **53**, 95–100.

Sullivan, M. X. (1907). The physiology of the digestive tract of elasmobranchs. *Bull. Bureau Fish.* **27**, 1–27.

Suyehiro, Y. (1942). A study on the digestive system and feeding habits of fish. *Jap. J. Zool.* **10**, 1–303.

Szarski, H. (1956). Causes of the absence of a stomach in Cyprinidae. *Bull. Acad. Pol. Sci. Cl. II Biol.* **4**, 155–156.

Takahashi, H., Sakamoto, T., and Narita, K. (2006). Cell proliferation and apoptosis in the anterior intestine of an amphibious, euryhaline mudskipper (*Periophthalmus modestus*). *J. Comp. Physiol. B* **176**, 463–468.

Teshima, K., and Hara, M. (1983). Epithelial cells of the intestine of the freshwater stingray *Potamotrygon magdalenae* taken from the Magdalena River, Colombia, South America. *Bull. Japan. Soc. Sci. Fish* **49**, 1665–1668.

Thacker, C. E. (2003). Molecular phylogeny of the gobioid fishes (Teleostei: Perciformes: Gobioidei). *Mol. Phylogenet. Evo.* **26**, 354–368.

Van Noorden, S., and Patent, G. J. (1980). Vasoactive Intestinal Polypeptide-like Immunoreactivity in Nerves of the Pancreatic Islet of the Teleost Fish, *Gillichthys mirabilis*. *Cell Tissue Res.* **212**, 139–146.

Verigina, I. A. (1974). The structure of the alimentary canal in some of the northern Blennioidei. I The alimentary canal of the Atlantic wolffish (*Anarrhichas lupus*). *Vopr. Ikhtiol.* **14**, 1098–1103.

Verigina, I. A., and Kobegenova, S. S. (1987). Structure of the digestive tract of stomachless fishes of the family Labridae (Perciformes). *Vopr. Ikhtiol.* **28**, 1030–1034.

Verigina, I. A. (1978). Structural features of the alimentary canal in connection with the lack of a stomach in the Black Sea blenny, *Blennius sanguinolentus*. *Vopr. Ikhtiol.* **17**, 964–968.

Wainwright, P. C., Turingan, R. G., and Brainerd, E. L. (1995). Functional Morphology of Pufferfish Inflation: Mechanism of the Buccal Pump. *Copeia* **1995**(3), 614–625.

Wainwright, P. C., and Turingan, R. G. (1997). Evolution of pufferfish inflation behavior. *Evo.* **51**, 506–518.

Wallace, K. N., and Pack, M. (2003). Unique and conserved aspects of gut development in zebrafish. *Dev. Biol.* **255**, 12–29.

Weisel, G. F. (1979). Histology of the feeding and digestive organs of the shovelnose sturgeon, *Scaphirhynchus platorynchus*. *Copeia* **1979**(3), 518–525.

Wentworth-Thompson, D. (1910). *Historia Animalium*. The works of Aristotle translated into English, pp. 633. Clarendon Press, Oxford.

Western, J. R. H. (1971). Feeding and digestion in two cottid fishes, the freshwater *Cottus gobio* L. and the marine *Enophrys bubalis* (Euphrasen). *J. Fish Biol.* **3**, 225–246.

Western, J. R. H., and Jennings, J. B. (1970). Histochemical demonstration of hydrochloric acid in the gastric tubules of teleosts using an in vivo Prussian blue technique. *Comp. Biochem. Physiol.* **35**, 879–884.

Wetherbee, B. M., and Gruber, S. H. (1993). Absorption efficiency of the lemon shark *Negaprion brevirostris* at varying rates of energy intake. *Copeia* **1993**, 416–425.

Williams, E. M., and Eddy, F. B. (1986). Chloride uptake in freshwater teleosts and its relationship to nitrite uptake and toxicity. *J. Comp. Physiol. B* **156**, 867–872.

Williams, E. M., and Eddy, F. B. (1988). Anion transport, chloride cell number and nitrite-induced methaemoglobinaemia in rainbow trout (*Salmo gairdneri*) and carp (*Cyprinus carpio*). *Aqua. Toxicol.* **13**, 29–42.

Wilson, M. V. H., and Caldwell, M. W. (1993). New Silurian and Devonian forktailed "thelodonts" are jawless vertebrates with stomachs and deep bodies. *Nature* **361**, 442–444.

Yamamoto, M., and Hirano, T. (1978). Morphological changes in the esophageal epithelium of the eel, *Anguilla japonica*, during adaptation to seawater. *Cell Tissue Res.* **192**, 25–38.

Yamamoto, T. (1966). An electron microscope study of the columnar epithelial cell in the intestine of fresh water teleosts: goldfish (*Carassius auratus*) and rainbow trout (*Salmo irideus*). *Zeit. Zekkfirsch.* **72**, 66–87.

Yao, X., and Forte, J. G. (2003). Cell biology of acid secretion by the parietal cell. *Annual Review of Physiology* **65**, 103–131.

Yasugi, S. (1987). Pepsinogen-like immunoreactivity among vertebrates: occurrence of common antigenicity to an anti-chicken pepsinogen antiserum in stomach gland cells of vertebrates. *Comp. Biochem. Physiol.* **86B**, 675–680.

Yasugi, S., and Mizuno, T. (2008). Molecular analysis of endoderm regionalization. *Dev. Grow. Diff.* **50**, S79–S96.

Yasugi, S., Matsunaga, T., and Mizuno, T. (1988). Presence of pepsinogens immunoreactive to anti-embryonic chicken pepsinogen antiserum in fish stomachs: possible ancestral molecules of chymosin of higher vertebrates. *Comp. Biochem. Physiol.* **91A**, 565–569.

Young, R. T., and Fox, D. L. (1936). The structure and functional of the gut in surf perches (Embiotocidæ) with reference to their carotenoid metabolism. *Biological Bulletin* **71**, 217–237.

Youson, J. H. (1981). The alimentary canal. In: *The Biology of Lampreys*, pp. 95–189. Academic Press, New York.

Yuen, B. B. H., Wong, C. K. C., Woo, N. Y. S., and Au, D. (2007). Induction and recovery of morphofunctional changes in the intestine of juvenile carnivorous fish (*Epinephelus coioides*) upon exposure to foodborne benzo[a]pyrene. *Aqua. Toxicol.* **82**, 181–194.

ADDITIONAL READING

Clements, K. D., and Bellwood, D. R. (1988). A comparison of the feeding mechanisms of two herbivorous labroid fishes, the temperate *Odax pullus* and the tropical. Scams rubroviolaceus. *Aust. J. Mar. fw. Res* **39**, 87–107.

MacLeod, M. G. (1978). Effects of salinity and starvation on the alimentary canal anatomy of the rainbow trout *Salmo gairdneri* Richardson. *J. Fish Biol.* **12**, 71–79.

Press, C. M., and Evensen, O. (1999). The morphology of the immune system in teleost fishes. *Fish & Shellfish Immunology* **9**, 309–318.

FEEDING, DIGESTION AND ABSORPTION OF NUTRIENTS

ANNE MARIE BAKKE
CHRIS GLOVER
ÅSHILD KROGDAHL

1. INTRODUCTION

The main function of the alimentary tract of any animal is the acquisition of food with subsequent assimiliation of vital nutrients. The natural diet of fishes varies tremendously between fish species and their natural habitats. Likewise, the structure of the digestive tract varies in the vastly diverse order Pisces. Despite this, the functional characteristics of nutrient assimilation

The Multifunctional Gut of Fish: Volume 30
FISH PHYSIOLOGY

show general consistency and include: capturing of food; puncturing, crushing, grinding and/or mixing; secretion of digestive enzymes and other components; and digestion and absorption of nutrients. Other functions of the gastrointestinal tract are intrinsically associated with nutrient assimilation from food, including osmoregulation; secretion of hormones not only involved in regulation of digestion, but also metabolism and other bodily functions; and protection of the organisms from alien compounds and pathogens that may reach the alimentary tract with the food and water. Thus, the digestive tract represents a critical interface between the internal and external environments of the fish.

Digestion is the process of modifying and/or hydrolyzing feed and food polymers into molecules and elements that can be absorbed across the intestinal wall. Understanding of the digestive physiology of fish is a field still in its infancy. However, large amounts of knowledge have been generated over the last three to four decades reflecting the increasing importance of fish as production animals. Evolutionary aspects of digestive physiology have also sparked researchers' curiosity.

This chapter summarizes the current state of knowledge, or lack thereof, highlighting the similarities and differences of fish digestive physiology relative to that of the more well-studied mammalian systems. Important references, especially review aricles, are noted to aid readers in finding more detailed information as needed. Feeding strategies and development of digestive processes in fish larvae and juveniles (Portella and Dabrowski 2008; Zambonino-Infante et al. 2008), as well as physiological responses to changes in ingredient composition of formulated feeds for cultured fish species (Bakke-McKellep and Refstie 2008) have been recently and extensively reviewed and will therefore not be exhaustively summarized in this chapter.

2. FEEDING STRATEGIES

Knowledge of diet and feeding habits is essential for the understanding of various aspects of fish biology and for developing aquaculture feeds and feeding methods. During the last several decades more than a hundred publications have documented the stomach contents from different fish species. The range of nutrient sources of fish is diverse, ranging from detritus, phytoplankton, zooplankton, micro- and macroalgae, aquatic plants, aquatic plants and meiofauna to insects, crustaceans, mollusks, shellfish, fish, birds and mammals. From these studies fish have been categorized as herbivores, omnivores, or carnivores/piscivores. Some studies add insectivorous and detrivorous groups. In general, omnivores with a preference for carnivory are the most numerous group. Many fish species are opportunistic feeders, feeding on whatever food they can find in their habitat.

The geographical location, ecology and climate will account for the available biotope, the prey and also the possible predators (Starck 1999; Clements et al. 2009). For many fish species, the choice of nutrient sources varies during their lifecycles from hatch to mature stages, from season to season, and also throughout the day. As larvae, most fish species are considered carnivores, dependent on zooplankton for normal development. Tilapia (*Oreochromis mossambicus*) may switch seasonally between zoophagy, phytophagy and detrivory (Maitipe and De Silva 1985). Moreover, fish may feed intensively in some periods and fast in others. For example, anadromous salmonids feed while at sea and cease feeding as they reach rivers for spawning. Rapid adaptation to variations in diet quality and availability is possible because the gastrointestinal tract is functionally dynamic and the epithelial cells are quickly renewed.

A study of 18 carnivorous fish from the same body of water in Australia showed clear species differences in selection of food organisms (Platell and Potter 2001). Each species showed size-related changes in diet as they grew, ranging from small amphipods, mysids and copepods, to the ingestion of larger prey such as polychaetes, caridean decapods, isopods and small teleosts. The differences in mouth size and morphology could sometimes account for diet differences. However, this was not always the case. Large fish with large mouths do not always eat large food. The basking shark (*Cetorhinus maximus*), the world's second largest fish, is a filter feeder living on plankton (Sims 2008).

Similar observations were made in a study of fish in a reef area in South Africa. Most fish species observed in the study had specialized on certain food items, thereby occupying their own "dietary niche" within the ecosystem (Lechanteur and Griffiths 2003). Related species, such as Pacific salmon species, feed on nutrient sources at different levels in the food chain (Clements et al. 2009; Johnson and Schindler 2009). Chinook (*Oncorhynchus tschawytscha*) and coho (*O. kisutch*) seemed to feed at higher trophic levels than pink (*O. gorbuscha*), chum (*O. keta*) and sockeye (*O. nerka*) salmon. Although some overlap between niches will occur, this strategy aids in reducing competition between species for food.

Fish species vary greatly in the way they capture and transfer food to the gastrointestinal tract. In some fish, digestion is initiated in the mouth cavity by physical breakdown of prey animals. The great white shark (*Carcharodon carcharias*) and other sharks, use their teeth and jaws to hold and tear the prey with strong movements of the head before swallowing the pieces. Some, such as salmonids and bottom-feeding fish species, swallow large volumes of water along with their food. Whereas others, such as the basking shark, assemble plankton by filtering large amounts of water through the gills, straining out the food with gill rakers.

Despite differences in natural diets, fish, as other animals of both higher and lower phyla, show the ability to select food and utilize its components to fulfill their energy and nutrient requirements (Dabrowski 1993; Jobling 1995; Choat and Clements 1998; de la Higuera 2001; Houlihan et al. 2001; Raubenheimer et al. 2009). For piscivorous fishes, nutrient requirements, especially regarding protein and lipids, are relatively easily met, with well-balanced amino and fatty acid compositions in their diets. Herbivorous fish species, on the other hand, may have more difficulty in meeting their amino acid/protein requirements when feeding on plants with low protein content and an unbalanced amino acid profile. However, herbivorous fish species may be supplied with nutrients from symbiotic microbiota, either directly as sources of nutrients themselves or indirectly by microbial fermentation of food components otherwise non-degradable by the fishes' endogenous enzymes; for example supplying absorbable carbohydrates and short chain fatty acids (Jobling 1995; Kihara and Sakata 1997; Clements et al. 2009). The role of microbial fermentation may be variable among herbivorous fish species, as indicated by recent investigations, in which only low short-chain fatty acid concentrations were found in the intestines of different minnow species (family Cyprinidae) with varying dietary affinity (German et al. 2010), as well as wood-eating (*Panaque nocturnus*, *P.* cf. *nigrolineatus* "Maraon," and *Hypostomus pyrineusi*) and detritivorous (*Pterygoplichthys disjunctivus*) catfishes (German and Bittong 2009). But the former study in minnows did demonstrate that some gastrointestinal physiological differences exist between related taxa that have different feeding preferences: the herbivores had longer intestinal tracts with higher activities of the carbohydrases amylase and laminarase, whereas the carnivores had higher chitinase activities. Interestingly, no differences in trypsin or lipase activities were found, possibly reflecting a strategy to efficiently digest vital dietary protein and lipids. Thus, physiological adaptations to diet type can be observed but are more appropriately compared between closely related species with different diets than distantly related species with similar diets (Dabrowski 1993; Raubenheimer et al. 2009; German et al. 2010).

Feeding strategies employed in fish cultivation take into account the natural feeding strategies during the species' lifecycle, including biorhythms and meal frequency, as well as the hierarchy between fish. In addition, aspects regarding fish growth rate and body composition, influence of environmental temperature and salinity, and feed efficiency are also important factors (Houlihan et al. 2001). Formulated feed characteristics, such as pellet sinking rate, size, consistency, taste and color, must be suitable. Appropriate grow-out feeding technology is also important. Feeding technology in aquaculture has developed from the early manual feeding procedures for fish in small units to today's automated feeding systems controlling feeding rate and frequency, taking into account feed waste, photoperiod, light intensity and water current in accordance with growth capacity and appetite of fish kept in large cages, even in very exposed areas of the ocean. The different production technologies in ponds, race ways, tanks or net pens, with or without recirculation, greatly influence what is an optimal feeding strategy in aquaculture (Houlihan et al. 2001).

3. SECRETION

The secretions of the alimentary tract, including the mouth, stomach, intestines, anus, liver and pancreas, are vital for (1) aiding passage and mixing of food, as well as providing digestive enzymes and other components necessary for digestion of the chyme, (2) solubilizing nutrients to ensure optimal digestion and delivery of nutrients for absorption by the intestinal mucosa, (3) protecting the intestinal mucosa from harsh dietary components as well as from the alimentary tract's own acid, alkaline and enzyme secretions, and (4) protecting the entire organism against microbes and chemicals that may be detrimental to the animal's health and well-being (Shephard 1994). Secretions include water, ions, digestive enzymes (see Table 2.1), proteins such as mucins and humoral agents, as well as other biologically active compounds. Secretion of the antioxidant ascorbic acid into the proximal intestine has been suggested in common carp (*Cyprinus carpio*), although the specific site from which it is secreted and its significance is not known (Dabrowski 1990). Levels and types of secreted components can be specific for the various regions of the alimentary tract and give rise to conditions optimal for their various functions. Neural, humoral, mechanical and chemical stimuli derived from the meal mediate secretory responses (see Chapters 7 and 8; Holmgren and Olsson 2009). In addition to aiding in nutrient digestion and absorption, water and ion secretions are vital for osmoregulation in fish (see Chapters 4 and 5).

3.1. Mucus

Mucus is secreted from specialized mucus-producing cells dispersed in the mucosal lining all along the alimentary canal and contains mainly water, ions and mucins (Shephard 1994). Mucins are highly glycated proteins with great

Table 2.1

Principal digestive enzymes*. The corresponding proenzymes are shown in parentheses

Source	Enzyme	Substrate	Specificity or products
Stomach	Pepsins (pepsinogens)	Proteins and polypeptides	Peptide bonds adjacent to aromatic amino acids
Exocrine pancreas	Trypsin (trypsinogen)	Proteins and polypeptides	Peptide bonds adjacent to arginine or lysine
	Chymotrypsins (chymotrypsinogens)	Proteins and polypeptides	Peptide bonds adjacent to aromatic amino acids
	Elastase (proelastase)	Elastin, some other proteins	Peptide bonds adjacent to aliphatic amino acids
	Carboxypeptidase A (procarboxypeptidase A)	Proteins and polypeptides	Carboxy terminal amino acids that have aromatic or branched aliphatic side chains
	Carboxypeptidase B (procarboxypeptidase B)	Proteins and polypeptides	Carboxy terminal amino acids that have basic side chains
	Colipase (procolipase)	Fat droplets	Binds to bile salt–triglyceride–water interface, making anchor for lipase
	Pancreatic lipase	Triglycerides	Monoglycerides and fatty acids
	Cholesteryl ester hydrolase	Cholesteryl esters	Cholesterol and fatty acids
	Pancreatic α-amylase	Starch	1,4,α linkages, producing α-limit dextrins, maltotriose, and maltose
	Ribonuclease	RNA	Nucleotides
	Deoxyribonuclease	DNA	Nucleotides
	Phospholipase A (prophospholipase A)	Phospholipids	Fatty acids, lysophospholipids
Intestinal mucosa	Enterokinase	Trypsinogen	Trypsin
	Aminopeptidases	Polypeptides	N-terminal amino acid from peptide
	Dipeptidases	Dipeptides	Two amino acids
	Glucoamylase	Maltose, maltotriose	Glucose
	Sucrase	Sucrose	Fructose and glucose
	Nuclease and related enzymes	Nucleic acids	Pentoses and purine and pyrimidine bases
Cytoplasm of mucosal cells	Various peptidases	Di-, tri-, and tetrapeptides	Amino acids

*Adapted from Krogdahl and Sundby (1999) and Ganong (2009).

water-holding capacity. The number of mucus-secreting cells varies between the buccal, esophageal, gastric and intestinal sections and putatively so does the flow of mucus, from low in the mouth region to high in intestinal regions (Martin and Blaber 1984; Kupermann and Kuz'mina 1994; Tibbetts 1997; Sklan et al. 2004; Abate et al. 2006). However, this type of gradient may vary with factors such as species, feeding habits, diet composition, and disease conditions (Olsen et al. 2007; Diaz et al. 2008; Dezfuli et al. 2009; Manjakasy et al. 2009). The appearance of the first functional goblet cells in juvenile fish and larvae appears to coincide with the first exogenous feeding.

The adherent mucus of the gastric and intestinal mucosa contains glycoconjugates that may be neutral, acidic or basic, or a combination of these. The quality of gut mucosubstances can vary with factors such as fish species, developmental stage and intestinal region (Shephard 1994; Tibbetts 1997; Domeneghini et al. 1998; Scocco et al. 1998; Arellano et al. 1999; Neuhaus et al. 2007; Diaz et al. 2008; Cao and Wang 2009; Schroers et al. 2009). The mucus may also contain immunoglobulins (Bøgwald et al. 1994; Abelli et al. 2004; Swan et al. 2008) and possibly other antimicrobial components such as enzymes and peptides. These factors aid in protecting the mucosa from potentially harmful dietary and environmental noxes, as well as providing optimal conditions for efficient nutrient digestion and absorption (see Sections 4.3, 5.6 and 5.7).

3.2. Foregut and Stomach

In stomachless fish, the foregut leads directly into the intestine and apparently secretes mucus but not digestive components (Buddington and Kuz'mina 2000a; Logothetis et al. 2001; Horn et al. 2006; Manjakasy et al. 2009). The stomach in fish with a secretory stomach is responsible for storage and initial physical and enzymatic breakdown of the diet. The most anterior cardiac region is generally not secretory, whereas the central fundic and distal pyloric regions are equipped with simple and branched gastric glands (Anderson 1986; Buddington and Kuz'mina 2000b). These are lined with goblet cells secreting mucus (Morrison and Wright 1999), as well as oxynticopeptic cells in the base of the glands, which in fish secrete both concentrated hydrochloric acid (HCl) from apically located tubules and pepsinogen from more basally located zymogen granules (Norris et al. 1973; Bomgren et al. 1998; Douglas et al. 1999; Buddington and Kuz'mina 2000b; Gawlicka et al. 2001; Mazlan and Grove 2004).

Secreted HCl contributes to the initiation of digestion by denaturing protein and converting the inactive zymogen pepsinogen into the active proteolytic enzyme pepsin (Yufera et al. 2004; Wu et al. 2009). Stomach pH may thus vary between 1 and 5 depending on species, developmental stage, feeding state and time after a meal. Between meals, stomach pH remains low in some species, indicating continuous secretion of acid, but increases in other species, showing intermittent acid secretion (Pérez-Jiménez et al. 2009). In the transition from the larval to juvenile stage in fishes, the gastric glands become functional, coinciding with the expression of genes for pepsinogen and the proton pump (Douglas et al. 1999; Yufera et al. 2004; Darias et al. 2007).

Interestingly, some fish appear to be capable of producing and secreting chitinase from the gastric mucosa, as well as from that of the intestine (Fange et al. 1979; Divakaran et al. 1999; Gutowska et al. 2004; Ikeda et al. 2009), an adaptation that can aid in breaking down chitin-containing exoskeletons of their prey. Although the contribution of exogenous chitinase from the diet, e.g. from prey animals, is not clear, genomic studies show the presence of a gene coding for a chitinase-like protein in the pufferfish *Takifugu rubripes* (Altschul et al. 1997). This indicates that at least some fish may indeed have an endogenous chitinase production.

3.3. Intestine and Accessory Organs

The food entering the intestine, subsequently known as chyme, is mixed with secretions from the intestine as well as the accessory digestive organs—the pancreas and liver/gall bladder—the latter via ducts, the pancreatic and bile ducts, respectively. The secretions include components involved in digestion (see Sections 3.3.1 and 3.3.2) as well as electrolytes, most notably bicarbonate, which neutralizes the acidic pH of the chyme coming from the stomach so that the digestive enzymes in the intestine can perform at a pH close to their optimums. The pH in the intestinal content of fishes appears to be more alkaline than in mammals, ranging from 7 to 9 (Deguara et al. 2003; Fard et al. 2007), suggesting bicarbonate secretion may be higher in fishes than in mammals. One reason may be that bicarbonate is also important in osmoregulation by mediating ion and water absorption in the intestine, at least in marine species (see Chapters 4 and 5). The relative contribution of bicarbonate secretion from the pancreas, pancreatic ducts, bile, and/or intestine is not known (Wilson and Grosell 2003; Grosell and Genz 2006). In marine fishes, bicarbonate from the intestinal epithelial cells is secreted by an apically located Cl^-/HCO_3^- exchanger (Grosell et al. 2001), rather than the basolaterally located exchanger in mammals. The alkaline pH may increase along the intestinal tract with a slight drop towards the anus, possibly due to an increase in content of short-chain fatty acids produced by microbial fermentation as shown for both young and adult Senegal sole (*Solea senegalensis*; Yufera and Darias 2007). Table 2.2 shows pH along the intestinal tract of Atlantic cod (*Gadus morhua*; Albrektsen et al. 2009).

Table 2.2
pH of gut contents along the digestive tract of Atlantic cod (*Gadus morhua*)*

	Stomach	PR	MI1	MI2	MI3	DI
Average	2.6	7.5	7.8	8.0	8.1	7.0
Standard deviation	0.92	0.35	0.20	0.16	0.18	0.39
Minimum	0.8	4.6	7.1	7.6	7.2	5.9
Maximum	4.9	8.0	8.2	8.3	8.4	8.0

*The results are averages of measurements on 160 cod of about 300 g (Albrektsen et al. 2009). PR: Pyloric region; MI1: first 1/3 of the mid intestine; MI2: mid 1/3 of the mid intestine; MI3: distal 1/3 of the mid intestine; DI: distal intestine.

3.3.1. PANCREAS

Although not well characterized functionally, the exocrine pancreas of various fish species, whether discrete or diffuse in its anatomical structure and location, contains acinar cells with zymogen granules which produce and store digestive enzymes (Kurokawa and Suzuki 1995; for review see Krogdahl and Sundby 1999). The zymogen granules and/or intestinal content of at least some fish species have been shown to contain the pancreatic enzymes or enzymatic activities corresponding to lipase, co-lipase, phospholipase, α-amylase, the proteolytic enzymes trypsin, chymotrypsin, elastase, carboxypeptidases A and B, as well as DNAase and RNAase (Kurokawa and Suzuki 1995; Pivnenko et al. 1997; Krogdahl and Sundby 1999; Kurtovic et al. 2009). Some important characteristics of some enzymes studied so far are given in Table 2.1.

From the pancreatic tissue, these enzymes are released into ductules which converge into pancreatic ducts, finely structured and numerous in fish with diffusely located pancreatic tissue. The ducts release the enzymes into the lumen of the pyloric ceca and/or proximal intestine, or into the bile duct(s) (Kurokawa and Suzuki 1995; Krogdahl and Sundby 1999; Morrison et al. 2004). The proteolytic enzymes and co-lipase are secreted as pro-enzymes that are activated in the intestinal lumen, whereas the lipases and α-amylase are released in active forms. The cascade of events that lead to activation of the pro-enzymes is initiated by enterokinase, which is secreted from intestinal cells. Enterokinase activates trypsinogen to form trypsin (Ogiwara and Takahashi 2007), which in turn activates the other pro-proteases (Fig. 2.1; see review by Krogdahl and Sundby 1999).

There is some uncertainty as to which lipolytic enzymes predominate in fish. The various pancreatic lipases found in higher vertebrates may be more limited in number and/or their functions modified in fish. Colipase activity has been found in rainbow trout (*Oncorhynchus mykiss*; Leger et al. 1979) and the elasmobranch dogfish (*Squalus acanthius*; Sternby et al. 1984),

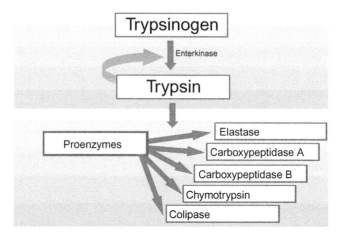

Fig. 2.1. Illustration of the activation of proenzymes secreted from the pancreas. Enterokinase from the intestinal mucosa activates trypsinogen to trypsin, which in turn activates the other proenzymes. Trypsin also shows autoactivation. Design: F. Venold.

whereas in other fish species such as Atlantic cod it has not been isolated despite efforts to do so (Gjellesvik et al. 1989, 1992). Several studies have attempted to verify the existence of an sn-1,3-specific pancreatic lipase, but this has been elusive (Olsen and Ringø 1997; Tocher 2003; Gottsche et al. 2005). Lipases purified or identified from red sea bream (*Pagrus major*) hepatopancreas and winter flounder (*Pseudopleuronectes americanus*) pancreatic tissue exhibit a bile-salt dependency (Iijima et al. 1998; Murray et al. 2003), indicating that bile salt-dependent carboxyl ester lipase (CEL) is secreted from pancreatic tissue of at least some species. In their review, Kurtovic et al. (2009) conclude that fish digestive lipases are either of the co-lipase-dependent pancreatic lipase (PL) or CEL type, and that PL may be present mainly in freshwater fish and CEL in marine species. However, some research indicates that in some fish species the wall of the digestive tract from the foregut to the distal-most regions may also be a source of lipases (Tocher 2003), since highest lipase activity was observed in the proximal intestinal region of most fish and in the distal region of others.

The amounts and activities of various secreted pancreatic enzymes appear to differ with species and/or their natural dietary preferences, although the diffusely located pancreatic tissue of many species makes such studies challenging. However, nutrient delivery into the intestinal lumen is the most important stimulus of exocrine pancreatic secretion in fish as in mammals (see review by Krogdahl and Sundby 1999). The nutrient composition and digestibility of the diet also differentially influences secretion of specific enzymes and other factors. For example, a diet containing a high level of protein, protein with low digestibility, and/or components that inhibit proteases, i.e. the antinutritional factors known as trypsin inhibitors in plant feedstuffs, has been shown to stimulate the pancreas to deliver a secretion with higher levels of trypsin in Atlantic salmon (*Salmo salar* L.; Olli et al. 1994; Krogdahl et al. 1999, 2003) and European sea bass (*Dicentrarchus labrax*; Péres et al. 1996, 1998). The relationship between dietary nutrient levels and corresponding enzyme secretions regarding lipids and carbohydrates appear to be more complicated (see reviews by Krogdahl and Sundby 1999; Morais et al. 2007; Zambonino Infante and Cahu 2007; Bogevik et al. 2009). Piscivorous species, such as Atlantic salmon, rainbow trout and sea bream (*Pagrus pagrus*) appear to have the capacity to secrete only low levels of amylase. In Atlantic salmon, amylase secretion is apparently not significantly up-regulated by carbohydrate content in the diet (Krogdahl et al. 2005; Frøystad et al. 2006). This is similar to the situation in some strictly carnivorous mammals.

3.3.2. LIVER

Bile acids, which are acidic steroids with powerful detergent properties, are produced in the hepatocytes and secreted from the liver/gall bladder via the bile duct(s). They aid digestion by emulsifying dietary lipids and fat-soluble vitamins and thereby allow for efficient action of lipases (see Section 3.3.1) and formation of micelles (see Section 4.3). The primary bile acids are cholic and chenodeoxycholic acids, which are formed from cholesterol. A large proportion of these are conjugated to taurine to form taurocholic acid, taurolithocholic acid and taurochenodeoxycholic acid, and some to glycine to form glycocholate in the few fish species studied to date (Haslewood 1967; Une et al. 1991; Velez et al. 2009). These primary bile acids may be further modified in the intestinal lumen by bacterial enzymes to form secondary bile acids, although the extent and significance of this in fish is largely unknown. Numerous studies have demonstrated that bile is also a medium for the excretion of many metabolites of endogenous and exogenous substances from the blood and liver in fishes.

The gall bladder is emptied by signals indicating that chyme is entering the intestine. Cholecytokinin (CCK) secreted via neural signals from endocrine cells lining the intestine may play a part in signaling both gall bladder and pancreatic secretion (see Chapters 7 and 8; Holmgren and Olsson 2009). During the course of fasting, the gall bladder becomes fuller and the bile more concentrated and darker in color. After a meal, it is usually more or less empty and light in color.

In many fishes, activities and amounts of the pancreatic enzymes and bile acids in the chyme decrease as the intestinal content moves distally toward the anus (see review by Krogdahl and Sundby 1999), presumably due to their recycling as in higher vertebrates. However, the extent of enterohepatic and enteropancreatic recycling of intact bile acids and pancreatic enzymes in fish is not known. At least some may be broken down into constituent parts, which are subsequently absorbed by the intestine.

4. DIGESTION

Digestion is the process of hydrolysis and solubilization of ingested nutrient polymers into molecules and elements suitable for transport across the intestinal wall. The digestive enzymes secreted from the stomach and exocrine pancreas are of major importance for enzymatic hydrolysis of complex food polymers, such as proteins, fats and carbohydrates, into smaller fragments. The resulting smaller fragments are further digested at the epithelium of the intestinal tract by the enzymes located in the brush border membrane of the enterocytes, releasing molecules small enough for absorption, i.e. small peptides and amino acids, monosaccharides, and fatty acids, respectively. This process is summarized in Fig. 2.2. However,

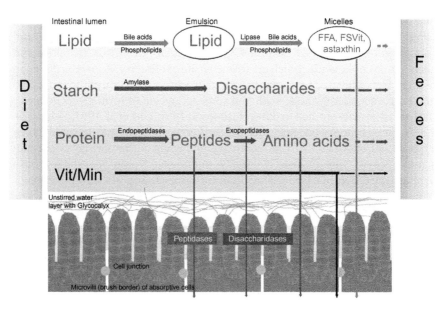

Fig. 2.2. Schematic drawing of the digestive processes along the digestive tract of fish. The location of various enzymes and other digestive components and the respective processes in the lumen as opposed to the intestinal mucosa are indicated. FFA: free fatty acids, FSVit: fat soluble vitamins. Design: F. Venold.

the contribution of exogenous digestive enzymes present in the natural diet to total digestive capacity has most likely been largely underestimated. A recent review focuses on exogenous contributions to digestion in fishes (see Kuz'mina 2008) and this chapter will therefore focus on endogenous gastrointestinal digestion processes. Considering the importance of providing cultured fish with highly digestible formulated feeds for rapid, cost-efficient fish growth and low waste released to the environment, the vast majority of the investigations on digestive processes and factors that affect nutrient digestibility have been carried out on production fish.

4.1. Digestive Enzyme Characteristics

Fish appear to have a digestive enzyme apparatus qualitatively similar to that of other animals with very similar substrate specificities across taxonomic groups (see Table 2.1). Although molecular characterizations are now being published with increasing frequency (see examples: Frøystad et al. 2006; Psochiou et al. 2007; Manchado et al. 2008; Kurtovic et al. 2009), knowledge is still limited regarding more specific characteristics of various digestive enzymes for most fish species. As in other animals, species-specific isoforms of the various enzymes exist with differences in, for example, molecular weights, specific activities, pH optima, and efficiencies towards different bonds (Pivnenko et al. 1997; Asgeirsson and Cekan 2006; Ogiwara and Takahashi 2007). Fish enzymes typically show higher specific activity and substrate affinities than in homoeothermic animals, presumably representing an evolutionary adaptation to function at lower temperatures (Pivnenko et al. 1997; Asgeirsson and Cekan 2006; Klomklao et al. 2006; Ogiwara and Takahashi 2007; Desrosiers et al. 2008; Jellouli et al. 2009). Trypsin from Atlantic cod, for example, has 17 times higher catalytic efficiency than bovine trypsin when measured at the same temperature range (Asgeirsson and Cekan 2006).

4.2. Protein Digestion

In fish species with stomachs, the low pH from HCl secretion denatures most of the proteins as they are solubilized, opening the structure for easier access by the proteolytic enzyme pepsin. Pepsinogen and pepsin from several fish species have been characterized (Wu et al. 2009). The enzyme seems to be present in fish in more than one form and the different forms show different activation rates, pH optima (varying between 1 and 5), specific activities and substrate specificities. Pepsins are endopeptidases, i.e. they hydrolyze peptide bonds, with a high affinity for hydrophobic bonds involving amino acids such as tyrosine and phenylalanine. The partial hydrolysis of the proteins increases solubility and dissolution of other food components, and prepares the diet—after this stage called chyme—for entry into the intestine through the pyloric sphincter.

Proteins and peptides entering the intestine, with or without prior processing in a stomach, are diluted and dissolved in alkaline secretions from the liver, pancreas and/or gut wall. The actions of the pancreatic endopeptidases trypsin, chymotrypsin and elastases I and II as well as the exopeptidases carboxypeptidase A and B result in a mixture of free amino acids and smaller peptides (see Fig. 2.2). The final steps of peptide hydrolysis take place at the brush border of the enterocytes by aminopeptidases or by intracellular peptidases following peptide transport across the membrane. However, some proteins and peptides entering the intestine either from the diet or with the gastrointestinal or pancreatic secretions may resist proteolysis and reach the distal intestine more or less intact.

Digestibility of protein from various diet sources varies among fish species. It is most often estimated in feeding trials as total apparent digestibility of protein (ADP) according to the following equation:

$$ADP = \left(1 - \left(P_{feces}/M_{feces}\right)/\left(P_{diet}/M_{diet}\right)\right) * 100$$

where P = protein concentration and M = indigestible marker concentration in the indicated material. Differences in ADP can be observed even in phylogenetically related species such as Atlantic salmon and rainbow trout (Krogdahl et al. 2004). Rainbow trout generally show higher ADP than Atlantic salmon. Greater variation, however, is seen between protein sources than between species (Adamidou et al. 2009; Borghesi et al. 2009; Tiril et al. 2009). Processing of ingredients before or during preparation of formulated feeds also affects digestibilities. For some proteins, heat treatment is necessary and increases digestibility by inactivating anti-nutrients. On the other hand, excessive heat treatment can reduce protein digestibility (Tran et al. 2008; Azaza et al. 2009; Tiril et al. 2009). Also, other factors such as the developmental stage of the fish, feeding rate, and environmental temperature may influence protein digestibility (Refstie et al. 2006; Venou et al. 2009).

4.3. Lipid Digestion

Lipids in natural fish food comprise mainly triglycerides, phospholipids, waxes, and free fatty acids, which are characterized by a high content of highly unsaturated fatty acids (Leaver et al. 2008). Efficient lipid digestion requires emulsifiers in the mixture of food—mainly proteins and phospholipids—as well as from endogenous bile acid and phospholipid secretion in the proximal part of the digestive tract. The emulsifiers orient themselves on the surface of lipid droplets that form as dietary lipid is released during the physical, chemical and enzymatic degradation of the food. If the emulsifying capacity is deficient, the digestion of released lipids may be hindered (Baeverfjord et al. 2006).

The main source of lipolytic enzymes in fish is the acinar cells of the exocrine pancreas (see Section 3.3.1; Kurtovic et al. 2009). Lipase activity differs between fish species as illustrated by the difference between even related species such as Atlantic salmon and rainbow trout (Bogevik et al. 2008). Active fishes such as mackerel (family Scombridae) and scup (family Sparidae) are among the species that have especially high activities (Kurtovic et al. 2009). Knowledge of characteristics and specificities of fish lipases is far from complete, as summarized in Section 3.3.1. Kurtovic et al. (2009) conclude that freshwater fishes may have mainly co-lipase-dependent pancreatic lipase (PL) whereas marine fishes have bile-acid-dependent carboxyl ester lipase (CEL). PL has higher specificity and digestive efficiency for triglycerides than the CEL. The latter hydrolyses a broader range of lipids including wax esters. Some studies have reported this lipase to be sn-1,3-specific (Tocher and Sargent 1984; Gjellesvik et al. 1989) which, concomitant with the metabolic advantage of re-esterification of monoacylglycerols within the enterocytes, opens up the possibility that bile-salt-dependent CEL in fish may actually possess sn-1,3-specific hydrolytic activity.

Fish lipases in general show higher affinity for glycerides with long, highly unsaturated fatty acids in contrast to mammalian lipases, which show highest activity towards ester bonds with fatty acids of chain length <20 carbons (Gjellesvik 1991; Gjellesvik et al. 1994). Also in contrast to mammals, at least some fishes seem to have the ability to hydrolyze wax esters, although at lower rates than triglycerides and phospholipids (Tocher and Sargent 1984; Olsen et al. 2004). Fish hydrolyze phospholipids quite efficiently. However, whether a specific phopholipase plays an important role in lipid digestion in fish is debated (Tocher 2003).

With great variation in lipid sources and lipid level in food organisms it would be expected that fish adapt enzyme level accordingly. A study with Atlantic salmon fed diets with varying proportions of fish oil and wax esters showed an ability to up-regulate both lipase activity and bile salt concentration with higher dietary levels of the less digestible wax esters compared to highly digestible fish oil (Bogevik et al. 2009).

Whether monoglycerols are the main products of lipolytic action in fishes, as in monogastric mammals, is not known. Results of a study of digestive processes in Atlantic salmon, summarized in Fig. 2.3, indicate that lipolysis in this species mainly produces free fatty acids (Krogdahl unpublished). Thus it would be expected from these results that most dietary fatty acids are absorbed as free fatty acids. However, the enzymes

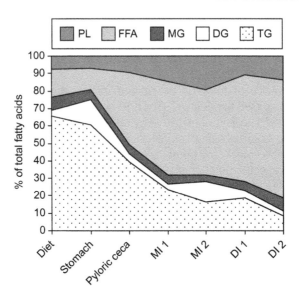

Fig. 2.3. Presence of fatty acids in various lipid classes as measured in the diet and along the intestinal tract of Atlantic salmon (0.5 kg body weight). The fish were fed a diet based on high-quality fish meal and marine oils (Krogdahl unpublished). PL: phospholipids; FFA: free fatty acids; MG: monoacyltriglycerides; DG: diacylglycerides; TG: triacylglycerides; MI1: proximal half of the mid intestine; MI2: distal half of the mid intestine; DI1: proximal half of the distal intestine; DI2: distal half of the distal intestine.

responsible for synthesis of triglycerides from absorbed fatty acids in the enterocytes seem to preferentially use monoglycerides as substrate, rather than glycerol (Oxley et al. 2007), suggesting that monoglyceride absorption may be important for optimal assimilation efficiency of triglycerides. The further fate of fatty acids in fish is not well understood but presumed to occur as in mammals (Tocher 2003). Thus, fatty acids liberated by lipolysis are picked up by primary micelles which turn into mixed micelles. Mixed micelles then become capable of assimilating more hydrophobic compounds such as fat-soluble vitamins and cholesterol esters. Mixed micelles purportedly disintegrate when they reach the so-called unstirred water layer of the intestinal mucosa. The mucus/water layer has a pH slightly lower than that of chyme and the difference facilitates the disintegration. The fatty acids are released from the micelles and absorbed (see Section 5.2).

Lipid digestibility has been reported in a great number of studies with cultivated fish species and generally lies in the range of 85–99%. The work has founded the basis for the development of prediction models for dietary lipid digestibility (Hua and Bureau 2009; Sales 2009). However, lipid digestibility varies between species and with developmental stage, feeding rate, water temperature, lipid source and its melting point, and diet composition (Røsjø et al. 2000; Hansen et al. 2008; Sales 2009). For example, in Atlantic salmon, digestibility of the individual dietary fatty acids decreases with increasing chain length and increases with increasing degree of desaturation (Røsjø et al. 2000). Further work is needed to understand the intricacies of lipid digestion and its regulation in fishes. Such knowledge would most likely lead to more optimal utilization of valuable, limited resources such as fish oil in formulated feeds for farmed fish.

4.4. Carbohydrate Digestion

Carbohydrates in natural fish diets and formulated feeds range from the highly soluble and digestible mono-, di- and oligosaccharides, glycogen and starch, to only marginally soluble and digestible chitin, hemicelluloses and celluloses. Fish species vary greatly in their capacity to digest and absorb even soluble carbohydrates. Some may have developed intestinal structures, functions and microbiota that enable hydrolysis of a greater variety of carbohydrates, although this appears to be variable even among herbivorous species (see Chapter 1 and Section 2 of this chapter). Fish have two categories of endogenous enzymes involved in carbohydrate digestion, pancreatic α-amylase and disaccharidases in the brush border membrane of the intestinal epithelial cells.

Pancreatic α-amylase hydrolyzes α(1-4) glucoside-linkages producing maltose and branched oligosaccharides from glycogen and starch. As starch is an uncommon dietary component for fish in the wild, the main role of the enzyme is most likely to hydrolyze glycogen in prey animals. The ability of fish to produce amylase is a great benefit for the aquaculture industry, which uses starch not only as a cheap energy source, but also as an efficient binder in formulated diets. Amylase activity varies greatly between fish species, but is generally highest in herbivorous species. For example, common carp has 150 times higher specific activity of intestinal amylase than the carnivorous eel (*Anguilla anguilla*; Hidalgo et al. 1999). Among carnivores, great differences in amylase activities have also been observed (Frøystad et al. 2006). However, the presence of amylase activity in the digestive tract may originate from not only the pancreas, but also from food organisms and/or gut microbiota (see review by Kuz'mina 2008). Amylase secretion and/or activity seems to be regulated according to intake of starch, at least in herbivorous and omnivorous species (see review by Krogdahl et al. 2005). The results are less clear regarding carnivores. In a study with rainbow trout, amylase activity decreased with increasing dietary starch level (Spannhof and Plantikow 1983). The seemingly inhibitory effect may, however, have been the result of amylase inhibitors in the diet overriding a possible stimulatory effect of dietary starch (Krogdahl et al. 2005).

Amylases from several fish species have been characterized regarding cDNA sequence showing a high degree of conservation (reviewed by Krogdahl et al. 2005). The sequence of Atlantic salmon amylase deviates especially from other fish amylases, showing a deletion in the peptide chain that may explain its inability to hydrolyze starch efficiently. It also shows a modification of an amino acid sequence of a signal peptide, which may affect import into the endoplasmatic reticulum during protein synthesis and result in low levels of amylase synthesis (Frøystad et al. 2006). Most amylases have two pH optima between 4 and 9, and different isoforms within a species may exist (Krogdahl et al. 2005).

Chitinase is a carbohydrase hydrolyzing β(1-4) bonds of the poly-saccharide n-acetylglucosamine, a major component of the exoskeleton of arthropods and cuticles of annelids and mollusks. Therefore, for many fish species chitin is a common food component. As discussed in Section 2, the origin of this enzyme may differ from species to species.

Polysaccharidase activity catalyzing cellulose degradation has been measured in the gut contents of several fish species (Das and Tripathi 1991; German and Bittong 2009). Both the intestine and hepatopancreas of the grass carp (*Ctenopharyngodon idella*) show cellulase activity and dietary cellulose level affected the activity significantly (Das and Tripathi 1991). Supplementation with antibiotics reduced the cellulase activity by one-third, indicating that the gut microbiota may, at least partly, be the source of cellulase activity. Microbial fermentation of undigested organic material such as cellulose is well documented in some fish species, and apparently of negligible importance in others (see Section 2). Short-chain fatty acids—acetate, propionate and some butyrate—are important products of carbohydrate fermentation and have been detected in the plasma (Seeto et al. 1996). The importance of short-chain fatty acids as energy sources for fish is not established. The role may be minor for most species (Krogdahl et al. 2005) but likely of at least some significance for fish living on plant material of low nutrient density.

The products of amylase and other carbohydrase activities are di- and oligosaccharides. Fish food may also contain sucrose and trehalose. These low-molecular-weight carbohydrates are further hydrolyzed by saccharidases located in the brush border membrane of enterocytes (see Fig. 2.2). In salmonids, hydrolysis of maltose is several times faster than that of other disaccharides (Krogdahl et al. 2003). The disaccharidase activities seem higher in herbivorous than carnivorous species (Ugolev and Kuz'mina 1994). Whether disaccharidase activities are regulated according to intake is not clear. The scientific literature gives conflicting information. Some studies indicate that at least some omnivorous and carnivorous species do not increase disaccharidase activities according to carbohydrate intake. However, a comparative study with rainbow trout and Atlantic salmon showed that both species increased disaccharidase activities up to 100% with increasing starch intake (Krogdahl et al. 2004). The reason for the conflicting results may be the unit of expression used in the different studies. Total enzymatic capacity of the intestinal tract per unit of body weight is recommended since fish may respond to increased levels of dietary carbohydrate by enlarging the intestine. Expressing enzyme activities as specific activity (related to amount of protein) or activity per weight of tissue may not reveal the effects on total enzymatic capacity.

Dietary starch digestibility has been analyzed in many studies. Variable experimental conditions make comparisons difficult and species differences have been specifically addressed in only a few studies. As mirrored by data on amylase activities, herbivorous carp show greater ability to digest starch than carnivorous rainbow trout, which in turn is more efficient than Atlantic salmon (Yamamoto et al. 2001; Krogdahl et al. 2004). Many species of fish show decreasing starch digestibilities with increasing dietary level, indicating only limited ability to increase digestive capacity to increased demand (reviewed by Krogdahl et al. 2005). Dietary lipid content as well as quality may also influence starch digestibility. Interestingly, short-term exposure of high dietary starch levels during early life may permanently influence carbohydrate digestive capacity in piscivorous rainbow trout (Geurden et al. 2007). Such feeding strategies can potentially increase the ability of the fish to utilize this cheap energy source and have protein- and lipid-sparing effects. In any case, present knowledge of carbohydrate digestibility is not sufficient to understand the great species variation in digestive efficiencies. Most studies are of applied character of little explanatory value regarding physiological responses and regulation.

5. ABSORPTION

Absorption of smaller, solubilized nutrients that have been released during digestion of dietary polymers are subsequently transported or otherwise absorbed across the apical (brush border) membrane (BBM) of the enterocytes lining the post-gastric alimentary tract, exit the cells across their basolateral (BL) membrane, and subsequently enter the circulatory system. Knowledge of absorptive mechanisms in fish gut is still rudimentary compared to that of mammals. However, general mechanisms appear to be conserved in fishes. Nutrients can enter (or exit) absorptive cells following a concentration gradient by simple diffusion or via the paracellular route. The contribution of the latter route has been considered negligible in fishes (Ferraris et al. 1990; Oxley et al. 2007). Nutrients can also enter the cell by specialized protein transporters, which are more or less specific for discrete chemical classes of nutrients and carry the nutrient across the cell membrane independent of substrate concentration gradient. Since there are limited numbers of transporters present in the membrane, transporter-mediated absorption is saturated at higher substrate concentrations. Depending on substrate, transport of macronutrients such as amino acids, peptides, monosaccharides and lipids may occur via either ion-dependent or facilitative, ion-independent mechanisms, or both. Ion-coupled transport can be dependent on ions such as Na^+, Cl^-, K^+ or H^+. Ion gradients are maintained by energy-requiring pumps or exchangers.

As is the case for the digestive enzymes, various nutrient transporters appear to have been conserved during evolution and thus, when investigated, have generally been found in fishes (see Table 2.3 and review by Collie and Ferraris 1995). Sequences encoding the Na^+/glucose symporter SGLT1 (Coady et al. 1990; Pajor et al. 1992) and the di- and tripeptide transporter PEPT1 (Verri et al. 2003) are examples. Any phylogenetic differences appear to be in substrate specificity and affinity constants, ion-to-nutrient coupling ratios, and transport rates. While fish and other poikilotherms absorb nutrients more slowly, their transporters have a higher affinity for nutrients than transporters in mammalian intestine. Besides temperature differences, these differences have been ascribed to lower nutrient concentrations in the intestinal lumen, transporter numbers (site density) per unit intestine, and total intestinal surface area in fish as compared to mammals (Karasov and Diamond 1981; Ferraris et al. 1989; Collie and Ferraris 1995). These observations appear to reflect the lower metabolic needs and energetic costs for locomotion in fish (Karasov et al. 1985).

Some studies of nutrient absorption by the fish intestine suggest transport mechanisms exist that do not exhibit saturation kinetics even at the highest concentrations utilized (Ferraris and Ahearn 1984; Bakke-McKellep et al. 2000; Nordrum et al. 2000a). Although the possibility that this non-saturable transport component is carrier-mediated is not decisively established, studies of neutral amino acid transport in sea bass (Balocco et al. 1993), *Boop salpa* (Boge et al. 2002) and dipeptide transport in tilapia (Thamotharan et al. 1996a) suggest that there may be ion-dependent low-affinity transporters for glycine and dipeptides. Christensen (1990) proposed that low-affinity transporters may exist for all amino acids. This remains to be established in higher vertebrates as well as in fishes. More research is needed to investigate the various absorptive pathways of different nutrients and their respective contribution to total nutrient assimilation in fishes.

Species and environmental differences such as temperature and water salinity influence nutrient absorption, as do differences between intestinal regions (Ferraris and Ahearn 1984; Collie 1985; Buddington et al. 1987; Collie

Table 2.3

Putative transport systems for hexoses and amino acids/peptides (as defined for mammalian systems in the first column) described and/or found expressed in the intestine of some fishes. Systems described in higher vertebrates but not in fishes are also indicated

System	Substrates	Location	Ion-dependency	Species	References
Hexoses					
SGLT1	D-glucose D-galactose	AM	Na$^+$	Gilthead sea bream *Sparus aurata*	Sala-Rabanal et al. 2004
				Atlantic salmon *Salmo salar*	Bakke-McKellep et al. 2008
				Asian weatherloach *Misgurnus anguillicaudatus*	Goncalves et al. 2007
				Pacific copper rockfish *Sebastes caurinus*	Ahearn et al. 1992
GLUT2	?	Transcripts found in intestinal tissue	?	Rainbow trout *Oncorhynchus mykiss*	Krasnov et al. 2001; Panserat et al. 2001
				Atlantic cod *Gadus morhua*	Hall et al. 2006
GLUT4	?	Transcripts found in intestinal tissue	?	Brown trout *Salmo trutta*	Planas et al. 2000
				Atlantic cod	Hall et al. 2006
Amino acids and peptides					
A	–	–	–	–	–
ASC	–	–	–	–	–
B^0	Neutral amino acids	AM	Na$^+$, Cl$^-$?	European eel *Anguilla anguilla*	Storelli et al. 1989
				Boops salpa	Boge et al. 2002
				European sea bass *Dicentrarchus labrax*	Balocco et al. 1993
b$^{0,+}$	–	–	–	–	

(Continued)

Table 2.3 (*continued*)

System	Substrates	Location	Ion-dependency	Species	References
IMINO/PAT	Imino acids (possibly regulated by ala)	AM	Na^+	European eel	Storelli et al. 1989; Vilella et al. 1989a
X_{AG}^-	Acidic amino acids	AM	Na^+	European eel	Storelli et al. 1989
X_c^-	–	–	–	–	–
y^+	Basic amino acids	AM		European eel	Storelli et al. 1989
Y^+L	–	–	–	–	–
L	Ala, gly, lys	AM	? (not Na^+)	European eel	Storelli et al. 1989
T	–	–	–	–	–
PepT1	Dipeptides	AM	H^+	Tilapia *Oreochromis mossambicus*	Thamotharan et al. 1996a
				Zebrafish *Danio rerio*	Verri et al. 2003
				Rockfish *Sebastes caurinus*	Thamotharan et al. 1996a
				European eel	Verri et al. 2000
				Icefish *Chionodraco hamatus*	Maffia et al. 2003
				European sea bass	Hakim et al. 2009; Sangaletti et al. 2009; Terova et al. 2009
				Asian weatherloach	Goncalves et al. 2007
PepT2	Dipeptides	AM	H^+	Zebrafish	Romano et al. 2006
Peptide transporter	Dipeptides	BM	H^+	Tilapia	Thamotharan et al. 1996b

AM: apical membrane; BM: basolateral membrane.

and Ferraris 1995; Lionetto et al. 1996; Buddington et al. 1997; Houpe et al. 1997; Bakke-McKellep et al. 2000; Nordrum et al. 2000a; Bakke et al. 2010) and differences caused by variations in natural dietary preferences, diet formulation and availability (Buddington and Hilton 1987; Nordrum et al. 2000a; Berge et al. 2004; Jutfelt et al. 2007; Terova et al. 2009). General agreement exists that net nutrient influx is higher per unit intestine in freshwater-adapted fish than in fish in seawater and increases with water temperature. The differences are ascribed to changes in electrophysiological characteristics and fluidity of the intestinal brush border membrane, respectively. A general trend of amino acid transporters having higher substrate affinities (lower K_ms) in the intestines of herbivorous and omnivorous fish species compared to those of transporters in carnivores has been suggested. As indicated for carbohydrate digestive capacity (see Section 4.4), maximum rates (V_{max}) of glucose transport in herbivores is generally greater than in carnivores (Ferraris and Ahearn 1984; Buddington et al. 1987). The authors suggest that this may reflect an adaptation to macronutrient levels in the respective natural diets and "facilitate the transport of the more common nutrient via a greater number of carriers or a higher transport rate, while assuring the absorption of the less common nutrient through high affinity processes" (Ferraris and Ahearn 1984).

Absorption of nutrients as studied *in vitro* generally appears to occur at a faster rate in tissues from the proximal intestinal regions compared to more distal regions in most (Collie 1985; Buddington et al. 1987; Dabrowski 1990; Bakke-McKellep et al. 2000; Jutfelt et al. 2007) but not all fish species studied—for example, sturgeon (*Acipensar* spp.; Buddington et al., 1987; Bakke et al. 2010). *In vivo* apparent absorption, as assessed by sequential nutrient digestibility along the intestinal tract or by histomorphological studies, confirms this and reveals that the proximal regions generally contribute more to nutrient absorption than more distal regions (Diaz et al. 1997; Olsen et al. 1999; Nordrum et al. 2000b; Hernandez-Blazquez et al. 2006). The pyloric ceca of Atlantic salmon and rainbow trout have been found to absorb up to 70% of nutrients *in vivo* (Buddington and Diamond 1987; Krogdahl et al. 1999; Nordrum et al. 2000b; Denstadli et al. 2004).

An interesting characteristic of the intestine of many fishes studied is that nearly the entire length of the post-gastric alimentary tract is capable of active nutrient transport (Ferraris and Ahearn 1984; Collie 1985; Bakke-McKellep et al. 2000). The enterocytes of even the most distal region are equipped with a brush border membrane in salmonids (Ezeasor and Stokoe 1981; Ingh et al. 1991; Murray et al. 1996), and is therefore not analogous to the mammalian colon. The permeability of this region appears to be higher for small water-soluble molecules than more proximal regions (Collie 1985; Schep et al. 1997; Bakke-McKellep et al. 2000; Jutfelt et al. 2007). The absorptive cells of this region also have numerous supranuclear absorptive vacuoles in salmonids that indicate that the distal intestine of some fish is important in the uptake of intact macromolecules (McLean and Ash 1987; McLean and Donaldson, 1990; Sire and Vernier 1992; Sire et al. 1992).

5.1. Protein/Peptide/Amino Acid Absorption

The carnivorous European eel is the teleostean species in which nutrient transport processes have been most systematically studied. As indicated in Table 2.3, investigators suggest the presence of at least four distinct Na^+-dependent amino acid transport systems in the eel (Storelli et al. 1989): an acidic, a basic and a neutral amino acid transport system, analogous to the mammalian systems X_{AG}^-, y^+ and B^0, respectively, along with a system responsible for proline and N-methylated amino acids (similar to IMINO and/or PAT systems). The latter system appears to differ substantially from that of mammals as it is totally inhibited by alanine, possibly by non-competitive inhibition (Vilella et al. 1989a), and partially by phenylalanine (Storelli et al. 1989). Na^+-independent transport systems have been found for alanine, glycine and lysine, suggesting the presence of a system analogous to the mammalian L system for neutral and basic amino acids, while proline and glutamate do not appear to be transported by such processes (Storelli et al. 1989). L-histidine appears to be transported by a highly specific transporter since a variety of other amino acids present did not affect transport (Glover and Wood 2008b). Substrate specificity of amino acid transporters may, however, vary between fish species (reviewed by Collie and Ferraris 1995).

The transport of peptides by the low-affinity/high-capacity H^+-dependent peptide transporters PepT1 and high-affinity PepT2 has also been demonstrated in various fishes (Thamotharan et al. 1996a,b; Maffia et al. 1997, 2003; Verri et al. 2000, 2003; Romano et al. 2006; Goncalves et al. 2007; Hakim et al. 2009; Sangaletti et al. 2009; Terova et al. 2009). Thus, peptide absorption followed by hydrolysis to amino acids within the cells or externalization as intact peptides comprises another possibility that may greatly contribute to the total amino acid assimilation (see review by Verri et al. 2010).

Free amino acids and small peptides seem to be absorbed earlier in the digestive process than intact proteins (Ambardekar et al. 2009). Absorption of larger peptides and/or proteins, most likely by endocytosis, has been demonstrated in distal intestinal regions of several fishes (McLean and Ash 1987; McLean and Donaldson 1990; Sire and Vernier 1992; Sire et al. 1992; Concha et al. 2002) including prion proteins in rainbow trout (Valle et al. 2008). Concha et al. (2002) suggested that intestinal fatty-acid-binding protein (I-FABP) may be involved by binding the peptide/protein before endocytosis. The relevance of protein absorption for fish is not entirely clear, but a source of amino acids as well as antigen presenting for mucosal immunological responses, as indicated by the successful use of some oral vaccines for fishes (Quentel et al. 2007), have been suggested.

5.2. Lipid Absorption

The absorption of the various lipid classes in fish is not well understood but is presumed to occur as in mammals (Tocher 2003). Fatty acids released from the micelles are thought to be absorbed by diffusion or facilitated transport. The involvement of transporters, as has been demonstrated in mammals (see review by Iqbal and Hussein 2009), has not been verified in fish. However, intestinal fatty acid binding protein (I-FABP), transcripts of which have been found to be expressed in zebrafish (*Danio rerio*) and common carp, may be involved in internalization of fatty acids (Andre et al. 2000; Concha et al. 2002) as well as macromolecular proteins in fish enterocytes (Concha et al. 2002).

The rate of lipid digestion and absorption in fishes is considered to be slower than that of mammals (Morais et al. 2005a). Although lipid absorption processes in fish mainly occurred in the proximal regions of the intestine and pyloric ceca (Diaz et al. 1997; Olsen et al. 1999; Røsjø et al. 2000; Denstadli et al. 2004; Hernandez-Blazquez et al. 2006), this may depend on lipid class, chain length and degree of saturation. Short- and medium-chain fatty acids, with their relatively high water solubility, are absorbed rapidly in the most proximal part of the intestine, leading to the hypothesis that they may not be incorporated into micelles (Røsjø et al. 2000; Denstadli et al. 2004). Conversely, saturated long-chain fatty acids, with their high degree of hydrophobicity and lower micellar solubility, may not easily reach the brush border and may therefore not be as readily absorbed as fatty acids with a similar chain length but with a lower degree of saturation. Free (non-esterified) fatty acids and phosphatidylcholine appear to be absorbed more readily than triacylglycerols (Morais et al. 2005a,b; Oxley et al. 2007).

Lipid droplets accumulate in the supranuclear space of enterocytes (Sire et al. 1981; Fontagne et al. 1998; Olsen et al. 1999; Hernandez-Blazquez et al. 2006), are metabolized (i.e. re-esterified into triacylglycerides), and packed into smaller lipoprotein particles similar to chylomicrons in mammals (Caballero et al. 2003), which appear to exit the enterocytes by exocytosis (Hernandez-Blazquez et al. 2006). In the intercellular spaces and lamina propria of the intestinal mucosa, the lipid particles vary in size and appear as very-low-density lipoproteins (VLDL) and chylomicrons

(Caballero et al. 2003). In most fishes, lipids are assumed to be carried via the circulatory system to other organs.

Lipid droplet accumulation in the enterocytes and number of lipoprotein particles in the intercellular spaces are affected by lipid source. High levels of dietary neutral lipid appear to cause an accumulation of large lipid droplets in the enterocytes of various species (Diaz et al. 1997; Fontagne et al. 1998; Olsen et al. 1999; Caballero et al. 2003). It has been suggested that this not only reduces intracellular metabolism and consequently lipid transport across the basolateral membrane, but also limits further absorption of lipid across the brush border membrane, as has been demonstrated in mammals. This helps explain lower lipid digestibility at high dietary lipid levels (Morais et al. 2005b). In piscivorous species such as gilthead sea bream (*Sparus aurata*) and Atlantic salmon, intracellular lipid accumulation in fish fed plant oils increased dose-dependently compared to control fish fed fish oil as the sole lipid source (Olsen et al. 1999, 2003; Caballero et al. 2003).

Dietary phospholipids, as important components of lipoproteins, have been found to greatly affect lipid digestion, absorption, intracellular metabolism and transcellular transport in cultured fish, especially during early stages of development (Geurden et al. 1998, 2008; Liu et al. 2002; Tocher et al. 2008). Especially phosphatidylcholine plays an active role in the ability of the enterocytes to efficiently export dietary neutral lipid as demonstrated by histological examination of intestines from carp and salmonids fed phosphatidylcholine-devoid diets (Fontagne et al. 1998; Olsen et al. 1999, 2003).

5.3. Carbohydrate Absorption

In various fish species investigated, the monosaccharides D-glucose and D-galactose appear to be transported by the same transporter across the brush border membrane, SGLT1, and in a similar manner as in mammals (see Table 2.3; Storelli et al. 1986; Buddington et al. 1987; Reshkin and Ahearn 1987a,b; Ahearn et al. 1992; Collie and Ferraris 1995; Maffia et al. 1996; Soengas and Moon 1998; Sala-Rabanal et al. 2004; Geurden et al. 2007; Goncalves et al. 2007; Bakke-McKellep et al. 2008). A putative facilitative transporter for D-fructose such as GLUT5 has not been demonstrated in fish, although fructose appears to be absorbed by the intestines of the fishes investigated: bream (*Abramis brama*) and common carp (Golovanova 1993). GLUT2 transcripts have been detected in intestinal tissue of some fishes (Krasnov et al. 2001; Panserat et al. 2001; Hall et al. 2006), though their exact location and characteristics have apparently not been investigated. A BBM-located facilitative transporter such as GLUT2 may play a role in fructose absorption, as has been suggested in mice (Gouyon et al. 2003). This remains to be investigated in fish, as is its involvement in basolateral efflux of monosaccharides.

5.4. Synthesis and Absorption of Vitamins

Vitamins have not been shown to be synthesized *in vivo* in any teleost fishes studied to date, although phylogenetically older Actinopterigian fishes, sharks and possibly stingrays and lungfish have the ability to synthesize ascorbic acid (Dabrowski 1994; Moreau and Dabrowski 2000; Fracalossi et al. 2001; Cho et al. 2007). Therefore it is presumed that the diet must supply most fishes' vitamin requirements.

Present knowledge on the mechanisms of vitamin absorption by the GI tract in fish is very limited. Of the fat-soluble vitamins (A, D, E and K) and carotenoids, retinol (vitamin A) is one of the few that has been studied in a fish species. Its *in vitro* absorption into intact intestinal tissue (proximal region) of juvenile sunshine bass (*Morone chrysops* × *M. saxatilis*) was via a saturable, transporter-mediated mechanism (Buddington et al. 2002), as has been demonstrated in mammals.

For water-soluble vitamins, transporter-mediated uptake has been suggested for ascorbic acid, inositol, riboflavin, biotin and folic acid, whereas nicotinamide appeared to occur by a non-saturable mechanism in the limited number of fish species studied (Vilella et al. 1989b; Rose and Chou 1990; Buddington et al. 1993; Casirola et al. 1995; Casirola and Ferraris 1997). This reflects mechanisms demonstrated in mammals. An exception was observed for folic acid absorption in channel catfish (*Ictalurus punctatus*), which appeared to be absorbed by diffusion (Casirola et al. 1995). Therefore, some differences in absorption mechanisms may exist from species to species.

5.5. Pigment Absorption

Dietary lipid-soluble carotenoids such as astaxanthin and canthaxanthin, which are important for the characteristic red pigmentation of salmonid flesh, are apparently not absorbed efficiently. Increased dietary lipid level has a positive effect on astaxanthin deposition in the muscle, possibly due to increased mixed micellar incorporation and, subsequently, eased transfer to the enterocytes for absorption (Torrissen et al. 1990; Choubert et al. 1991). This is apparently further modulated by lipid class, degree of saturation and/or chain lengths (Olsen et al. 2005). Mechanisms of absorption are not known.

5.6. Mineral and Trace Element (Metal) Absorption

Minerals are important components of structural and physiological processes. Elements such as calcium, phosphorus and magnesium are required at high levels for bone and scale formation and also have critical roles in intracellular processes such as cell signaling and energy metabolism. Metal micronutrients are often only required in trace amounts, but are critical as co-factors for cellular respiration, oxygen transport, and enzyme catalysis. The maintenance of appropriate levels of minerals in the body depends on acquisition from the environment via absorptive processes across epithelia. In this respect, a fundamental difference exists between fishes and mammals: while the gut is the sole absorptive epithelium for their acquisition in mammals, fish are also capable of absorbing certain minerals and metals from the water via the gills. There is a significant body of literature demonstrating interplay between these two epithelia, such that absorption at the gill can impact uptake across the gut (see Bury et al. 2003).

The gastrointestinal absorption of calcium, magnesium, phosphorus and iodide, and the metal micronutrients iron, copper, zinc and nickel will be briefly discussed. The absence of mechanistic information regarding the absorption of trace elements such as manganese, cobalt, molybdenum, chromium and selenium currently precludes their discussion. This review will not cover the monovalent mineral ions of sodium, potassium and chloride, as these will be discussed elsewhere in light of their critical roles in salt and water homeostasis (see Chapters 4 and 5).

5.6.1. PHOSPHORUS

The digestive tract is the major absorptive epithelium for phosphorus, largely owing to the limited levels and bioavailability of phosphorus in water. This is especially true of marine and brackish waters where high concentrations of divalent cations lead to the formation of insoluble phosphate compounds. This also has consequences for uptake in marine fish gut, as drinking seawater is likely to restrict gastrointestinal phosphorus bioavailability.

Absorption of phosphorus in fish gut has been well characterized at both a molecular and physiological level. A sodium-phosphorus co-transporter (Na-P_i-II; SLC34) was first cloned from winter flounder (Werner et al. 1994). Originally identified in the kidney, this transporter was also found to operate at the apical surface of the enterocytes (Kohl et al. 1996). Piscine Na-P_i-II exhibited properties similar to its mammalian homologue, with a distinct pH dependence (Forster et al. 1997).

From studies in a number of fish it is apparent that significant differences in transporter expression, distribution, regulation and function exist between different species (Sugiura 2009). Expression of Na-P_i-II transporters is greatest in the proximal intestine in tilapia and carp. Rainbow trout express two distinct Na-P_i-II transporter isoforms, one that is primarily localized to the pyloric ceca (PC) and another that is more prevalent in the mid intestine. The transporter distribution is consistent with functional assays that show these regions display highest phosphorus uptake rates (Avila et al. 2000; Sugiura and Ferraris 2004a). The intestinal form responds to dietary phosphorus content, but the PC form does not (Sugiura and Ferraris 2004a). The PC transporter is sodium-independent, and characterized as a high-affinity, low-capacity facilitator compared to the intestinal form. Transport in the PC, however, seems less important than diffusion for total absorption (Sugiura and Ferraris 2004a). Zebrafish also express two Na-P_i-II transporters, which can also be classified as high- and low-affinity forms. As opposed to the situation in rainbow trout, the zebrafish transporters exhibit overlapping distribution along the digestive tract (Nalbant et al. 1999; Graham et al. 2003). Differences between species likely reflect variations in phylogeny (i.e. gene copy number and isoforms), ontogeny and environment (e.g. freshwater versus marine). Vitamin D 24-hydroxylase and vitamin D receptor may influence gut phosphorus uptake and indicate endocrine control of phosphorus homeostasis (e.g. Sugiura and Ferraris 2004b), although details in fishes are largely lacking.

5.6.2. CALCIUM

Assimilation from the diet plays a relatively minor role in calcium absorption, with the bulk of uptake achieved by the gills in fishes (Flik and Verbost 1993). The exception is when calcium demand is high, e.g. during gonad development prior to spawning or skeletal development, or when water levels are low (Sundell and Björnsson 1988; Guerreiro et al. 2002). Marine fish absorb a greater proportion of calcium through the digestive tract than freshwater species, primarily due to intake of seawater rather than increased acquisition from the diet. Intestinal uptake ranges from 20 to 70% of total calcium uptake for adult marine fish (Schoenmakers et al. 1993; Guerreiro et al. 2002). Intestinal calcium absorption is apparently regulated in euryhaline fish by decreasing activities of putative calcium transporters when moving from freshwater to seawater (Schoenmakers et al. 1993).

The mechanism of calcium absorption in the gut of fish is only poorly understood. At the apical surface, calcium appears to be absorbed via a low-affinity process (Klaren et al. 1993). A P_2 purinoceptor-responsive transporter operating well below saturation (Klaren et al. 1997) and an L-type calcium channel (Larsson et al. 1998) have been hypothesized as mediators of calcium absorption in fish enterocytes. Molecular evidence exists for the presence of an epithelial calcium channel (ECaC) in the intestine of pufferfish (Qiu and

Hogstrand 2004) and zebrafish (Pan et al. 2005), but only low levels of transcription are detectable and there is not yet evidence of its cellular location. Basolateral transport of calcium is achieved by the action of a Ca^{2+}-ATPase and/or a sodium–calcium exchanger (NCX). Recently, expression of an NCX transporter has been detected in the intestine of zebrafish (On et al. 2009). Although the mechanistic understanding of calcium uptake is not well developed, it is clear that significant differences between species are likely to exist (Flik and Verbost 1993).

There is significant endocrine control of intestinal calcium uptake also in fish. Stanniocalcin (Sundell et al. 1992), vitamin D (Larsson et al. 2002), parathyroid hormone-like protein (Fuentes et al. 2006) and vitamin A (only in concert with vitamin D; Ørnsrud et al. 2009), or their active metabolites, have all been described as having regulatory effects on gastrointestinal calcium assimilation.

5.6.3. MAGNESIUM

The intestinal absorption of magnesium in fish is not well understood. Dietary uptake appears to be regulated to maintain whole body magnesium homeostasis (Bijvelds et al. 1996a). Freshwater fish are likely to have more efficient mechanisms of magnesium uptake given the relatively lower level of magnesium in freshwater, although this has not been specifically examined (Bijvelds et al. 1998). Recent evidence suggests that the stomach is the primary region of magnesium absorption (Bucking and Wood 2007).

Mechanistically, apical uptake of magnesium is mediated by an unidentified transport protein in tilapia, which facilitates the uptake down a favorable electrochemical gradient (Bijvelds et al. 2001). Basolateral transfer is achieved by an anion-coupled process suggestive of a symporter (Bijvelds et al. 1996b). Mechanisms of magnesium absorption have not been studied in other fishes.

5.6.4. IODIDE

There is insufficient mechanistic information regarding gastrointestinal absorption of iodide. It is, however, believed that branchial absorption is the most important route of iodide uptake, although evidence exists for gastrointestinal assimilation via a sodium-iodide symporter in the gut of larval fish (Moren et al. 2008).

5.6.5. IRON

The uptake of iron across the gastrointestinal tract of fish is highly dependent upon the chemical environment in the intestine. A reducing environment that favors the formation of ferrous (Fe^{2+}) iron over ferric (Fe^{3+}) significantly stimulates absorption (Bury et al. 2001; Kwong and

Niyogi 2008), suggesting a divalent metal ion transporter similar to the divalent metal transporter-1 (DMT-1) in mammalian intestines (Conrad and Umbriet 2002). Recent physiological (Kwong and Niyogi 2009) and molecular evidence (see Table 2.4) supports the presence of such a

Table 2.4
Metal micronutrient transporters characterized in piscine gastrointestinal tract

Transporter	Evidence for metal transport in fish	Fish species	References
SLC11A2 (divalent metal/ cation transporter 1; DMT1; DCT1; NRAMP)	Responsive to dietary iron Transporter of iron in expression vectors	Zebrafish (*Danio rerio*) Rainbow trout (*Oncorhyncus mykiss*)	Donovan et al. 2002 Bury and Grosell 2003 Cooper et al. 2006b Craig et al. 2009
SLC31A1 (copper transporter 1; CTR1)	Responsive to waterborne and dietary copper and dietary iron	Zebrafish Sea bream (*Sparus aurata*)	Mackenzie et al. 2004 Minghetti et al. 2008 Craig et al. 2009
ATP7A (basolateral Cu-ATPase; Menkes transporter)	Responsive to waterborne copper and dietary iron	Zebrafish	Craig et al. 2009
SLC40A (Ferroportin; FPN; iron regulated transporter; IREG)	Responsive to waterborne copper and dietary iron Transporter of iron in expression vectors	Zebrafish	Donovan et al. 2000 Cooper et al. 2006b Craig et al. 2009
SLC39A[1] (Zrt/IRT-like protein; ZIP)	Some ZIP genes are responsive to waterborne and/or dietary zinc Transporter of zinc in expression vectors	Zebrafish Rainbow trout Pufferfish (*Takifugu rubripes*)	Feeney et al. 2005 Qiu and Hogstrand 2005 Qiu et al. 2005 Qiu et al. 2007
SLC30A[2] (Zinc transporter; ZnT)	Some ZnT genes are responsive to waterborne and/or dietary zinc Transporter (efflux) of zinc in expression vectors	Zebrafish Pufferfish	Feeney et al. 2005 Balesaria and Hogstrand 2006

[1]A total of 12 ZIP genes have been identified in fish by bioinformatic approaches (Feeney et al. 2005), with studies in zebrafish (7 of the 8 genes examined; Feeney et al. 2005) showing a gastrointestinal expression of these genes.
[2]A total of eight ZnT genes have been identified in fish by bioinformatic approaches (Feeney et al. 2005), with studies in zebrafish (4 of the 7 genes examined; Feeney et al. 2005) and pufferfish (ZnT1; Balesaria and Hogstrand 2006) showing a gastrointestinal expression of these genes.

transporter in fish. The preference for Fe^{2+} as the preferred substrate is further supported by the presence of ferric reductase in teleost intestine (Carriquiriborde et al. 2004). Basolateral transfer is achieved by ferroportin in zebrafish (Donovan et al. 2000). This is supported by accumulation of iron in the intestinal mucosa of zebrafish mutants lacking a functional ferroportin gene (Fraenkel et al. 2005).

The intestinal environment of marine fish, however, may be unfavorable for iron absorption. The intestinal/pancreatic secretion of bicarbonate ion into the gut lumen (Wilson et al. 2002) can result in the formation of insoluble iron carbonates that limit iron bioavailability. Binding of iron and other positively charged metals to mucus and other chelating ligands (see Section 5.7) may partially hinder formation of insoluble carbonates and promote absorption. A further obstacle for iron absorption is the alkaline gut environment, which may not favor the actions of a proton-coupled DMT-1. DMT-1 transporter isolated from the gill of rainbow trout, however, shows that iron influx rates at pH 7.4 were equivalent to those at acidic pH (Cooper et al. 2007). If this functional property is similar for intestinal DMT-1, i.e. that DMT-1 of marine fish is not H^{+}-coupled as is the case in mammals, it may help explain how the gastrointestinal epithelium of marine fish is capable of absorbing iron at neutral or alkaline pH.

The transcript levels of both DMT-1 and ferroportin increase in response to reduced dietary iron levels in zebrafish (Cooper et al. 2006b; Craig et al. 2009). As in mammals, the uptake of iron in fish is homeostatically regulated, capable of being stimulated by experimentally induced anemia (Bury et al. 2001).

5.6.6. COPPER

Copper absorption is probably the best studied of all metal micronutrient transport pathways in the piscine intestine. A mechanistic model for copper uptake that has attempted to assimilate the often contrasting physiological and pharmacological data has recently been proposed (Nadella et al. 2007). Copper uptake is favored by high luminal sodium, confirmed by both *in vitro* (Nadella et al. 2006b) and *in vivo* (Kjoss et al. 2005) studies. This effect may, however, be indirect (Handy et al. 2002) by stimulating sodium/hydrogen exchange mechanisms at the apical membrane and subsequently acidifying the apical microenvironment, which in turn facilitates uptake through the copper-proton symporters DMT-1 and/or copper transporter-1 (CTR1). The latter is a putative apical copper transporter in mammalian intestine (Sharp 2002). Support for this apical uptake pathway in fishes follows from the observation that hypercapnia, which stimulates apical sodium-proton exchange, promotes copper uptake (Nadella et al. 2007). Basolateral transport of copper is proposed to occur via a Cu-ATPase (Menkes protein), similar to the mechanism in mammalian gut (Nadella et al. 2007). Molecular components of this model have all been identified in fish gut (see Table 2.4). Evidence also exists for a basolateral copper/anion symporter (Handy et al. 2000). Transport kinetics is summarized in Table 2.5.

Recent evidence suggests the stomach of fish may play an important role in absorption of copper. An indirect role for the stomach in metal micronutrient uptake in higher vertebrates is well documented, with acid secretion demonstrated to increase metal bioavailability (Whitehead et al. 1996). In addition, organic ligands in gastric secretions are believed to prevent metal hydroxypolymerization, maintaining their bioavailability for subsequent absorption (Whitehead et al. 1996). There is, however, growing support for a direct absorptive role of the stomach in gastrointestinal copper uptake in fish. In rainbow trout, as much as 25% of total absorbed copper *in vitro* may occur in the stomach (Nadella et al. 2006b), and *in vivo* the copper uptake rate in the stomach was equivalent to that in the mid and distal regions of the intestine (Nadella et al. 2006a). It appears that the stomach may also be an important route of uptake for other nutritive metals, particularly nickel (Ojo and Wood 2007). It is unclear how efficiently fish species lacking stomachs and therefore acid digestion, with its favorable effects on metal bioavailability, assimilate metals.

Although copper can exist in two redox states (Cu^+ and Cu^{2+}), absorption is largely independent of which species is presented. This suggests that transport pathways are equally adept at handling either species, or implies the presence of copper reductase activity in the gut (Nadella et al. 2007).

5.6.7. ZINC

The uptake of zinc across the gut of fish has been investigated in both freshwater and marine teleosts using a range of *in vivo* and *in vitro* techniques. *In vivo*, both saturable and diffusive components of zinc uptake in marine teleosts can be discerned (Shears and Fletcher 1983; Glover et al. 2003a), while only saturable uptake is observed in freshwater teleosts (Glover and Hogstrand 2002b; see Table 2.5 for kinetic parameters). *In vitro*, however, both saturable and diffusive components were observed for freshwater fish (Glover et al. 2003b), suggesting that zinc-induced secretion of mucus in the freshwater gut occluded diffusive uptake. Differences related to methodology and environment may explain observed differences in zinc absorption in fish.

The study of mechanisms of zinc absorption in fish has been aided by a marine zinc-hyperaccumulating species, the squirrelfish (*Holocentrus adscensionis*). In addition to a capacity to safely store potentially toxic levels of zinc (Hogstrand et al. 1996), female squirrelfish have an enhanced intestinal zinc uptake capacity compared to males and other tested marine

Table 2.5

Kinetic parameters of the saturable component of gastrointestinal metal micronutrient uptake

Metal	Species	Method	Affinity (K_m)	Capacity (J_{max})	Reference
Copper	Rainbow trout (*Oncorhynchus mykiss*)	*In vitro* gut sac	19–32 µM	12–17 pmol cm^{-2} h^{-1}	Nadella et al. 2006b
		Isolated intestinal cells	216 µM	18 pmol mg protein^{-1} min^{-1}	Burke and Handy 2005
Zinc	Rainbow trout	*In vivo* perfusion	309 µM	933 nmol kg^{-1} h^{-1}	Glover and Hogstrand 2002b
		Brush-border membrane vesicles	57 µM	1,867 nmol mg protein^{-1} min^{-1}	Glover et al. 2003b
	Yellow perch (*Perca flavescens*)	*In vitro* gut sac	265–374 µM	0.9–1.1 nmol cm^{-2} h^{-1}	Niyogi et al. 2007
Nickel	Rainbow trout	*In vitro* gut sac	11–42 µM	53 –215 pmol cm^{-2} h^{-1}	Leonard et al. 2009

Although saturable uptake has been demonstrated for intestinal iron uptake in fish the nature of these experiments has to date precluded calculation of kinetic parameters.

species (Glover et al. 2003a). Molecular characterization of the structural components of gastrointestinal absorption suggests that basic elements of intestinal zinc uptake are conserved between piscine and mammalian systems, namely homologous apical (Zrt/IRT-like protein; ZIP) and basolateral (Zinc transporter; ZnT) transport proteins (see Table 2.4).

Gut chemistry also plays an important role in zinc uptake, with organic composition of the gut lumen having a particularly prominent role in modifying absorption (see Section 5.7). The roles of inorganic ions in zinc uptake are less well defined. Calcium has been shown to have an inhibitory or a stimulatory effect depending on concentrations used (Glover and Hogstrand 2003; Ojo and Wood 2008). This effect appears to be mediated at the apical epithelium, but is not consistent with interaction at an epithelial calcium channel (Glover et al. 2004).

5.6.8. NICKEL

Saturable absorption of nickel has been demonstrated *in vitro* in rainbow trout with substrate affinity constants in the same order of magnitude as those determined for zinc and copper, as well as those for nickel uptake in mammalian intestines (Leonard et al. 2009). At high nickel levels (>60 μM), a linear, diffusive mode of absorption has been characterised. *In vivo* nickel is efficiently absorbed by the fish intestine, with 50% assimilation from a commercial fish meal (Leonard et al. 2009) and is under homeostatic control (e.g. Ptashynski and Klaverkamp 2002). Yet, there is still some contention as to whether nickel is an essential micronutrient in fish (Muyssen et al. 2004; Chowdhury et al. 2008). Its biochemical role remains elusive and no nickel-dependent enzymes have been isolated (Phipps et al. 2002). Both magnesium and calcium are non-competitive inhibitors of nickel uptake *in vitro* (Leonard et al. 2009).

5.7. Interactions between Nutrients and Other Intestinal Components

Pathways of absorption, particularly those of nutrients that possess similar physicochemical properties, are often shared. Consequently the absorption of one nutrient can be impacted by the presence in the gut lumen of competing nutrients. This is an effect that has been described in the intestines of higher vertebrates as well as fish for both amino acids (e.g. Berge et al. 1999, 2004; Rosas et al. 2008) and metal micronutrients (e.g. Glover and Hogstrand 2003; Ojo and Wood 2008; Kwong and Niyogi 2009; Ojo et al. 2009). Most amino acid transporters translocate their substrate on the basis of a physicochemical attribute shared among many amino acids, e.g. charge at physiological pH. Consequently transport depends not only on the presence and concentration of a given amino acid, but also on the other substrates available. Similarly many metal transporters also appear to be promiscuous, capable of binding and transporting multiple metals. Thus, by virtue of similar ionic radii, some metals can utilize transport pathways usually devoted to macronutrient ions (Bury et al. 2003).

The interactions described above are predictable based on the chemical similarity between the nutrients involved. However, there is also evidence that physicochemically dissimilar nutrients can significantly influence nutrient uptake. For example, dietary phosphate and fiber, and the anti-nutrient phytate, at least in some studies, have a demonstrated negative effect on the bioavailability of zinc and other minerals (e.g. Richardson et al. 1985; Clearwater et al. 2002). The presence of high levels of ascorbate (vitamin C) in aquacultural feeds promotes ferrous iron (Fe^{2+}) speciation, and is likely to enhance iron absorption (Cooper et al. 2006a). Further effects of gut chemistry on trace metal absorption have been summarized above (see Sections 5.6.1–5.6.8).

Intestinal absorption of iron and other positively charged metals in various fishes is favored by the presence of gut mucus (Bury et al. 2001; Cooper et al. 2007; Ojo and Wood 2007). The polyanionic nature of mucus at the pH encountered in the gut lumen effectively binds metals. The actions of apical ion transporters such as sodium–hydrogen exchangers and/or proton pumps, acidify the mucosal-adjacent mucus, and create an environment that disfavors the continued binding of the metal by the mucus layer (Whitehead et al. 1996). This effectively generates high levels of metal in close proximity to the transporters. Thus, mucus appears to have a significant role in metal absorption in the gut and is a factor that is often overlooked by more refined techniques for examining metal uptake in the gut.

Perhaps the best-studied nutrient interaction in fish gastrointestinal absorption is that between metal micronutrients and amino acids. Using an *in vivo* gut perfusion protocol, histidine and cysteine promoted zinc uptake in rainbow trout in freshwater (Glover and Hogstrand 2002a). *In vitro*, a similar effect of histidine was observed on copper uptake (Nadella et al. 2006b). Both these amino acids have a high affinity for binding zinc. However, taurine, an amino acid with limited zinc-binding activity, also significantly influenced zinc uptake (Glover and Hogstrand 2002a). The organic composition of the gut also appears to be critical in determining copper uptake (Blanchard et al. 2009). The availability of amino acids and other small organic ligands with the capacity to bind copper are thought to facilitate uptake, potentially through pathways of absorption that remain relatively unaffected by inorganic gut chemistry. Thus many nutrient interactions at the level of absorption involve pathways that are far more complicated than simple competition between compounds with shared physicochemical properties for tranport.

Two mechanistic hypotheses exist to explain the findings of positive effects of chelating amino acids on metal micronutrient uptake. The first suggests binding by amino acids maintains metal solubility, and enhances presentation to the apical metal transporters (donor effect). The second suggests that this facilitation extends to the actual transport of a metal-amino acid chelate (transport effect). These hypotheses were recently tested for copper and histidine uptake across the brush border membrane of rainbow trout intestine (Glover and Wood 2008a,b). The concentration-dependent kinetics of copper and histidine when assayed together was significantly different than when examined individually. This effect was not simply a consequence of altered luminal concentrations of unchelated substrates. It was suggested that the presence of a copper-histidine chelate created an additional pathway for absorption, possibly via amino acid tranporters. This is supported by *in vitro* data indicating that methionine and selenomethionine compete for the same absorption pathways in intact intestinal tissue of green sturgeon (*Acipenser medirostris*; Bakke et al. 2010), as has been demonstrated in mammalian systems.

The importance of this effect was recently highlighted by a study of copper uptake in killifish (*Fundulus heteroclitus*; Blanchard et al. 2009). These authors suggested that the transport of copper chelated to amino acids was the dominant pathway for total copper assimilation. This study has important implications for many of the mechanistic investigations of metal nutrient absorption in fish gut. In real-world scenarios, metal speciation is likely to be dominated by the presence of organic chelates. Studies of nutrient absorption traditionally focus on single nutrients in simplified inorganic matrices, and/or carefully controlled simple mixtures. Clearly such studies do not adequately reflect the complexity of the numerous nutrient interactions that may occur in natural diets. Nevertheless, developing a mechanistic understanding of individual nutrient absorption pathways in experimental conditions better enables interpretation of mixture effects, and under-standing of processes such as acclimation and homeostatic regulation of gastrointestinal uptake. Such basic research will also provide knowledge that can lead to decreased losses of valuable dietary components and undesirable effluents to the environment, contributing to the consumer acceptance and sustainability of the aquaculture industry.

REFERENCES

Abate, F., Germana, G. P., De Carlos, F., Montalbano, G., Laura, R., Levanti, M. B., and Germana, A. (2006). The oral cavity of the adult zebrafish (*Danio rerio*). *Anat. Histol. Embryol.* **35**, 299–304.

Abelli, L., Coscia, M. R., De Santis, A., Zeni, C., and Oreste, U. (2004). Evidence for hepato-biliary transport of immunoglobulin in the antarctic teleost fish *Trematomus bernacchii*. *Dev. Comp. Immunol.* **29**, 431–442.

Adamidou, S., Nengas, I., Henry, M., Grigorakis, K., Rigos, G., Nikolopoulou, D., Kotzamanis, Y., Bell, G. J., and Jauncey, K. (2009). Growth, feed utilization, health and organoleptic characteristics of European seabass (*Dicentrarchus labrax*) fed extruded diets including low and high levels of three different legumes. *Aquaculture* **293**, 263–271.

Ahearn, G. A., Behnke, R. D., Zonno, V., and Storelli, C. (1992). Kinetic heterogeneity of Na-D-glucose cotransport in teleost gastrointestinal tract. *Am. J. Physiol.* **263**, R1018–R1023.

Albrektsen, S., Krogdahl, Å., Nortvedt, R., Sandnes, K. and Hillestad, M. (2009). Salmon bone in cod diet. In: *Rubin Report*. Nr. 178, pp. 1–22. Rubin, Trondheim.

Altschul, S. F., Madden, T. L., Schaffer, A. A., Zhang, J. H., Zhang, Z., Miller, W., and Lipman, D. J. (1997). Gapped BLAST and PSI-BLAST: a new generation of protein database search programs. *Nucleic Acids Res.* **25**, 3389–3402.

Ambardekar, A., Reigh, R. C., and Williams, M. B. (2009). Absorption of amino acids from intact dietary proteins and purified amino acid supplements follows different time-courses in channel catfish (*Ictalurus punctatus*). *Aquaculture* **291**, 179–187.

Anderson, T. A. (1986). Histological and cytological structure of the gastrointestinal tract of the luderick, *Girella tricuspidata* (pisces, kyphosidae), in relation to diet. *J. Morphol.* **190**, 109–119.

Andre, M., Ando, S., Ballagny, C., Durliat, M., Poupard, G., Briancon, C., and Babin, P. J. (2000). Intestinal fatty acid binding protein gene expression reveals the cephalocaudal patterning during zebrafish gut morphogenesis. *Int. J. Devel. Biol.* **44**, 249–252.

Arellano, J., Dinis, M. T., and Sarasquete, C. (1999). Histomorphological and histochemical characteristics of the intestine of the Senegal sole, *Solea senegalensis*. *Eur. J. Histochem.* **43**, 121–133.

Asgeirsson, B., and Cekan, P. (2006). Microscopic rate-constants for substrate binding and acylation in cold-adaptation of trypsin I from Atlantic cod. *FEBS Lett.* **580**, 4639–4644.

Avila, E. M., Tu, H., Basantes, S., and Ferraris, R. P. (2000). Dietary phosphorus regulates intestinal transport and plasma concentrations of phosphate in rainbow trout. *J. Comp. Physiol. B* **170**, 201–209.

Azaza, M. S., Kammoun, W., Abdelmouleh, A., and Kraiem, M. M. (2009). Growth performance, feed utilization, and body composition of Nile tilapia (*Oreochromis niloticus* L.) fed with differently heated soybean-meal-based diets. *Aquacult. Int.* **17**, 507–521.

Baeverfjord, G., Refstie, S., Krogedal, P., and Åsgard, T. (2006). Low feed pellet water stability and fluctuating water salinity cause separation and accumulation of dietary oil in the stomach of rainbow trout (*Oncorhynchus mykiss*). *Aquaculture* **261**, 1335–1345.

Bakke, A. M., Tashjian, D. H., Wang, C. F., Lee, S. H., Bai, S. C., and Hung, S. S. O. (2010). Competition between selenomethionine and methionine absorption in the intestinal tract of green sturgeon (*Acipenser medirostris*). *Aq. Toxicol.* **96**, 62–69.

Bakke-McKellep, A. M., and Refstie, S. (2008). Alternative protein sources and digestive function alterations in teleost fishes. In: *Feeding and Digestive Functions of Fishes* (J. E. P. Cyrino, D. P. Bureau and B. G. Kapoor, eds), pp. 445–478. Science Publishers, Enfield.

Bakke-McKellep, A. M., Nordrum, S., Krogdahl, Å., and Buddington, R. K. (2000). Absorption of glucose, amino acids, and dipeptides by the intestines of Atlantic salmon (*Salmo salar* L.). *Fish Physiol. Biochem.* **22**, 33–44.

Bakke-McKellep, A. M., Sanden, M., Danieli, A., Acierno, R., Hemre, G.-I., Maffia, M., and Krogdahl, Å. (2008). Atlantic salmon (*Salmo salar* L.) parr fed genetically modified soybeans and maize: histological, digestive, metabolic, and immunological investigations. *Res. Vet. Sci.* **84**, 395–408.

Balesaria, S., and Hogstrand, C. (2006). Identification, cloning and characterization of a plasma membrane zinc efflux transporter, TrZnT-1, from Fugu pufferfish (*Takifugu rubripes*). *Biochem. J.* **394**, 485–493.

Balocco, C., Boge, G., and Roche, H. (1993). Neutral amino-acid-transport by marine fish intestine—role of the side-chain. *J. Comp. Physiol. B* **163**, 340–347.

Berge, G. E., Bakke-McKellep, A. M., and Lied, E. (1999). In vitro uptake and interaction between arginine and lysine in the intestine of Atlantic salmon (*Salmo salar*). *Aquaculture* **179**, 181–193.

Berge, G. E., Goodman, M., Espe, M., and Lied, E. (2004). Intestinal absorption of amino acids in fish: kinetics and interaction of the in vitro uptake of L-methionine in Atlantic salmon (*Salmo salar* L.). *Aquaculture* **229**, 265–273.

Bijvelds, M. J. C., Flik, G., Kolar, Z. I., and Wendelaar Bonga, S. E. (1996a). Uptake, distribution and excretion of magnesium in *Oreochromis mossambicus*: dependence on magnesium in diet and water. *Fish Physiol. Biochem.* **15**, 287–298.

Bijvelds, M. J. C., Kolar, Z. I., Wendelaar Bonga, S. E., and Flik, G. (1996b). Magnesium transport across the basolateral plasma membrane of the fish enterocyte. *J. Membr. Biol.* **154**, 217–225.

Bijvelds, M. J. C., Van der Velden, J. A., Kolar, Z. I., and Flik, G. (1998). Magnesium transport in freshwater teleosts. *J. Exp. Biol.* **201**, 1981–1990.

Bijvelds, M. J. C., Kolar, Z. I., and Flik, G. (2001). Electrodiffusive magnesium transport across the intestinal brush border membrane of tilapia (*Oreochromis mossambicus*). *Eur. J. Biochem.* **268**, 2867–2872.

Blanchard, J., Brix, K., and Grosell, M. (2009). Subcellular fractionation of Cu exposed oysters, *Crassostrea virginica*, and Cu accumulation from a biologically incorporated Cu rich oyster diet in *Fundulus heteroclitus* in fresh and sea water. *Comp. Biochem. Physiol.* **149C**, 531–537.

Boge, G., Roche, H., and Balocco, C. (2002). Amino acid transport by intestinal brush border vesicles of a marine fish, *Boops salpa*. *Comp. Biochem. Physiol. B* **131**, 19–26.

Bogevik, A. S., Tocher, D. R., Waagbø, R., and Olsen, R. E. (2008). Triacylglycerol-, wax ester- and sterol ester-hydrolases in midgut of Atlantic salmon (*Salmo salar*). *Aquacult. Nutr.* **14**, 93–98.

Bogevik, A. S., Tocher, D. R., Langmyhr, E., Waagbø, R., and Olsen, R. E. (2009). Atlantic salmon (*Salmo salar*) postsmolts adapt lipid digestion according to elevated dietary wax esters from *Calanus finmarchicus*. *Aquacult. Nutr.* **15**, 94–103.

Bøgwald, J., Stensvåg, K., Stuge, T. B., and Jørgensen, T.Ø. (1994). Tissue localisation and immune responses in Atlantic salmon, *Salmo salar* L., after oral administration of *Aeromonas salmonicida*, *Vibrio anguillarum* and *Vibrio salmonicidia* antigens. *Fish Shellfish Immunol.* **4**, 353–368.

Bomgren, P., Einarsson, S., and Jonsson, A. C. (1998). Similarities and differences in oxynticopeptic cell ultrastructure of one marine teleost, *Gadus morhua*, and one freshwater teleost, *Oncorhynchus mykiss*, during basal and histamine-stimulated phases of acid secretion. *Fish Physiol. Biochem.* **18**, 285–296.

Borghesi, R., Dairiki, J. K., and Cyrino, J. E. P. (2009). Apparent digestibility coefficients of selected feed ingredients for dourado *Salminus brasiliensis*. *Aquacult. Nutr.* **15**, 453–458.

Buddington, R. K., and Diamond, J. M. (1987). Pyloric ceca of fish: a "new" absorptive organ. *Am. J. Physiol.* **259**, G65–G76.

Buddington, R. K., and Hilton, J. W. (1987). Intestinal adaptations of rainbow trout to changes in dietary carbohydrate. *Am. J. Physiol.* **253**, G489–G496.

Buddington, R. K. and Kuz'mina, V. (2000a). Digestive system. In: *The Laboratory Fish* (G. K. Ostrander, ed.), Part 3 Gross Functional Anatomy, pp. 173–179. Academic Press, San Diego.

Buddington, R. K. and Kuz'mina, V. (2000b). Digestive system. In: *The Laboratory Fish* (G. K. Ostrander, ed.), Part 4 Microscopic Functional Anatomy, pp. 379–384. Academic Press, San Diego.

Buddington, R. K., Chen, J. W., and Diamond, J. M. (1987). Genetic and phenotypic adaptation of intestinal nutrient transport to diet in fish. *J. Physiol.* **393**, 261–281.

Buddington, R. K., Puchal, A. A., Houpe, K. L., and Diehl, W. J. (1993). Hydrolysis and absorption of two monophosphate derivatives of ascorbic acid by channel catfish *Ictalurus punctatus* intestine. *Aquaculture* **114**, 317–326.

Buddington, R. K., Krogdahl, Å., and Bakke-McKellep, A. M. (1997). The intestines of carnivorous fish: structure and functions and the relations with diet. *Acta Physiol. Scand.* **161**(Suppl. 638), 67–80.

Buddington, R. K., Buddington, K. K., Deng, D.-F., Hemre, G.-I., and Wilson, R. P. (2002). A high retinol dietary intake increases its apical absorption by the proximal small intestine of juvenile sunshine bass (*Morone chrysops* × *M. saxatilis*). *J. Nutr.* **132**, 2713–2716.

Bucking, C. P., and Wood, C. M. (2007). Gastrointestinal transport of Ca^{2+} and Mg^{2+} during the digestion of a single meal in the freshwater rainbow trout. *J. Comp. Physiol. B* **177**, 349–360.

Burke, J., and Handy, R. D. (2005). Sodium-sensitive and -insensitive copper accumulation by isolated intestinal cells of rainbow trout. *Oncorhynchus mykiss. J. Exp. Biol.* **208**, 391–407.

Bury, N., and Grosell, M. (2003). Iron acquisition by teleost fish. *Comp. Biochem. Physiol.* **135C**, 97–105.

Bury, N. R., Grosell, M., Wood, C. M., Hogstrand, C., Wilson, R. W., Rankin, J. C., Busk, M., Lecklin, T., and Jensen, F. B. (2001). Intestinal iron uptake in the European flounder (*Platichthys flesus*). *J. Exp. Biol.* **204**, 3779–3787.

Bury, N. R., Walker, P. A., and Glover, C. N. (2003). Nutritive metal uptake in teleost fish. *J. Exp. Biol.* **206**, 11–23.

Caballero, M. J., Izquierdo, M. S., Kjørsvik, E., Montero, D., Socorro, J., Fernández, A. J., and Rosenlund, G. (2003). Morphological aspects of intestinal cells from gilthead seabream (*Sparus aurata*) fed diets containing different lipid sources. *Aquaculture* **225**, 325–340.

Cao, X. J., and Wang, W. M. (2009). Histology and mucin histochemistry of the digestive tract of yellow catfish, *Pelteobagrus fulvidraco. Anat. Histol. Embryol.* **38**, 254–261.

Carriquiriborde, P., Handy, R. D., and Davies, S. J. (2004). Physiological modulation of iron metabolism in rainbow trout (*Oncorhynchus mykiss*) fed low and high iron diets. *J. Exp. Biol.* **207**, 75–86.

Casirola, D. M., and Ferraris, R. P. (1997). Intestinal absorption of water-soluble vitamins in rainbow trout (*Oncorhynchus mykiss*). *Comp. Biochem. Physiol. A* **116**, 273–279.

Casirola, D. M., Vinnakota, R. R., and Ferraris, R. P. (1995). Intestinal absorption of water-soluble vitamins in channel catfish (*Ictalurus punctatus*). *Am. J. Physiol. A* **269**, R490–R496.

Cho, Y. S., Douglas, S. E., Gallant, J. W., Kim, K. Y., Kim, D. S., and Nam, Y. K. (2007). Isolation and characterization of cDNA sequences of L-gulono-gamma-lactone oxidase, a key enzyme for biosynthesis of ascorbic acid, from extant primitive fish groups. *Comp. Biochem. Physiol.* **147**, 178–190.

Choubert, G., de la Nou, J., and Blanc, J.-M. (1991). Apparent digestibility of canthaxanthin in rainbow trout: effect of dietary fat level, antibiotics and number of pyloric caeca. *Aquaculture* **99**, 319–323.

Choat, J. H., and Clements, K. D. (1998). Vertebrate herbivores in marine and terrestrial environments: a nutritional ecology perspective. *Annu. Rev. Ecol. Syst.* **29**, 375–403.

Chowdhury, M. J., Bucking, C., and Wood, C. M. (2008). Pre-exposure to waterborne nickel downregulates gastrointestinal nickel uptake in rainbow trout: indirect evidence for nickel essentiality. *Environ. Sci. Technol.* **42**, 1359–1364.

Christensen, H. N. (1990). Role of amino acid transport and countertransport in nutrition and metabolism. *Physiol. Rev.* **70**, 43–77.

Clearwater, S. J., Farag, A. M., and Meyer, J. S. (2002). Bioavailability and toxicity of dietborne copper and zinc to fish. *Comp. Biochem. Physiol.* **132C**, 269–313.

Clements, K. D., Raubenheimer, D., and Choat, J. H. (2009). Nutritional ecology of marine herbivorous fishes: ten years on. *Func. Ecol.* **23**, 79–92.

Coady, M. J., Pajor, A. M., and Wright, E. M. (1990). Sequence homologies among intestinal and renal Na^+/glucose transporters. *Am. J. Physiol.* **259**, C605–C610.

Collie, N. L. (1985). Intestinal nutrient transport in coho salmon (*Oncorhynchus kisutch*) and the effects of development, starvation, and seawater adaptation. *J. Comp. Physiol. B* **156**, 163–174.

Collie, N. L., and Ferraris, R. P. (1995). Nutrient fluxes and regulation in fish intestine. In: *Metabolic Biochemistry (Biochemistry and Molecular Biology of Fishes Series)* (P. W. Hochachka and T. P. Mommsen, eds), Vol. 4, pp. 221–239. Elsevier Science, Amsterdam.

Concha, M. I., Santander, C. N., Villanueva, J., and Amthauer, R. (2002). Specific binding of the endocytosis tracer horseradish peroxidase to intestinal fatty acid-binding protein (I-FABP) in apical membranes of carp enterocytes. *J. Exp. Zool.* **293**, 541–550.

Conrad, M. E., and Umbreit, J. N. (2002). Pathways of iron absorption. *Blood Cells Mol. Dis.* **29**, 336–355.

Cooper, C. A., Bury, N. R., and Grosell, M. (2006a). The effects of pH and the iron redox state on iron uptake in the intestine of a marine teleost fish, gulf toadfish (*Opsanus beta*). *Comp. Biochem. Physiol.* **143A**, 292–298.

Cooper, C. A., Handy, R. D., and Bury, N. R. (2006b). The effects of dietary iron concentration on gastrointestinal and branchial assimilation of both iron and cadmium in zebrafish (*Danio rerio*). *Aquat. Toxicol.* **79**, 167–175.

Cooper, C. A., Shayeghi, M., Techau, M. E., Capdevila, D. M., MacKenzie, S., Durrant, C., and Bury, N. R. (2007). Analysis of the rainbow trout solute carrier 11 family reveals iron import ≤pH 7.4 and a functional isoform lacking transmembrane domains 11 and 12. *FEBS Lett.* **581**, 2599–2604.

Craig, P. M., Galus, M., Wood, C. M., and McClelland, G. B. (2009). Dietary iron alters waterborne copper-induced gene expression in soft water acclimated zebrafish (*Danio rerio*). *Am. J. Physiol.* **296**, R362–R373.

Dabrowski, K. (1990). Absorption of ascorbic-acid and ascorbic sulfate and ascorbate metabolism in common carp (*Cyprinus-carpio* L.). *J. Comp. Physiol. B* **160**, 549–561.

Dabrowski, K. (1993). Ecophysiological adaptations exist in nutrient requirements of fish: true or false? *Comp. Biochem. Physiol. A* **104**, 579–584.

Dabrowski, K. (1994). Primitive Actinopterigian fishes can synthesize ascorbic-acid. *Experientia* **50**, 745–748.

Darias, M. J., Murray, H. M., Gallant, J. W., Douglas, S. E., Yufera, M., and Martinez-Rodriguez, G. (2007). Ontogeny of pepsinogen and gastric proton pump expression in red porgy (*Pagrus pagrus*): determination of stomach functionality. *Aquaculture* **270**, 369–378.

Das, K. M., and Tripathi, S. D. (1991). Studies on the digestive enzymes of grass carp, *Ctenopharyngodon idella* (Val.). *Aquaculture* **92**, 21–32.

de la Higuera, M. (2001). Effects of nutritional factors and feed characteristics on feed intake. In: *Food Intake in Fish* (D. Houlihan, T. Boujard and M. Jobling, eds), pp. 250–268. Blackwell Science Ltd., Oxford.

Deguara, S., Jauncey, K., and Agius, C. (2003). Enzyme activities and pH variations in the digestive tract of gilthead sea bream. *J. Fish Biol.* **62**, 1033–1043.

Denstadli, V., Vegusdal, A., Krogdahl, Å., Bakke-McKellep, A. M., Berge, G. M., Holm, H., Hillestad, M., and Ruyter, B. (2004). Lipid absorption in different segments of the gastrointestinal tract of Atlantic salmon (*Salmo salar* L.). *Aquaculture* **240**, 385–398.

Desrosiers, V., Le Francois, N. R., and Blier, P. U. (2008). Trypsin-like enzyme from Atlantic wolffish (*Anarhichas lupus*) viscera: purification and characterization. *J. Aquat. Food Prod. Tech.* **17**, 11–26.

Dezfuli, B. S., Szekely, C., Giovinazzo, G., Hills, K., and Giari, L. (2009). Inflammatory response to parasitic helminths in the digestive tract of *Anguilla anguilla* (L.). *Aquaculture* **296**, 1–6.

Diaz, A. O., Garcia, A. M., Figueroa, D. E., and Goldemberg, A. L. (2008). The mucosa of the digestive tract in *Micropogonias furnieri*: a light and electron microscope approach. *Anat. Histol. Embryol.* **37**, 251–256.

Diaz, J. P., Guyot, E., Vigier, S., and Connes, R. (1997). First events in lipid absorption during post-embryonic development of the anterior intestine in gilt-head sea bream. *J. Fish Biol.* **51**, 180–192.

Divakaran, S., Kim, B. G., and Ostrowski, A. C. (1999). Digestive enzymes present in Pacific threadfin *Polydactylus sexfilis* (Bloch & Schneider 1801) and bluefin trevally *Caranx melampygus* (Cuvier 1833). *Aquacult. Res.* **30**, 781–787.

Domeneghini, C., Straini, R. P., and Veggetti, A. (1998). Gut glycoconjugates in *Sparus aurata* L. (Pisces, Teleostei). A comparative histochemical study in larval and adult ages. *Histol. Histopathol.* **13**, 359–372.

Donovan, A., Brownlie, A., Zhou, Y., Shepard, J., Pratt, S. J., Moynihan, J., Paw, B. H., Drejer, A., Barut, B., Zapata, A., Law, T. C., Brugnara, C., Kingsley, P. D., Palis, J., Fleming, M. D., Andrews, N. C., and Zon, L. I. (2000). Positional cloning of zebrafish ferroportin1 identifies a conserved vertebrate iron exporter. *Nature* **403**, 776–781.

Donovan, A., Brownlie, A., Dorschner, M. O., Zhou, Y., Pratt, S. J., Paw, B. H., Phillips, R. B., Thisse, C., Thisse, B., and Zon, L. I. (2002). The zebrafish mutant gene chardonnay (cdy) encodes divalent metal transporter 1 (DMT1). *Blood* **100**, 4655–4659.

Douglas, S., Gawlicka, A., Mandla, S., and Gallant, J. W. (1999). Ontogeny of the stomach in the winter flounder: characterization and expression of the pepsinogen and proton pump genes and determination of pepsin activity. *J. Fish Biol.* **55**, 897–915.

Ezeasor, D. N., and Stokoe, W. M. (1981). Light and electron microscopic studies on the absorptive cells of the intestine, caeca and rectum of the adult rainbow trout, *Salmo gairdneri*, Rich. *J. Fish Biol.* **18**, 527–544.

Fange, R., Lundblad, G., Lind, J., and Slettengren, K. (1979). Chitinolytic enzymes in the digestive-system of marine fishes. *Marine Biol.* **53**, 317–321.

Fard, M. R. S., Weisheit, C., and Poynton, S. L. (2007). Intestinal pH profile in rainbow trout *Oncorhynchus mykiss* and microhabitat preference of the flagellate *Spironucleus salmonis* (Diplomonadida). *Dis. Aquat. Org.* **76**, 241–249.

Feeney, G. P., Zheng, D. L., Kille, P., and Hogstrand, C. (2005). The phylogeny of teleost ZIP and ZnT zinc transporters and their tissue specific expression and response to zinc in zebrafish. *Biochim. Biophys. Acta—Gene Struct. Express* **1732**, 88–95.

Ferraris, R. P., and Ahearn, G. A. (1984). Sugar and amino-acid-transport in fish intestine. *Comp. Biochem. Physiol. A* **77**, 397–413.

Ferraris, R. P., Lee, P. P., and Diamond, J. M. (1989). Origin of regional and species differences in intestinal glucose uptake. *Am. J. Physiol.* **257**, G689–G697.

Ferraris, R. P., Yasharpour, S., Lloid, K. C. K., Mirzayan, R., and Diamond, J. M. (1990). Luminal glucose concentration in the gut under normal conditions. *Am. J. Physiol.* **259**, G820–G837.

Flik, G., and Verbost, P. M. (1993). Calcium transport in fish gills and intestine. *J. Exp. Biol.* **184**, 17–29.

Fontagne, S., Geurden, I., Escaffre, A. M., and Bergot, P. (1998). Histological changes induced by dietary phospholipids in intestine and liver of common carp (*Cyprinus carpio*) larvae. *Aquaculture* **161**, 213–223.

Forster, I. C., Wagner, C. A., Busch, A. E., Lang, F., Biber, J., Hernando, N., Murer, H., and Werner, A. (1997). Electrophysiological characterization of the flounder type II Na^+/P_i co-transporter (NaPi-5) expressed in *Xenopus laevis* oocytes. *J. Membr. Biol.* **160**, 9–25.

Fracalossi, D. M., Allen, M. E., Yuyama, L. K., and Oftedal, O. T. (2001). Ascorbic acid biosynthesis in Amazonian fishes. *Aquaculture* **192**, 321–332.

Fraenkel, P. G., Traver, D., Donovan, A., Zahrieh, D., and Zon, L. I. (2005). Ferroportin1 is required for normal iron cycling in zebrafish. *J. Clin. Invest.* **115**, 1532–1541.

Frøystad, M. K., Lilleeng, E., Sundby, A., and Krogdahl, Å. (2006). Cloning and characterization of alpha-amylase from Atlantic salmon (*Salmo salar* L.). *Comp. Biochem. Physiol. A* **145**, 479–492.

Fuentes, J., Figueiredo, J., Power, D. M., and Canário, A. V. M.. (2006). Parathyroid hormone-related protein regulates intestinal calcium transport in sea bream (*Sparus auratus*). *Am. J. Physiol.* (**291**), R1499–R1506.

Ganong, W. F. (2009). Digestion and absorption. In: *Review of Medical Physiology* (W. F. Ganong, ed.), 14th edn, pp. 398–407. Appelton & Lange, London.

Gawlicka, A., Leggiadro, C. T., Gallant, J. W., and Douglas, S. E. (2001). Cellular expression of the pepsinogen and gastric proton pump genes in the stomach of winter flounder as determined by in situ hybridization. *J. Fish Biol.* **58**, 529–536.

German, D. P., and Bittong, D. P. (2009). Digestive enzyme activities and gastrointestinal fermentation in wood-eating catfishes. *J. Comp. Physiol. B* **179**, 1025–1042.

German, D. P., Nagle, B. C., Villeda, J. M., Ruiz, A. M., Thomson, A. W., Balderas, S. C., and Evans, D. H. (2010). Evolution of herbivory in a carnivorous clade of minnows (Teleostei: Cyprinidae): effects on gut size and digestive physiology. *Physiol. Biochem. Zool.* **83**, 1–18.

Geurden, I., Bergot, P., Schwarz, L., and Sorgeloos, P. (1998). Relationship between dietary phospholipid class composition and neutral lipid absorption in postlarval turbot. *Fish Physiol. Biochem.* **19**, 217–228.

Geurden, I., Aramendi, M., Zambonino-Infante, J., and Panserat, S. (2007). Early feeding of carnivorous rainbow trout (*Oncorhynchus mykiss*) with a hyperglucidic diet during a short period: effect on dietary glucose utilization in juveniles. *Am. J. Physiol.* **292**, R2275–R2283.

Geurden, I., Kaushik, S., and Corraze, G. (2008). Dietary phosphatidylcholine affects postprandial plasma levels and digestibility of lipid in common carp (*Cyprinus carpio*). *Brit. J. Nutr.* **100**, 512–517.

Gjellesvik, D. R. (1991). Fatty acid specificity of bile salt-dependent lipase: enzyme recognition and super-substrate effects. *Biochim. Biophys. Acta—Lipids Lipid Metab.* **1086**, 167–172.

Gjellesvik, D. R., Raae, A. J., and Walther, B. T. (1989). Partial purification and characterization of a triglyceride lipase from cod (*Gadus morhua*). *Aquaculture* **79**, 177–184.

Gjellesvik, D. R., Lombardo, D., and Walther, B. T. (1992). Pancreatic bile salt dependent lipase from cod (*Gadus morhua*): purification and properties. *Biochim. Biophys. Acta— Lipids Lipid Metab.* **1124**, 123–134.

Gjellesvik, D. R., Lorens, J. B., and Male, R. (1994). Pancreatic carboxylester lipase from Atlantic salmon (*Salmo salar*) cDNA sequence and computer-assisted modelling of tertiary structure. *Eur. J. Biochem.* **226**, 603–612.

Glover, C. N., and Hogstrand, C. (2002a). Amino acid modulation of *in vivo* intestinal zinc absorption in freshwater rainbow trout. *J. Exp. Biol.* **205**, 151–158.

Glover, C. N., and Hogstrand, C. (2002b). *In vivo* characterisation of intestinal zinc uptake in freshwater rainbow trout. *J. Exp. Biol.* **205**, 141–150.

Glover, C. N., and Hogstrand, C. (2003). Effects of dissolved metals and other hydrominerals on *in vivo* intestinal zinc uptake in freshwater rainbow trout. *Aquat. Toxicol.* **62**, 281–293.

Glover, C. N., and Wood, C. M. (2008a). Absorption of copper and copper-histidine complexes across the apical surface of freshwater rainbow trout intestine. *J. Comp. Physiol. B* **178**, 101–109.

Glover, C. N., and Wood, C. M. (2008b). Histidine absorption across apical surfaces of freshwater rainbow trout intestine: mechanistic characterization and the influence of copper. *J. Membr. Biol.* **221**, 87–95.

Glover, C. N., Balesaria, S., Mayer, G. D., Thompson, E. D., Walsh, P. J., and Hogstrand, C. (2003a). Intestinal zinc uptake in two marine teleosts, squirrelfish (*Holocentrus adscensionis*) and gulf toadfish (*Opsanus beta*). *Physiol. Biochem. Zool.* **76**, 321–330.

Glover, C. N., Bury, N. R., and Hogstrand, C. (2003b). Zinc uptake across the apical membrane of freshwater rainbow trout intestine is mediated by high affinity, low affinity, and histidine-facilitated pathways. *Biochim. Biophys. Acta—Biomembr.* **1614**, 211–219.

Glover, C. N., Bury, N. R., and Hogstrand, C. (2004). Intestinal zinc uptake in freshwater rainbow trout: evidence for apical pathways associated with potassium efflux and modified by calcium. *Biochim. Biophys. Acta—Biomembr.* **1663**, 214–221.

Golovanova, I. L. (1993). Characteristics of carbohydrate transport in different parts of bream (*Abramis brama*) and carp (*Cyprinus carpio*) intestine. *J. Ichthyol.* **33**, 26–35.

Goncalves, A. F., Castro, L. F. C., Pereira-Wilson, C., Coimbra, J., and Wilson, J. M. (2007). Is there a compromise between nutrient uptake and gas exchange in the gut of *Misgurnus anguillicaudatus*, an intestinal air-breathing fish? *Comp. Biochem. Physiol. D* **2**, 345–355.

Gottsche, J. R., Nielsen, N. S., Nielsen, H. H., and Mu, H. L. (2005). Lipolysis of different oils using crude enzyme isolate from the intestinal tract of rainbow trout, *Oncorhynchus mykiss*. *Lipids* **40**, 1273–1279.

Gouyon, F., Caillaud, L., Carriere, V., Klein, C., Dalet, V., Citadelle, D., Kellett, G. L., Thorens, B., Leturque, A., and Brot-Laroche, E. (2003). Simple-sugar meals target GLUT2 at enterocyte apical membranes to improve sugar absorption: a study in GLUT2-null mice. *J. Physiol.—London* **552**, 823–832.

Graham, C., Nalbant, P., Schölermann, B., Hentschel, H., Kinne, R. K. H., and Werner, A. (2003). Characterization of a type IIb sodium-phosphate cotransporter from zebrafish (*Danio rerio*) kidney. *Am. J. Physiol.* **284**, F727–F736.

Grosell, M., and Genz, J. (2006). Ouabain-sensitive bicarbonate secretion and acid absorption by the marine teleost fish intestine play a role in osmoregulation. *Am. J. Physiol.* **291**, R1145–R1156.

Grosell, M., Laliberte, C. N., Wood, S., Jensen, F. B., and Wood, C. M. (2001). Intestinal HCO_3^- secretion in marine teleost fish: evidence for an apical rather than basolateral Cl^-/HCO_3^- exchanger. *Fish Physiol. Biochem.* **24**, 81–95.

Guerreiro, P. M., Fuentes, J., Canario, A. V. M., and Power, D. M. (2002). Calcium balance in sea bream (*Sparus aurata*): the effect of oestradiol-17β. *J. Endocrinol.* (**173**), 377–385.

Gutowska, M. A., Drazen, J. C., and Robison, B. H. (2004). Digestive chitinolytic activity in marine fishes of Monterey Bay, California. *Comp. Biochem. Physiol. A—Mol. Integr. Physiol* **139**, 351–358.

Hakim, Y., Harpaz, S., and Uni, Z. (2009). Expression of brush border enzymes and transporters in the intestine of European sea bass (*Dicentrarchus labrax*) following feed deprivation. *Aquaculture* **290**, 110–115.

Hall, J. R., Short, C. E., and Driedzic, W. R. (2006). Sequence of Atlantic cod (*Gadus morhua*) GLUT4, GLUT2, and GPDH: developmental stage expression, tissue expression and relationship to starvation-induced changes in blood glucose. *J. Exp. Biol.* **209**, 4490–4502.

Handy, R. D., Musonda, M. M., Phillips, C., and Falla, S. J. (2000). Mechanisms of gastrointestinal copper absorption in the African Walking Catfish: copper dose-effects and a novel anion-dependent pathway in the intestine. *J. Exp. Biol.* **203**, 2365–2377.

Handy, R. D., Eddy, F. B., and Baines, H. (2002). Sodium-dependent copper uptake across epithelia: a review of rationale with experimental evidence from gill and intestine. *Biochim. Biophys. Acta—Biomembr.* **1566**, 104–115.

Hansen, J. O., Berge, G. M., Hillestad, M., Krogdahl, Å., Galloway, T. F., Holm, H., Holm, J., and Ruyter, B. (2008). Apparent digestion and apparent retention of lipid and fatty acids in Atlantic cod (*Gadus morhua*) fed increasing dietary lipid levels. *Aquaculture* **284**, 159–166.

Haslewood, G. A. D. (1967). Bile salt evolution. *J. Lipid Res.* **8**, 535–550.

Hernandez-Blazquez, F. J., Guerra, R. R., Kfoury, J. R., Bombonato, P. P., Cogliati, B., and da Silva, J. R. M. C. (2006). Fat absorptive processes in the intestine of the Antarctic fish *Notothenia coriiceps* (Richardson, 1844). *Polar Biol.* **29**, 831–836.

Hidalgo, M. C., Urea, E., and Sanz, A. (1999). Comparative study of digestive enzymes in fish with different nutritional habits. Proteolytic and amylase activities. *Aquaculture* **170**, 267–283.

Hogstrand, C., Gassman, N. J., Popova, B., Wood, C. M., and Walsh, P. J. (1996). The physiology of massive zinc accumulation in the liver of female squirrelfish and its relationship to reproduction. *J. Exp. Biol.* **199**, 2543–2554.

Holmgren, S., and Olsson, C. (2009). The neuroendocrine regulation of gut function. In: *Fish Neuroendocrinology (Fish Physiology Series)* (N. J. Bernier, G. Van Der Kraak, A. P. Farrell and C. J. Brauner, eds), Vol. 28, pp. 467–512. Academic Press, San Diego.

Horn, M. H., Gawlicka, A. K., German, D. P., Logothetis, E. A., Cavanagh, J. W., and Boyle, K. S. (2006). Structure and function of the stomachless digestive system in three related species of New World silverside fishes (Atherinopsidae) representing herbivory, omnivory, and carnivory. *Marine Biol.* **149**, 1237–1245.

Houlihan, D., Boujard, T., and Jobling, M. (2001). *Food Intake in Fish*. Blackwell Science Ltd., Oxford.

Houpe, K. L., Malo, C., Oldham, P. B., and Buddington, R. K. (1997). Thermal modulation of channel catfish intestinal dimensions, BBM fluidity, and glucose transport. *Am. J. Physiol.* **270**, R1037–R1043.

Hua, K., and Bureau, D. P. (2009). Development of a model to estimate digestible lipid content of salmonid fish feeds. *Aquaculture* **286**, 271–276.

Iijima, N., Tanaka, S., and Ota, Y. (1998). Purification and characterization of bile salt-activated lipase from the hepatopancreas of red sea bream. *Pagrus major. Fish Physiol. Biochem.* **18**, 59–69.

Ikeda, M., Miyauchi, K., Mochizuki, A., and Matsumiya, M. (2009). Purification and characterization of chitinase from the stomach of silver croaker. *Pennahia argentatus. Protein Expres. Purif.* **65**, 214–222.

Ingh, T. S. G. A. M. van den, Krogdahl, Å., Olli, J., Hendricks, H. G. C. J. M., and Koninkx, J. F. J. G. (1991). Effects of soybean-containing diets on the proximal and distal intestine in Atlantic salmon (*Salmo salar*): a morphological study. *Aquaculture* **94**, 297–305.

Iqbal, J., and Hussain, M. M. (2009). Intestinal lipid absorption. *Am. J. Physiol.* **296**, E1183–E1194.

Jellouli, K., Bougatef, A., Daassi, D., Balti, R., Barkia, A., and Nasri, M. (2009). New alkaline trypsin from the intestine of Grey triggerfish (*Balistes capriscus*) with high activity at low temperature: purification and characterisation. *Food Chem.* **116**, 644–650.

Jobling, M. (1995). *Environmental Biology of Fishes.* Chapman & Hall, London.

Johnson, S. P., and Schindler, D. E. (2009). Trophic ecology of Pacific salmon (*Oncorhynchus* spp.) in the ocean: a synthesis of stable isotope research. *Ecol. Res.* **24**, 855–863.

Jutfelt, F., Olsen, R. E., Björnsson, B. T., and Sundell, K. (2007). Parr–smolt transformation and dietary vegetable lipids affect intestinal nutrient uptake, barrier function and plasma cortisol levels in Atlantic salmon. *Aquaculture* **273**, 298–311.

Karasov, W. H., and Diamond, J. M. (1981). Intestinal glucose transport varies with taxa and diet. *Amer. Zool.* **21**, 1030.

Karasov, W. H., Buddington, R. K., and Diamond, J. M. (1985). Adaptation of intestinal sugar and amino acid transport in vertebrate evolution. In: *Transport Processes, Iono- and Osmoregulation* (R. Gilles and M. Gilles-Baillien, eds), pp. 227–239. Springer Verlag, Berlin.

Kihara, M., and Sakata, T. (1997). Fermentation of dietary carbohydrates to short-chain fatty acids by gut microbes and its influence on intestinal morphology of a detritivorous teleost Tilapia (*Oreochromis niloticus*). *Comp. Biochem. Physiol. A* **118**, 1201–1207.

Kjoss, V. A., Grosell, M., and Wood, C. M. (2005). The influence of dietary Na on Cu accumulation in juvenile rainbow trout exposed to combined dietary and waterborne Cu in soft water. *Arch. Environ. Contam. Toxicol.* **49**, 520–527.

Klaren, P. H. M., Flik, G., Lock, R. A. C., and Wendelaar Bonga, S. E. (1993). Ca^{2+} transport across the intestinal brush border membranes of the cichlid teleost *Oreochromis mossambicus.* *J. Membr. Biol.* **132**, 157–166.

Klaren, P. H. M., Wendelaar Bonga, S. E., and Flik, G. (1997). Evidence for P_2 purinoceptor-mediated uptake of Ca^{2+} across a fish (*Oreochromis mossambicus*) intestinal brush border membrane. *Biochem. J.* **322**, 129–134.

Klomklao, S., Benjakul, S., Visessanguan, W., Kishimura, H., Simpson, B. K., and Saeki, H. (2006). Trypsins from yellowfin tuna (*Thunnus albacores*) spleen: purification and characterization. *Comp. Biochem. Physiol. B* **144**, 47–56.

Kohl, B., Herter, P., Huelseweh, B., Elger, M., Hentschel, H., Kinne, R. H., and Werner, A. (1996). Na-P_i cotransport in flounder: same transport system in kidney and intestine. *Am. J. Physiol.* **270**, F937–F944.

Krasnov, A., Teerijoki, H., and Molsa, H. (2001). Rainbow trout (*Oncorhynchus mykiss*) hepatic glucose transporter. *Biochim. Biophys. Acta* **1520**, 174–178.

Krogdahl, Å., and Sundby, A. (1999). Characteristics of pancreatic function in fish. In: *Biology of the Pancreas in Growing Animals* (S. G. Pierzynowski and R. Zabielski, eds), pp. 437–458. Elsevier Science, Amsterdam.

Krogdahl, Å., Nordrum, S., Sørensen, M., Brudeseth, L., and Røsjø, C. (1999). Effects of diet composition on apparent nutrient absorption along the intestinal tract and of subsequent fasting on mucosal disaccharidase activities and plasma nutrient concentration in Atlantic salmon *Salmo salar* L. *Aquacult. Nutr.* **5**, 121–133.

Krogdahl, Å., Bakke-McKellep, A. M., and Baeverfjord, G. (2003). Effects of graded levels of standard soybean meal on intestinal structure, mucosal enzyme activities, and pancreatic response in Atlantic salmon (*Salmo salar* L.). *Aquacult. Nutr.* **9**, 361–371.

Krogdahl, Å., Sundby, A., and Olli, J. J. (2004). Atlantic salmon (*Salmo salar*) and rainbow trout (*Oncorhynchus mykiss*) digest and metabolize nutrients differently. Effects of water salinity and dietary starch level. *Aquaculture* **229**, 335–360.

Krogdahl, Å., Hemre, G.-I., and Mommsen, T. P. (2005). Carbohydrates in fish nutrition: digestion and absorption in postlarval stages. *Aquacult. Nutr.* **11**, 103–122.

Kupermann, B. I., and Kuz'mina, V. V. (1994). The ultrastructure of the intestinal epithelium in fishes with different types of feeding. *J. Fish Biol.* **44**, 181–193.

Kurokawa, T., and Suzuki, T. (1995). Structure of the exocrine pancreas of flounder (*Paralichthys-olivaceus*)—immunological localization of zymogen granules in the digestive tract using anti-trypsinogen antibody. *J. Fish Biol.* **46**, 292–301.

Kurtovic, I., Marshall, S. N., Zhao, X., and Simpson, B. K. (2009). Lipases from mammals and fishes. *Rev. Fish. Sci.* **17**, 18–40.

Kuz'mina, V. V. (2008). Classical and modern concepts in fish digestion. In: *Feeding and Digestive Functions of Fishes* (J. E. P. Cyrino, D. P. Bureau and B. G. Kapoor, eds), pp. 85–154. Science Publishers, Enfield.

Kwong, R. W. M., and Niyogi, S. (2008). An *in vitro* examination of intestinal iron absorption in a freshwater teleost, rainbow trout (*Oncorhynchus mykiss*). *J. Comp. Physiol. B* **178**, 963–975.

Kwong, R. W. M., and Niyogi, S. (2009). The interactions of iron with other divalent metals in the intestinal tract of a freshwater teleost, rainbow trout (*Oncorhynchus mykiss*). *Comp. Biochem. Physiol.* **150C**, 442–449.

Larsson, D., Lundgren, T., and Sundell, K. (1998). Ca^{2+} uptake through voltage-gated L-type Ca^{2+} channels by polarized enterocytes from Atlantic cod. *Gadus morhua. J. Membr. Biol.* **164**, 229–237.

Larsson, D, Aksnes, L., Björnsson, B. T., Larsson, B., Lundgren, T., and Sundell, K. (2002). Antagonistic effects of $24R,25$-dihydroxyvitamin D_3 and 25-hydroxyvitamin D_3 on L-type Ca^{2+} channels and Na^+/Ca^{2+} exchange in enterocytes from Atlantic cod (*Gadus morhua*). *J. Mol. Endocrinol.* **28**, 53–68.

Leaver, M. J., Bautista, J. M., Bjornsson, B. T., Jonsson, E., Krey, G., Tocher, D. R., and Torstensen, B. E. (2008). Towards fish lipid nutrigenomics: current state and prospects for fin-fish aquaculture. *Rev. Fish. Sci.* **16**, 73–94.

Lechanteur, Y. A. R. G., and Griffiths, C. L. (2003). Diets of common suprabenthic reef fish in False Bay, South Africa. *Afr. Zool.* **38**, 213–227.

Leger, C., Ducruet, V., and Flanzy, J. (1979). Lipase et colipase de la truite arc-en-ciel. Quelques resultats recents. *Ann. Biol. Anim. Biochim. Biophys.* **19**, 825–832.

Leonard, E. M., Nadella, S. R., Bucking, C., and Wood, C. M. (2009). Characterization of dietary Ni uptake in the rainbow trout, *Oncorhynchus mykiss. Aquat. Toxicol.* **93**, 205–216.

Lionetto, M. G., Maffia, M., Vignes, F., Storelli, C., and Schettino, T. (1996). Differences in intestinal electrophysiological parameters and nutrient transport rates between eels (*Anguilla anguilla*) at yellow and silver stages. *J. Exp. Zool.* **275**, 399–405.

Liu, J., Caballero, M. J., Izquierdo, M. S., El-sayed Ali, T., Hernandez-Cruz, C. M., Valencia, A., and Fernandez-Palacios, H. (2002). Necessity of dietary lecithin and eicosapentaenoic acid for growth, survival, stress resistance and lipoprotein formation in gilthead sea bream. *Sparus aurata. Fish. Sci.* **68**, 1165–1172.

Logothetis, E. A., Horn, M. H., and Dickson, K. A. (2001). Gut morphology and function in *Atherinops affinis* (Teleostei: Atherinopsidae), a stomachless omnivore feeding on macroalgae. *J. Fish Biol.* **59**, 1298–1312.

Mackenzie, N. C., Brito, M., Reyes, A. E., and Allende, M. L. (2004). Cloning, expression pattern and essentiality of the high-affinity copper transporter 1 (ctr1) gene in zebrafish. *Gene* **328**, 113–120.

Maffia, M., Acierno, R., Cillo, E., and Storelli, C. (1996). Na^+-D-glucose cotransport by intestinal BBMVs of the Antarctic fish. *Trematomus bernacchii. Am. J. Physiol.* **271**, R1576–R1583.

Maffia, M., Verri, T., Danieli, A., Thamotharan, M., Pastore, M., Ahearn, G. A., and Storelli, C. (1997). H^+-glycyl-L-proline cotransport in brush-border membrane vesicles of eel (*Anguilla anguilla*) intestine. *Am. J. Physiol.* **272**, R217–R225.

Maffia, M., Rizzello, A., Acierno, R., Verri, T., Rollo, M., Danieli, A., Doring, F., Daniel, H., and Storelli, C. (2003). Characterisation of intestinal peptide transporter of the Antarctic haemoglobinless teleost *Chionodraco hamatus*. *J. Exp. Biol.* **206**, 705–714.

Maitipe, P., and De Silva, S. (1985). Switches between zoophagy, phytophagy and detritivory of *Sarotherodon mossambicus* populations in twelve Sri Lankan lakes. *J. Fish Biol.* **26**, 49–61.

Manchado, M., Infante, C., Asensio, E., Crespo, A., Zuasti, E., and Canavate, J. P. (2008). Molecular characterization and gene expression of six trypsinogens in the flatfish Senegalese sole (*Solea senegalensis* Kaup) during larval development and in tissues. *Comp. Biochem. Physiol. B* **149**, 334–344.

Manjakasy, J. M., Day, R. D., Kemp, A., and Tibbetts, I. R. (2009). Functional morphology of digestion in the stomachless, piscivorous needlefishes *Tylosurus gavialoides* and *Strongylura leiura ferox* (Teleostei: Beloniformes). *J. Morphol.* **270**, 1155–1165.

Martin, T. J., and Blaber, S. J. M. (1984). Morphology and histology of the alimentary tracts of ambassidae (Cuvier) (Teleostei) in relation to feeding. *J. Morphol.* **182**, 295–305.

Mazlan, A. G., and Grove, D. J. (2004). Quantification of gastric secretions in the wild whiting fed on natural prey in captivity. *J. Appl. Ichthyol.* **20**, 295–301.

McLean, E., and Ash, R. (1987). Intact protein (antigen) absorption in fishes: mechanisms and physiological significance. *J. Fish Biol. F* (Suppl. A), 219–223.

McLean, E., and Donaldson, E. M. (1990). Absorption of bioactive proteins by the gastrointestinal tract of fish: a review. *J. Aquat. Anim. Health* **2**, 1–11.

Minghetti, M., Leaver, M. J., Carpene, E., and George, S. G. (2008). Copper transporter 1, metallothionein and glutathione reductase genes are differentially expressed in tissues of sea bream (*Sparus aurata*) after exposure to dietary or waterborne copper. *Comp. Biochem. Physiol.* **147C**, 450–459.

Morais, S., Rojas-Garcia, C. R., Conceição, L. E. C., and Rønnestad, I. (2005a). Digestion and absorption of a pure triacylglycerol and a free fatty acid by *Clupea harengus* L. larvae. *J. Fish Biol.* **67**, 223–238.

Morais, S., Koven, W., Ronnestad, I., Dinis, M. T., and Conceicao, L. E. (2005b). Dietary protein:lipid ratio and lipid nature affects fatty acid absorption and metabolism in a teleost larva. *Brit. J. Nutr.* **93**, 813–820.

Morais, S., Conceicao, L. E., Ronnestad, I., Koven, W., Cahu, C., Zambonino Infante, J. L., and Dinis, M. T. (2007). Dietary neutral lipid level and source in marine fish larvae: effects on digestive physiology and food intake. *Aquaculture* **268**, 106–122.

Moreau, R., and Dabrowski, K. (2000). Biosynthesis of ascorbic acid by extant actinopterygians. *J. Fish Biol.* **57**, 733–745.

Moren, M., Sloth, J. J., and Hamre, K. (2008). Uptake of iodide from water in Atlantic halibut larvae (*Hippoglossus hippoglossus* L.). *Aquaculture* **285**, 174–178.

Morrison, C. M., and Wright, J. R. (1999). A study of the histology of the digestive tract of the Nile tilapia. *J. Fish Biol.* **54**, 597–606.

Morrison, C. M., Pohajdak, B., Tam, J., and Wright, J. R. (2004). Development of the islets, exocrine pancreas, and related ducts in the Nile tilapia, *Oreochromis niloticus* (Pisces: Cichlidae). *J. Morphol.* **261**, 377–389.

Murray, H. M., Wright, G. M., and Goff, G. P. (1996). A comparative histological and histochemical study of the post-gastric alimentary canal from three species of pleuronectid, the Atlantic halibut, the yellowtail flounder and the winter flounder. *J. Fish Biol.* **48**, 187–206.

Murray, H. M., Gallant, J. W., Perez-Casanova, J. C., Johnson, S. C., and Douglas, S. E. (2003). Ontogeny of lipase expression in winter flounder. *J. Fish Biol.* **62**, 816–833.

Muyssen, B. T. A., Brix, K. V., DeForest, D. K., and Janssen, C. R. (2004). Nickel essentiality and homeostasis in aquatic organisms. *Environ. Rev.* **12**, 113–131.

Nadella, S. R., Bucking, C., Grosell, M., and Wood, C. M. (2006a). Gastrointestinal assimilation of Cu during digestion of a single meal in the freshwater rainbow trout (*Oncorhynchus mykiss*). *Comp. Biochem. Physiol.* 143C, 394–401.

Nadella, S. R., Grosell, M., and Wood, C. M. (2006b). Physical characterization of high-affinity gastrointestinal Cu transport *in vitro* in freshwater rainbow trout. *Oncorhynchus mykiss. J. Comp. Physiol.* B 176, 793–806.

Nadella, S. R., Grosell, M., and Wood, C. M. (2007). Mechanisms of dietary Cu uptake in freshwater rainbow trout: evidence for Na-assisted Cu transport and a specific metal carrier in the intestine. *J. Comp. Physiol.* B 176, 793–806.

Nalbant, P., Boehmer, C., Dehmelt, L., Wehner, F., and Werner, A. (1999). Functional characterization of a Na^+-phosphate cotransporter (NaP_i-II) from zebrafish and identification of related transcripts. *J. Physiol.* 520, 79–89.

Neuhaus, H., van der Marel, M., Caspari, N., Meyer, W., Enss, M. L., and Steinhagen, D. (2007). Biochemical and histochemical effects of perorally applied endotoxin on intestinal mucin glycoproteins of the common carp. *Cyprinus carpio. Dis. Aquat. Org.* 77, 17–27.

Niyogi, S., Pyle, G. G., and Wood, C. M. (2007). Branchial versus intestinal zinc uptake in wild yellow perch (*Perca flavescens*) from reference and metal-contaminated aquatic ecosystems. *Can. J. Fish. Aquat. Sci.* 64, 1605–1613.

Nordrum, S., Bakke-McKellep, A. M., Krogdahl, Å., and Buddington, R. K. (2000a). Effects of soybean meal and salinity on intestinal transport of nutrients in Atlantic salmon (*Salmo salar* L.) and rainbow trout (*Oncorhynchus mykiss*). *Comp. Biochem. Physiol.* B 125, 317–335.

Nordrum, S., Krogdahl, Å., Røsjø, C., Olli, J. J., and Holm, H. (2000b). Effects of methionine, cysteine and medium chain tryglycerides on nutrient digestibility, absorption of amino acids along the intestinal tract and nutrient retention in Atlantic salmon (*Salmo salar* L.) under pair-feeding regime. *Aquaculture* 186, 341–360.

Norris, J. S., Norris, D. O., and Windell, J. T. (1973). Effect of stimulated meal size on gastric acid and pepsin secretory rates in bluegill (*Lepomis macrochirus*). *J. Fish. Res. Board Can.* 30, 201–204.

Ogiwara, K., and Takahashi, T. (2007). Specificity of the medaka enteropeptidase serine protease and its usefulness as a biotechnological tool for fusion-protein cleavage. *Proc. Natl. Acad. Sci. USA* 104, 7021–7026.

Ojo, A. A., and Wood, C. M. (2007). *In vitro* analysis of the bioavailability of six metals via the gastrointestinal tract of the rainbow trout (*Oncorhynchus mykiss*). *Aquat. Toxicol.* 83, 10–23.

Ojo, A. A., and Wood, C. M. (2008). *In vitro* characterization of cadmium and zinc uptake via the gastrointestinal tract of the rainbow trout (*Oncorhynchus mykiss*): interactive effects and the influence of calcium. *Aquat. Toxicol.* 89, 55–64.

Ojo, A. A., Nadella, S. R., and Wood, C. M. (2009). *In vitro* examination of interactions between copper and zinc uptake via gastrointestinal tract of the rainbow trout (*Oncorhynchus mykiss*). *Arch. Environ. Contam. Toxicol.* 56, 244–252.

Olli, J., Hjelmeland, K., and Krogdahl, Å. (1994). Soybean trypsin inhibitors in diets for Atlantic salmon (*Salmo salar*, L.): effects on nutrient digestibilities and trypsin in pyloric caeca homogenate and intestinal content. *Comp. Biochem. Physiol.* A 109, 923–928.

Olsen, R. E., and Ringø, E. (1997). Lipid digestibility in fish: a review. *Recent Res. Dev. Lipids Res.* 1, 199–265.

Olsen, R. E., Myklebust, R., Kaino, T., and Ringø, E. (1999). Lipid digestibility and ultrastructural changes in the enterocytes of Arctic char *Salvelinus alpinus* L. fed linseed oil and soybean lecithin. *Fish Physiol. Biochem.* 21, 35–44.

Olsen, R. E., Dragnes, B. T., Myklebust, R., and Ringø, E. (2003). Effect of soybean oil and soybean lecithin on intestinal lipid composition and lipid droplet accumulation of rainbow trout, *Oncorhynchus mykiss* Walbaum. *Fish Physiol. Biochem.* **29**, 181–192.

Olsen, R. E., Henderson, R. J., and Sountama, J. (2004). Atlantic salmon, *Salmo salar*, utilizes wax ester-rich oil from *Calanus finmarchicus* effectively. *Aquaculture* **240**, 433–449.

Olsen, R. E., Kiessling, A., Milley, J. E., Ross, N. W., and Lall, S. P. (2005). Lipid source, not bile acids, affects absorption of astaxanthin in Atlantic salmon. *Salmo salar*. *Aquaculture* **250**, 804–812.

Olsen, R. E., Hansen, A. C., Rosenlund, G., Hemre, G.-I., Mayhew, T. M., Knudsen, D. L., Eroldogan, O. T., Myklebust, R., and Karlsen, O. (2007). Total replacement of fish meal with plant proteins in diets for Atlantic cod (*Gadus morhua* L.) II—Health aspects. *Aquaculture* **272**, 612–624.

On, C., Marshall, C. R., Perry, S. F., Dinh Le, H., Yurkov, V., Omelchenko, A., Hnatowich, M., Hryshko, L. V., and Tibbits, G. F. (2009). Characterization of zebrafish (*Danio rerio*) NCX4: a novel NCX with distinct electrophysiological properties. *Am. J. Physiol.* **296**, C173–C181.

Ørnsrud, R., Lock, E. J., Glover, C. N., and Flik, G. (2009). Retinoic acid cross-talk with calcitriol activity in Atlantic salmon (*Salmo salar*). *J. Endocrin.* **202**, 473–482.

Oxley, A., Jutfelt, F., Sundell, K., and Olsen, R. E. (2007). Sn-2-monoacylglycerol, not glycerol, is preferentially utilised for triacylglycerol and phosphatidylcholine biosynthesis in Atlantic salmon (*Salmo salar* L.) intestine. *Comp. Biochem. Physiol. B* **146**, 115–123.

Pajor, A. M, Hirayama, B. A., and Wright, E. M. (1992). Molecular biology approaches to comparative study of sodium glucose cotransport. *Am. J. Physiol.* **263**, R489–R495.

Pan, T. C., Liao, B. K., Huang, C. J., Lin, L. Y., and Hwang, P. P. (2005). Epithelial Ca^{2+} channel expression and Ca^{2+} uptake in developing zebrafish. *Am. J. Physiol.* **279**, R1202–R1211.

Panserat, S., Plagnes-Juan, E., and Kaushik, S. (2001). Nutritional regulation and tissue specificity of gene expression for proteins involved in hepatic glucose metabolism in rainbow trout (*Oncorhynchus mykiss*). *J. Exp. Biol.* **204**, 2351–2360.

Péres, A., Cahu, C. L., Zambonino Infante, J. L., Le Gall, M. M., and Quazuguel, P. (1996). Amylase and trypsin responses to intake of dietary carbohydrate and protein depend on the developmental stage in sea bass (*Dicentrarchus labrax*) larvae. *Fish Physiol. Biochem.* **15**, 237–242.

Péres, A., Zambonino Infante, J. L., and Cahu, C. L. (1998). Dietary regulation of activities and mRNA levels of trypsin and amylase in sea bass (*Dicentrarchus labrax*) larvae. *Fish Physiol. Biochem.* **19**, 145–152.

Pérez-Jiménez, A., Cardenete, G., Morales, A. E., García-Alcázar, A., Abellán, E., and Hidalgo, M. C. (2009). Digestive enzymatic profile of *Dentex dentex* and response to different dietary formulations. *Comp. Biochem. Physiol. A* **154**, 157–164.

Phipps, T., Tank, S. L., Wirtz, J., Brewer, J., Coyner, A., Ortego, L. A., and Fairbrother, A. (2002). Essentiality of nickel and homeostatic mechanisms for its regulation in terrestrial organisms. *Environ. Rev.* **10**, 209–261.

Pivnenko, T. N., Epstein, L. M., and Okladnikova, S. V. (1997). Comparative analysis of substrate specificity of pancreatic serine proteinases of different origin. *J. Evol. Biochem. Physiol.* **33**, 540–544.

Planas, J. V., Capilla, E., and Gutiérrez, J. (2000). Molecular identification of a glucose transporter from fish muscle. *FEBS Lett.* **481**, 266–270.

Platell, M. E., and Potter, I. C. (2001). Partitioning of food resources amongst 18 abundant benthic carnivorous fish species in marine waters on the lower west coast of Australia. *J. Exp. Mar. Biol. Ecol.* **261**, 31–54.

Portella, M. C., and Dabrowski, K. (2008). Diets, physiology, biochemistry and digestive tract development of freshwater fish larvae. In: *Feeding and Digestive Functions of Fishes* (J. E. P. Cyrino, D. P. Bureau and B. G. Kapoor, eds), pp. 227–279. Science Publishers, Enfield.

Psochiou, E., Sarropoulou, E., Mamuris, Z., and Mouton, K. A. (2007). Sequence analysis and tissue expression pattern of *Sparus aurata* chymotrypsinogens and trypsinogen. *Comp. Biochem. Physiol. B* **147**, 367–377.

Ptashynski, M. D., and Klaverkamp, J. F. (2002). Accumulation and distribution of dietary nickel in lake whitefish (*Coregonus clupeaformis*). *Aquat. Toxicol.* **58**, 249–264.

Qiu, A. D., and Hogstrand, C. (2004). Functional characterisation and genomic analysis of an epithelial calcium channel (ECaC) from pufferfish *Fugu rubripes*. *Gene* **342**, 113–123.

Qiu, A. D., and Hogstrand, C. (2005). Functional expression of a low-affinity zinc uptake transporter (*FrZIP2*) from pufferfish (*Takifugu rubripes*) in MDCK cells. *Biochem. J.* **390**, 777–786.

Qiu, A., Shayeghi, M., and Hogstrand, C. (2005). Molecular cloning and functional characterization of a high-affinity zinc importer (*DrZIP1*) from zebrafish (*Danio rerio*). *Biochem. J.* **388**, 745–754.

Qiu, A., Glover, C. N., and Hogstrand, C. (2007). Regulation of branchial zinc uptake by 1α,25-(OH)$_2$D$_3$ in rainbow trout and associated changes in expression of ZIP1 and ECaC. *Aquat. Toxicol.* **84**, 142–152.

Quentel, C., Bremont, M., and Pouliquen, H. (2007). Farm fish vaccination. *Prod. Anim.* **20**, 233–238.

Raubenheimer, D., Simpson, S. J., and Mayntz, D. (2009). Nutrition, ecology and nutritional ecology: toward an integrated framework. *Func. Ecol.* **23**, 4–16.

Refstie, S., Landsverk, T., Bakke-McKellep, A. M., Ringø, E., Sundby, A., Shearer, K. D., and Krogdahl, Å. (2006). Digestive capacity, intestinal morphology, and microflora of 1-year and 2-year old Atlantic cod (*Gadus morhua*) fed standard or bioprocessed soybean meal. *Aquaculture* **261**, 269–284.

Reshkin, S. J., and Ahearn, G. A. (1987a). Basolateral glucose transport by intestine of teleost, *Oreochromis mossambicus*. *Am. J. Physiol.* **252**, R579–R586.

Reshkin, S. J., and Ahearn, G. A. (1987b). Intestinal glucose transport and salinity adaptation in a euryhaline teleost. *Am. J. Physiol.* **252**, R567–R578.

Richardson, N. L., Higgs, D. A., Beames, R. M., and McBride, J. R. (1985). Influence of dietary calcium, phosphorus, zinc and sodium phytate level on cataract incidence, growth and histopathology in juvenile Chinook salmon (*tshawytscha*). *J. Nutr.* **115**, 553–567.

Romano, A., Kottra, G., Barca, A., Tiso, N., Maffia, M., Argenton, F., Daniel, H., Storelli, C., and Verri, T. (2006). High-affinity peptide transporter PEPT2 (SLC15A2) of the zebrafish *Danio rerio*: functional properties, genomic organization, and expression analysis. *Physiol. Genom.* **24**, 207–217.

Rosas, A., Vazquez-Duhalt, R., Tinoco, R., Shimada, A., Dabramo, L. A., and Viana, M. T. (2008). Comparative intestinal absorption of amino acids in rainbow trout (*Oncorhynchus mykiss*), totoaba (*Totoaba macdonaldi*) and Pacific bluefin tuna (*Thunnus orientalis*). *Aquacult. Nutr.* **14**, 481–489.

Rose, R. C., and Chou, J. (1990). Intestinal absorption and metabolism of ascorbic acid in rainbow trout. *Am. J. Physiol.* **258**, R1238–R1241.

Røsjø, C., Nordrum, S., Olli, J. J., Krogdahl, Å., Ruyter, B., and Holm, H. (2000). Lipid digestibility and metabolism in Atlantic salmon (*Salmo salar*) fed medium-chain triglycerides. *Aquaculture* **190**, 65–76.

Sala-Rabanal, M., Gallardo, M. A., Sanchez, J., and Planas, J. M. (2004). Na-dependent D-glucose transport by intestinal brush border membrane vesicles from gilthead sea bream (*Sparus aurata*). *J. Membr. Biol.* **201**, 85–96.

Sales, J. (2009). Linear models to predict the digestible lipid content of fish diets. *Aquacult. Nutr.* **15**, 537–549.

Sangaletti, R., Terova, G., Peres, A., Bossi, E., Cora, S., and Saroglia, M. (2009). Functional expression of the oligopeptide transporter PepT1 from the sea bass (*Dicentrarchus labrax*). *Pflugers Arch.-Eur. J. Physiol.* **459**, 47–54.

Schep, L. J., Tucker, I. G., Young, G., and Butt, A. G. (1997). Regional permeability differences between the proximal and distal portions of the isolated salmonid posterior intestine. *J. Comp. Physiol. B* **167**, 370–377.

Schoenmakers, T. J. M., Verbost, P. M., Flik, G., and Wendelaar Bonga, S. E. (1993). Transcellular intestinal calcium transport in freshwater and seawater fish and its dependence on sodium/calcium exchange. *J. Exp. Biol.* **176**, 195–206.

Schroers, V., van der Marel, M., Neuhaus, H., and Steinhagen, D. (2009). Changes of intestinal mucus glycoproteins after peroral application of *Aeromonas hydrophila* to common carp (*Cyprinus carpio*). *Aquaculture* **288**, 184–189.

Scocco, P., Accili, D., Menghi, G., and Ceccarelli, P. (1998). Unusual glycoconjugates in the oesophagus of a tilapine polyhybrid. *J. Fish Biol.* **53**, 39–48.

Seeto, G. S., Veivers, P. C., Clements, K. D., and Slaytor, M. (1996). Carbohydrate utilisation by microbial symbionts in the marine herbivorous fishes *Odax cyanomelas* and *Crinodus lophodon*. *J. Comp. Physiol. B* **165**, 571–579.

Sharp, P. A. (2002). CTR1 and its role in body copper homeostasis. *J. Biochem. Cell Biol.* **35**, 288–291.

Shears, M. A., and Fletcher, G. L. (1983). Regulation of Zn^{2+} uptake from the gastrointestinal tract of a marine teleost, the winter flounder (*Pseudopleuronectes americanus*). *Can. J. Fish. Aquat. Sci.* **40**, 197–205.

Shephard, K. L. (1994). Functions for fish mucus. *Rev. Fish Biol. Fisher* **4**, 401–429.

Sims, D. W. (2008). Sieving a living: a review of the biology, ecology and conservation status of the plankton-feeding basking shark. *Cetorhinus maximus. Adv. Mar. Biol.* **54**, 171–220.

Sire, M.-F., and Vernier, J.-M. (1992). Intestinal absorption of protein in teleost fish. *Comp. Biochem. Physiol. A* **103**, 771–781.

Sire, M.-F., Lutton, C., and Vernier, J.-M. (1981). New views on intestinal absorption of lipids in teleostean fishes: an ultrastructural and biochemical study in the rainbow trout. *J. Lipid Res.* **22**, 81–94.

Sire, M.-F., Dorin, D., and Vernier, J.-M. (1992). Intestinal absorption of macromolecular proteins in rainbow trout. *Aquaculture* **100**, 234–235.

Sklan, D., Prag, T., and Lupatsch, I. (2004). Structure and function of the small intestine of the tilapia *Oreochromis niloticus* × *Oreochromis aureus* (Teleostei, Cichlidae). *Aquacult. Res.* **35**, 350–357.

Soengas, J. L., and Moon, T. W. (1998). Transport and metabolism of glucose in isolated enterocytes of the black bullhead *Ictalurus melas*: effects of diet and hormones. *J. Exp. Biol.* **201**, 3263–3273.

Spannhof, L., and Plantikow, H. (1983). Studies on carbohydrate digestion in rainbow trout. *Aquaculture* **30**, 95–108.

Starck, J. M. (1999). Structural flexibility of the gastro-intestinal tract of vertebrates— implications for evolutionary morphology. *Zool. Anzeiger* **238**, 87–101.

Sternby, B., Engstrom, A., and Hellman, U. (1984). Purification and characterization of pancreatic colipase from the dogfish (*Squalus acanthius*). *Biochim. Biophys. Acta, Protein Struct. Mol. Enzymol.* **789**, 159–163.

Storelli, C., Vilella, S., and Cassano, G. (1986). Na-dependent D-glucose and L-alanine transport in eel intestinal brush border membrane vesicles. *Am. J. Physiol.* **251**, R463–R469.

Storelli, C., Vilella, S., Romano, A., Maffia, M., and Cassano, G. (1989). Brush-border amino acid transport mechanisms in carnivorous eel intestine. *Am. J. Physiol.* **257**, R506–R510.

Sugiura, S. (2009). Identification of intestinal phosphate transporters in fishes and shellfishes. *Fish. Sci.* **75**, 99–108.

Sugiura, S. H., and Ferraris, R. P. (2004a). Contributions of different NaP$_i$ cotransporter isoforms to dietary regulation of P transport in the pyloric caeca and intestine of rainbow trout. *J. Exp. Biol.* **207**, 2055–2064.

Sugiura, S. H., and Ferraris, R. P. (2004b). Dietary phosphorus-responsive genes in the intestine, pyloric ceca, and kidney of rainbow trout. *Am. J. Physiol.* **287**, R541–R550.

Sundell, K., and Björnsson, B. T. (1988). Kinetics of calcium fluxes across the intestinal mucosa of the marine teleost, *Gadus morhua*, measured using an *in vitro* perfusion method. *J. Exp. Biol.* **140**, 171–186.

Sundell, K., Björnsson, B. T., Itoh, H., and Kawauchi, H. (1992). Chum salmon (*Oncorhynchus keta*) stanniocalcin inhibits *in vitro* intestinal calcium uptake in Atlantic cod (*Gadus morhua*). *J. Comp. Physiol. B* **162**, 489–495.

Swan, C. M., Lindstrom, N. M., and Cain, K. D. (2008). Identification of a localized mucosal immune response in rainbow trout, *Oncorhynchus mykiss* (Walbaum), following immunization with a protein-hapten antigen. *J. Fish Dis.* **31**, 383–393.

Terova, G., Cora, S., Verri, T., Rimoldi, S., Bernardini, G., and Saroglia, M. (2009). Impact of feed availability on PepT1 mRNA expression levels in sea bass (*Dicentrarchus labrax*). *Aquaculture* **294**, 288–299.

Thamotharan, M., Gomme, J., Zonno, V., Maffia, M., Storelli, C., and Ahearn, G. A. (1996a). Electrogenic, proton-coupled, intestinal dipeptide transport in herbivorous and carnivorous teleosts. *Am. J. Physiol.* **270**, R939–R947.

Thamotharan, M., Zonno, V., Storelli, C., and Ahearn, G. A. (1996b). Basolateral dipeptide transport by the intestine of the teleost *Oreochromis mossambicus*. *Am. J. Physiol.* **270**, R948–R954.

Tibbetts, I. R. (1997). The distribution and function of mucous cells and their secretions in the alimentary tract of *Arrhamphus sclerolepis krefftii*. *J. Fish Biol.* **50**, 809–820.

Tiril, S. U., Karayucel, I., Alagil, F., Dernekbasi, S., and Yagci, F. B. (2009). Evaluation of extruded chickpea, common bean and red lentil meals as protein source in diets for juvenile rainbow trout (*Oncorhynchus mykiss*). *J. Anim. Vet. Adv.* **8**, 2079–2086.

Tocher, D. R. (2003). Metabolism and functions of lipids and fatty acids in teleost fish. *Rev. Fish. Sci.* **11**, 107–184.

Tocher, D. R., and Sargent, J. R. (1984). Studies on triacylglycerol, wax ester and sterol ester hydrolases in intestinal ceca of rainbow trout (*Salmo-Gairdneri*) fed diets rich in triacylglycerols and wax esters. *Comp. Biochem. Physiol. B* **77**, 561–571.

Tocher, D. R., Bendiksen, E. A., Campbell, P. J., and Bell, J. G. (2008). The role of phospholipids in nutrition and metabolism of teleost fish. *Aquaculture* **280**, 21–34.

Torrissen, O. J., Hardy, R. W., Shearer, K. D., Scott, T. M., and Stone, F. E. (1990). Effects of dietary canthaxanthin level and lipid level on apparent digestibility coefficients for canthaxanthin in rainbow trout (*Oncorhynchus mykiss*). *Aquaculture* **88**, 351–362.

Tran, Q. D., Hendriks, W. H., and van der Poel, A. F. B. (2008). Effects of extrusion processing on nutrients in dry pet food. *J. Sci. Food Agricult.* **88**, 1487–1493.

Ugolev, A. M., and Kuz'mina, V. V. (1994). Fish enterocyte hydrolases—nutrition adaptations. *Comp. Biochem. Physiol. A* **107**, 187–193.

Une, M., Goto, T., Kihira, K., Kuramoto, T., Ki, H., Nakajima, T., and Hoshita, T (1991). Isolation and identification of bile salts conjugated with cysteinolic acid from bile of the red seabream, *Pagrosomus major. J. Lipid Res.* **32**, 1619–1623.

Valle, A. Z, Iriti, M., Faoro, F., Berti, C., and Ciappellano, S. (2008). In vivo prion protein intestinal uptake in fish. *Acta Pathol. Microbiol. Immunol. Scand.* **116**, 173–180.

Velez, Z., Hubbard, P. C., Welham, K., Hardege, J. D., Barata, E. N., and Canario, A. V. M. (2009). Identification, release and olfactory detection of bile salts in the intestinal fluid of the Senegalese sole (*Solea senegalensis*). *J. Comp. Physiol. A* (**195**), 691–698.

Venou, B., Alexis, M. N., Fountoulaki, E., and Haralabous, J. (2009). Performance factors, body composition and digestion characteristics of gilthead sea bream (*Sparus aurata*) fed pelleted or extruded diets. *Aquacult. Nutr.* **15**, 390–401.

Verri, T., Maffia, M., Danieli, A., Herget, M., Wenzel, U., Daniel, H., and Storelli, C. (2000). Characterization of the H^+/peptide cotransporter of eel intestinal brush-border membrane. *J. Exp. Biol.* **203**, 2991–3001.

Verri, T., Kottra, G., Romano, A., Tiso, N., Peric, M., Maffia, M., Boll, M., Argenton, F., Daniel, H., and Storelli, C. (2003). Molecular and functional characterisation of the zebrafish (*Danio rerio*) PEPT1-type peptide transporter. *FEBS Letters* **549**, 115–122.

Verri, T., Romano, A., Barca, A., Kottra, G., Daniel, H. and Storelli, C. (2010). Transport of di- and tripeptides in teleost fish intestine. *Aquacult. Res.* Doi: 10.1111/j.1365-2109.2009.02270.x.

Vilella, S., Ahearn, G. A., Cassano, G., and Storelli, C. (1989a). How many Na^+-dependent carriers for L-alanine and L-proline in the eel intestine? Studies with brush border membrane vesicles. *Biochim. Biophys. Acta* **984**, 188–192.

Vilella, S., Reshkin, S. J., Storelli, C., and Ahearn, G. A. (1989b). Brush-border inositol transport by intestines of carnivorous and herbivorous teleosts. *Am. J. Physiol.* **256**, G501–G508.

Werner, A., Murer, H., and Kinne, R. K. H. (1994). Cloning and expression of a renal Na-P_i cotransport system from flounder. *Am. J. Physiol.* **267**, F311–F317.

Whitehead, M. W., Thompson, R. P. H., and Powell, J. J. (1996). Regulation of metal absorption in the gastrointestinal tract. *Gut* **39**, (625–628).

Wilson, R. W., and Grosell, M. (2003). Intestinal bicarbonate secretion in marine teleost fish—source of bicarbonate, pH sensitivity, and consequences for whole animal acid–base and calcium homeostasis. *Biochim. Biophys. Acta—Biomembr.* **1618**, 163–174.

Wilson, R. W., Wilson, J. M., and Grosell, M. (2002). Intestinal bicarbonate secretion by marine teleost fish—why and how? *Biochim. Biophys. Acta—Biomembr.* **1566**, 182–193.

Wu, T., Sun, L. C., Du, C. H., Cai, Q. F., Zhang, Q. B., Su, W. J., and Cao, M. J. (2009). Identification of pepsinogens and pepsins from the stomach of European eel (*Anguilla anguilla*). *Food Chem.* **115**, 137–142.

Yamamoto, T., Shima, T., Furuita, H., Suzuki, N., and Shiraishi, M. (2001). Nutrient digestibility values of a test diet determined by manual feeding and self-feeding in rainbow trout and common carp. *Fish. Sci.* **67**, 355–357.

Yufera, M., and Darias, M. J. (2007). Changes in the gastrointestinal pH from larvae to adult in Senegal sole (*Solea senegalensis*). *Aquaculture* **267**, 94–99.

Yufera, M., Fernandez-Diaz, C., Vidaurreta, A., Cara, J. B., and Moyano, F. J. (2004). Gastrointestinal pH and development of the acid digestion in larvae and early juveniles of *Sparus aurata* (Pisces: Teleostei). *Marine Biol.* **144**, 863–869.

Zambonino Infante, J. L., and Cahu, C. L. (2007). Dietary modulation of some digestive enzymes and metabolic processes in developing marine fish: applications to diet formulation. *Aquaculture* **268**, 98–105.

Zambonino-Infante, J. L., Gisbert, E., Sarasquete, C., Navarro, I., Gutierrez, J., and Cahu, C. L. (2008). Ontogeny and physiology of the digestive system of marine fish larvae. In: *Feeding and Digestive Functions of Fishes* (J. E. P. Cyrino, D. P. Bureau and B. G. Kapoor, eds), pp. 281–348. Science Publishers, Enfield.

3

BARRIER FUNCTION AND IMMUNOLOGY

KENNETH CAIN
CHRISTINE SWAN

Barrier and immunological functions of the gut in fish are discussed in this chapter. The gut is known to be an important site often targeted by invading pathogenic microbes, and constant exposure to aquatic microbes requires defense mechanisms to be present in the gut that act rapidly to limit infection. These include many physical, chemical, and cellular components that prevent infection by presenting a barrier to invasion; however, innate and adaptive immune responses can be initiated in the gastrointestinal tract as part of a mucosal immune response. The mucosal immune system in fish includes not only the gut but also mucosal tissues associated with the gills and skin. In mammals, this system is well developed and exposure to pathogens or immunization with antigens delivered to one mucosal site results in antigen-specific antibody production at other mucosal sites. Fish are capable of mounting similar responses via the gut and skin, but the mucosal system in fish functions in a more primitive manner than in mammals. Protective functions of the gut of fish involve inhibition of colonization by microbes at mucosal sites, rapid elimination of pathogens through innate responses, and development of adaptive immunity following antigen uptake and processing.

The Multifunctional Gut of Fish: Volume 30
FISH PHYSIOLOGY

1. INTRODUCTION

The gastrointestinal (GI) tract of fish is primarily thought of as the site where digestion of food and absorption of nutrients takes place; however, the GI tract is also an important immunological site and plays a primary role in protecting the animal from various pathogenic insults. The components of the fish gut include the esophagus, stomach, pyloric ceca, and the anterior and posterior intestine. The gut acts as a physical and chemical barrier, thus providing a first line of defense against invading organisms entering the body via feed intake or through ingestion of water. Cells within the GI tract in fish produce a range of chemical substances which can enhance the barrier function, contribute to innate defenses, or activate processes that lead to an adaptive immune response. The posterior portion of the GI tract can serve as a site where uptake of macromolecules and foreign antigens occurs, which leads to antigen processing and the development of an adaptive immune response and long-term immunity. Unlike mammals, in fish this occurs in the absence of structures such as Peyer's patches. Mucosal immunity has been shown to exist in fish and plays a distinct role in both innate and adaptive immune responses. In mammals, this system is highly developed and a strong localized mucosal immune response can be elicited following antigen uptake at mucosal sites, such as those of the nasal, GI and respiratory tracts among others. Antigen uptake at the mucosal sites results in the production of a distinct immunoglobulin isotype (IgA) that is transported to mucosal sites throughout the body and provides immediate defense to pathogen invasion. Since the GI tract, gills, and entire body surface of fish are covered with mucus, mucosal immunity involves not only the GI tract and internal surfaces of the gut, but appears integrated with the responses of the cutaneous mucosal surfaces of the skin and gills. However, the mechanisms that link these sites or the existence of a localized immune response in teleosts are not completely understood. What is apparent is that the fish gut is a key site involved in immunoprotection and this site is affected by pathogenic as well as non-pathogenic microbes. Although limited information is available for fish, the presence of natural gut microbes appears to contribute to immune function and fish health through a variety of mechanisms. To begin to understand the role the gut plays in immune function, an overview of the fish immune system is presented and contrasted to the mammalian system, mucosal immunity in fish is discussed, and the potential effect of non-pathogenic gut microbes common in fish is discussed in the context of fish health and immunity.

2. THE IMMUNE SYSTEM IN FISH

To understand the function of the gut and other mucosal sites in the context of both barrier function and immunology, the fish immune system as a whole and how an immune response is elicited are briefly reviewed. As with any other vertebrate, fish have the ability to respond to foreign or non-self substances (antigens) with innate or immediate non-specific responses as well as adaptive or specific responses leading to antigen recognition and memory. Both innate and adaptive immunity involve cellular and humoral components that work together to target and eliminate invading microorganisms or other foreign substances (Balfry 1997; Magnadottir 2006).

2.1. Innate Response

An innate immune response is a critical non-specific action that is considered a first line of defense to pathogen invasion. The innate arm of the immune system functions in all vertebrates and invertebrates. Cells involved in innate immunity can often recognize common molecular structures on pathogens or certain products they release through interactions with pattern-recognition receptors on the cell surface (Dixon and Stet 2001; Magnadottir 2006). This cellular response requires no antigenic priming and is either immediate or can be induced within a short time following an encounter with the foreign antigen (Ellis 2001; Watts et al. 2001). It has been suggested that some aspects of the innate response in fish may be stronger than that in mammals and may contribute a greater proportion of overall immunity (Ellis 2001). The reason for this has been suggested to be due to a slower more temperature-dependent requirement for an antigen-specific adaptive response to develop in fish (Ellis 2001; Magnadottir 2006).

Cells associated primarily with the innate immune system include phagocytes such as macrophages, neutrophils and other granulocytes, non-specific cytotoxic cells and epithelial cells; however, other cell types such as pillar cells as well as B lymphocytes have been shown to have phagocytic abilities (Chilmonczyk and Monge 1980; Li et al. 2006). Bacterial pathogens encountered by macrophages and neutrophils are phagocytized and engulfed within the cell where destruction and digestion occur (Secombes 1996; Neumann et al. 2001). Degradation of phagocytized bacterial pathogens can occur via respiratory burst activity resulting in the production of toxic free radicals along with hydrogen peroxide, superoxide anion as well as nitric oxide (Secombes 1996; Ellis 1999; Neumann et al. 2001). Phagocytes may contain proteases, lipases, phosphatases as well as other proteins and peptides that have antimicrobial properties and act directly on pathogens (Neumann et al. 2001). Macrophages are known to produce lysozyme, and it has been shown that production of lysozymes by macrophages is enhanced following exposure to bacterial LPS or yeast β-glucans (Paulsen et al. 2001). Non-specific cytotoxic cells (NCC) have been identified in fish such as rainbow trout *Oncorhynchus mykiss* (Zimmerman et al. 2004) and catfish *Ictalurus punctatus*

(Shen et al. 2004) and are believed to be precursors to mammalian natural killer (NK) cells. Fish NCC cells appear to function in a similar manner to NK cells in mammals and can lyse target cells through apoptosis.

Inflammation is an important part of the innate response and in mammals causes swelling, redness, heat and pain. Typically, an influx of phagocytic cells occurs at a site that is affected following damage or pathogen entry. Inflammation can be initiated when chemical signals are released following binding of cellular receptors with carbohydrate antigens such as lipopolysaccharide (LPS) and peptidoglycans (Watts et al. 2001). Additionally, enzymes and cellular proteins that are released from damaged host cells can trigger phagocyte influx and inflammation (Goldsby et al. 2003). This response in teleosts appears similar to that in mammals in that inflammatory cells migrate rapidly to sites affected by pathogens such as viruses, bacteria, fungi (Gerwick et al. 2002) and parasites (vanMuiswinkel 1995).

Many humoral factors have been identified and associated with innate immunity in fish. These include components such as complement, interferon, transferrin, C-reactive protein (vanMuiswinkel 1995), lectins, and lysozymes (Secombes 1996). In mammals, interferon acts to defend against viruses by signaling the production and release of antiviral proteins such as Mx proteins (Robertsen 2006). Mx proteins are also found in fish and appear to have a similar function. Atlantic salmon *Salmo salar* Mx proteins have been demonstrated to have antiviral properties and studies with Japanese flounder *Paralichthys olivaceus* have demonstrated similar properties (Caipang et al. 2003; Larsen et al. 2004). In addition to humoral factors discussed above, interactions and signaling of proteins associated with the complement system are important in both innate and adaptive immune responses. The complement system involves a biological cascade that acts to eliminate pathogens from the animal. It is a key function of the innate immune response but it can also be activated by the adaptive immune system. Many proteins are associated with the complement system and it is activated by classical, alternative or mannose-binding lectin pathways (Boshra et al. 2006). Once this system is activated, it triggers the cleavage of specific proteins that creates a cascade of events resulting in amplification of numerous responses involved in inflammation, phagocytosis, lysis of foreign cells, opsonization, and formation of membrane attack complexes (Holland and Lambris 2002; Boshra et al. 2006). There are at least 25 proteins involved in the complement pathway. The C3 protein molecule plays a primary role in this cascade following activation of a complement pathway (Boshra et al. 2006). The C3 protein appears to be present in fish and C3 homologues have been isolated from a range of teleost fish including rainbow trout (Sunyer et al. 1996), carp *Cyprinus carpio* (Nakao et al. 2000),

Japanese medaka fish *Oryzias latipes* (Kuroda et al. 2000) and sea bream *Sparus aurata* (Sunyer et al. 1997).

Transferrin is an important blood glycoprotein involved in binding of iron. Transferrin sequesters iron from bacteria that require it, resulting in an inhibition in bacterial growth and damage *in vivo* (Weinberg 1974; Ellis 1999). Other molecules such as lectins act to bind sugars. Pentraxin lectins, serum amyloid proteins, and C-reactive proteins are involved in the acute phase response and are associated with complement activation, opsonization, and phagocytosis in mammals (Arason 1996; Magnadottir 2006). Pentraxin-like proteins have been found in several teleost species including rainbow trout, carp (Cartwright et al. 2004), halibut *Hippoglossus hippoglossus*, cod *Gadus morhua*, wolffish *Anarhichas lupus*, Atlantic salmon (Lund and Olafsen 1998), snapper *Pagrus auratus* (Cook et al. 2003), and plaice *Pleuronectes platessa* (Jensen et al. 1995). The specific role pentraxins play in immune defenses and the acute phase response are not clear, but they have been shown to either increase or decrease in fish serum during an infection (Kodama et al. 1989; Szalai et al. 1994; Jensen et al. 1997; Lund and Olafsen 1999). Lysozyme, an enzyme that has bactericidal activity, targets the peptidoglycan layer of bacterial cell walls and is capable of causing lysis (Magnadottir 2006). Lysozyme is common and has been isolated from mucus, serum, eggs, and lymphoid tissue of fish (Fange et al. 1976; Grinde et al. 1988; Yousif et al. 1994). Although some molecules described above can be associated with adaptive immunity as well as innate responses, many are activated immediately and act on pathogens directly to decrease the possibility of a pathogen causing damage and establishing a serious infection.

2.2. Adaptive Response

Adaptive immunity is critical to long-term protection from specific pathogens. In comparison to the innate response, it is much slower to develop but can be directed at antigens specific to a pathogen and results in immunologic memory to that antigen. Adaptive responses involve both cellular and humoral components and only occur in vertebrates (Du Pasquier 2001). Fish have the ability to mount an adaptive response, but there are some differences when compared to a mammalian response. In mammals, the innate response helps to activate the adaptive immune response, allowing these two systems to work together to provide enhanced protection from pathogens (Watts et al. 2001; Re and Strominger 2004). This is also believed to occur in fish, but the adaptive immune system can take more than two weeks to manifest in fish (Kollner and Kotterba 2002). Protection via the production of antigen-specific antibody occurs in a manner similar to that observed in mammalian systems, and the teleost immune system has many of the same attributes as the mammalian system. A number of cells and molecules are important to this response and include, but are not limited to, B lymphocytes (Kaattari 1992), T-cell receptors (Hordvik et al. 1996; Partula et al. 1996a,b; Wilson et al. 1998), immunoglobulin (Ig) (Warr 1995), complement (Holland and Lambris 2002) and natural killer-like cells (Shen et al. 2004).

It is known that the primary cell types important for adaptive response are B and T lymphocytes and antigen presenting cells (APCs). It is thought that neutrophils and macrophages are the primary APCs in fish but other cell types may play a role in antigen presentation. Once a pathogen or other foreign substance is encountered by the cells it is generally phagocytized, digested and antigenic epitopes are presented on the cell surface in association with the major histocompatibility complex (MHC) (Vallejo et al. 1992). These epitopes are then recognized by B cells alone or by T-cell receptors on T cells in the context of MHC. This results in cytokine activation, which in turn activates additional T and B cells. B cells will proliferate and begin to differentiate resulting in the production of antibody-secreting plasma cells and memory B cells (Goldsby et al. 2003). B cells and circulating antibody are well documented in fish; however, it has been more difficult to identify T cells. T cells have been identified in sea bass *Dicentrarchus labrax* following immunostaining with a monoclonal anti-body specific for thymocytes and peripheral T cells (Scapigliati et al. 2000). The development of such markers in other teleosts has been difficult, but the presence of T cells is generally accepted due to the large proportion of lymphocytes remaining following staining for B cells and the identifica-tion of the T-cell receptor (TcR) gene in a number of species. A few important species that the TcR gene has been identified in include rainbow trout (Partula et al. 1995, 1996a), Atlantic salmon (Hordvik et al. 2004), turbot *Scophthalmus maximus* (Taylor et al. 2005), and catfish (Wilson et al. 1998). Early evidence for the presence of T cells in fish was suggested due to the occurrence of graft rejection and delayed type hypersensitivity reactions, a response known to be strongly linked to T cells (Nakanishi et al. 1999).

Key components of the adaptive immune response are activated and regulated by numerous humoral factors produced in cells and tissues. These components include many cytokines, tumor necrosis factor, and chemo-kines. However, the primary humoral factor important in the adaptive immune response is antibody or immunoglobulin (Ig). Immunoglobulin is expressed on the surface of B cells, but can also be secreted by the B cells and by plasma B cells. Fish and mammals differ in the types of anti-body produced. Mammals are known to have five different isotypes of antibody (IgM, IgD, IgE, IgG and IgA) and each of these is linked with a different biological function (Bengten et al. 2000). IgG is predominant in the serum and can protect fetal tissues, activate complement, and enhance phagocytosis. IgE is important in allergic reactions and plays a role in hypersensitivity reactions (Snider et al. 1994; Goldsby et al. 2003), and while IgD is highly conserved in mammals, its specific immunological role has not been determined (Frazer and Capra 1999). IgA, when found in its polymeric form, is linked primarily to mucosal immunity and is found in mammalian mucosal secretions of the gastrointestinal, genital and respiratory tracts (Underdown and Mestecky 1994; Stratton et al. 1999; Bengten et al. 2000; Johansen et al. 2000). IgM is highly conserved and appears to have been the first Ig to develop during evolution. It is the initial antibody produced by immature B cells, and when a primary immune response is initiated, IgM is produced followed by isotype switching to other Ig molecules (Frazer and Capra 1999). Along with IgA, mammals produce a pentameric secretory IgM that plays an important accessory role in mammalian mucosal immunity.

In teleost fish, IgM is the principal antibody produced in the serum (Wilson and Warr 1992; Warr 1995; Bengten et al. 2000) but other isotypes including IgD, IgT, IgZ and a chimeric IgM-IgZ have been identified in various species. The gene encoding IgD has been identified in channel catfish (Wilson et al. 1997), Atlantic salmon (Hordvik et al. 1999), Atlantic cod (Stenvik and Jørgensen 2000), Japanese flounder *Paralichthys olivaceus* (Hirono et al. 2003) and Fugu *Takifugu rubripes* (Saha et al. 2004). Other genes in the Ig family that have been identified in fish include IgT in rainbow trout (Hansen et al. 2005) and stickleback (Gambon-Deza et al. 2010), IgZ in zebrafish *Danio rerio* and mandarin fish *Siniperca chuatsi* (Tian et al. 2009) and the chimeric IgM-IgZ identified in common carp (Savan et al. 2005). In teleost fish, IgM is found in the serum and mucosa associated with the gut, skin and gills (Davidson et al. 1997). IgM in the serum of fish is generally found as a tetramer with a molecular weight of ~800 kDa. Each of the four monomeric subunits of tetrameric IgM is comprised of two heavy chains (H chain) at approximately 70 kDa and two light chains (L chain) between 22 and 25 kDa (Wilson and Warr 1992; Kaattari et al. 1998). Eight different covalent forms of teleost IgM have been described by Kaattari et al. (1998), and each can form non-covalent bonds between H chains in a myriad of different fashions resulting in numerous heterogeneous (redox) forms of IgM (Kaattari et al. 1998). These redox forms appear to be species-dependent and it is speculated that antibody diversity may arise from bonding differences between H chains and this could serve to functionally enhance IgM in lieu of the limited Ig isotype repertoire in teleost fish (Kaattari et al. 1998) (Table 3.1).

Table 3.1

General comparison of antibody isotype differences between mammals and fish

Antibody isotype	Mammals	Teleosts	Comments
IgM	+ (pentamer)	+ (tetramer)	Numerous redox forms exist in fish
IgD	+	+	Gene encoding IgD found in a number of fish species
IgG	+	−	Primary serum antibody in mammals
IgE	+	−	Active in allergic and hypersensitivity reactions
IgA	+	−	Primary mucosal antibody in mammals
IgT	−	+	Gene identified in trout and stickleback belonging to Ig family
IgZ	−	+	Gene identified in zebrafish and mandarin fish belonging to Ig family
IgM-IgZ (chimeric)	−	+	Gene identified in carp belonging to Ig family

Note: (+) = present; (−) = not identified.

3. MUCOSAL IMMUNITY IN FISH: THE GUT AS A FIRST LINE OF DEFENSE

3.1. Barrier Function and Innate Immunity

It is well established that the mucosal layer of the gut creates physical, chemical, and cellular barriers to pathogen invasion. Mucus covers all external surfaces, gills, and all internal surfaces of the gut in fish (Shephard 1994). Epithelial tissues secrete mucus which provides a protective replaceable outer barrier against pathogen attachment (Ellis 2001). In general, fish lacking scales often produce greater amounts of mucus in response to external stresses, which is believed to enhance barrier function by preventing physical contact and interaction with aquatic pathogens. Epithelial and mucosal surfaces can inhibit bacteria, viruses and other pathogens from attaching and entering host tissues (vanMuiswinkel 1995; Ellis 1999; Press and Evensen 1999; Watts et al. 2001; Goldsby et al. 2003). These surface barriers include the skin, mucus, and the epithelium covering the gill, buchal cavity and intestine. Mucus in the gut provides a similar function to that of skin mucus, but there are additional complex interactions associated with nutrient uptake from food, production of various factors that aid in digestion, establishment of normal microflora within the environment of the gut, and regulation of immune defense mechanisms.

Beyond the direct physical barrier, mucus can provide a chemical barrier that acts in a variety of ways that may also inhibit colonization or invasion by infectious agents, or contribute to innate and adaptive immune responses. Mucus is composed primarily of glycoproteins (Fletcher and Grant 1968), but a number of other substances have been identified and include such chemical compounds as cytokines (Lindenstrom et al. 2003), peptides (Cole et al. 1997; Fernandes et al. 2004b), lysozyme (Fernandes et al. 2004a), lipoprotein (Concha et al. 2003), complement (Dalmo et al. 1997), lectins (Itami et al. 1993; Tsutsui et al. 2003, 2005), proteases (Aranishi and Mano 2000) and antibodies (Cain et al. 2000; Hatten et al. 2001; LaFrentz et al. 2002). Many of these compounds work to defend against pathogens either directly or indirectly, and mucus collected from the GI tract and skin of fish has been shown to have natural anti-microbial activity. This activity can be targeted to bacterial as well as viral pathogens. Cain et al. (1996) demonstrated that mucus collected from the skin and lower GI tract of pathogen-free uninfected rainbow trout was capable of neutralizing infectious hematopoietic necrosis virus (IHNV) *in vitro*. Similarly, early work by Harrell et al. (1976) has shown that vibriostatic and vibriocidal activity is present in skin and gut mucus from uninfected trout.

Additional innate responses can be activated and further function to prevent pathogen entry or rapidly eliminate a pathogen. During the innate response a number of humoral and cellular factors are activated. These can act immediately to clear an infection but are often involved in the initiation and development of an adaptive immune response leading to longer-term protection. The intestinal epithelium itself consists of a surface that is composed of a diversity of cell types that are important for immunity. These cells consist of intestinal epithelial cells (IECs), macrophages, intraepithelial lymphocytes (IELs), and goblet cells (GCs) that actively produce and secrete mucus (Komatsu et al. 2009). These cell types play different roles, but IECs typically form a tight junction that prevents pathogen entry and therefore act to physically exclude organisms (Tsukita et al. 2001). Chemical signals are known to be produced by IECs in mammals, and these cells are considered to be immunocompetent. Cytokines and chemokines produced by IECs include IL-7, IL-8, and IL-15, which results in the activation of neutrophil chemotaxis as well as macrophage and IEL activation (Watanabe et al. 1995; Hirose et al. 1998; Pitman and Blumberg 2000). In fish, the involvement of IECs in immunity is only just now beginning to be understood. Recent work by Komatsu et al. (2009) demonstrated that IECs in rainbow trout could produce the inflammatory cytokines IL-1β and TNFα2 when stimulated with live *Aeromonas salmonicida*, but interestingly, exposure to killed *A. salmonicida*, conditioned media, or live *Escherichia coli* did not result in up-regulation of these cytokines. It was speculated that the up-regulation of cytokine production plays a defensive role in combating furunculosis, the disease caused by *A. salmonicida*, since the GI tract has been implicated as the primary infection site for this pathogen.

Other humoral factors such as interferon (IFNγ) and migration inhibiting factor (MIF) have been shown to be produced from excised portions of the anterior and posterior intestine of rainbow trout when incubated with *Candida albicans* (Jirillo et al. 2007). This resulted in enhancement of phagocytosis of *C. albicans* by macrophages due to IFNγ production, but MIF production led to inhibition of macrophage movement in agarose. Taken together, the authors suggested that the gut-associated lymphoid tissue of trout was capable of mounting a type IV hypersensitivity response following challenge with microbial antigens.

3.2. Adaptive Responses

As in mammals, the mucosal surfaces of teleosts are major portals of entry for pathogens. Fish live in a pathogen-rich environment and mucosal surfaces are continually in contact with possible infectious agents. The mucosal immune system in fish functions to exclude pathogens through innate immunity and by creating barriers as described above; however, the immunological function of the gut includes an ability to elicit an adaptive response at mucosal sites. Our current level of understanding of the gut immune response in fish can be attributed to research aimed at developing vaccines for aquaculture. In particular, work that has investigated practical vaccine delivery strategies such as oral delivery via the feed has improved our understanding of mucosal immunity.

In mammals there is a distinct localized mucosal immune system that is separate from the systemic immune system and protects the animal against pathogenic entry at these points. Mucosal associated lymphoid tissues (MALT) are secondary lymphoid tissues found in both the gastrointestinal and respiratory tracts that produce copious amounts of IgA and lesser amounts of IgM. MALT includes tonsils and adenoids in the bronchial region (nasal-associated lymphoreticular tissue or NALT) and Peyer's patches in the gastrointestinal region (gut-associated lymphoreticular tissues or GALT). MALT is considered to be the inductive site for antigen recognition and acts in the uptake, processing and presentation of antigens for stimulation of mucosal immune responses at effector or remote mucosal sites. Antigens presented at MALT are taken up by specific cells in the epithelium called microfold (M) cells and are then moved through the lumen into inductive regions such as Peyer's patches or other specialized regions (lymph nodes, tonsils, appendix and adenoids) (Holmgren and Czerkinsky 2005). Peyer's patches contain germinal centers and naïve B and T lymphocytes which are stimulated upon antigenic encounter. Activated B and T cells disperse via the lymph throughout the body to diffuse effector regions in the mucosa (Stratton et al. 1999; Holmgren and Czerkinsky 2005) where they act to protect the host from pathogenic infection.

Because antigen can be presented at one mucosal site and produce antibody and T-cell involvement at other remote effector sites, the effectiveness of mucosal immunization has been studied extensively, but only in mammals. It has been shown that mucosal immunization can produce antigen-specific antibody at induced sites like Peyer's patches, tonsils and adenoids, but also at other remote mucosal sites, giving rise to the idea of a common or localized mucosal immune system (McGhee et al. 1999; Stratton et al. 1999; Holmgren and Czerkinsky 2005). Antigen introduced at mucosal sites can also stimulate systemic immunity, but generally the introduction of antigens parenterally (by injection) does not induce protective mucosal immunity (McGhee and Kiyono 1999; McGhee et al. 1999; Stratton et al. 1999).

As described earlier, IgA is the main antibody in mammalian mucosal secretions and is normally found in monomeric form in the serum, but in the mucus it is found as dimers and tetramers. A secondary antibody in mucosal secretion, IgM, is found as a pentamer. Both IgA and IgM are expressed in mucus in conjunction with two components, J chain and a secretory piece (Underdown and Mestecky 1994; Bengten et al. 2000; Johansen et al. 2000; Goldsby et al. 2003). The J chain is a small polypeptide of ~15 kDa that is disulfide bonded to IgA and IgM. Attachment of J chain allows for the formation of polymeric IgA and IgM and is required for the binding of both Igs with a polymeric Ig receptor (pIgR) on the basolateral surface of mucosal epithelial cells. Once bound, the polymeric IgA and IgM are endocytosed and transported to the lumen and finally the apical surface where the pIgR is cleaved to form a secretory piece. The secretory piece is covalently bound to IgA, making this Ig quite stable in acidic environments. IgM is not covalently bound and is less stable in acidic secretions. The two components, J chain and the polymeric pIgR working together, allow the polymeric forms of IgA and IgM to cross cell membranes and be presented in the mucosa (Johansen et al. 2000; Goldsby et al. 2003). IgA and IgM effectively bind antigens due to their polymeric form and increased avidity (Goldsby et al. 2003).

In fish, the main antibody produced is a tetrameric IgM, rather than the pentameric form found in mammals. The J chain is an important component in mammalian IgA and IgM, and has been noted in birds, amphibians and some fish such as catfish and lamprey (Underdown and Mestecky 1994). The J chain has been identified in serum from nurse sharks *Ginglymostoma cirratum* (see McCumber and Clem 1976; Hohman et al. 2003) and catfish (Weinheimer et al. 1971; Mestecky et al. 1975). The appearance of a J chain-like polypeptide (11 kDa) in rainbow trout serum (Sanchez et al. 1989) and leopard shark *Triakis semifasciata* serum (Klaus et al. 1971) has also been reported. However, other research indicates that J chain is not present in longnose gar *Lepisosteus osseus* (see Mestecky et al. 1975), paddlefish *Polydon spathula* (see Weinheimer et al. 1971), pike *Esox lucius* (see Clerx et al. 1980) or in salmonids (Voss et al. 1980; Kobayashi et al. 1982; Warr 1983). J chain is generally considered to be absent and not associated with most fish IgM molecules (Flajnik 1996; Magnadottir 1998) and the presence of J chain in the mucosa of rainbow trout has not been documented.

Although there is substantial information documenting the presence and induction of IgM in the serum and mucus of teleost fish, the relationship between serum and mucosal IgM molecules has not been extensively studied. Research suggests that the composition of IgM collected from serum and mucus of rainbow trout may not be structurally identical and heterogeneous forms of IgM may exist (Cain et al. 2000). Cain et al. (2000) showed molecular mass differences between serum and mucus Ig after purification over a mannose-binding protein column. Another study suggested structural diversity of serum and mucus Ig based on the differential binding of monoclonal antibodies to those Igs (Rombout et al. 1993b). Recent studies have shown that mucous antibodies appeared to be heterogeneous to serum Ig because of the presence of more monomers and dimers in the mucus instead of the disulfide bonded tetramers which make up the bulk of serum Ig (Bromage et al. 2006). Mucosal Ig also showed specificity to an anti-trout IgM monoclonal antibody that recognized the heavy chain at about 100 kDa whereas serum Ig had no specificity at that region (Bromage et al. 2006).

The ability of the intestine to act as a site of antigen uptake is considered a key component to developing a protective immune response. In the intestine of fish, antigens are taken up by intestinal epithelial cells and transported to APCs that stimulate B and T lymphocytes (Rombout et al. 1986; Rombout and Van Den Berg 1989; Joosten et al. 1996). There is no evidence that fish possess specialized tissues such as Peyer's patches to act as antigen retrieval centers (Rombout et al. 1993a) but it has been shown that the hindgut of some teleost species is populated with diffuse GALT (Abelli et al. 1997) including macrophages, plasma cell and lymphoid cells, and this GALT has been shown to be the site of particulate or soluble antigen absorption (Rombout et al. 1985, 1986; Georgopoulou and Vernier 1986; Rombout and Van Den Berg 1989; Abelli et al. 1997; Petrie and Ellis 2006). It has been suggested that the fish intestine is immunocompetent irrespective of the presence of such structures (Hart et al. 1988). Structurally, it is known that the anterior portion of the GI tract of fish is primarily involved with the absorption of lipid and proteins, while the posterior portion can take up macromolecular proteins (Vernier 1990) and this appears to be the primary site where antigen uptake occurs and an immune response is initiated.

While it has been shown that mammals have an antigen-specific antibody response in mucosal tissues following mucosal vaccination (Mestecky et al. 1994; Stratton et al. 1999), this response in fish is not well understood. Most studies look at antibody production following immunization at mucosal sites; however, it has recently been shown that anal administration of antigens in carp could also induce cell-mediated responses in systemically derived leucocytes (Sato et al. 2005). How the mucosal and systemic immune systems in fish are separated is unclear, and studies disagree as to whether teleost fish possess a distinct mucosal immune system capable of producing antigen-specific antibody locally as is the case in mammals. However, studies do agree that specific antibody is present at mucosal sites in some species following anal intubation (Rombout et al. 1986; Cain et al. 2000; Vervarcke et al. 2005), immersion or waterborne infection treatments (Lobb 1987; Maki and Dickerson 2003), and oral vaccination (Lin et al. 2000). Interestingly, serum or plasma antibody titers have been reported to be elevated in fish following oral or anal immunization as well (Jenkins et al. 1994; Joosten et al. 1997). In catfish, it has been suggested that there is a localized response to *Ichthyophthirius multifiliis* "Ich" after immunization or infection (Maki and Dickerson 2003; Sigh et al. 2004). It has also been shown that skin excised from immunized fish produced antibodies that immobilized or weakened *Ichthyophthirius multifiliis* theronts in culture (Xu and Klesius 2002; Xu et al. 2002). More recently, antibody to *Flavobacterium columnare* was found in explanted skin tissue from infected channel catfish (Shoemaker et al. 2005). The relationship of this cutaneous response in catfish to intestinal immunity is not clear. An analysis of immune cells of the intestinal tract of catfish using flow cytometry and histology found that neutrophils were the primary cells with less than 6% B cells present. Other immune cells were also present, but the authors suggested that the specific immune response in the mucosa of catfish may be limited and that this species may rely more on innate responses in the gut since few cells associated with adaptive immunity were identified (Hebert et al. 2002). This may hold true for other species as well and explain some of the limitations to the development of effective oral vaccines in aquaculture. Detection of antibody in serum or external mucus following oral or anal immunization is often limited in comparison with injection delivery of antigens. In trout, it was found that fish immunized peranally (p.a.) with the hapten-carrier antigen FITC-KLH were able to develop an anti-FITC antibody response in their serum but were non-responsive to the KLH carrier. Whereas fish immunized interperitoneally (i.p.) had a strong anti-FITC and anti-KLH response. This lack of responsive KLH-specific B cells in the intestine was not a result of systemic tolerance and was thought to reflect either a restriction in the repertoire of gut-associated B cells or induction of an anergic state in the KLH specific population when compared to peripheral B cells (Jones et al. 1999). In another study, rainbow trout were immunized i.p. (with and without adjuvant) and p.a. with FITC-KLH and although serum anti-FITC responses were detected in p.a. immunized fish, they were limited in comparison to i.p. responses. Furthermore, anti-FITC antibody was detected in cutaneous mucus of trout following p.a. immunization, but only at week 6 and titers were lower than those from i.p. immunized fish (Cain et al. 2000). Although delivery of antigen to the GI tract resulted in mucosal antibody, the comparison of the magnitude of this response to i.p. immunization supports the idea of a limited B-cell repertoire in the gut or a lack of efficient antigen uptake. The production of antigen-specific antibodies in the cutaneous mucus of fish has been further documented in rainbow trout following i.p. immunization with the pathogen *Flavobacterium psychrophium* (LaFrentz et al. 2002) and in white sturgeon *Acipenser transmontanus* following i.p. administration of FITC-KLH (Drennan et al. 2007). However, oral immunization of carp *Carassius auratus gibelio* with *Aeromonas hydrophila* ghosts (empty bacterial cell envelopes) was recently shown to elicit antigen-specific antibodies in the intestinal mucosa and fish were protected from disease challenge (Tu et al. 2010). These and other studies showing that mucosal antibody can be elicited following delivery of antigens to the gut suggest that a distinct localized mucosal immune system that functions in a similar manner to that found in mammals may be present in fish. However, it appears much more limited and recent studies in rainbow trout suggest that unlike mammals, B cells in fish can mature systemically and then migrate to mucosal tissues where they produce antibodies locally (Swan et al. 2008). This would explain the heightened mucosal response following i.p. immunization compared to p.a. administration in fish. Swan et al. (2008) found that leukocytes isolated from spleen and blood of fish immunized i.p. with FITC-KLH produced significantly elevated anti-FITC antibody levels at week 10 when cultured *in vitro*. Interestingly, this was also true for excised skin, intestine and gill tissues, which also exhibited significantly elevated anti-FITC responses indicating localized production of antibody in the mucosa from tissue-specific B cells. This localized mucosal immune response was only elicited following i.p. immunization and was not observed if antigens were delivered directly to the GI tract through p.a. immunization (Swan et al. 2008). Based on much of the work to date, it is likely that adaptive responses develop more effectively in some fish species than in others. However, it seems safe to conclude that the production of a localized response is incomplete in teleosts and the primitive response observed likely represents the evolutionary precursor to the more advanced mucosal responses found in mammals.

4. NATURAL GUT MICROBES AND THEIR ROLE IN IMMUNITY

Until recently, little attention has been given to the presence of normal endogenous gut microbes and how they may affect the health of fish. However, investigations into probiotics and their potential to improve health have shown that the gut microbiota can be an important component of the mucosal barrier. Such microbes are known to affect colonization of pathogenic microorganisms through chemical inhibition or competitive exclusion. This in itself likely enhances the overall barrier function of the gut, but in many animals the establishment of normal microbial flora in the gut is known to contribute to pathogen exclusion and health maintenance (Balcazar et al. 2006b). It has been shown that bacteria are the primary microorganism that invade and constitute the microbiota of the gut in healthy fish (Spanggaard et al. 2000; Pond et al. 2006), but yeast has been reported in some cases (Andlid et al. 1998; Gatesoupe 2007). Interestingly, in larval fish it has been shown that bacteria will colonize the gut immediately after hatching and that this may be complete within hours (Gomez and Balcazar 2008). This colonization affects gene expression in the GI tract and is thought to contribute to a favorable habitat for the colonizing bacterial flora that then lead to prevention of other bacterial colonization (Balcazar et al. 2006a). The microbiota of the GI tract must be considered along with other chemical factors, acids, bile salts and enzymes that contribute to creating an unfavorable environment for many pathogens. If a balance between these many components is established in the GI tract then protection from invading pathogens can be enhanced. However, when conditions are altered or bacterial or other pathogens breach this first line of defense, then other humoral and cellular factors of the immune system are triggered. Our understanding of the complex actions and the ability of true normal bacterial flora to develop in fish is limited, but it can be speculated that if beneficial bacteria can be introduced early on then this may lead to disease protection down the road. This is the approach being taken in selecting probiotic strains of bacteria for use in aquaculture and appears to have great potential as a disease management tool.

5. CONCLUSION

Although the immune system and responses observed in fish are limited in function when compared to mammals, it is apparent that each component of this system is important and contributes to disease protection. The gut in fish is often the initial site where pathogens are encountered and is considered a primary portal of entry. Therefore, various physical and chemical barriers exist to limit pathogen entry. This barrier function works in conjunction with innate and adaptive immune responses that together make up the mucosal immune system in fish. Although study of this system and gut immunology in fish is limited, evidence supports the gut as a primary site of immunological importance. Mucosal immunity in fish may be primitive in comparison to mammals, but it is obvious that innate and adaptive responses can be elicited at these sites. Many immune cells are present in the gut mucosa and activation of these cells can lead to the development of antigen-specific antibody. However, the ability to effectively trigger adaptive responses following antigen delivery to the gut appears somewhat limited. This observation is supported by the fact that many antigens are degraded as they pass through the gut, and that antigen uptake is much less efficient in the GI tract when compared to delivery via injection or even immersion. The decreased antigen uptake also points to the effective function of the gut to eliminate pathogen entry through barrier function and chemical inhibition. If antigens or pathogenic microbes reach the lower GI tract, then antigen uptake can occur and both innate and adaptive responses are initiated. However, the contribution these responses make to disease protection is not clear, but it is apparent that the gut of fish is routinely exposed to many microbes, both non-pathogenic and pathogenic. This and other evidence presented in this chapter suggests that barrier function and initiation of early innate responses in the gut of fish are central to disease protection and that adaptive responses contribute to protection, but most likely to a lesser degree.

REFERENCES

Abelli, L., Picchietti, S., Romano, N., Mastrolia, L., and Scapigliati, G. (1997). Immunohistochemistry of gut-associated lymphoid tissue of the sea bass *Dicentrarchus labrax* (L.). *Fish Shellfish Immunol.* **7**, 235–245.

Andlid, T., Vazquez-Juarez, R., and Gustafsson, L. (1998). Yeasts isolated from the intestine of rainbow trout adhere to and grow in intestinal mucus. *Mol. Mar. Biol. Biotechnol.* **7**, 115–126.

Aranishi, F., and Mano, N. (2000). Antibacterial cathepsins in different types of ambicoloured Japanese flounder skin. *Fish Shellfish Immunol.* **10**, 87–89.

Arason, G. J. (1996). Lectins as defence molecules in vertebrates and invertebrates. *Fish Shellfish Immunol.* **6**, 277–289.

Balcazar, J., de Blas, I., Ruiz-Zarzuela, I., Cunningham, D., Vendrell, D., and Múzquiz, J. L. (2006a). The role of probiotics in aquaculture. *Vet. Microbiol.* **114**, 173–186.

Balcazar, J., Decamp, O., Vendrell, D., de Blas, I., and Ruiz-Zarzuela, I. (2006b). Health and nutritional properties of probiotics in fish and shellfish. *Microb. Ecol. Health Dis.* **18**, 65–70.

Balfry, S. K. (1997). The non-specific immune system and innate disease resistance in different strains of teleost fish. *Animal Science.* University of British Columbia, Vancouver.

Bengten, E., Wilson, M., Miller, N., Clem, L. W., Pilstrom, L., and Warr, G. W. (2000). Immunoglobulin isotypes: structure, function and genetics. *Current Topics in Microbiology and Immunology* **248**, 189–219.

Boshra, H., Li, J., and Sunyer, J. O. (2006). Recent advances on the complement system of teleost fish. *Fish Shellfish Immunol.* **20**, 239–262.

Bromage, E. S., Ye, J., and Kaattari, S. L. (2006). Antibody structural variation in rainbow trout fluids. *Comp. Biochem. Physiol. B Biochem. Mol. Biol.* **143**, 61–69.

Cain, K. D., LaPatra, S. E., Baldwin, T. J., Shewmaker, B., Jones, J., and Ristow, S. S. (1996). Characterization of mucosal immunity in rainbow trout *Oncorhynchus mykiss* challenged with infectious hematopoietic necrosis virus: identification of antiviral activity. *Dis. Aquat. Org.* **27**, 161–172.

Cain, K. D., Jones, D. R., and Raison, R. L. (2000). Characterisation of mucosal and systemic immune responses in rainbow trout (*Oncorhynchus mykiss*) using surface plasmon resonance. *Fish Shellfish Immunol.* **10**, 651–666.

Caipang, C. M. A., Hirono, I., and Aoki, T. (2003). In vitro inhibition of fish rhabdoviruses by Japanese flounder, *Paralichthys olivaceus* Mx. *Virology* **317**, 373–382.

Cartwright, J. R., Tharia, H. A., Burns, I., Shrive, A. K., Hoole, D., and Greenhough, T. J. (2004). Isolation and characterisation of pentraxin-like serum proteins from the common carp *Cyprinus carpio*. *Dev. Comp. Immunol.* **28**, 113–125.

Chilmonczyk, S., and Monge, D. (1980). Rainbow trout gill pillar cell—demonstration of inert particle phagocytosis and involvement in viral infection. *J. Reticuloendoth. Soc.* **28**(4), 327–332.

Clerx, J. P., Castel, A., Bol, J. F., and Gerwig, G. J. (1980). Isolation and characterization of the immunoglobin of pike (*Esox lucius* L.). *Vet. Immunol. Immunopathol.* **1**, 125–144.

Cole, A. M., Weis, P., and Diamond, G. (1997). Isolation and characterization of pleurocidin, an antimicrobial peptide in the skin secretions of winter flounder. *J. Biol. Chem.* **272**, 12008–12013.

Concha, M. I., Molina, S., Oyarzun, C., Villanueva, J., and Amthauer, R. (2003). Local expression of apolipoprotein A-I gene and a possible role for HDL in primary defence in the carp skin. *Fish Shellfish Immunol.* **14**, 259–273.

Cook, M. T., Hayball, P. J., Birdseye, L., Bagley, C., Nowak, B. F., and Hayball, J. D. (2003). Isolation and partial characterization of a pentraxin-like protein with complement-fixing activity from snapper (*Pagrus auratus*, Sparidae) serum. *Dev. Comp. Immunol.* **27**, 579–588.

Dalmo, R. A., Ingebrigtsen, K., and Bogwald, J. (1997). Non-specific defence mechanisms in fish, with particular reference to the reticuloendothelial system (RES). *J. Fish Dis.* **20**, 241–273.

Davidson, G. A., Lin, S. H., Secombes, C. J., and Ellis, A. E. (1997). Detection of specific and "constitutive" antibody secreting cells in the gills, head, kidney and peripheral blood leucocytes of dab (*Limanda limanda*). *Vet. Immunol. Immunopathol.* **58**, 363–374.

Dixon, B., and Stet, R. J. M. (2001). The relationship between major histocompatibility receptors and innate immunity in teleost fish. *Dev. Comp. Immunol.* **25**, 683–699.

Drennan, J., LaPatra, S., Swan, C., Ireland, S., and Cain, K. D. (2007). Characterization of serum and mucosal antibody responses in white sturgeon (*Acipenser transmontanus* Richardson) following immunization with WSIV and a protein hapten antigen. *Fish Shellfish Immunol.* **23**, 657–669.

Du Pasquier, L. (2001). The immune system of invertebrates and vertebrates. *Comparative Biochemistry and Physiology Part B: Biochemistry and Molecular Biology* **129**, 1–15.

Ellis, A. (1999). Immunity to bacteria in fish. *Fish Shellfish Immunol.* **9**, 291–308.

Ellis, A. E. (2001). Innate host defense mechanisms of fish against viruses and bacteria. *Dev. Comp. Immunol.* **25**, 827–839.

Fange, R., Lundblad, G., and Lind, J. (1976). Lysozyme and chitinase in blood and lymphomyeloid tissues of marine fish. *Marine Biology* **36**, 277–282.

Fernandes, J. M. O., Kemp, G. D., and Smith, V. J. (2004a). Two novel muramidases from skin mucosa of rainbow trout (*Oncorhynchus mykiss*). *Comparative Biochemistry and Physiology Part B: Biochemistry and Molecular Biology* **138**, 53–64.

Fernandes, J. M. O., Molle, G., Kemp, G. D., and Smith, V. J. (2004b). Isolation and characterisation of oncorhyncin II, a histone H1-derived antimicrobial peptide from skin secretions of rainbow trout, *Oncorhynchus mykiss*. *Dev. Comp. Immunol.* **28**, 127–138.

Flajnik, M. F. (1996). The immune system of ectothermic vertebrates. *Vet. Immunol. Immunopathol.* **54**, 145–150.

Fletcher, T., and Grant, P. (1968). Glycoproteins in the external mucous secretions of the plaice, *Pleuronectes platessa*, and other fishes. *Biochem. J.* **106**, 12.

Frazer, J. K., and Capra, J. D. (1999). Immunoglobulins: structure and function. In: *Fundamental Immunology* (W. E. Paul, ed.), pp. 37–74. Lippincott-Raven, Philadelphia.

Gambon-Deza, F., Sanchez-Espinel, C., and Magadan-Mompo, S. (2010). Presence of a unique IgT on the IGH locus in three-spined stickleback fish (*Gasterosteus aculeatus*) and the very recent generation of a repertoire of VH genes. *Dev. Comp. Immunol.* **34**(2), 114–122.

Gatesoupe, F. (2007). Live yeasts in the gut: natural occurrence, dietary introduction, and their effects on fish health and development. *Aquaculture* **267**, 20–30.

Georgopoulou, U., and Vernier, J.-M. (1986). Local immunological response in the posterior intestinal segment of the rainbow trout after oral administration of macromolecules. *Dev. Comp. Immunol.* **10**, 529–537.

Gerwick, L., Steinhauer, R., LaPatra, S., Sandell, T., Ortuno, J., Hajiseyedjavadi, N., and Bayne, C. J. (2002). The acute phase response of rainbow trout (*Oncorhynchus mykiss*) plasma proteins to viral, bacterial and fungal inflammatory agents. *Fish Shellfish Immunol.* **12**, 229–242.

Goldsby, R. A., Kindt, T. J., Osborne, B. A., and Kuby, J. (2003). *Immunology,* pp. 551. W.H. Freeman and Company, New York.

Gomez, G., and Balcazar, J. (2008). A review on the interactions between gut microbiota and innate immunity of fish. *FEMS Immunol. Med. Microbiol.* **52**, 145–154.

Grinde, B., Lie, O., Poppe, T., and Salte, R. (1988). Species and individual variation in lysozyme activity in fish of interest in aquaculture. *Aquaculture* **68**, 299–304.

Hansen, J. D., Landis, E. D., and Phillips, R. B. (2005). Discovery of a unique Ig heavy-chain isotype (IgT) in rainbow trout: implications for a distinctive B cell developmental pathway in teleost fish. *Proc. Natl. Acad. Sci. USA* **102**, 6919–6924.

Harrell, L. W., Etlinger, H. M., and Hodgins, H. O. (1976). Humoral factors important in resistance of salmonid fish to bacterial disease. II. Anti-*Vibrio anguillarum* activity in mucus and observations on complement. *Aquaculture* **7**, 363–370.

Hart, S., Wrathwell, A., Harris, J., and Grayson, T. (1988). Gut immunology in fish: a review. *Dev. Comp. Immunol.* **12**, 453–480.

Hatten, F., Fredriksen, A., Hordvik, I., and Endresen, C. (2001). Presence of IgM in cutaneous mucus, but not in gut mucus of Atlantic salmon, *Salmo salar*. Serum IgM is rapidly degraded when added to gut mucus. *Fish Shellfish Immunol.* **11**, 257–268.

Hebert, P., Ainsworth, A. J., and Boyd, B. (2002). Histological enzyme and flow cytometric analysis of channel catfish intestinal tract immune cells. *Dev. Comp. Immunol.* **26**, 53–62.

Hirono, I., Nam, B.-H., Enomoto, J., Uchino, K., and Aoki, T. (2003). Cloning and characterisation of a cDNA encoding Japanese flounder *Paralichthys olivaceus* IgD. *Fish Shellfish Immunol.* **15**, 63–70.

Hirose, K., Suzuki, H., Nishimura, H., Mitani, A., Washizu, J., Matsuguchi, T., and Yoshikai, Y. (1998). Interleukin-15 may be responsible for early activation of intestinal intraepithelial lymphocytes after oral infection with *Listeria monocytogenes* in rat. *Infect. Immun.* **66**(12), 5677–5683.

Hohman, V. S., Stewart, S. E., Rumfelt, L. L., Greenberg, A. S., Avila, D. W., Flajnik, M. F., and Steiner, L. A. (2003). J chain in the nurse shark: implications for function in a lower vertebrate. *J. Immunol.* **170**, 6016–6023.

Holland, M. C. H., and Lambris, J. D. (2002). The complement system in teleosts. *Fish Shellfish Immunol.* **12**, 399–420.

Holmgren, J., and Czerkinsky, C. (2005). Mucosal immunity and vaccines. *Nature Medicine* **11**, 45–53.

Hordvik, I., Jacob, A. L., Charlemagne, J., and Endresen, C. (1996). Cloning of T-cell antigen receptor beta chain cDNAs from Atlantic salmon (*Salmo salar*). *Immunogenetics* **45**, 9–14.

Hordvik, I., Thevarajan, J., Samdal, I., Bastani, N., and Krossoy, B. (1999). Molecular cloning and phylogenetic analysis of the Atlantic salmon immunoglobulin D gene. *Scandinavian J. Immunol.* **50**, 202–210.

Hordvik, I., Torvund, J., Moore, L., and Endresen, C. (2004). Structure and organization of the T cell receptor alpha chain genes in Atlantic salmon. *Mol. Immunol.* **41**, 553–559.

Itami, T., Ishida, Y., Endo, F., Kawazoe, N., and Takahashi, Y. (1993). Haemagglutinins in the skin mucus of ayu. *Fish Pathol.* **28**, 41–47.

Jirillo, F., Passantino, G., Massaro, M. A., Cianciotta, A., Crasto, A., Perillo, A., Passantino, L., and Jirillo, E. (2007). In vitro elicitation of intestinal immune responses in teleost fish: evidence for a type IV hypersensitivity reaction in rainbow trout. *Immunopharmacol. Immunotoxicol.* **29**(1), 69–80.

Jenkins, P. G., Wrathmell, A. B., Harris, J. E., and Pulsford, A. L. (1994). Systemic and mucosal immune responses to enterically delivered antigen in *Oreochromis mossambicus*. *Fish Shellfish Immunol.* **4**, 255–271.

Jensen, L. E., Petersen, T. E., Thiel, S., and Jensenius, J. C. (1995). Isolation of a pentraxin-like protein from rainbow trout serum. *Dev. Comp. Immunol.* **19**, 305–314.

Jensen, L. E., Hiney, M. P., Shields, D. C., Uhlar, C. M., Lindsay, A. J., and Whitehead, A. S. (1997). Acute phase proteins in salmonids: evolutionary analyses and acute phase response. *J. Immunol.* **158**, 384–392.

Johansen, F. E., Braathen, R., and Brandtzaeg, P. (2000). Role of J chain in secretory immunoglobulin formation. *Scandinavian J. Immunol.* **52**, 240–248.

Jones, D. R., Hannan, C. M., Russell-Jones, G. J., and Raison, R. L. (1999). Selective B cell non-responsiveness in the gut of the rainbow trout (*Oncorhynchus mykiss*). *Aquaculture* **172**, 29–39.

Joosten, P. H. M., Kruijer, W. J., and Rombout, J. H. W. M. (1996). Anal immunisation of carp and rainbow trout with different fractions of a *Vibrio anguillarum* bacterin. *Fish Shellfish Immunol.* **6**, 541–551.

Joosten, P. H. M., Tiemersma, E., Threels, A., Caumartin-Dhieux, C., and Rombout, J. H. W. M. (1997). Oral vaccination of fish against *Vibrio anguillarum* using alginate microparticles. *Fish Shellfish Immunol.* **7**, 471–485.

Kaattari, S. L. (1992). Fish B lymphocytes: defining their form and function. *Annu. Rev. Fish Dis.* **2**, 161–180.

Kaattari, S., Evans, D., and Klemer, J. (1998). Varied redox forms of teleost IgM: an alternative to isotypic diversity? *Immunological Reviews* **166**, 133–142.

Klaus, G. G., Halpern, M. S., Koshland, M. E., and Goodman, J. W. (1971). A polypeptide chain from leopard shark 19S immunoglobulin analogous to mammalian J chain. *J. Immunol.* **107**, 1785–1787.

Kobayashi, K., Hara, A., Takano, K., and Hirai, H. (1982). Studies on subunit components of immunoglobulin M from a bony fish, the chum salmon (*Oncorhynchus keta*). *Mol. Immunol.* **19**, 95–103.

Kodama, H., Yamada, F., Murai, T., Nakanishi, Y., Mikami, T., and Izawa, H. (1989). Activation of trout macrophages and production of CRP after immunization with *Vibrio anguillarum*. *Dev. Comp. Immunol.* **13**, 123–132.

Kollner, B., and Kotterba, G. (2002). Temperature dependent activation of leucocyte populations of rainbow trout, *Oncorhynchus mykiss*, after intraperitoneal immunisation with *Aeromonas salmonicida*. *Fish Shellfish Immunol.* **12**, 35–48.

Komatsu, K., Tsutsui, S., Hino, K., Araki, K., Yoshiura, Y., Yamamoto, A., Nakamura, O., and Watanabe, T. (2009). Expression profiles of cytokines release in intestinal epithelial cells of the rainbow trout, *Oncorhynchus mykiss*, in response to bacterial infection. *Dev. Comp. Immunol.* **33**, 499–506.

Kuroda, N., Naruse, K., Shima, A., Nonaka, M., and Sasaki, M. (2000). Molecular cloning and linkage analysis of complement C3 and C4 genes of the Japanese medaka fish. *Immunogenetics* **51**, 117–128.

LaFrentz, B. R., LaPatra, S. E., Jones, G. R., Congleton, J. L., Sun, B., and Cain, K. D. (2002). Characterization of serum and mucosal antibody responses and relative per cent survival in rainbow trout (*Oncorhynchus mykiss*) following immunization and challenge with *Flavobacterium psychrophilum*. *J. Fish Dis* **25**, 703–713.

Larsen, R., Rokenes, T. P., and Robertsen, B. (2004). Inhibition of infectious pancreatic necrosis virus replication by Atlantic salmon Mx1 protein. *Journal of Virology* **78**, 7938–7944.

Li, J., Barreda, D. R., Zhang, Y.-A., Boshra, H., Gelman, A. E., LaPatra, S., Tort, L., and Sunyer, J. O. (2006). B lymphocytes from early vertebrates have potent phagocytic and microbicidal abilities. *Nature Immunology* **7**(10), 1116–1124.

Lin, S. H., Davidson, G. A., Secombes, C. J., and Ellis, A. E. (2000). Use of a lipid-emulsion carrier for immunisation of dab (*Limanda limanda*) by bath and oral routes: an assessment of systemic and mucosal antibody responses. *Aquaculture* **181**, 11–24.

Lindenstrom, T., Buchmann, K., and Secombes, C. J. (2003). *Gyrodactylus derjavini* infection elicits IL-1β expression in rainbow trout skin. *Fish Shellfish Immunol.* **15**, 107–115.

Lobb, C. J. (1987). Secretory immunity induced in catfish, *Ictalurus punctatus*, following bath immunization. *Dev. Comp. Immunol.* **11**, 727–738.

Lund, V., and Olafsen, J. A. (1998). A comparative study of pentraxin-like proteins in different fish species. *Dev. Comp. Immunol.* **22**, 185–194.

Lund, V., and Olafsen, J. A. (1999). Changes in serum concentration of a serum amyloid P-like pentraxin in Atlantic salmon, *Salmo salar* L., during infection and inflammation. *Dev. Comp. Immunol.* **23**, 61–70.

Magnadottir, B. (1998). Comparison of immunoglobulin (IgM) from four fish species. *Buvisindi* **12**, 47–59.

Magnadottir, B. (2006). Innate immunity of fish (overview). *Fish Shellfish Immunol.* **20**, 137–151.

Maki, J. L., and Dickerson, H. W. (2003). Systemic and cutaneous mucus antibody responses of channel catfish immunized against the protozoan parasite, *Ichthyophthirius multifiliis*. *Clin. Diagn. Lab. Immunol.* **10**, 876–881.

McCumber, L. J., and Clem, L. W. (1976). A comparative study of J chain structure and stoichiometry in human and nurse shark IgM. *Immunochemistry* **13**, 479–484.

McGhee, J. R., and Kiyono, H. (1999). The mucosal immune system. In: *Fundamental Immunology* (W. E. Paul, ed.), pp. 909–945. Lippincott-Raven, Philadelphia.

McGhee, J. R., Czerkinsky, C., and Mestecky, J. (1999). Mucosal vaccines: an overview. In: *Mucosal Immunology* (P. L. Ogra, M. E. Lamm, J. Bienenstock, J. Mestecky, W. Strober and J.R. McGhee, eds), pp. 741–757. Academic Press, San Diego.

Mestecky, J., Kulhavy, R., Schrohenloher, R. E., Tomana, M., and Wright, G. P. (1975). Identification and properties of J chain isolated from catfish macroglobulin. *J. Immunol.* **115**, 993–997.

Mestecky, J., Abraham, R., and Ogra, P. L. (1994). Common mucosal immune system and strategies for the development of vaccines effective at the mucosal surfaces. In: *Handbook of Mucosal Immunology* (P. L. Ogra, J. Mestecky, M. E. Lamm, W. Strober, J. R. McGhee and J. Bienenstock, eds), pp. 357–372. Academic Press, San Diego.

Nakanishi, T., Aoyagi, K., Xia, C., Dijkstra, J. M., and Ototake, M. (1999). Specific cell-mediated immunity in fish. *Vet. Immunol. Immunopathol.* **72**, 101–109.

Nakao, M., Mutsuro, J., Obo, R., Fujiki, K., Nonaka, M., and Yano, T. (2000). Molecular cloning and protein analysis of divergent forms of the complement component C3 from a bony fish, the common carp (*Cyprinus carpio*): presence of variants lacking the catalytic histidine. *Eur. J. Immunol.* **30**, 858–866.

Neumann, N. F., Stafford, J. L., Barreda, D., Ainsworth, A. J., and Belosevic, M. (2001). Antimicrobial mechanisms of fish phagocytes and their role in host defense. *Dev. Comp. Immunol.* **25**, 807–825.

Partula, S., Guerra, A. d., Fellah, J., and Charlemagne, J. (1995). Structure and diversity of the T cell antigen receptor beta-chain in a teleost fish. *J. Immunol.* **155**, 699–706.

Partula, S., Guerra, A. d., Fellah, J., and Charlemagne, J. (1996a). Structure and diversity of the TCR alpha-chain in a teleost fish. *J. Immunol.* **157**, 207–212.

Partula, S., Schwager, J., Timmusk, S., Pilstrom, L., and Charlemagne, J. (1996b). A second immunoglobulin light chain isotype in the rainbow trout. *Immunogenetics* **45**, 44–51.

Paulsen, S., Engstad, R., and Robertsen, B. (2001). Enhanced lysozyme production in Atlantic salmon (*Salmo salar* L.) macrophages treated with yeast B-glucan and bacterial lipopolysaccharide. *Fish Shellfish Immunol.* **11**, 23–37.

Petrie, A. G., and Ellis, A. E. (2006). Evidence of particulate uptake by the gut of Atlantic salmon (*Salmo salar* L.). *Fish Shellfish Immunol.* **20**, 660–664.

Pitman, R. S., and Blumberg, R. S. (2000). First line of defense: the role of the intestinal epithelium as an active component of the mucosal immune system. *J. Gastroenterol.* **35**(1), 805–814.

Pond, M., Stone, D., and Alderman, D. (2006). Comparison of conventional and molecular techniques to investigate the intestinal micorflora of rainbow trout (*Oncorhynchus mykiss*). *Aquaculture* **261**, 194–203.

Press, C., and Evensen, O. (1999). The morphology of the immune system in teleost fishes. *Fish Shellfish Immunol.* **9**, 309–318.

Re, F., and Strominger, J. L. (2004). Heterogeneity of TLR-induced responses in dendritic cells: from innate to adaptive immunity. *Immunobiology* **209**, 191–198.

Robertsen, B. (2006). The interferon system of teleost fish. *Fish Shellfish Immunol.* **20**, 172–191.

Rombout, J. W. H. M., Blok, L. J., Lamers, C. H. J., and Egberts, E. (1986). Immunization of carp (*Cyprinus carpio*) with a *Vibrio anguillarum* bacterin: indications for a common mucosal immune system. *Dev. Comp. Immunol.* **10**, 341–351.

Rombout, J. H. W. M., and Van Den Berg, A. A. (1989). Immunological importance of the second gut segment of carp. I. Uptake and processing of antigens by epithelial cells and macrophages. *J. Fish Biol.* **35**, 13–22.

Rombout, J. H. W. M., Lamers, C. H. J., Helfrich, M. H., Dekker, A., and Taverne-Thiele, J. J. (1985). Uptake and transport of intact macromolecules in the intestinal epithelium of carp (*Cyprinus carpio* L.). *Cell Tissue Res.* **239**, 519–530.

Rombout, J. H. W. M., Taverne-Thiele, A. J., and Villena, M. I. (1993a). The gut-associated lymphoid tissue (GALT) of carp (*Cyprinus carpio* L.): an immunocytochemical analysis. *Dev. Comp. Immunol.* **17**, 55–66.

Rombout, J. H. W. M., Taverne, N., van de Kamp, M., and Taverne-Thiele, A. J. (1993b). Differences in mucus and serum immunoglobulin of carp (*Cyprinus carpio* L.). *Dev. Comp. Immunol.* **17**, 309–317.

Saha, N. R., Suetake, H., Kiyoshi, K., and Suzuki, Y. (2004). Fugu immunoglobulin D: a highly unusual gene with unprecedented duplications in its constant region. *Immunogenetics* **56**, 438–447.

Sanchez, C., Dominguez, J., and Coll, J. (1989). Immunoglobulin heterogeneity in the rainbow trout, *Salmo gardineri* Richardson. *J. Fish Dis.* **12**, 459–465.

Sato, A., Somamoto, T., Yokooka, H., and Okamoto, N. (2005). Systemic priming of alloreactive cytotoxic cells in carp, following anal administration of allogeneic cell antigens. *Fish Shellfish Immunol.* **19**, 43–52.

Savan, R., Aman, A., Nakao, M., Watanuki, H., and Sakai, M. (2005). Discovery of a novel immunoglobulin heavy chain gene chimera from common carp (*Cyprinus carpio* L.). *Immunogenetics* **57**, 458–463.

Scapigliati, G., Romano, N., Abelli, L., Meloni, S., Ficca, A., Buonocore, F., Bird, S., and Secombes, C. (2000). Immunopurification of T-cells from sea bass *Dicentrarchus labrax* (L.). *Fish Shellfish Immunol.* **10**, 329–341.

Secombes, C. J. (1996). The nonspecific immune system: cellular defenses. In: *The Fish Immune System* (G. Iwama and T. Nakanishi, eds), pp. 63–157. Academic Press, San Diego.

Shen, L., Stuge, T. B., Bengten, E., Wilson, M., Chinchar, V. G., Naftel, J. P., Bernanke, J. M., Clem, L. W., and Miller, N. W. (2004). Identification and characterization of clonal NK-like cells from channel catfish (*Ictalurus punctatus*). *Dev. Comp. Immunol.* **28**, 139–152.

Shephard, K. L. (1994). Functions for fish mucus. *Reviews in Fish Biology and Fisheries* **4**(4), 401–429.

Shoemaker, C. A., Xu, D.-H., Shelby, R. A., and Klesius, P. H. (2005). Detection of cutaneous antibodies against *Flavobacterium columnare* in channel catfish, *Ictalurus punctatus* (Rafinesque). *Aquacult. Res.* **36**, 813–818.

Sigh, J., Lindenstrom, T., and Buchmann, K. (2004). The parasitic ciliate *Ichthyophthirius multifiliis* induces expression of immune relevant genes in rainbow trout, *Oncorhynchus mykiss* (Walbaum). *J. Fish Dis.* **27**, 409–417.

Snider, D., Marshall, J., Perdue, M., and Liang, H. (1994). Production of IgE antibody and allergic sensitization of intestinal and peripheral tissues after oral immunization with protein Ag and cholera toxin. *J. Immunol.* **153**, 647–657.

Spanggaard, B., Huber, I., Nielsen, J., Nielsen, T., Appel, K., and Gram, L. (2000). The microflora of rainbow trout intestine: a comparison of traditional and molecular identification. *Aquaculture* **182**, 1–15.

Stenvik, J., and Jørgensen, T. Ø. (2000). Immunoglobulin D (IgD) of Atlantic cod has a unique structure. *Immunogenetics* **51**, 452–461.

Stratton, K. R., Durch, J. S., and Lawrence, R. S. (1999). *Vaccines for the 21st Century, a Tool for Decisionmaking.* National Academy Press, Washington DC, pp. 460.

Sunyer, J. O., Zarkadis, I. K., Sahu, A., and Lambris, J. D. (1996). Multiple forms of complement C3 in trout that differ in binding to complement activators. *Proc. Natl. Acad. Sci. USA* **93**, 8546–8551.

Sunyer, J. O., Tort, L., and Lambris, J. D. (1997). Structural C3 diversity in fish: characterization of five forms of C3 in the diploid fish *Sparus aurata*. *J. Immunol.* **158**, 2813–2821.

Swan, C., Lindstrom, N., and Cain, K. D. (2008). Identification of a localized mucosal immune response in rainbow trout, *Oncorhynchus mykiss* (Walbaum), following immunization with a protein-hapten antigen. *J. Fish Dis.* **31**, 383–393.

Szalai, A., Bly, J., and Clem, L. (1994). Changes in serum concentrations of channel catfish (*Ictalurus punctatus* Rafinesque) phosphorylcholine-reactive protein (PRP) in response to inflammatory agents, low temperature-shock and infection by the fungus *Saprolegnia* sp. *Fish Shellfish Immunol.* **4**, 323–336.

Taylor, I. S., Adam, B., Veverkova, M., Tatner, M. F., Low, C., Secombes, C., and Birkbeck, T. H. (2005). T-cell antigen receptor genes in turbot (*Scophthalmus maximus* L.). *Fish Shellfish Immunol.* **18**, 445–448.

Tian, J., Sun, B., Luo, Y., Zhang, Y., and Nie, P. (2009). Distribution of IgM, IgD and IgZ in mandarin fish, *Siniperca chuatsi* lymphoid tissues and their transcriptional changes after *Flavobacterium columnare* stimulation. *Aquaculture* **288**, 14–21.

Tsukita, S., Furuse, M., and Itoh, M. (2001). Multifunctional strands in tight junctions. *Nat. Rev. Mol. Cell Biol.* **2**(4), 285–293.

Tsutsui, S., Tasumi, S., Suetake, H., and Suzuki, Y. (2003). Lectins homologous to those of monocotyledonous plants in the skin mucus and intestine of pufferfish, *Fugu rubripes*. *J. Biol. Chem.* **278**, 20882–20889.

Tsutsui, S., Tasumi, S., Suetake, H., Kikuchi, K., and Suzuki, Y. (2005). Demonstration of the mucosal lectins in the epithelial cells of internal and external body surface tissues in pufferfish (*Fugu rubripes*). *Dev. Comp. Immunol.* **29**, 243–253.

Tu, F., Chu, W., Zhuang, X., and Lu, C. (2010). Effect of oral immunization with *Aeromonas hydrophila* ghosts on protection against experimental fish infection. *Lett. Appl. Microbiol.* **50**, 13–17.

Underdown, B. J., and Mestecky, J. (1994). Mucosal immunoglobulins. In: *Handbook of Mucosal Immunology* (P. L. Ogra, J. Mestecky, M. E. Lamm, W. Strober, J. R. McGhee and J. Bienenstock, eds), pp. 79–97. Academic Press, San Diego.

Vallejo, A. N., Miller, N. W., Harvey, N. E., Cuchens, M. A., Warr, G. W., and Clem, L. W. (1992). Cellular pathway(s) of antigen processing and presentation in fish APC: endosomal involvement and cell-free antigen presentation. *Dev. Immunol.* **3**, 51–65.

vanMuiswinkel, W. B. (1995). The piscine immune system: innate and acquired immunity. In: *Fish Diseases and Disorders: Protozoan and Metazoan Infections* (P. T. K Woo, ed.), vol. 1, pp. 729–750. CABI Publishing, New York.

Vernier, J. (1990). Intestinal ultrastructure in relation to lipid and protein absorption in teleost fish. *Comp. Physiol.* **5**, 166–175.

Vervarcke, S., Ollevier, F., Kinget, R., and Michoel, A. (2005). Mucosal response in African catfish after administration of *Vibrio anguillarum* O2 antigens via different routes. *Fish Shellfish Immunol.* **18**, 125–133.

Voss, E. W., Jr., Groberg, W. J., Jr., and Fryer, J. L. (1980). Metabolism of coho salmon Ig. Catabolic rate of coho salmon tetrameric Ig in serum. *Mol. Immunol.* **17**, 445–452.

Warr, G. W. (1983). Immunoglobulin of the toadfish, *Spheroides Glaber. Comp. Biochem. Physiol. B* **76**, 507–514.

Warr, G. W. (1995). The immunoglobulin genes of fish. *Dev. Comp. Immunol.* **19**, 1–12.

Watanabe, M., Ueno, Y., Yajima, T., Iwao, Y., Tsuchiya, M., Ishikawa, H., Aiso, S., Hibi, T., and Ishii, H. (1995). Interleukin 7 is produced by human intestinal epithelial cells and regulates the proliferation of intestinal mucosal lymphocytes. *J. Clin. Invest.* **95**(6), 2945–2953.

Watts, M., Munday, B. L., and Burke, C. M. (2001). Immune responses of teleost fish. *Australian Veterinary Journal* **79**, 570–574.

Weinberg, E. D. (1974). Iron and susceptibility to infectious disease. *Science* **184**, 952–956.

Weinheimer, P. F., Mestecky, J., and Acton, R. T. (1971). Species distribution of J chain. *J. Immunol.* **107**, 1211–1212.

Wilson, M. R., and Warr, G. W. (1992). Fish immunoglobulins and the genes that encode them. *Annu. Rev. Fish Dis.* **2**, 201–221.

Wilson, M., Bengtén, E., Miller, N. W., Clem, L. W., Pasquier, L. D., and Warr, G. W. (1997). A novel chimeric Ig heavy chain from a teleost fish shares similarities to IgD. *Proc. Natl. Acad. Sci. USA* **94**, 4593–4597.

Wilson, M. R., Zhou, H., Bengten, E., Clem, L. W., Stuge, T. B., Warr, G. W., and Miller, N. W. (1998). T-cell receptors in channel catfish: structure and expression of TCR α and β genes. *Mol. Immunol.* **35**, 545–557.

Xu, D.-H., and Klesius, P. H. (2002). Antibody mediated immune response against *Ichthyophthirius multifiliis* using excised skin from channel catfish, *Ictalurus punctatus* (Rafinesque), immune to *Ichthyophthirius*. *J. Fish Dis.* **25**, 299–306.

Xu, D. H., Klesius, P. H., and Shelby, R. A. (2002). Cutaneous antibodies in excised skin from channel catfish, *Ictalurus punctatus* Rafinesque, immune to *Ichthyophthirius multifiliis*. *J. Fish Dis.* **25**, 45–52.

Yousif, A. N., Albright, L. J., and Evelyn, T. (1994). *In vitro* evidence for the antibacterial role of lysozyme in salmonid eggs. *Dis. Aquat. Org.* **19**, 15–19.

Zimmerman, A., Evenhuis, J., Thorgaard, G., and Ristow, S. (2004). A single major chromosomal region controls natural killer cell-like activity in rainbow trout. *Immunogenetics* **55**, 825–835.

4

THE ROLE OF THE GASTROINTESTINAL TRACT IN SALT AND WATER BALANCE

MARTIN GROSELL

Fish living in marine and freshwater environments face contrasting challenges in terms of maintaining salt and water balance. The present chapter summarizes and reviews much of what is known about intestinal water absorption, which is essential for survival of fish in marine environments, and the impacts of intestinal processes on integrative organismal physiology. Transcellular Cl^- and Na^+ absorption via parallel co-transport systems and anion exchange drive water absorption across the intestinal epithelium and leave behind high concentrations of Mg^{2+} and SO_4^{2-} in the intestinal lumen resulting in fluids isotonic to the blood plasma but of unique chemical composition. High rates of Cl^-/HCO_3^- exchange by the intestinal epithelium render the intestinal fluids alkaline and high in HCO_3^- and result in $CaCO_3$ formation in the intestinal lumen. Rectal excretion of these precipitates and rectal fluids rich in HCO_3^- comprise a substantial base loss which along with salt gain from water absorption are compensated for at the gill where both NaCl and acid are excreted. Suggested directions for future studies in the area of intestinal contribution to osmoregulation are provided.

The Multifunctional Gut of Fish: Volume 30
FISH PHYSIOLOGY

1. INTRODUCTION

Marine teleost fish and lampreys constantly experience a diffusive water loss to their relatively concentrated environment and are forced to ingest seawater and perform solute coupled water absorption in the intestine to maintain water balance (Grosell 2006). Marine elasmobranchs rely much less on drinking as they are iso-osmotic or slightly hyper-osmotic compared to seawater, but nevertheless drink small volumes of seawater (De Boeck et al. 2001; Grosell and Taylor 2007) and are capable of intestinal salt and water absorption (Anderson et al. 2002, 2007; Grosell and Taylor 2007). In both cases, ion transport across the intestinal epithelium is critical for water absorption in a hypersaline environment and leads to salt gain, which must be compensated for by active salt excretion (Marshall and Grosell 2005).

In contrast, freshwater teleosts, lampreys and elasmobranchs gain water from their dilute environment and do not rely on intestinal water absorption for salt and water balance (Marshall and Grosell 2005). However, the freshwater fish intestine nevertheless plays a role in ionoregulation by contributing to uptake of Na^+ and Cl^-, a process which likely acts to compensate for the diffusional loss of these ions across the gill to the hyposaline environment. Of particular importance is the intestinal uptake, from dietary sources, as the sole means for replacing diffusive Cl^- loss to the freshwater environment in a number of euryhaline and freshwater teleost fish that lack branchial Cl^- uptake (Tomasso and Grosell 2004). In addition, recent research has revealed a significant intestinal contribution to Na^+ homeostasis in freshwater fish through uptake from dietary sources (see Chapter 5). The emphasis of the present chapter is on the direct role of the intestine in osmoregulation in fasting marine and euryhaline fish.

2. DRINKING

2.1. Drinking Rates

Since drinking and intestinal water absorption by marine teleosts was demonstrated in 1930 (Smith 1930), it has been established that drinking rate is greatly influenced by ambient salinity such that euryhaline fish display 10–50-fold higher drinking rates in seawater than they do in freshwater (Malvin et al. 1980; Carrick and Balment 1983; Perrot et al. 1992; Lin et al. 2002). Generally, marine and seawater-acclimated euryhaline fish drink 1–5 ml/kg/h (Shehadeh and Gordon 1969; Carrick and Balment 1983; Perrot et al. 1992; Fuentes et al. 1996; Fuentes and Eddy 1996, 1997a,b;

Grosell and Jensen 1999; Takei and Tsuchida 2000; Lin et al. 2002; Grosell et al. 2004). A change in drinking rate presents the most immediate level at which fluid absorption is controlled by the gastrointestinal tract and thus, control of drinking is central to marine fish salt and water balance.

2.2. Sensing External Salinity and Control of Drinking

In addition to ambient salinity, a range of factors influence drinking rate (see Marshall and Grosell 2005 for a summary). The renin-angiotensin system (RAS) is clearly involved in regulation of drinking in teleosts (Carrick and Balment 1983; Perrot et al. 1992; Fuentes et al. 1996b; Fuentes and Eddy 1998; Takei and Tsuchida 2000), with the terminally active angiotensin II (ANGII) acting on regions in the medulla oblongata (the so-called swallowing center) to initiate water ingestion (Takei et al. 1979, 1988; Takei and Tsuchida 2000). Reduced blood volume and reduced blood pressure, as a result of vasodilation, potently stimulate drinking (Hirano 1974; Fuentes and Eddy 1996), a response that appears to be mediated through the RAS system via elevated ANGII (Fuentes and Eddy 1996). In contrast, elevated plasma osmotic pressure acts to reduce rather than stimulate drinking rate. This response seems counterintuitive and it cannot be excluded that hypervolemia resulting from NaCl infusion might have masked the effect of osmotic pressure increases alone (Hirano 1974; Takei et al. 1988).

Bradykinin (BK), the active component of the kallikrein-kini (KK) system, and atrial natriuretic peptide (ANP) both inhibit drinking. They interact with RAS by altering plasma ANGII levels (Gardiner et al. 1993; Tsuchida and Takei 1998; Conlon 1999; Takei et al. 2001), although it is unknown at present whether the effects of BK and ANP are mediated solely through RAS or if they act directly or in some other way on the drinking reflex.

While RAS appears to respond to hypotension and hypovolemia to control drinking rate in marine and euryhaline fish, elegant studies performed on Japanese eel revealed that seawater exposure stimulates a drinking response within minutes, much faster than can be accounted for by hypovolemia or hypotension caused by dehydration (Hirano 1974; Balment et al. 2003). Furthermore, it has been demonstrated that distension of the stomach and intestine, as well as increased intestinal salt concentrations act through a gastrointestinal negative feedback system to reduce drinking (Hirano 1974; Ando and Nagashima 1996). It cannot be excluded that the gastrointestinal distension feedback system operates via the RAS system, but the immediate response to elevated ambient salinity and intestinal lumen salt concentrations precede any effects on blood volume and blood pressure that might result from hypersaline exposure. It thus appears that sensory systems capable of detecting ambient and intestinal lumen salt concentrations affect drinking reflexes in ways distinct from RAS/ANGII. Ion replacement studies demonstrated that Cl^- is the trigger for externally stimulated drinking and for the negative feedback resulting from elevated intestinal lumen salt concentrations (Hirano 1974; Ando and Nagashima 1996). The calcium sensing receptor (CaR) has been proposed to act as a salinity sensor in fish (Nearing et al. 2002). However, while there is support for a role of CaR in osmosensing of extracellular fluids in elasmobranchs and effects of CaRs agonists have been demonstrated in teleost osmoregulatory organs, CaRs have not yet been implicated in sensing of external osmotic pressure or ion concentrations, or in mediating changes in drinking rate.

The reader is referred also to Chapter 7 for discussions of endocrine control of drinking.

3. GASTROINTESTINAL PROCESSING OF INGESTED SEAWATER

Imbibed seawater undergoes substantial processing to allow for intestinal water absorption and this leaves intestinal fluids with a unique chemistry, one that is much different from the ingested seawater (Grosell et al. 2001; Marshall and Grosell 2005). Overall, most studies agree that 70–85% of the ingested volume of seawater is absorbed with the remaining volume being excreted via the rectum (Hickman 1968; Shehadeh and Gordon 1969; Wilson et al. 1996; Genz et al. 2008). However, >95% of the NaCl contained in the ingested seawater is absorbed by the intestinal tract; this NaCl absorption facilitates water absorption (Genz et al. 2008). When considering the concentrations of NaCl in seawater (~450 mM) and in anterior intestinal fluids (<100 mM in most cases) of fasted marine fish (Grosell et al. 2001; Marshall and Grosell 2005; Grosell 2006), it becomes clear that the majority of the NaCl uptake occurs in the proximal part of the gastrointestinal tract. The gastric epithelium is generally assumed not to be active with respect to ion absorption, certainly in fasted fish, but fluids in the stomach of unfed fish are much less concentrated than seawater (Smith 1930; Kirsch and Meister 1982; Parmelee and Renfro 1983; Wilson et al. 1996) illustrating the important role of the esophagus on desalinization of ingested seawater.

3.1. Esophageal Desalinization

Mass balance estimates of intestinal tract handling of seawater suggests that ~50% of the NaCl absorbed across the gastrointestinal tracts occurs by the epithelium of the relatively short esophagus and that absorption occurs in a 1:1 ratio of Na^+:Cl^- (Grosell 2006). Such high rates of NaCl absorption combined with low water permeability of the esophageal epithelium (Hirano and Mayer-Gostan 1976; Parmelee and Renfro 1983) explains the marked reduction in the osmotic pressure of the ingested seawater (~1,000 mOsm) as it travels to the stomach (~500 mOsm) (Smith 1930; Kirsch and Meister 1982; Parmelee and Renfro 1983; Wilson et al. 1996). Based on such studies of stomach fluid composition it appears that the esophagus is relatively impermeable to ions other than Na^+ and Cl^-. Both passive (downhill electrochemical gradients) and active absorption of Na^+ and Cl^- occurs in the esophagus driven at least in part by Na^+/K^+-ATPase activity, and is insensitive to furosemide but sensitive to amiloride in the few species examined (Hirano and Mayer-Gostan 1976; Parmelee and Renfro 1983). These observations rule out the involvement of $Na^+:Cl^-$ (NC) and $Na^+:K^+:Cl^-$ (NKCC) co-transporters, but imply that either Na^+/H^+ exchangers (NHEs) or sodium channels function in esophageal salt uptake. In addition to the uncertainty regarding Na^+ uptake pathways, nothing is known about the absorption of Cl^- across this epithelium. Clearly, transport properties of this important osmoregulatory organ beg further study which will be greatly facilitated with the use of modern molecular techniques.

3.2. Intestinal Processing

3.2.1. ABSORPTION

Overall the intestine of marine teleost fish absorbs water to combat diffusive water loss mainly at the gill and renal water loss via modest release of iso-osmotic urine. Water absorption is tightly linked to absorption of monovalent ions (Skadhauge 1974; Mackay and Janicki 1978; Usher et al. 1991) and the concentrations of Na^+ and Cl^- in fluids that are absorbed by the intestine generally exceed those of the plasma (Grosell 2006). Thus, the net salt gain at the gut adds to that occurring at the respiratory surface, with the cumulative salt gain being compensated for predominantly by branchial secretion (Evans et al. 2005; Evans and Claiborne 2008). Refer to Fig. 4.1 for a comprehensive marine teleost intestinal epithelium transport model.

3.2.1.1. Na^+. Entry of Na^+ from the intestinal lumen proceeds down an electrochemical gradient across the apical membrane into the intestinal epithelial cells via at least two co-transport pathways, NKCC2 (the absorptive form) and NC (Frizzell et al. 1979; Field et al. 1980; Musch et al. 1982; Halm et al. 1985a,b; Cutler and Cramb 2002). Considering the inward-directed Na^+ gradient, Na^+ uptake across the apical membrane could occur via Na^+ channels but this possibility remains to be documented. Furthermore, considering the outward-directed H^+ gradient across the

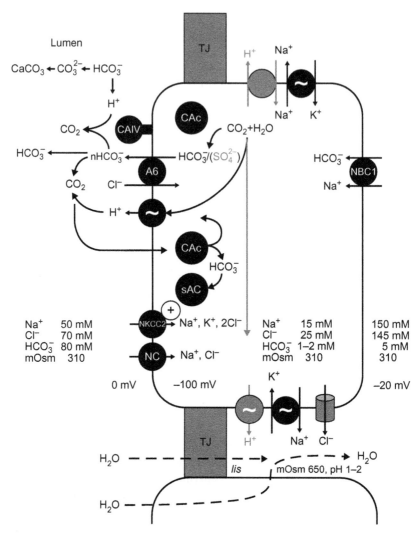

Fig. 4.1. Accepted (solid black) and putative (gray) transport processes in the intestinal epithelium of marine teleost fish (redrawn from Grosell et al. 2009b). Water transport, transcellular and/or paracellular (dotted lines), is driven by active NaCl absorption providing a hyperosmotic coupling compartment in the lateral interspace (*lis*). Apical Na^+ entry via co-transporters (NKCC2 and NC) and extrusion across the basolateral membrane via Na^+/K^+-ATPase accounts for transepithelial Na^+ movement. Entry of Cl^- across the apical membrane occurs via both co-transporters and Cl^-/HCO_3^- exchange conducted by the SLC26a6 anion exchanger while Cl^- exits the cell via basolateral anion channels. Cellular HCO_3^- for apical anion exchange is provided in part by HCO_3^- entry across the apical membrane via NBC1 and in part by hydration of endogenous CO_2. Cytosolic carbonic anhydrase (CAc) found mainly in the apical region of the enterocytes facilitates CO_2 hydration. Protons arising from CO_2

apical membrane (lumen pH is typically > 8.0) apical NHEs could possibly contribute to Na^+ entry. Indeed, studies on rainbow trout (*Oncorhynchus mykiss*) show a dramatic increase in intestinal mRNA expression of the apical NHE3 isoform following transfer from freshwater to 65% seawater (Grosell et al. 2007), suggesting a role for this transporter in seawater osmoregulation. However, attempts to demonstrate a role for NHE3 in apical H^+ extrusion in seawater-acclimated rainbow trout showed no reduction of acid following amiloride (10^{-4} M) addition. These negative results may not necessarily rule out intestinal NHE3 in seawater osmoregulation by rainbow trout because the sensitivity of this transporter to amiloride, although assumed, has yet to be demonstrated.

Transport of Na^+ across the basolateral membrane occurs via Na^+/K^+-ATPase (NKA), which is richly abundant in basal and lateral membranes of the columnar enterocytes (Grosell et al. 2007) and shows the highest NKA activity of the three teleost osmoregulatory tissues (gill, kidney and intestine) (Grosell et al. 1999; Hogstrand et al. 1999). Intestinal NKA mRNA expression and activity is in many cases higher in seawater-acclimated euryhaline fish than in freshwater-acclimated individuals, which has also been observed for NKCC intestinal mRNA levels (Jampol and Epstein 1970; Colin et al. 1985; Madsen et al. 1994; Fuentes and Eddy 1997a; Kelly et al. 1999; Seidelin et al. 2000; Grosell et al. 2007). The activity of the electrogenic NKA contributes to a cytosolic negative membrane potential of ~ -100 mV and ~ -85 mV for the apical and basolateral membrane,

---

hydration are extruded mainly across the basolateral membrane by an Na^+-dependent pathway and possibly by vacuolar H^+ pumps. Some H^+ extrusion occurs across the apical membrane via H^+-pumps and masks some of the apical HCO_3^- secretion by HCO_3^- dehydration in the intestinal lumen yielding molecular CO_2. This molecular CO_2 may diffuse back into the enterocytes for rehydration and continued apical anion exchange. Furthermore, molecular CO_2 from this reaction is rehydrated in the enterocytes and resulting HCO_3^- is sensed by soluble adenylyl cyclase (sAC) which appears to stimulate ion absorption via NKCC2 (+). Conversion of HCO_3^- to CO_2 in the intestinal lumen is facilitated by membrane-bound carbonic anhydrase, CAIV and possibly other isoforms, a process which consumes H^+ and thereby contributes to luminal alkalinization and CO_3^{2-} formation. Titration of luminal HCO_3^- and formation of CO_3^{2-} facilitates formation of $CaCO_3$ precipitates to reduce luminal osmotic pressure and thus water absorption. SLC26a6, the electrogenic anion exchanger, exports $nHCO_3^-$ in exchange for $1Cl^-$ and its activity is therefore stimulated by the hyperpolarizing effect of the H^+-pump. The apical electrogenic $nHCO_3^-/Cl^-$ exchanger (SLC26a6) and electrogenic H^+-pump constitutes a transport metabolon perhaps accounting for the apparently active secretion of HCO_3^- and the uphill movement of Cl^- across the apical membrane. Note that SLC26a6 may also operate to secrete SO_4^{2-} across the apical membrane. The indicated values for osmotic pressure and pH in the absorbed fluids are based on measured net movements of H_2O and electrolytes including H^+s but the degree of hypertonicity and acidity in *lis* likely are much less than indicated due to rapid equilibration with sub-epithelial fluid compartments. See text for further details.

respectively (Loretz 1995), and a low cytosolic Na^+ concentration of about 15 mM (Zuidema et al. 1986). In addition to the favorable electrochemical gradient for Na^+ entry across the apical membrane (luminal Na^+ is typically 50–100 mM), NKA also drives Na^+-coupled transport processes across the basolateral membrane. Evidence for the presence of basolateral H^+ extrusion via an Na^+-dependent pathway comes from observations of reduced apical HCO_3^- secretion (see below) following serosal application of ouabain and removal of serosal Na^+ (Grosell and Genz 2006). In addition to a possible NHE in the basolateral membrane, recent studies demonstrate the presence of a basolateral Na^+:HCO_3^- co-transporter (NBC1) allowing for basolateral uptake of HCO_3^- and Na^+, driven by NKA. Despite the presence of these basolateral Na^+ importers, the overall net transport for the marine teleost intestinal epithelium is in the direction of Na^+ absorption.

3.2.1.2. K^+. As for Na^+, the net transport for K^+ is in the direction of absorption. In fact, considering that ingested seawater contains 10 mM K^+, that intestinal fluids typically contain <5 mM (Grosell et al. 2001) and that most of the ingested water is absorbed, it is clear that the majority of ingested K^+ is absorbed by the intestinal tract. Undoubtedly, NKCC2 contributes to K^+ absorption across the apical membrane. Cytosolic K^+ (87 mM) (Duffey 1979) is far above electrochemical equilibrium in the epithelial cells and has been documented to exit across the basolateral membrane via K^+:Cl^- transport (Smith et al. 1980). In addition, an apical Ba^{3+}-sensitive K^+ channel allows for K^+ secretion (Musch et al. 1982). It is possible that the apical K^+ channel serves K^+ recycling across the apical membrane to facilitate NaCl absorption via NKCC2. Owing to the difference between concentrations of Na^+, Cl^- and K^+ in seawater, as well as in intestinal fluids, K^+ would clearly be limiting for intestinal salt absorption via NKCC2 in the absence of apical K^+ channels.

3.2.1.3. Cl^-. The net absorption of Cl^-, like Na^+, is against a transepithelial concentration gradient. The transepithelial potential (TEP) of the marine fish intestine under *"in vivo"*-like conditions is about −15 to −20 mV (blood side negative) and both Na^+ and Cl^- concentrations are typically <100 mM in intestinal fluids of unfed fish (Wilson et al. 2002; Wilson and Grosell 2003; Grosell and Genz 2006). However, in contrast to Na^+, which enters the enterocytes across the apical membrane down an electrochemical gradient, Cl^- uptake across the apical membrane is against an electrochemical gradient. Cytosolic Cl^- concentrations, however, are above electrochemical equilibrium and Cl^- exits enterocytes across the basolateral membrane via K^+:Cl^- co-transport (see K^+ section above) and

Cl^- channels, including a CFTR paralogue (Loretz and Fourtner 1988; Marshall et al. 2002).

The electrochemical gradient against Cl^- entry across the apical membrane is overcome by Na^+-coupled co-transport via NC and NKCC2 (see Na^+ section above), which contributes significantly to Cl^- uptake. However, net Cl^- absorption is far greater than net Na^+ absorption across the marine teleost intestinal epithelium as first demonstrated for the southern flounder (Hickman 1968) and since confirmed for all other examined species (Shehadeh and Gordon 1969; Pickering and Morris 1973; Skadhauge 1974; Grosell et al. 1999, 2001, 2005; Marshall et al. 2002). When considering the sum of net Na^+ and K^+ uptake it becomes clear that even NKCC2 cannot account for the excess Cl^- (compared to Na^+) uptake. Indeed, Cl^- uptake persists in the absence of Na^+ uptake, although at reduced rates, demonstrating Na^+-independent Cl^- uptake pathway(s) (Grosell et al. 1999, 2001).

3.2.1.3.1. Anion Exchange. Strong correlations between net Cl^- uptake and apparent HCO_3^- secretion across the intestinal epithelium of seawater-acclimated rainbow trout provided the first evidence for Cl^-/HCO_3^- exchange being a significant contributor to intestinal Cl^- uptake (Shehadeh and Gordon 1969; Wilson et al. 1996). Later observations of continued Cl^- uptake either in the absence of Na^+ in luminal fluids, or when Na^+ uptake was abolished by exposure of the intestinal epithelium to silver, provided further support for Cl^- uptake via anion exchange (Grosell et al. 1999, 2001). The Cl^- uptake observed despite inhibition of Na^+ uptake was concurrent with substantial intestinal HCO_3^- secretion rates and the magnitude of Cl^- uptake was matched closely by HCO_3^- secretion rates. Furthermore, under control conditions in the absence of silver, Cl^- uptake in excess of Na^+ uptake was again matched closely by rates of HCO_3^- secretion (Grosell et al. 1999).

Recent advances have identified a member of the SLC26 gene family, namely SLC26a6, as a highly likely candidate for the anion exchange protein involved in intestinal Cl^- absorption by marine teleosts (Kurita et al. 2008; Grosell et al. 2009b). Studies on a euryhaline pufferfish, mefugu (*Takifugu obscures*), utilizing the genomic information available for its close relative, the tiger puffer (*Tagifugu rubripes*), identified an SLC26a6 isoform as a candidate. Of a high number of genes coding for HCO_3^- transporting proteins, the SLC26a6 isoform, as well as an NBC1 (a SLC4 member) coding gene, showed robustly elevated mRNA expression following transfer from freshwater to seawater (Kurita et al. 2008). A possibly unique feature of SLC26a6 is that it appears to be electrogenic ($nHCO_3^-/Cl^-$ exchange) although this is the subject of controversy and likely differs among species

(Chernova et al. 2005; Clark et al. 2008). Such a stoichiometry of net export of negative charge means that anion exchange by SLC26a6 is facilitated by the apical membrane potential and possibly can account for movement of Cl^- and HCO_3^- against electrochemical gradients. This rationale formed the basis for a parallel attempt, using targeted cloning of SLC26a6 from the gulf toadfish, to identify the anion exchange protein involved in marine teleost intestinal Cl^- uptake (Grosell et al. 2009b). Expression of SLC26a6 in intestinal tissue is high in both mefugu and the gulf toadfish. In the latter case SLC26a6 expression is highest in the anterior and mid-region of the intestine, which are responsible for the majority of anion exchange and show elevated transcription following salinity challenge. Furthermore, SLC26a6 from both species exhibits electrogenic Cl^- uptake in accord with the suggested stoichiometry (Kurita et al. 2008; Grosell et al. 2009b). Taken together, the above observations from a euryhaline and a stenohaline marine fish indicate that SLC26a6 may be a widespread anion exchange protein involved in marine fish osmoregulation, although other anion exchange proteins are present and may contribute to Cl^- uptake (Grosell unpublished observations).

A link between Cl^- uptake and H^+ extrusion for freshwater fish gills and amphibian skin is well established and is assumed to be associated with titration of boundary layer HCO_3^- and accumulation of cellular HCO_3^- as a result of apical H^+ extrusion (Larsen et al. 1992; Fenwick et al. 1999; Jensen et al. 2002, 2003; Boisen et al. 2003). Recent observations of apical H^+ pump (V-type) localization, elevated proton pump mRNA expression, and bafilomycin-sensitive acid-extrusion demonstrate the activity of an apical proton pump in euryhaline and marine fish (see H^+ section below) and suggest a role in osmoregulation (Grosell et al. 2007, 2009a,b). The combination of apical H^+-pump activity and an electrogenic SLC26a6 may constitute a transport metabolon in which anion exchange is facilitated not only by the above-mentioned impacts of H^+ extrusion on HCO_3^- gradients but also facilitated by the hyperpolarizing effect of H^+ pump activity (Grosell et al. 2009b). Activity of the H^+-pump hyperpolarizes the apical membrane which in turn provides an additional driving force for apical $nHCO_3^-/Cl^-$ exchange via SLC26a6. While the coupling of H^+-pump activity and electrogenic SLC26a6 provides an explanation for anion exchange and thereby Cl^- uptake across the marine teleost intestinal epithelium, it may also account for Cl^- uptake by freshwater organisms against unfavorable electrochemical gradients. Indeed, SLC26a6 has been identified as one of the anion exchange proteins involved in Cl^- uptake by zebrafish (Bayaa et al. 2009; Perry et al. 2009), although the electrogenicity of the zebrafish SLC26a6 remains to be determined.

3.2.1.4. Water. The salt absorption described above provides the driving force for osmotic water movement across the intestinal epithelium despite the absence of net osmotic gradients. A number of studies have shown little if any osmotic gradient across the intestinal epithelium, but it appears that the lateral interspace (*lis*) between enterocytes may act as a coupling compartment for salt and water absorption (Fig. 4.1). The tonicity of fluids absorbed by the marine teleost intestinal epithelium is considerably higher than the osmotic pressure in the intestinal lumen or the extracellular fluids, which can account for water absorption even against a slight osmotic gradient (Grosell 2006). Salt transport across the basolateral membrane renders fluids in the lateral interspace hyperosmotic, which will drive water absorption (Fig. 4.1). Water absorption in the direction from the lumen to the blood side of the epithelium is likely ensured by barrier properties of the tight junctions resulting in dispersal of salt, and thus water, from *lis* to the blood side of the epithelium.

3.2.1.4.1. Paracellular versus Transcellular Water Transport—A Role for Aquaporins? At present it is unknown if water enters *lis* by moving from the lumen across the tight junctions or via a transcellular path. Of the 13 known aquaporins (AQPs) a putative role in water transport by the marine teleost intestine has been examined for AQPe, AQP1 and AQP3 (Lignot et al. 2002; Aoki et al. 2003; Martinez et al. 2005a; Cutler et al. 2007; Raldua et al. 2008). These aquaporins are all expressed in the intestinal tract and at least AQP1 may play a role in water absorption. Evidence for a direct role of AQP1 in transcellular intestinal water absorption is provided by (i) elevated mRNA expression in intestinal tissue from euryhaline fish following seawater transfer, (ii) elevated AQP1 protein concentrations in intestinal tissue from seawater-acclimated fish and (iii) apical and lateral localization of the AQP1 protein (Aoki et al. 2003; Martinez et al. 2005a,c; Raldua et al. 2008). The fish AQP1 confers water permeability in expression systems, but direct support for intestinal water absorption by AQP1 is lacking due in part to the absence of suitable pharmacological tools for aquaporins in general.

While the observations for AQP1 are highly supportive of a direct role in water absorption, a similar response (elevated expression following seawater transfer) has been reported in the esophagus (Martinez et al. 2005b), a tissue known for its lack of water permeability in seawater-acclimated fish. Thus, mRNA expression and protein abundance of aquaporins are not necessarily associated with direct transcellular water movements. Thus, it is possible that the transcriptional responses and the elevated protein abundance may be related to sensing of osmotic pressure or cellular volume as in other organs (Watanabe et al. 2009), rather than transepithelial water movement. Clearly further studies are required to address this issue.

3.2.2. SECRETION AND EXCLUSION

Despite substantial absorption of Na^+, Cl^-, K^+ and water, the intestinal epithelium of marine teleost fish exhibits several secretory functions, and on a net basis is highly efficient in excluding divalent ions even in the presence of considerable lumen-blood gradients. Several of these secretory processes are directly related to osmoregulation by facilitating water absorption and exclusion of divalent ions. These processes minimize the demand for renal excretion and thus urinary fluid loss, thereby contributing to osmoregulation by marine teleost fish.

3.2.2.1. HCO_3^-. High total CO_2 concentrations and alkaline conditions (evident from the phenol red reaction) in intestinal fluids were first reported 80 years ago for three different starved, as well as fed, marine teleost species (Smith 1930). These early observations excluded a strict role of intestinal HCO_3^- secretion in digestion. Later, intestinal fluid pH was reported to be higher in seawater-acclimated fish than in freshwater-acclimated conspecifics, suggesting a role of luminal alkalinization in osmoregulation (Cordier and Maurice 1956; Shehadeh and Gordon 1969). This suggested role was further supported by observations of reduced amounts of precipitated CO_3^{2-} complexes (see Ca^{2+} section) in toadfish held in 25% seawater compared to seawater (Walsh et al. 1991), by highly alkaline rectal fluids in coho salmon smolts but circum-neutral in post-smolt individuals 24 hours post-seawater transfer (Kerstetter and White 1994), and finally by comparing intestinal fluid HCO_3^- concentrations between freshwater- and seawater-acclimated fish (Wilson 1999; Grosell and Taylor 2007). More recent studies (Genz et al. 2008) found that total intestinal base secretion as measured by rectal excretions was greatly reduced in gulf toadfish acclimated to 9 ppt (iso-osmotic), while greatly elevated in fish acclimated to 50 ppt when compared to control fish in seawater (35 ppt).

The link between Cl^- absorption and HCO_3^- secretion (see Cl^- section above) was first suggested four decades ago based on correlations between Cl^- absorption and luminal alkalinization and has since been extensively documented through ion replacement studies (Dixon and Loretz 1986; Ando and Subramanyam 1990; Wilson et al. 1996; Grosell et al. 2001, 2005), DIDS sensitivity (Dixon and Loretz 1986; Ando and Subramanyam 1990; Grosell and Jensen 1999; Grosell et al. 2001, 2009a), and most recently by functional characterization of SLC26a6 gene products (see Cl^- section above) in euryhaline as well as stenohaline marine fish.

Intestinal HCO_3^- secretion by marine fish is temperature dependent (Grosell and Genz 2006) but is generally high, in the order of 0.5 $\mu mol/cm^2/h$ under resting conditions and in the presence of physiological levels of serosal

HCO_3^- (Dixon and Loretz 1986; Ando and Subramanyam 1990; Grosell and Jensen 1999; Furukawa et al. 2000; Grosell et al. 2001, 2005, 2009a,b; Wilson et al. 2002; Grosell and Genz 2006; Taylor and Grosell 2009; Taylor et al. 2010) and are comparable to rates observed for resting duodenal HCO_3^- secretion in mammals at 37°C (Hogan et al. 1997; Clarke and Harline 1998; Tuo and Isenberg 2003; Tuo et al. 2004). The high HCO_3^- secretion rates result in concentrations in intestinal fluids of marine fish reaching levels of up to 100 mM (Wilson et al. 2002; Grosell 2006; Grosell et al. 2009b). The observations of extremely high HCO_3^- secretion rates led to multiple investigations of the HCO_3^- source for secretion and the source of energy for thermodynamic uphill transport of HCO_3^-.

3.2.2.1.1. Source and Energy for Intestinal HCO_3^- Secretion. Both transepithelial HCO_3^- transport from the blood to the intestinal lumen and hydration of endogenous epithelial CO_2 provide cellular substrate for apical Cl^-/HCO_3^- exchange although the relative contribution of these two sources differs among species. In most species examined to date, both HCO_3^- sources contribute significantly to overall intestinal HCO_3^- secretion (Dixon and Loretz 1986; Ando and Subramanyam 1990; Wilson and Grosell 2003; Grosell et al. 2005; Grosell and Genz 2006). However, seawater-acclimated killifish, *Fundulus heteroclitus*, rely mainly on transepithelial HCO_3^- movement (Genz and Grosell unpublished), while secretion by seawater-acclimated rainbow trout is sustained by hydration of endogenous CO_2 alone (Grosell et al. 2009a).

At least for mefugu and the gulf toadfish, HCO_3^- entry from the extracellular fluids to the intestinal epithelium is by a basolateral, electrogenic NBCe1, which is fueled by the electrochemical Na^+-gradient (Kurita et al. 2008; Taylor et al. 2010). For both species, NBC1 mRNA expression is increased following salinity transfer illustrating an important and possibly rate-limiting step in intestinal HCO_3^- secretion by this transporter. The gulf toadfish NBCe1 exhibits an affinity constant of ~10 mM HCO_3^- (Taylor et al. 2010), which means that HCO_3^- transport by the protein *in vivo* will scale with serosal HCO_3^- concentrations and possibly be important for systemic acid–base balance regulation (see Chapter 6).

Hydration of endogenous CO_2 to provide cytosolic substrate for apical anion exchange is clearly facilitated by the enzyme carbonic anhydrase (CA), as illustrated by inhibition of HCO_3^- secretion during application of permeant pharmacological agents (Dixon and Loretz 1986; Wilson et al. 1996; Grosell et al. 2005, 2009a). In addition, elevated mRNA expression of the cytosolic CAc isoform in rainbow trout following acclimation to 65% SW and elevated CAc mRNA expression in the gulf toadfish following exposure to 60 ppt, as well as elevated CA activity in the cytosolic fraction of the intestinal epithelium in both species, points to a role of this CA isoform in HCO_3^- secretion and thus osmoregulation (Grosell et al. 2007; Sattin et al. 2010). The CAc isoform localizes to the apical region of the enterocytes without being directly associated with membrane fractions of the epithelial cells (Grosell et al. 2007). The mechanism by which CAc is concentrated in the apical region is unknown but it cannot be dismissed that CAc interacts directly with SLC26a6 or other membrane-associated proteins. However, it is clear that the localization of CAc in the apical region may confer high HCO_3^- secretion rates by facilitating CO_2 hydration in the immediate vicinity of the SLC26a6 export protein, which in turn would facilitate the continued hydration of CO_2.

High HCO_3^- secretion rates, even in the absence of serosal HCO_3^- when endogenous CO_2 is the sole source of HCO_3^-, indicate that the intestinal epithelium of marine teleosts is a metabolically active tissue. In isolated toadfish intestinal tissue under *in vivo*-like conditions, oxygen consumption rate is 9 µmol/g intestinal tissue/h (three-fold higher than whole toadfish values (Gilmour et al. 1998)), which is similar to HCO_3^- secretion rates in the absence of HCO_3^- in serosal salines (Taylor and Grosell 2009). These observations not only confirm that the intestinal epithelium is metabolically active and thus produces sufficient amounts of CO_2 to sustain high HCO_3^- secretion rates, but also demonstrates that the majority of the metabolic CO_2 produced is hydrated and secreted as HCO_3^-.

The high CO_2 hydration rate by the intestinal tissue generates H^+s at a high rate, which must be secreted from the cell to maintain cellular acid–base balance and to allow for continued CO_2 hydration. The bulk of H^+ secretion occurs across the basolateral membrane such that the tissue exhibits overall high rates of base secretion across the apical membrane into the lumen. Indeed, net acid extrusion across the basolateral membrane has been measured directly by two different techniques (Grosell and Genz 2006; Grosell and Taylor 2007). Intestinal HCO_3^- secretion by the gulf toadfish is dependent on serosal Na^+, even in the absence of serosal HCO_3^-, and is sensitive to ouabain, strongly suggesting that H^+ extrusion across the basolateral membrane occurs via Na^+/H^+ exchange (NHE) and is important for continued CO_2 hydration by the enterocytes (Grosell and Genz 2006).

The link between HCO_3^- secretion and electrochemical Na^+ gradients via NBC1 and NHE, and the inhibition of HCO_3^- secretion by application of ouabain, illustrates that part of the energy required for apparently active HCO_3^- secretion is provided by basolateral NKA. In addition, observations of basolateral localization of H^+-pumps (Grosell et al. 2007) may suggest that H^+ extrusion across the basolateral membrane may occur via this ATPase, although functional evidence for such a role for the H^+-pump remains to be observed. Furthermore, apical H^+-pump activity promotes apical anion exchange as discussed below, and therefore contributes energy to the active secretion of HCO_3^- and Cl^- uptake (Grosell et al. 2009a).

3.2.2.2. H^+. Considering the high rates of CO_2 hydration, H^+-pump expression and activity in the intestinal tissue was hypothesized and demonstrated for rainbow trout acclimated to 65% seawater. Predictions of strictly basolateral localization in this base-secreting tissue was, however, not supported as substantial apical, as well as basolateral staining was observed in immunohistochemical studies (Grosell et al. 2007). These observations led to the demonstration of apical H^+ extrusion in seawater-acclimated rainbow trout and the gulf toadfish by exposing the apical membrane to the H^+-pump inhibitor, bafilomycin (Grosell et al. 2009a,b). In both cases, proton pump inhibition resulted in an increase in apparent base secretion of 10–25% demonstrating that simultaneous apical HCO_3^- and H^+-secretion combine to a net base secretion, which is lower than HCO_3^- secretion via anion exchange. The 10–25% reduction of net base secretion owing to H^+-pump activity is likely an underestimate of the H^+-pump contribution since inhibition of the H^+-pump is likely to reduce cellular HCO_3^- formation and HCO_3^- secretion (Grosell et al. 2009a,b).

In addition to the functional importance of H^+ extrusion for continued HCO_3^- formation from hydration of CO_2, apical H^+-pump activity may aid apical anion exchange by reducing extracellular HCO_3^- concentrations near the apical membrane and by membrane hyperpolarization. Hyperpolarization of the apical membrane will facilitate anion exchange via the electrogenic SLC26a6 (see Cl^- section). In addition, the titration of HCO_3^- by H^+ secretion across the apical membrane likely results in high partial pressure of CO_2 near the apical surface which may diffuse across the apical membrane and allow for further cellular CO_2 hydration. Such a recycling of CO_2 would allow for continued Cl^- absorption without excessive accumulation of the HCO_3^- osmolyte in the intestinal lumen and would thus favor further water absorption. Titration of luminal HCO_3^- is possibly facilitated by extracellular CA activity. Recent studies have revealed that apical membrane-bound CA-IV (extracellular) is present in rainbow trout intestinal epithelium and that transcription, as well as membrane-bound CA activity, is increased during acclimation to 65% seawater (Grosell et al. 2007, 2009a).

A final functional significance of apical H^+ extrusion is that it will act to reduce luminal osmotic pressure by removal of HCO_3^-. Plasma and intestinal fluid osmotic pressures are tightly correlated and any reduction in luminal osmotic pressure will therefore manifest as a beneficial reduction on plasma osmolality. Titration of luminal HCO_3^- by apical H^+-secretion may seem counter to luminal $CaCO_3$ formation (see below) but since luminal CO_3^{2-} concentrations *in vivo* exceed Ca^{2+} concentrations, total CO_2, HCO_3^- or CO_3^{2-} concentrations are not limiting for $CaCO_3$ formation (Grosell et al. 2009b).

3.2.2.3. Ca^{2+}. Intestinal fluid Ca^{2+} concentrations are typically < 5 mM in marine teleost fish as compared to 10 mM in seawater (Marshall and Grosell 2005). This drop in Ca^{2+} concentrations occurs despite water absorption and is the result of both intestinal Ca^{2+} uptake and formation of $CaCO_3$ in the intestinal lumen. The high luminal HCO_3^- concentrations and high pH in intestinal fluids are conducive for $CaCO_3$ formation and precipitates are visible in the intestinal lumen of unfed marine fish. These $CaCO_3$ precipitates are excreted with the rectal fluids and have recently been shown to contribute significantly to oceanic $CaCO_3$ production (Wilson et al. 2009). Reports on intestinal Ca^{2+} uptake in marine fish differ significantly with respect to the observed fraction of ingested Ca^{2+} being absorbed. Original work on the southern flounder was interpreted to show that 68.5 % of ingested Ca^{2+} was absorbed but did not take into account the amount of Ca^{2+} lost as $CaCO_3$ precipitates via rectal fluids (Hickman 1968). More recently, experiments on the European flounder revealed a much lower fractional Ca^{2+} absorption of only 20% taking into account the amount of Ca^{2+} lost as precipitated $CaCO_3$ (Wilson and Grosell 2003). However, the later study was an intestinal perfusion experiment with perfusion rates exceeding normal drinking rates and the Ca^{2+} "intake" therefore was super-physiological. The very low fractional Ca^{2+} uptake in these experiments could be an underestimate of physiological fractional uptake due to artificially high Ca^{2+} perfusion. Most recently, in experiments on toadfish allowed to ingest seawater naturally, where Ca^{2+} lost as $CaCO_3$ was accounted for, revealed an approximately 80% fractional Ca^{2+} uptake (Genz et al. 2008). Thus, while the study by Hickman likely overestimates fractional Ca^{2+} absorption because $CaCO_3$ precipitates were not considered, and the study by Wilson and Grosell potentially may underestimate fractional absorption due to high perfusion rates, it seems clear that differences likely exist among species with respect to intestinal Ca^{2+} handling.

Regardless of these species differences $CaCO_3$ formation in the intestinal lumen is a consistent phenomenon among marine teleosts, and precipitates accounting for a significant fraction of Ca^{2+} in intestinal and rectal fluids (~25%; Wilson and Grosell 2003; Genz et al. 2008). Furthermore, the magnitude of this precipitate formation is sufficient to reduce osmotic pressure by as much as 70 mOsm as Ca^{2+} and HCO_3^- are removed from solution. A reduction in osmotic pressure of this magnitude is significant for water absorption and thus osmoregulation in itself (Wilson et al. 2002). Interestingly, the activity of an extracellular, membrane-bound carbonic anhydrase isoform (presumably CA-IV) likely facilitates $CaCO_3$ formation in the intestinal lumen (Grosell et al. 2009b).

3.2.2.4. Mg^{2+}. Luminal concentrations of Mg^{2+} in the intestine of marine fish are typically 120–150 mM due to selective absorption of Na^+, Cl^- and water (Grosell et al. 2001; Marshall and Grosell 2005; Grosell 2006). As plasma Mg^{2+} concentrations are < 1 mM, the gradient, from the intestinal lumen (~200-fold) and from the water (~ > 50-fold) to the extracellular fluids is greater for Mg^{2+} than for any other ion. Despite this, very little is known about Mg^{2+} handling by osmoregulatory tissues of marine teleosts. The majority (~90%) of Mg^{2+} ingested with seawater passes through the intestinal track unabsorbed (Hickman 1968; Wilson and Grosell 2003; Genz et al. 2008). However, some Mg^{2+} absorption is unavoidable owing to the substantial gradient across the intestinal epithelium. Limited Mg^{2+} uptake at the gill may also occur. The resulting Mg^{2+} excess is cleared by renal excretion as evident from high urine Mg^{2+} concentrations (McDonald 2006; McDonald and Grosell 2006), but as for possible intestinal Mg^{2+} transport, virtually nothing is known about the cellular mechanism of Mg^{2+} transport by renal tubules in marine teleosts.

3.2.2.5. SO_4^{2-}. Intestinal net SO_4^{2-} absorption is also highly limited and intestinal fluid SO_4^{2-} concentrations are therefore typically 100–120 mM (Marshall and Grosell 2005; McDonald and Grosell 2006). Uptake, although limited when considering the gradient across the intestinal epithelium, is unavoidable and excess Mg^{2+} and SO_4^{2-} are eliminated by urinary excretion (McDonald 2006; McDonald and Grosell 2006). It seems that the limited intestinal SO_4^{2-} uptake is not simply a function of passive barrier functions. Elegant studies have shown that the intestinal epithelium of the winter flounder exhibits SO_4^{2-} secretion via anion exchange for Cl^- uptake to reduce overall net SO_4^{2-} uptake (Pelis and Renfro 2003). The nature of the anion exchanger responsible for SO_4^{2-} secretion by the intestinal epithelium of the winter flounder remains unknown; but recent advances in our understanding of renal SO_4^{2-} handling may shed some light on this issue. The SLC26a6 isoform was recently implicated in renal SO_4^{2-} secretion in a euryhaline pufferfish (Kato et al. 2009). The SLC26a6 isoform is capable of SO_4^{2-} secretion under *in vivo*-like conditions and operates to secrete SO_4^{2-} into renal tubules against considerable gradients in exchange for Cl^-. In marine fish, renal tubule and intestinal lumen fluid chemistry is similar with respect to SO_4^{2-} and Cl^- concentrations (McDonald and Grosell 2006), so it is likely that the SLC26a6 isoform which is also expressed in the intestine is involved in SO_4^{2-} as well as HCO_3^- secretion.

3.3. Integrative Aspects

A high number of individual transport processes involved in osmo-regulation are observed and discussed above, but these processes operate in an integrative and coordinate manner to ensure salt and water balance, as well as acid–base balance (see Chapter 6). Furthermore, many of these processes are influenced by other physiological processes including feeding (see Chapter 5).

3.3.1. THE GASTROINTESTINAL TRACT

Absorption of Cl^- by the intestinal epithelium occurs by multiple pathways including two co-transporters and at least one anion exchange isoform. Although direct evidence is lacking, it seems likely that Na^+ uptake, which occurs via co-transporters, likely also occurs via NHEs and possibly via Na^+-channels. The reasons for such apparent redundancy in Cl^- and Na^+ uptake pathways are unknown, but may reflect a need for NaCl absorption under a wide range of conditions, and thus subject to thermodynamical constraints imposed by the multiple functions of the intestinal tissue.

At least for Cl^- absorption, some insight is emerging and illustrating the apparent coordination between co-transporter activity and anion exchange. First, it appears that Cl^- absorption via anion exchange is minimal when luminal Na^+ and Cl^- concentrations are high and thus favors absorption by co-transport pathways (Grosell and Taylor 2007). When luminal Na^+ and Cl^- concentrations are reduced and Mg^{2+} and SO_4^{2-} concentrations are increased to in vivo-like concentrations, Cl^- absorption via anion exchange is activated. While activation of anion exchange under conditions of low luminal Cl^- allow for continued Cl^- absorption via the electrogenic SLC26a6 exchanger operating alongside an apical H^+-pump, it is unknown at present how the intestinal epithelium senses the change in luminal chemistry to activate Cl^-/HCO_3^- exchange via SLC26a6. More recently, the opposite interaction between anion exchange and co-transporters was demonstrated. Elevated luminal HCO_3^-, which impairs further Cl^- uptake via anion exchange, potently stimulates Cl^- absorption via co-transport pathways (Tresguerres et al. 2010). The activation of co-transport by luminal HCO_3^- is mediated by soluble adenylyl cyclase (sAC), as illustrated by application of 4-catechol estrogen (CE, a specific sAC inhibitor). Furthermore, the link between luminal HCO_3^- and activation of NKCC2 in the apical membrane seems to involve CA and the apical H^+-pump since application of etoxzolamide (CA inhibitor) and bafilomycin mutes the response of NKCC2 to HCO_3^-. Based on these observations, a model involving apical H^+ extrusion leading to elevated PCO_2 near the apical surface, diffusion of CO_2 into the cell and finally rehydration of CO_2 to form HCO_3^- was proposed. sAC is effectively an HCO_3^- sensor (Chen et al. 2000) and appears to stimulate NKCC2 in response to elevated cytosolic HCO_3^- via a cAMP-dependent mechanism yet to be determined (Tresguerres et al. 2010).

Secretion of HCO_3^- is necessary for Cl^- absorption via SLC26a6 under low luminal Cl^- conditions and results in accumulation of luminal HCO_3^-. While HCO_3^- secretion promotes the formation of $CaCO_3$, and thus reduced luminal osmotic pressure, HCO_3^- accumulates to very high concentrations in the intestinal lumen. Thus while Cl^- absorption via anion exchange and formation of $CaCO_3$ acts to promote water absorption, accumulation of HCO_3^- in the intestinal fluids exerts osmotic pressure and may therefore ultimately impair water absorption. Recent findings of apical H^+ extrusion suggest that the potential problem of excessive HCO_3^- accumulation may be avoided by titration of luminal HCO_3^-, which ultimately allows for the recycling of molecular CO_2, reduced luminal osmotic pressure and continued water absorption (Grosell et al. 2009b). The need on the one hand to maintain luminal HCO_3^- levels sufficient for $CaCO_3$ formation and on the other hand to avoid elevated osmotic pressure from excessive accumulation likely means that HCO_3^- secretion directly or HCO_3^- accumulation indirectly via altered H^+ extrusion is under regulatory control. One might predict that high luminal Ca^{2+} combined with high osmotic pressure would promote HCO_3^- secretion to facilitate $CaCO_3$ formation, while low luminal Ca^{2+} combined with high osmotic pressure might result in reduced HCO_3^- secretion or enhanced apical H^+ extrusion. Indeed, net HCO_3^- secretion at least in the European flounder is stimulated by elevated luminal Ca^{2+}. However, it remains to be investigated how net HCO_3^- secretion is controlled and to what extent apical H^+ extrusion rather than HCO_3^- secretion is the point of regulation.

3.3.2. THE INTACT FISH

It has long been recognized that the NaCl gained across the intestinal epithelium to promote water absorption is extruded across the gills. However, regardless of the magnitude of apical H^+ extrusion, the intestinal epithelium exhibits net base secretion or net acid uptake. Whether intestinal HCO_3^- secretion occurs as transepithelial HCO_3^- movement or results from hydration of endogenous CO_2 with most of the resulting H^+s extruded across the basolateral membrane, the net outcome of acid gain is the same. Fluid absorption across the intestinal epithelium therefore results in NaCl gain as well as H^+ gain. The rate of H^+ gain amounts to a theoretical pH in the absorbed fluids of <1.5 (Grosell 2006; Grosell and Taylor 2007). It should be noted, however, that such low pH does not occur anywhere in the intestinal epithelium of associated extracellular fluid compartments due to rapid equilibration with larger fluid compartments and buffer capacity of these compartments. Nevertheless, a substantial acid gain is associated with intestinal water absorption and is matched by a compensatory branchial acid extrusion (Genz et al. 2008). The often observed transient acidosis in euryhaline fish transferred from freshwater to seawater (Wilkes and Mcmahon 1986; Nonnotte and Truchot 1990; Maxime et al. 1991) likely reflects the onset of intestinal base secretion and a delayed compensatory brachial acid extrusion response.

While the intestinal gain of Na^+, Cl^- and H^+ associated with water absorption is eliminated at the gill it is generally assumed that divalent cations and SO_4^{2-} are extruded mainly by the kidney (Marshall and Grosell 2005). A complete overview of mass balance for salt, acid–base and water exchange between marine fish and their environment is challenging to compile because three different organs are intimately involved. To date, simultaneous measurements of all relevant branchial, intestinal and renal contributions have not been performed. An early and impressive account of intestinal and renal contributions for the winter flounder did not consider transfer of acid–base equivalents (Hickman 1968), while later examinations have measured these parameters for intestinal and branchial fluxes (Wilson and Grosell 2003; Genz et al. 2008), but not the renal contribution. Even though simultaneous measurements of all branchial, renal and intestinal transport processes including drinking rates have not been performed on the same individuals, most of these parameters have been measured in gulf toadfish. Measurements of a full range of osmoregulatory relevant fluxes include branchial Na^+ and Cl^- extrusion rates (Kormanik and Evans 1982; Grosell et al. 2010), rectal and branchial acid–base flux measurements, drinking rates, rectal fluid, ionic and $CaCO_3$ output (Genz et al. 2008), as well as urinary excretions and bladder urine ionic composition (McDonald and Grosell 2006; Genz and Grosell 2010). A summary of these observations is presented in Fig. 4.2.

Overall the mass balance emerging from the compiled measurement from toadfish agrees well with a recent estimate based on average values from a range of marine species (Grosell 2007), and also with an early description of mass balance for the winter flounder (Hickman 1968). Full agreement among the three mass balance estimates apply to Na^+, Cl^- and K^+, for which $>95\%$ of ingested ions are absorbed along the gastrointestinal tract. Of these monovalent ions, a few percent are lost with renal excretions, while the rest is extruded across the gills. Similar agreement exists for handling of SO_4^{2-} and Mg^{2+}, which are largely unabsorbed by the intestine and show little if any branchial uptake or excretion and are excreted renally. For Ca^{2+}, a discrepancy exists between on the one hand the mass balance presented for

	Ingestion	Branchial			Rectal	Renal
		Efflux	Influx	Net flux		
Volume	2.6 ml/kg/h			−1.8 ml/kg/h	−0.7 ml/kg/h	−0.1 ml/kg/h
Na^+	1092	−7364	6296	−1068	−19	−5
Cl^-	1274	−21597	20394	−1203	−47	−24
Mg^{2+}	130			5	−120	−15
SO_4^{2-}	78			−9	−63	−6
Ca^{2+}	26			−21.5	−4	−0.5
K^+	26			−24.5	−1	−0.5
HCO_3^- eqv	6			230	−68	−0.3

Fig. 4.2. Whole animal balance of water, acid–base equivalents and main electrolytes (μmol/kg/h) in the gulf toadfish acclimated to natural seawater. Positive values indicate gain by the fish while negative values indicate loss and excretion. Values in **bold** represent measured values, values in *italics* represent estimated values (based on assumed Ca^{2+} and K^+ concentrations in urine of 5 mM and urine HCO_3^- concentrations of 3 mM). With exception of branchial net flux of HCO_3^- equivalents (in red) all branchial flux values are calculated from ingestion rates, and rectal and renal efflux rates assuming steady state with respect to salt and water balance. Branchial net flux of HCO_3^- equivalents (uptake) are the sum of measured titratable non-intestinal acid excretion and ammonia excretion, with ammonia efflux contributing approximately 25% to the overall net acid excretion (Genz et al. 2008). Branchial unidirectional efflux rates for Na^+ and Cl^- were measured recently in toadfish in seawater (Grosell et al. 2010) and are in agreement with earlier reports from this species and other marine teleosts (Kormanik and Evans 1982; Grosell and Wood 2001). Branchial influx rates were calculated from the corresponding measured efflux rate and the calculated net flux rates. Drinking rates and loss via rectal fluids as well as renal losses were all measured (Blanchard and Grosell 2006; Genz et al. 2008; Genz and Grosell 2010) and ingestion of various ions was calculated from the measured drinking rate and an average assumed seawater composition. Toadfish photo: unknown source. See Section 3.3.2. for further details. See color plate section.

toadfish and the winter flounder and on the other hand for an "average" marine teleost. For both toadfish and winter flounder it appears that the majority of the ingested Ca^{2+} is assimilated (70–80%) and that renal excretion is a modest contribution compared to branchial excretion (it is assumed that extra-renal excretion is largely by the gills). This observation, first made by Hickman in 1968 and now evident also from toadfish data, contrasts the common view that all divalent ions are eliminated by renal excretion and clearly studies are required to further investigate the role of the gills in Ca^{2+} excretion.

Substantial rectal base secretion, limited renal contribution and a compensatory branchial acid excretion are newly recognized and as such were not discussed by Hickman but are becoming accepted as general phenomena among marine teleost fish. The gulf toadfish kidney is aglomerular and urine is formed strictly by secretion. However, the mass balance presented in Fig. 4.2 likely represents even teleosts with glomeruli as urine flow and composition generally are indistinguishable among these two groups of fishes (Beyenbach and Liu 1996; Beyenbach 2004).

4. FUTURE DIRECTIONS

1. Sensing of external, luminal and internal osmotic pressure and/or salinity resulting in altered behavior and function of individual osmoregulatory organs remains poorly understood. While some evidence indicates that the calcium-sensing receptor (CaR) may be involved the only documented cases for control of organ function pertaining to osmoregulation are the intestine in marine teleosts and the rectal gland in elasmobranchs (Nearing et al. 2002). Observations of drinking reflex being responsive in part to ambient Cl^- and Cl^- in the intestinal lumen suggest that sensors other than CaR are involved but the nature of this Cl^- sensor remains unknown, clearly offering an interesting area for further research.

2. The influence of hypervolemia on drinking and the putative interactions between RAS and BK and ANP (see Section 2.2) begs examination.

3. Owing to only a few studies, esophageal salt absorption is poorly understood and deserves further attention. The recent advances in our knowledge of the molecular identity of many transporters involved in salt transport should assist future efforts towards a better understanding of salt absorption by this quantitatively important osmoregulatory organ.

4. It is most likely that additional transporters are present and function in the intestine to serve osmoregulation. The NC co-transporter, for example, is poorly characterized in the marine teleost intestinal epithelium. Furthermore, some evidence strongly suggests a role for NHE and anion exchangers in addition to SLC26a6 in intestinal transport of marine teleosts and it is clearly possible that Na-channels operate to facilitate Na^+ uptake by the intestine. The highly effective exclusion of Mg^{2+} by the intestinal epithelium may point to an apical Mg^{2+} export function analogous to the proposed SO_4^{2-} secretion by SLC26a6, yet the putative nature of such a transporter is unknown.

5. Control of paracellular permeability of osmoregulatory tissues is undoubtedly important for osmoregulation and acclimation to changing salinities. Recent identification of the family of claudins as well as occludin as tight junction proteins with barrier functions combined with observations of their regulation in response to changes in salinity suggest that these tight junction proteins may play an important role in osmoregulation (Seidelin et al. 2000; Bagherie-Lachidan et al. 2008; Chasiotis and Kelly 2008; Tipsmark et al. 2008a,b,c, 2009; Chasiotis et al. 2009). The putative role of claudins and occludin in osmoregulation offers an exciting area of research for fish physiologists. However, this research area is also challenging since as many as 24 claudins exist in mammals, mostly with distinct functional properties (Angelow et al. 2008; Findley and Koval 2009). Furthermore, genome duplication events in the teleost lineage mean that fish possess as many as 56 individual claudin proteins (Loh et al. 2004).

6. Although it appears to be accepted that intestinal HCO_3^- secretion does not play a dynamic role in acid–base balance in unfed fish, it clearly is of significance for postprandial acid–base balance (see Chapter 5) and recent characterization of NBCe1 in toadfish (Taylor et al. 2010) strongly implies that this transporter may be responsive to systemic acid–base balance.

7. Finally, indirect evidence (Section 3.3.2 and Fig. 4.2) seems to suggest that the gill may play a so far unrecognized role in Ca^{2+} excretion which warrants further examination.

REFERENCES

Anderson, W. G., Takei, Y., and Hazon, N. (2002). Osmotic and volaemic effects on drinking rate in elasmobranch fish. *J. Exp. Biol.* **205**, 1115–1122.

Anderson, W. G., Taylor, J. R., Good, J. P., Hazon, N., and Grosell, M. (2007). Body fluid volume regulation in elasmobranch fish. *Comp. Biochem. Physiol.* **148**, 3–13.

Ando, M., and Nagashima, K. (1996). Intestinal Na^+ and Cl^- levels control drinking behavior in the seawater-adapted eel, *Anguilla japonica. J. Exp. Biol.* **199**, 711–716.

Ando, M., and Subramanyam, M. V. V. (1990). Bicarbonate transport systems in the intestine of the seawater eel. *J. Exp. Biol.* (**150**), 381–394.

Angelow, S., Ahlstrom, R., and Yu, A. S. (2008). Biology of claudins. *Am. J. Physiol. Renal Physiol.* **295**, F867–F876.

Aoki, M., Kaneko, T., Katoh, F., Hasegawa, S., Tsutsui, N., and Aida, K. (2003). Intestinal water absorption through aquaporin 1 expressed in the apical membrane of mucosal epithelial cells in seawater-adapted Japanese eel. *J. Exp. Biol.* **206**, 3495–3505.

Bagherie-Lachidan, M., Wright, S. I., and Kelly, S. P. (2008). Claudin-3 tight junction proteins in *Tetraodon nigroviridis*: cloning, tissue-specific expression, and a role in hydromineral balance. *Am. J. Physiol. Regul. Integr. Comp. Physiol.* **294**, R1638–R1647.

Balment, R. J., Warne, J. M., and Takei, Y. (2003). Isolation, synthesis, and biological activity of flounder [Asn(1), Ile(5), Thr(9)] angiotensin I. *Gen. Comp. Endocrinol.* **130**, 92–98.

Bayaa, M., Vulesevic, B., Esbaugh, A., Braun, M., Ekker, M. E., Grosell, M., and Perry, S. F. (2009). The involvement of SLC26 anion transporters in chloride uptake in zebrafish (*Danio rerio*) larvae. *J. Exp. Biol.* **212**, 3283–3295.

Beyenbach, K. W. (2004). Kidneys sans glomeruli. *Am. J. Physiol. Renal Physiol.* **286**, F811–F827.

Beyenbach, K. W., and Liu, P. L. F. (1996). Mechanism of fluid secretion common to aglomerular and glomerular kidneys. *Kidney International* **49**, 1543–1548.

Blanchard, J., and Grosell, M. (2006). Copper toxicity across salinities from freshwater to seawater in the euryhaline fish, *Fundulus heteroclitus*: is copper an ionoregulatory toxicant in high salinites? *Aquat. Tox.* **80**, 131–139.

Boisen, A. M. Z., Amstrup, J., Novak, I., and Grosell, M. (2003). Sodium and chloride transport in soft water and hard water acclimated zebrafish (*Danie rerio*). *Biochim. Biophys. Acta* **1618**, 207–218.

Carrick, S., and Balment, R. J. (1983). The renin-angiotensin system and drinking in the Euryhaline flounder, *Platichthys flesus. Gen. Comp. Endocrinol.* **51**, 423–433.

Chasiotis, H., and Kelly, S. P. (2008). Occludin immunolocalization and protein expression in goldfish. *J. Exp. Biol.* **211**, 1524–1534.

Chasiotis, H., Effendi, J. C., and Kelly, S. P. (2009). Occludin expression in goldfish held in ion-poor water. *J. Comp. Physiol. B* **179**, 145–154.

Chen, Y., Cann, M. J., Litvin, T. N., Iourgenko, V., Sinclair, M. L., Levin, L. R., and Buck, J. (2000). Soluble adenylyl cyclase as an evolutionarily conserved bicarbonate sensor. *Science* **289**, 625–628.

Chernova, M. N., Jiang, L. W., Friedman, D. J., Darman, R. B., Lohi, H., Kere, J., Vandorpe, D. H., and Alper, S. L. (2005). Functional comparison of mouse slc26a6 anion exchanger with human SLC26A6 polypeptide variants—differences in anion selectivity, regulation, and electrogenicity. *J. Biol. Chem.* **280**, 8564–8580.

Clark, J. S., Vandorpe, D. H., Chernova, M. N., Heneghan, J. F., Stewart, A. K., and Alper, S. L. (2008). Species differences in Cl^- affinity and in electrogenicity of SLC26A6-mediated oxalate/Cl^- exchange correlate with the distinct human and mouse susceptibilities to nephrolithiasis. *J. Physiol.—Lond.* **586**, 1291–1306.

Clarke, L. L., and Harline, M. C. (1998). Dual role of CFTR in cAMP-stimulated HCO_3^- secretion across murine duodenum. *Am. J. Physiol.* **274**, G718–G726.

Colin, D. A., Nonnotte, G., Leray, C., and Nonnotte, L. (1985). Na^+ transport and enzyme activities in the intestine of the freshwater and sea-water adapted trout (*Salmo gairdneri* R). *Comp. Biochem. Physiol. A* **81**, 695–698.

Conlon, J. M. (1999). Bradykinin and its receptors in non-mammalian vertebrates. *Regulat. Pept.* **79**, 71–81.

Cordier, D., and Maurice, A. (1956). Etude sur l'absorption intestinale des sucres chez l'anguille (*Anguilla vulgaris, L.*) vivant dans l'eau de mer ou dans l'eau douce. *C. r. Séanc. Soc. Biol.* **150**, 1957–1959.

Cutler, C. P., and Cramb, G. (2002). Two isoforms of the $Na^+/K^+/2CI(^-)$ cotransporter are expressed in the European eel (*Anguilla anguilla*). *Biochim. Biophys. Acta—Biomembranes* **1566**, 92–103.

Cutler, C. P., Martinez, A. S., and Cramb, G. (2007). The role of aquaporin 3 in teleost fish. *Comp. Biochem. Physiol. A Mol. Integr. Physiol.* **148**, 82–91.

De Boeck, G., Grosell, M., and Wood, C. (2001). Sensitivity of the spiny dogfish (*Squalus acanthias*) to waterborne silver exposure. *Aquat. Toxicol.* **54**, 261–275.

Dixon, J. M., and Loretz, C. A. (1986). Luminal alkalinization in the intestine of the goby. *J. Comp. Physiol.* **156**, 803–811.

Duffey, M. E. (1979). Intracellular chloride activities and active chloride absorption in the intestinal epithelium of the winter flounder. *J. Membrane Biol.* **50**, 331–341.

Evans, D. H., and Claiborne, J. B. (2008). Osmotic and ionic regulation in fishes. In: *Osmotic and ionic regulation: cells and animals.* (D. Evans, ed.). CRC Press, Taylor and Francis Group, Boca Raton, FL, USA.

Evans, D. H., Piermarini, P. M., and Choe, K. P. (2005). The multifunctional fish gill: dominant site of gas exchange, osmoregulation, acid–base regulation, and excretion of nitrogenous waste. *Physiolog. Rev.* **85**, 97–177.

Fenwick, J. C., Wendelaar Bonga, S. E., and Gert, F. (1999). *In vivo* bafilomycin-sensitive Na$^+$ uptake in young freshwater fish. *J. Exp. Biol.* **202**, 3659–3666.

Field, M., Smith, P. L., and Bolton, J. E. (1980). Ion transport across the isolated intestinal mucosa of the winter flounder *Pseudopleuronectes americanus*: II. Effects of cyclic AMP. *J. Membrane Biol.* **53**, 157–163.

Findley, M. K., and Koval, M. (2009). Regulation and roles for claudin-family tight junction proteins. *IUBMB. Life* **61**, 431–437.

Frizzell, R. A., Smith, P. L., Field, M., and Vosburgh, E. (1979). Coupled sodium-chloride influx across brush border of flounder intestine. *J. Membrane Biol.* **46**, 27–39.

Fuentes, J., and Eddy, F. B. (1996). Drinking in freshwater-adapted rainbow trout fry, *Oncorhynchus mykiss* (Walbaum), in response to angiotensin I, angiotensin II, angiotensin-converting enzyme inhibition, and receptor blockade. *Physiolog. Zool.* **69**, 1555–1569.

Fuentes, J., and Eddy, F. B. (1997a). Drinking in Atlantic salmon presmolts and smolts in response to growth hormone and salinity. *Comp. Biochem. Physiol.* **117A**, 487–491.

Fuentes, J., and Eddy, F. B. (1997b). Effect of manipulation of the renin-angiotensin system in control of drinking in juvenile Atlantic salmon (*Salmo salar* L) in fresh water and after transfer to sea water. *J. Comp. Physiol.* **167B**, 438–443.

Fuentes, J., and Eddy, F. B. (1998). Cardiovascular responses in vivo to angiotensin II and the peptide antagonist saralasin in rainbow trout *Oncorhynchus mykiss*. *J. Exp. Biol.* **201**, 267–272.

Fuentes, J., Bury, N. R., Carroll, S., and Eddy, F. B. (1996a). Drinking in Atlantic salmon presmolts (*Salmo salar* L.) and juvenile rainbow trout (*Oncorhynchus mykiss* Walbaum) in response to cortisol and sea water challenge. *Aquaculture* **141**, 129–137.

Fuentes, J., McGeer, J. C., and Eddy, F. B. (1996b). Drinking rate in juvenile Atlantic salmon, *Salmo salar* L fry in response to a nitric oxide donor, sodium nitroprusside and an inhibitor of angiotensin converting enzyme, enalapril. *Fish Physiol. Biochem.* **15**, 65–69.

Furukawa, O., Kawauchi, S., Miorelli, T., and Takeuchi, K. (2000). Stimulation by nitric oxide of HCO$_3^-$ secretion in bull frog duodenum in vitro—roles of cyclooxygenase-1 and prostaglandins. *Med. Sci. Monitor* **6**, 454–459.

Gardiner, S. M., Kemp, P. A., and Bennett, T. (1993). Differential-effects of captopril on regional hemodynamic-responses to angiotensin-I and bradykinin in conscious rats. *Br. J. Pharmacol.* **108**, 769–775.

Genz, J. and Grosell, M. (2010). Magnesium sulfate limits teleost survival in hypersaline waters. In preparation.

Genz, J., Taylor, J. R., and Grosell, M. (2008). Effects of salinity on intestinal bicarbonate secretion and compensatory regulation of acid–base balance in *Opsanus beta*. *J. Exp. Biol.* **211**, 2327–2335.

Gilmour, K. M., Perry, S. F., Wood, C. M., Henry, R. P., Laurent, P., Part, P., and Walsh, P. J. (1998). Nitrogen excretion and the cardiorespiratory physiology of the gulf toadfish, *Opsanus beta*. *Physiolog. Zool.* **71**, 492–505.

Grosell, M. (2006). Intestinal anion exchange in marine fish osmoregulation. *J. Exp. Biol.* **209**, 2813–2827.

Grosell, M. (2007). Intestinal transport processes in marine fish osmoregulation. In: *Fish Osmoregulation* (B. Baldisserotto, J. M. Mancera and B. G. Kapoor, eds), pp. 332–357. Science Publisher, USA.

Grosell, M., and Genz, J. (2006). Ouabain sensitive bicarbonate secretion and acid absorption by the marine fish intestine play a role in osmoregulation. *Am. J. Physiol.* **291**, R1145–R1156.

Grosell, M., and Jensen, F. B. (1999). NO_2^- uptake and HCO_3^- excretion in the intestine of the European flounder (*Platichthys flesus*). *J. Exp. Biol.* **202**, 2103–2110.

Grosell, M., Laliberte, C. N., Wood, S., Jensen, F. B., and Wood, C. M. (2001). Intestinal HCO_3^- secretion in marine teleost fish: evidence for an apical rather than a basolateral Cl^-/HCO_3^- exchanger. *Fish Physiol. Biochem.* **24**, 81–95.

Grosell, M., De Boeck, G., Johannsson, O., and Wood, C. M. (1999). The effects of silver on intestinal ion and acid–base regulation in the marine teleost fish, *Papophrys vetulus*. *Comp. Biochem. Physiol. C.* **124**, 259–270.

Grosell, M., and Wood, C. M. (2001). Branchial versus intestinal silver toxicity and uptake in the marine teleost. *Parophrys vetulus. J. Comp. Physiol.* **171B**, 585–594.

Grosell, M., Gilmour, K. M., and Perry, S. F. (2007). Intestinal carbonic anhydrase, bicarbonate, and proton carriers play a role in the acclimation of rainbow trout to seawater. *Am. J. Physiol.* **293**, R2099–R2111.

Grosell, M., McDonald, M. D., Walsh, P. J., and Wood, C. M. (2004). Effects of prolonged copper exposure in the marine gulf toadfish (*Opsanus beta*). II. Drinking rate, copper accumulation and Na^+/K^+-ATPase activity in osmoregulatory tissues. *Aquat. Tox.* **68**, 263–275.

Grosell, M., Wood, C. M., Wilson, R. W., Bury, N. R., Hogstrand, C., Rankin, J. C., and Jensen, F. B. (2005). Bicarbonate secretion plays a role in chloride and water absorption of the European flounder intestine. *Am. J. Physiol.* **288**, R936–R946.

Grosell, M., Genz, J., Taylor, J. R., Perry, S. F., and Gilmour, K. M. (2009a). The involvement of H^+-ATPase and carbonic anhydrase in intestinal HCO_3^- secretion on seawater-acclimated rainbow trout. *J. Exp. Biol.* **212**, 1940–1948.

Grosell, M., Mager, E. M., Williams, C., and Taylor, J. R. (2009b). High rates of HCO_{3-} secretion and Cl^- absorption against adverse gradients in the marine teleost intestine: the involvement of an electrogenic anion exchanger and H^+-pump metabolon? *J. Exp. Biol.* **212**, 1684–1696.

Grosell, M., McDonald, M. D., Wood, C. M. and Walsh, P. J. (2010). Prolonged copper exposure impairs acid–base balance, branchial Na^+ and Cl^- extrusion, nitrogenous waste excretion, intestinal transport physiology and affects ^{110m}Ag toxicokinetics in the gulf toadfish, *Opsanus beta*. In preparation.

Grosell, M., and Taylor, J. R. (2007). Intestinal anion exchange in teleost water balance. *Comp. Biochem. Physiol.* **148A**, 14–22.

Halm, D. R., Krasny, E. J., and Frizzell, R. A. (1985a). Electrophysiology of flounder intestinal mucosa. I. Conductance of cellular and paracellular pathways. *J. Gen. Physiol.* **85**, 843–864.

Halm, D. R., Krasny, E. J., and Frizzell, R. A. (1985b). Electrophysiology of flounder intestinal mucosa. II. Relation of the electrical potential profile to coupled NaCl absorption. *J. Gen. Physiol.* **85**, 865–883.

Hickman, C. P. (1968). Ingestion, intestinal absorption, and elimination of seawater and salts in the southern flounder, *Paralichthys lethostigma. Can. J. Zool.* **46**, 457–466.

Hirano, T. (1974). Some factors regulating water intake by the eel, *Anguilla japonica. J. Exp. Biol.* **61**, 737–747.

Hirano, T., and Mayer-Gostan, N. (1976). Eel esophagus as an osmoregulatory organ. *Proc. Natl. Acad. Sci.* **73**, 1348–1350.

Hogan, D. L., Crombie, D. L., Isenberg, J. I., Svendsen, P., DeMuckadell, O. B. S., and Ainsworth, M. A. (1997). Acid-stimulated duodenal bicarbonate secretion involves a CFTR-mediated transport pathway in mice. *Gastroenterology* **113**, 533–541.

Hogstrand, C., Ferguson, E. A., Galvez, F., Shaw, J. R., Webb, N. A., and Wood, C. M. (1999). Physiology of acute silver toxicity in the starry flounder (*Platichthys stellatus*) in seawater. *J. Comp. Physiol.* **169B**, 461–473.

Jampol, L. M., and Epstein, F. H. (1970). Sodium-potassium-activated adenosine triphosphate and osmotic regulation by fishes. *Am. J. Physiol.* **218**, 607–611.

Jensen, L. J., Willumsen, N. J., and Larsen, E. H. (2002). Proton pump activity is required for active uptake of chloride in isolated amphibian skin exposed to freshwater. *J. Comp. Physiol.* **176B**, 503–511.

Jensen, L. J., Willumsen, N. J., Amstrup, J., and Hviid Larsen, E. H. (2003). Proton pump-driven cutaneous chloride uptake in anuran amphibia. *Biochim. Biophys. Acta* **1618**, 120–132.

Kato, A., Chang, M. H., Kurita, Y., Nakada, T., Ogoshi, M., Nakazato, T., Doi, H., Hirose, S., and Romero, M. F. (2009). Identification of renal transporters involved in sulfate excretion in marine teleost fish. *Am. J. Physiol. Regul. Integr. Comp. Physiol.* **297**, R1647–R1659.

Kelly, S. P., Chow, I. N. K., and Woo, N. Y. S. (1999). Effects of prolactin and growth hormone on strategies of hypoosmotic adaptation in a marine teleost, *Sparus sarba. Gen. Comp. Endocrinol.* **113**, 9–22.

Kerstetter, T. H., and White, R. J. (1994). Changes in intestinal water absoprtion in coho salmon during short-term seawater adaptation: a developmental study. *Aquaculture* **121**, 171–180.

Kirsch, R., and Meister, M. F. (1982). Progressive processing of ingested water in the gut of seawater teleost. *J. Exp. Biol.* **98**, 67–81.

Kormanik, G. A., and Evans, D. H. (1982). The relation of Na and Cl extrusion in *Opsanus beta*, the Gulf toadfish, acclimated to seawater. *J. Exp. Zool.* **224**, 187–194.

Kurita, Y., Nakada, T., Kato, A., Doi, H., Mistry, A. C., Chang, M. H., Romero, M. F., and Hirose, S. (2008). Identification of intestinal bicarbonate transporters involved in formation of carbonate precipitates to stimulate water absorption in marine teleost fish. *Am. J. Physiol.* **294**, R1402–R1412.

Larsen, E. H., Willumsen, N. J., and Christoffersen, B. C. (1992). Role of proton pump of mitochondria-rich cells for active-transport of chloride-ions in toad skin epithelium. *J. Physiol.* **450**, 203–216.

Lignot, J. H., Cutler, C. P., Hazon, N., and Cramb, G. (2002). Immunolocalisation of aquaporin 3 in the gill and the gastrointestinal tract of the European eel *Anguilla anguilla* (L.). *J. Exp. Biol.* **205**, 2653–2663.

Lin, L.-Y., Weng, C.-F., and Hwang, P.-P. (2002). Regulation of drinking rate in euryhaline Tilapia larvae (*Oreochromis mossambicus*) during salinity challenges. *Phys. Biochem. Zool.* **74**, 171–177.

Loh, Y. H., Christoffels, A., Brenner, S., Hunziker, W., and Venkatesh, B. (2004). Extensive expansion of the claudin gene family in the teleost fish, *Fugu rubripes. Genome Res.* **14**, 1248–1257.

Loretz, C. A. (1995). Electrophysiology of ion transport in the teleost intestinal cells. In: *Cellular and Molecular Approaches to Fish Ionic Regulation, Fish Physiology* (C. M. Wood and T. J. Shuttleworth, eds.) **14**, 25–56.

Loretz, C. A., and Fourtner, C. R. (1988). Functional-characterization of a voltage-gated anion channel from teleost fish intestinal epithelium. *J. Exp. Biol.* **136**, 383–403.

Mackay, W. C., and Janicki, R. (1978). Changes in the eel intestine during seawater adaptation. *Comp. Biochem. Physiol.* **62A**, 757–761.

Madsen, S. S., Mccormick, S. D., Young, G., Endersen, J. S., Nishioka, R. S., and Bern, H. S. (1994). Physiology of seawater acclimation in the striped bass, *Morone saxatilis* (Walbaum). *Fish Physiol. Biochem.* **13**, 1–11.

Malvin, R. L., Schiff, D., and Eiger, S. (1980). Angiotensin and drinking rates in the euryhaline killifish. *Am. J. Physiol.* **239**, R31–R34.

Marshall, W. S., and Grosell, M. (2005). Ion transport, osmoregulation and acid–base balance. In: *Physiology of Fishes* (D. Evans and J. B. Claiborne, eds), 3rd edition. CRC Press.

Marshall, W. S., Howard, J. A., Cozzi, R. R. F., and Lynch, E. M. (2002). NaCl and fluid secretion by the intestine of the teleost *Fundulus heteroclitus*: involvement of CFTR. *J. Exp. Biol.* **205**, 745–758.

Martinez, A.-S., Cutler, C. P., Wilson, G. D., Phillips, C., Hazon, N., and Cramb, G. (2005a). Cloning and expression of three aquaporin homologues from the European eel (*Anguilla anguilla*): effects of seawater acclimation and cortisol treatment on renal expression. *Biol. Cell.* **97**, 615–627.

Martinez, A.-S., Wilson, G., Phillips, C., Cutler, C., Hazon, N., and Cramb, G. (2005b). Effects of cortisol on aquaporin expression in the esophagus of the European eel, *Anguilla anguilla*. *Ann. NY Acad. Sci.* **1040**, 395–398.

Martinez, A. S., Cutler, C. P., Wilson, G. D., Phillips, C., Hazon, N., and Cramb, G. (2005c). Regulation of expression of two aquaporin homologs in the intestine of the European eel: effects of seawater acclimation and cortisol treatment. *Am. J. Physiol. Regul. Integr. Comp. Physiol.* **288**, R1733–R1743.

Maxime, V., Pennec, J. P., and Peyraud, C. (1991). Effects of direct transfer from fresh-water to seawater on respiratory and circulatory variables and acid–base status in rainbow-trout. *J. Comp. Physiol.* **161B**, 557–568.

McDonald, M. D. (2006). Renal contribution to teleost fish osmoregulation. In: *Fish Osmoregulation* (B. Bladisserotto, J. M. Mancera and B. G. Kapoor, eds.). Science publishers, USA.

McDonald, M. D., and Grosell, M. (2006). Maintaining osmotic balance with an aglomerular kidney. *Comp. Biochem. Physiol. A—Mol. Integr. Physiol.* **143**, 447–458.

Musch, M. W., Orellana, S. A., Kimberg, L. S., Field, M., Halm, D. R., Krasny, E. J., and Frizzell, R. A. (1982). Na^+-K^+-$2Cl^-$ co-transport in the intestine of a marine teleost. *Nature* **300**, 351–353.

Nearing, J., Betka, M., Quinn, S., Hentschel, H., Elger, M., Baum, M., Bai, M., Chattopadyay, N., and Brown, E. M. (2002). Polyvalent cation receptor proteins (CaRs) are salinity sensors in fish. *PNAS* **99**, 9231–9236.

Nonnotte, G., and Truchot, J. P. (1990). Time course of extracellular acid-base adjustments under hypoosmotic or hyperosmotic conditions in the euryhaline fish *Platichthys flesus*. *J. Fish Biol.* **36**, 181–190.

Parmelee, J. T., and Renfro, J. L. (1983). Esophageal desalination of seawater in flounder: role of active sodium transport. *Am. J. Physiol.* **245**, R888–R893.

Pelis, R. M., and Renfro, J. L. (2003). Active sulfate secretion by the intestine of winter flounder is through exchange for luminal chloride. *Am. J. Physiol.* **284**, R380–R388.

Perrot, M. N., Grierson, N., Hazon, N., and Balment, R. J. (1992). Drinking behavior in sea water and fresh water teleosts: the role of the renin-angiotensin system. *Fish Physiol. Biochem.* **10**, 161–168.

Perry, S. F., Vulesevic, B., Grosell, M., and Bayaa, M. (2009). Evidence that SLC26 anion transporters mediate branchial chloride uptake in adult zebrafish (*Danio rerio*). *Am. J. Physiol. Regul. Integr. Comp. Physiol.* **297**, R988–R997.

Pickering, A. D., and Morris, R. (1973). Localization of ion-transport in the intestine of the migrating river lamprey, *Lametra fluviatilis* L. *J. Exp. Biol.* **58**, 165–176.

Raldua, D., Otero, D., Fabra, M., and Cerda, J. (2008). Differential localization and regulation of two aquaporin-1 homologs in the intestinal epithelia of the marine teleost, *Sparus aurata*. *Am. J. Physiol. Regul. Integr. Comp. Physiol.* **294**, R993–1003.

Sattin, G., Mager E.M., Beltramini, M. and Grosell, M. (2010). Cytosolic carbonic anhydrase in the gulf toadfish is important for tolerance to hypersalinity. *Comp. Biochem. Physiol.* In press.

Seidelin, M., Madsen, S. S., Blenstrup, H., and Tipsmark, C. K. (2000). Time-course changes in the expression of the Na^+, K^+-ATPase in gills and pyloric caeca of brown trout (*Salmo trutta*) during acclimation to seawater. *Phys. Biochem. Zool.* **73**, 446–453.

Shehadeh, Z. H., and Gordon, M. S. (1969). The role of the intestine in salinity adaptation of the rainbow trout, *Salmo gairdneri*. *Comp. Biochem. Physiol* **30**, 397–418.

Skadhauge, E. (1974). Coupling of transmural flows of NaCl and water in the intestine of the eel (*Anguilla anguilla*). *J. Exp. Biol.* **60**, 535–546.

Smith, C. P., Smith, P. L., Welsh, M. J., Frizzell, R. A., Orellana, S. A., and Field, M. (1980). Potassium transport by the intestine of the winter flounder *Pseudopleuronectes americanus*: evidence for KCl co-transport. *Bull. Mt. Desert. Island Biol. Lab.* **20**, 92–96.

Smith, H. W. (1930). The absorption and excretion of water and salts by marine teleosts. *Am. J. Physiol.* **93**, 480–505.

Takei, Y., and Tsuchida, T. (2000). Role of the renin-angiotensin system in drinking of the seawater-adapted eels *Anguilla japonica*: reevaluation. *Am. J. Physiol.* **279**, R1105–R1111.

Takei, Y., Hirano, T., and Kobayashi, H. (1979). Angiotensin and water intake in the Japanese eel, *Anguilla japonica*. *Gen. Comp. Endocrinol.* **38**, 446–475.

Takei, Y., Okubo, J., and Yamaguchi, K. (1988). Effect of cellular dehydration on drinking and plasma angiotensin II level in the eel, *Anguilla japonica*. *Zool. Sci.* **5**, 43–51.

Takei, Y., Tsuchida, T., Li, Z. H., and Conlon, J. M. (2001). Antidipsogenic effects of eel bradykinins in the eel *Anguilla japonica*. *Am. J. Physiol.* **281**, R1090–R1096.

Taylor, J. R., and Grosell, M. (2009). The intestinal response to feeding in seawater gulf toadfish, *Opsanus beta*, includes elevated base secretion and increased epithelial oxygen consumption. *J. Exp. Biol.* **212**, 3873–3881.

Taylor, J. R., Mager, E. M., and Grosell, M. (2010). Basolateral NBCe1 plays a rate-limiting role in transepithelial intestinal HCO_3^- secretion serving marine fish osmoregulation. *J. Exp. Biol.* **213**, 459–468.

Tipsmark, C. K., Baltzegar, D. A., Ozden, O., Grubb, B. J., and Borski, R. J. (2008a). Salinity regulates claudin mRNA and protein expression in the teleost gill. *Am. J. Physiol. Regul. Integr. Comp. Physiol.* **294**, R1004–R1014.

Tipsmark, C. K., Kiilerich, P., Nilsen, T. O., Ebbesson, L. O., Stefansson, S. O., and Madsen, S. S. (2008b). Branchial expression patterns of claudin isoforms in Atlantic salmon during seawater acclimation and smoltification. *Am. J. Physiol. Regul. Integr. Comp. Physiol.* **294**, R1563–R1574.

Tipsmark, C. K., Luckenbach, J. A., Madsen, S. S., Kiilerich, P., and Borski, R. J. (2008c). Osmoregulation and expression of ion transport proteins and putative claudins in the gill of southern flounder (*Paralichthys lethostigma*). *Comp. Biochem. Physiol. A Mol. Integr. Physiol.* **150**, 265–273.

Tipsmark, C. K., Jorgensen, C., Brande-Lavridsen, N., Engelund, M., Olesen, J. H., and Madsen, S. S. (2009). Effects of cortisol, growth hormone and prolactin on gill claudin expression in Atlantic salmon. *Gen. Comp. Endocrinol.* **163**, 270–277.

Tomasso, J. R., and Grosell, M. (2004). Physiological basis for large differences in resistance to nitrite among freshwater and freshwater acclimated euryhaline fishes. *Environ. Sci. Technol.* **39**, 98–102.

Tresguerres, M., Levin, R., Buck, J. and Grosell, M. (2010). Interaction between HCO_3^- secretion and NaCl absorption in the marine teleost intestine is mediated by soluble adenylyl cyclase. *Am. J. Physiol.* Submitted.

Tsuchida, T., and Takei, Y. (1998). Effects of homologous atrial natriuretic peptide on drinking and plasma ANG II level in eels. *Am. J. Physiol.* **44**, R1605–R1610.

Tuo, B. G., and Isenberg, J. I. (2003). Effect of 5-hydroxytryptamine on duodenal mucosal bicarbonate secretion in mice. *Gastroenterology* **125**, 805–814.

Tuo, B. G., Sellers, Z., Paulus, P., Barrett, K. E., and Isenberg, J. I. (2004). 5-HT induces duodenal mucosal bicarbonate secretion via cAMP- and Ca^{2+}-dependent signaling pathways and 5-HT4 receptors in mice. *Am. J. Physiol.* **286**, G444–G451.

Usher, M. L., Talbot, C., and Eddy, F. B. (1991). Intestinal water transport in juvenile Atlantic salmon (*Salmo salar* L.) during smolting and following transfer to seawater. *Comp. Biochem. Physiol.* **100A**, 813–818.

Walsh, P. J., Blackwelder, P., Gill, K. A., Danulat, E., and Mommsen, T. P. (1991). Carbonate deposits in marine fish intestines: a new source of biomineralization. *Limnol. Oceanogr.* **36**, 1227–1232.

Watanabe, S., Hirano, T., Grau, E. G., and Kaneko, T. (2009). Osmosensitivity of prolactin cells is enhanced by the water channel aquaporin-3 in a euryhaline Mozambique tilapia (*Oreochromis mossambicus*). *Am. J. Physiol. Regul. Integr. Comp. Physiol.* **296**, R446–R453.

Wilkes, P. R. H., and Mcmahon, B. R. (1986). Responses of a stenohaline fresh-water teleost (*Catostomus-Commersoni*) to hypersaline exposure. 1. The dependence of plasma pH and bicarbonate concentration on electrolyte regulation. *J. Exp. Biol.* **121**, 77–94.

Wilson, R. W. (1999). A novel role for the gut of seawater teleosts in acid–base balance. *Regulation of Acid–Base Status in Animals and Plants, SEB seminar series*, **68**, pp. 257–274. Cambridge University Press, Cambridge.

Wilson, R. W., and Grosell, M. (2003). Intestinal bicarbonate secretion in marine teleost fish— source of bicarbonate, pH sensitivity, and consequence for whole animal acid–base and divalent cation homeostasis. *Biochim. Biophys. Acta* **1618**, 163–193.

Wilson, R. W., Gilmour, K., Henry, R., and Wood, C. (1996). Intestinal base excretion in the seawater-adapted rainbow trout: a role in acid–base balance? *J. Exp. Biol.* **199**, 2331–2343.

Wilson, R. W., Wilson, J. M., and Grosell, M. (2002). Intestinal bicarbonate secretion by marine teleost fish—why and how? *Biochim. Biophys. Acta* **1566**, 182–193.

Wilson, R. W., Millero, F. J., Taylor, J. R., Walsh, P. J., Christensen, V., Jennings, S., and Grosell, M. (2009). Contribution of fish to the marine inorganic carbon cycle. *Science* **323**, 359–362.

Zuidema, T., Kamermans, M., and Vanheukelom, J. S. (1986). Influence of glucose-absorption on ion activities in cells and submucosal space in goldfish intestine. *Pflugers Archiv.— European J. Physiol.* **407**, 292–298.

5

THE ROLE OF FEEDING IN SALT AND WATER BALANCE

CHRIS M. WOOD
CAROL BUCKING

The importance of salt and water absorption from the food by the gastrointestinal tract and its impact on iono- and osmoregulation have been largely overlooked by fish physiologists. The present review aims to correct this situation. Techniques for overcoming the practical difficulties of studying ionoregulation in feeding fish are critically assessed, and ion and water contents of a range of diets are surveyed. In freshwater fish, the quantities of most major electrolytes ingested via a normal ration far exceed

The Multifunctional Gut of Fish: Volume 30
FISH PHYSIOLOGY

those transported from the water by the gills, but net absorption rates of specific ions vary greatly with a range of influences, including complex interactions involving mucins and bile salts. The stomach plays a key role in absorption, while net secretion generally occurs in the anterior intestinal region due to biliary and pancreatic discharge. Dry commercial diets help minimize water uptake, while high NaCl diets have marked effects on plasma composition, branchial ion fluxes, and renal function. In seawater teleosts, Na^+, K^+, Cl^- and water absorption from the food are superimposed on ion and water transport from drinking; Ca^{2+} and Mg^{2+} are largely excluded. Effects of feeding on the ionoregulatory functions of gills and kidney remain to be investigated. Marine elasmobranchs constitute a special case due to their urea-based osmoregulary strategy; feeding causes marked stimulation of urea synthesis, urea secretion into the chyme, and rectal gland metabolism. Future directions are highlighted.

1. INTRODUCTION

It is generally thought that salt and water balance in freshwater and seawater fish proceeds via opposing strategies. In freshwater the gills and kidney are the major organs involved for salt uptake and water loss, respectively. In seawater the gastrointestinal tract and gills are utilized for water uptake and salt unloading, respectively. A survey of many comparative physiology textbooks and review articles on fish osmoregulation would suggest that the gastro-intestinal tract plays little role in salt and water balance in freshwater fish, yet a major role in seawater fish. However, this conclusion simply reflects the state of most science in the field, which devolves from the penchant of comparative physiologists to simplify their experimental conditions and standardize their experimental animals through fasting. Working with feeding fish is a messy business—the annoying problem of defecation into the water for example—and feeding is well known to have immense, time-dependent postprandial effects on oxygen consumption (e.g. Secor 2009) and nitrogen metabolism (e.g. Wood 2001a). Therefore, the vast majority of experimental studies in the literature have been performed on fish which have been fasted long enough to ensure that the gastrointestinal tract has been voided and the stimulatory effects of feeding on metabolism have abated. Of course, this protocol ignores one of the fundamental functions of the GI tract and reveals only part of the story.

In nature, most fish feed either opportunistically or on a diurnal cycle and in aquaculture, fish are fed regularly. A meal, either another fish or a commercial diet made from fish meal, contains about 3,000–12,000 μmol kg^{-1} of each of Na^+, Cl^-, K^+, Ca^{2+}, and Mg^{2+}, and up to 90 ml kg^{-1} of water, assuming a daily intake of about 3% dry matter (relative to wet body weight). If all these ions were to be absorbed, the gastrointestinal uptake rates (125–500 µmol kg^{-1} h^{-1}) would far exceed the net uptake rates of these same ions measured at the gills in fasted freshwater fish (generally less than 50 µmol kg^{-1} h^{-1}). Similarly water absorption through the tract (up to 3.8 ml kg^{-1} h^{-1}) would be comparable to urine flow rates of fasted freshwater teleosts, which are traditionally assumed to match branchial water uptake rates. In seawater this intake of Na$^+$, Cl$^-$, and water is exceeded by intake through other routes (gills, drinking) during fasting. However, the intake of calcium is above and beyond those in the fasted condition. Considering calcium's role in water uptake (Grosell 2006), the consequences of calcium deficiency may be considerable. Combining the function of the gastrointestinal tract in digestion with salt and water balance was unexplored until a small "thought-experiment" paper by Smith et al. (1989), which became a watershed. These workers estimated the ingestion rates of Na$^+$ over the year for a typical freshwater salmonid, by using temperature-dependent bioenergetic models for food intake in the wild, commercial recommendations for temperature-dependent feeding rates in aquaculture, and measured Na$^+$ concentrations of invertebrates representative of salmonid diets in nature or commercial feed pellets in aquaculture. These estimates were compared to branchial Na$^+$ influx and efflux rates, temperature-extrapolated from the measurements of Wood et al. (1984) at $13°C$ on fasted rainbow trout. The analysis indicated that in winter, when feeding was very low, net Na$^+$ uptake at the gills dominated, but in summer, uptake via the gut could potentially equal or exceed uptake via the gills (Fig. 5.1).

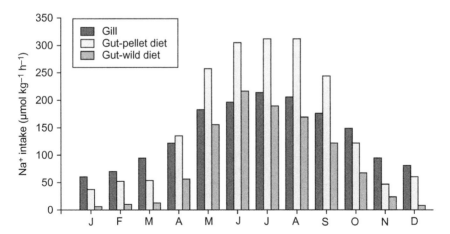

Fig. 5.1. Estimated seasonal rates of Na$^+$ uptake from the water via the gills (black bars), Na$^+$ uptake via the gut from a diet of commercial pellets (white bars), and Na$^+$ uptake via the gut from a wild invertebrate diet, for salmonid parr living in freshwater in Scotland. Data recalculated from Smith et al. (1989). See original reference for assumptions.

Apart from a few early exceptions such as Smith et al. (1989), it is only in the last decade that this oversight as to the potential contribution of the gut to ionoregulation has been recognized. The focus of this chapter will be on those investigations, most of them quite recent, which have addressed how the two key functions of the gastrointestinal tract (feeding/digestion/assimilation and ionoregulation/osmoregulation) are integrated. Our emphasis will be at the level of the whole organism; the reader is referred to Chapters 2, 4 and 6 for cellular and molecular detail on the transport processes that may be involved. We will show that when freshwater fish are allowed to feed, the role of the gastrointestinal tract in salt and water balance is far more important than previously believed, and when seawater fish are allowed to feed, the role is far more complex than previously believed.

2. PRACTICAL DIFFICULTIES AND SOLUTIONS FOR IONO/OSMOREGULATORY STUDIES ON FEEDING FISH

2.1. Getting the Fish to Feed under Conditions Which Allow the Study of Ionoregulation

Standard *in vivo* ionoregulatory methodology relies heavily on isolating fish in small-volume chambers. This practice allows for the measurement of ion flux rates, drinking rates, and even blood, urine and rectal fluid collection by indwelling catheters. Unfortunately, fish will almost never feed spontaneously under these unnatural conditions, although Taylor et al. (2007) succeeded in this approach for flux measurements with non-cannulated European flounder. Practical solutions include the following:

2.1.1. SYNCHRONIZED FEEDING FOLLOWED BY SERIAL SACRIFICE

The simplest approach places groups of fish on a fixed daily feeding regime, preferably satiation feeding. Fish are then sacrificed at various times after the meal for sampling of chyme composition and plasma parameters (Fig. 5.2A). Early examples focused on rainbow trout (Hille 1984; Dabrowski et al. 1986) while recent examples include the detailed studies on gulf toadfish (Taylor and Grosell 2006a) and spiny dogfish (Wood et al. 2007b). Serial sacrifice is also useful with fish that feed almost continuously in the wild (Bergman et al. 2003).

2.1.2. SYNCHRONIZED FEEDING FOLLOWED BY ISOLATION

A modification of the above methodology involves transferring the fish to individual chambers for flux measurements at various times post-feeding

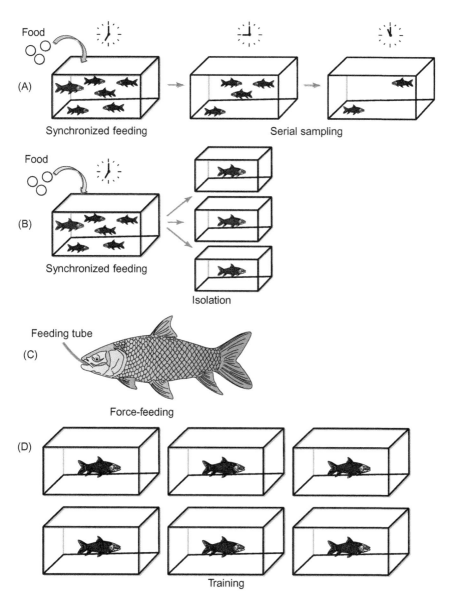

Fig. 5.2. An illustration of four techniques which can be used to facilitate the feeding of fish under conditions which allow the study of ionoregulation. See text, Sections 2.1–2.4 for details.

(Fig. 5.2B). This approach has produced useful information on postprandial acid–base and osmolyte fluxes in dogfish (Kajimura et al. 2006, 2008; Wood et al. 2007b) and trout (Smith et al. 1995; Bucking and Wood 2008; Cooper and Wilson 2008). However, the sudden isolation is stressful, so unfed control fish that are handled in a similar manner are necessary. Cannulation is not recommended because anesthesia will often cause the fish to vomit or defecate.

2.1.3. FORCE-FEEDING

Administration of food by gavage (Fig. 5.2C) while the fish are under anesthesia (e.g. Ruohonen et al. 1997) or simply restrained (Bucking et al. 2009) offers the advantage that an exact meal size may be given, but carries considerable disadvantages. An improvement on this technique is to implant a feeding tube into the stomach several days prior to the experiment (e.g. Clearwater et al. 2000; Wood et al. 2005, 2009), allow the animal to recover from anesthesia and surgery, and then to infuse a food slurry down the tube. Although this approach is probably less stressful, it again carries the problem of lack of meal anticipation. Cooper and Wilson (2008) reported that postprandial disturbances in blood ions and acid–base status were greater in trout fed by stomach-tube than in trout which naturally consumed the same ration. For both techniques it is essential that sham-fed fish subjected to the same disturbances be included in the experimental design.

2.1.4. PRIOR TRAINING

The best approach is to habituate the fish to the experimental conditions, be it isolation with or without implanted cannulae, and train them to feed under these conditions (Fig. 5.2D). Beamish and Thomas (1984) is a classic early example of this approach. Individual rainbow trout were fitted with indwelling urinary bladder catheters, then housed in chambers that allowed collection of feces, and trained to take pellets introduced through a feeding port. This approach allowed separate collection of branchial, fecal and urinary N-compounds in fish fed diets of different compositions for up to 8 weeks. Bucking and Wood (2008) similarly trained adult trout to feed in individual chambers, then fasted the fish for 1 week before fitting them with dorsal aorta catheters. After recovery from surgery, these cannulated trout readily consumed a large meal, and serial blood sampling from the arterial catheter allowed analysis of postprandial changes in plasma composition. Wood et al. (2010) trained killifish to eat individually in small chambers so as to analyze postprandial acid–base fluxes to the external water.

2.2. Quantifying Food Consumption and the Absorption of Electrolytes from the Food

When fish are fed in groups, the ration to the entire group is known, but widely varying amounts may be consumed by each individual animal. Additionally there is the problem of quantifying absorption of ions and fluid from the food. Simple measurements of chyme composition are of limited value, because large fluid fluxes may dilute or concentrate the salts, and the mass of the solid fraction will change as nutrients are absorbed.

2.2.1. FORCE-FEEDING OR INDIVIDUAL FEEDING COMBINED WITH RECTAL CATHETERS

As noted above, these approaches allow exact measurements of ration, and if the entire rectal output can be collected, the net fluid and ion absorption (or secretion) across the tract associated with the meal could be calculated. To our knowledge, this has never been done, though Cooper and Wilson (2008) did administer food by stomach tube to trout fitted with rectal catheters, using the latter approach to avoid fecal contamination of the external water. Rectal catheters have been used in several studies where there was a need to collect rectal effluent continuously from fasted fish (e.g. Shehadeh and Gordon 1969; Wilson et al. 1996; Grosell and Wood 2001; Wilson et al. 2002; Wilson and Grosell 2003). Fecal traps, as often used in nutritional research (Cho et al. 1982), are not recommended, because of the loss of readily soluble ions to the water (e.g. Sugiura et al. 1998a,b).

2.2.2. VOLUNTARY FEEDING COMBINED WITH THE USE OF AN INERT MARKER

An approach that will overcome all these problems is the inclusion of an inert marker in the food, one which can be easily quantified in individual fish (as a measure of consumption) and as it passes down the tract (as a reference for absorption and secretion calculations, and gut passage time measurement). Characteristics of ideal markers include the following: (i) they must not affect the palatability of the food; (ii) they must be neither absorbed nor secreted by the tract; (iii) they must be easily and specifically quantifiable, preferably by non-invasive techniques; and (iv) they must move through the tract at the same rate as the food/chyme.

The traditional marker used in many nutritional studies is chromic oxide (Cr_2O_3; Austreng 1978; reviewed by Wood 1995). However, it cannot be measured in the tract by non-invasive means, and there is some evidence that it may not move through the tract at the same rate as the chyme (see Dabrowski and Dabrowska 1981, for example), it may alter carbohydrate utilization (e.g. Shiau and Chen 1993), it may impact digestibility (e.g. DeSilva et al. 1997), and it may be slightly absorbed (Sugiura et al. 1998a). More recently, there has been a trend to replace it with yttrium oxide (Y_2O_3; Sugiura et al. 1998b).

For almost 40 years, X-radiographic studies of feeding have been the approach of choice for physiological studies. The fish may be quickly (often non-terminally) X-rayed at any time post-feeding, and the amount and location of the marker in the tract quantified. Early studies used radio-opaque barium sulfate ($BaSO_4$) but it tends to make the feed unpalatable (e.g. Edwards 1971, 1973; Ross and Jauncey 1981). Radio-opaque iron powder (Talbot and Higgins 1983; Salman and Eddy 1988a, 1990) and leaded glass ballotini beads are preferable, and the latter have been widely used since their introduction into fish feeding physiology by D. F. Houlihan's laboratory (e.g. McCarthy et al. 1992, 1993; Carter et al. 1993). Bucking and Wood (2006a) demonstrated that the beads moved down the tract of rainbow trout after a meal at almost the same rate as a marker for the fluid phase of the chyme, radiolabeled polyethylene glycol (M.W. $= 4,000$; [^{3}H]PEG-4000). This technique has now been used to calculate the specific net fluxes of water (Bucking and Wood 2006a), Na^+, K^+, Cl^-, Ca^{2+}, and Mg^{2+} (Bucking and Wood 2006b, 2007; Bucking et al. 2010b), and the nutrient metals Cu (Nadella et al. 2006) and Ni (Leonard et al. 2009) in various segments of the tract of rainbow trout after a single satiation meal. The reader is referred to these studies for calculation details. In brief, comparison of the number of beads in the fish with their concentration in the original food provides a direct measure of individual consumption, and comparison (in fish sacrificed at various times after feeding) of the ratio of ions or water to the beads in the chyme with the comparable ratio in the previous segment allows calculation of net secretory and/or absorptive fluxes in each segment of the tract, as well as across the entire length of the tract.

2.3. *In vitro* to *in vivo* Comparisons

While this chapter focuses on *in vivo* studies, for mechanistic analyses there is often a need for *in vitro* studies. *In vitro* preparations have been widely used to analyze osmoregulatory transport in the gut from fasted animals (reviewed by Loretz 1995; Grosell 2006). However, we are aware of no studies which have applied subcellular analytical techniques, such as brush border membrane vesicles or various patch clamp approaches, to fish gut tissue harvested following a meal. Indeed, only very recently has feeding been considered using two useful tissue level approaches (gut sacs, Ussing chambers) which will be briefly addressed here in the hope that they will be used in future feeding studies. For *in vitro* approaches, a prime concern lies in duplicating the composition of the postprandial chyme in the "inside" mucosal compartment, and in interpreting results from preparations which are deprived of all natural hormonal and neural controls as well as the blood flow and mechanical stimulation changes which normally occur during digestion.

2.3.1. GUT SACS

In most fish, each segment of the gastrointestinal tract can easily be made into a tied-off sac preparation filled with saline of known composition. When incubated in oxygenated serosal saline, ion transport can be followed for several hours using radiotracer or cold techniques, and fluid transport can be followed gravimetrically or by monitoring changes in concentration of an impermeable marker. PEG-4000 is not recommended due to evidence of transport in the trout (Bogé et al. 1988) and killifish (C. M. Wood and M. Grosell unpublished results). The great advantages of this approach are simplicity and scale, but to date it has been used in only four feeding-related investigations. Baldisserotto et al. (2006) and Klinck et al. (2009) characterized changes in Ca and Cd transport in gut sacs from trout subjected to different pre-feeding regimes with diets containing these two antagonistic metals. Bucking et al. (2009) and Wood et al. (2010) used this technique to demonstrate large changes in intestinal fluid, Cl^-, and HCO_3^- transport after feeding in rainbow trout and killifish, with substantial differences between freshwater- and seawater-acclimated animals. Practically, there seems no reason why true chyme could not be used in the mucosal compartment in such experiments, but we are aware of no studies which have done this.

A variation on this approach is the everted gut sac in which uptake from now "outside" mucosal fluid into the tissue is measured (Karasov and Diamond 1983). It appears to be particularly suitable for monitoring transport in the many fine pyloric cecae which are present in some fish species (Buddington and Diamond 1986, 1987). Sugiura and Ferraris (2004a) used this technique to show that the pyloric cecae could account for up to 89% of the absorption of inorganic phosphate from the diet in freshwater rainbow trout.

2.3.2. USSING CHAMBERS

This classic technique allows precise control of both mucosal and serosal composition, and the continuous recording of additional parameters (e.g. conductance, transepithelial potential, short circuit current, net acid–base-flux, O_2 consumption) as well as ion fluxes for many hours. The only disadvantages are that fluid flux is not easily measurable, and fewer preparations can be run in most laboratories because the approach is equipment-intensive. In the single Ussing chamber-based feeding study to date, Taylor and Grosell (2009) demonstrated substantial increases in intestinal HCO_3^- transport and O_2 consumption in post-feeding preparations from the marine gulf toadfish.

2.4. Fish Larvae—A Special Case

Due to their size, fish larvae present an additional challenge for studying feeding. Ingestion of commercially prepared diets is limited, and therefore the majority of larviculture depends on live feeds. Unfortunately the compositions of live feeds are difficult to manipulate, and incorporation of inert markers into living tissues is problematic. A majority of larval studies conducted to date use tracers instead of inert markers, usually stable isotopes of carbon (^{13}C) or nitrogen (^{15}N) or radioactive isotopes (^{14}C or ^{35}S). These tracers are added to living prey so as to be incorporated into proteins, fatty acids, carbohydrates, etc. (e.g. Morais et al. 2004; Tonheim et al. 2004). The tracers can also be added directly to mechanically prepared meals as labeled amino acids or fatty acids, etc. (e.g. Morais and Conceicao 2009). More recently, inert markers incorporated into commercial diets have been attempted (Hansen et al. 2009). Additionally, fish larvae can be force-fed various diets. Ronnestad et al. (2001) provides an excellent technical review of the *in vivo* study of digestion in fish larvae using a tube feeding technique involving lightly anesthetized fish. To date, studies focusing on digestion have concentrated primarily on larval nutrient assimilation, and little is currently known about the role of feeding in larval salt and water balance. It is conceivable that salt and water tracer studies could be attempted, using the same techniques. Conceicao et al. (2008) provide a comprehensive review of the area.

3. ION LEVELS IN DIETS AND RATION LEVELS

Most pelleted commercial feeds are based in whole or in part on dried, processed animal or plant material, so the most obvious differences from natural food are in the reduced water content (~5–10% versus 70–80%) and associated elevations in Na^+, K^+, Cl^- concentrations. Apart from this, it is difficult to make any generalizations. For nutritionists, concerns are largely focused on trace elements and phosphorus; the latter is required in large amounts (about 0.6% or about 200 mmol kg^{-1} of diet) for good growth (Lall 2002). Most studies do not report the concentrations of those electrolytes in the diet which are of major concern to ionoregulatory physiologists, and many do not report ration levels. Table 5.1 surveys some recent studies which have, with ration (when reported), tabulated on a wet mass food to wet mass fish basis. We have also included some of our own analyses on commercial feeds.

Table 5.1

A survey of the ionic and water contents of various natural and commercial diets used in feeding studies focused on ionoregulation/osmoregulation. Ration levels are also given when reported

Consumer	Food	Environment	Reference	H_2O (ml kg^{-1})	Na^+ (mmol kg^{-1})	Cl^- (mmol kg^{-1})	Mg^{2+} (mmol kg^{-1})	Ca^{2+} (mmol kg^{-1})	K^+ (mmol kg^{-1})	Ration (% body mass)
Opsanus beta	*Loligo forbesi*	SW	Taylor and Grosell (2006a)	~750	121	141	21	1	61	9.0
Opsanus beta	*Sardinia pilchardus*	SW	Taylor and Grosell (2006a)	~750	123	117	30	404	70	9.0
Squalus acanthias	*Merluccius productus*	SW	Wood et al. (2007b)	802	56	46	16	73	113	5.5
Perca flavescens	Aquatic invertebrates (reference lake)	FW	Klinck et al. (2007)	~800	34	7	13	238	–	–
Perca flavescens	Aquatic invertebrates (metal contaminated lake)	FW	Klinck et al. (2007)	~800	41	7	12	12	–	–
Oncorhynchus mykiss	Commercial salmonid diet	FW	Bucking and Wood (2006 a,b, 2007)	61	216	187	109	194	97	3.1
Oncorhynchus mykiss	Formulated low Na^+ diet	FW	D'Cruz and Wood (1998)	~100	78	79	–	–	–	0.6
Oncorhynchus mykiss	Spiked high Na^+ diet	FW	Kjoss et al. (2005)	85	2050	1950	–	–	–	4.0

(Continued)

CHRIS M. WOOD AND CAROL BUCKING

Table 5.1 (*continued*)

Consumer	Food	Environment	Reference	H_2O (ml kg^{-1})	Na$^+$ (mmol kg^{-1})	Cl$^-$ (mmol kg^{-1})	Mg^{2+} (mmol kg^{-1})	Ca^{2+} (mmol kg^{-1})	K$^+$ (mmol kg^{-1})	Ration (% body mass)
Oncorhynchus mykiss	Spiked high Ca$^+$ diet	FW	Baldisserotto et al. (2006)	~100	303	–	–	1615	–	3.0
Fundulus heteroclitus	Commercial pellets	FW & SW	Wood et al. (2010)	40	230	170	77	628	326	1.5%
Goldfish	Commercial goldfish food	FW	Wood (unpublished)	70	308	243	107	406	220	–
Tilapia	Commercial cichlid flakes	FW	Wood (unpublished)	50	242	184	47	444	146	–
				40	382	344	83	291	255	–
Algavores	Commercial Spirulina flakes	FW	Wood (unpublished)	70	495	428	93	232	238	–
				60	131	92	98	219	146	–

The unifying feature of these data is their diversity. Note, for example, the large differences in ionic composition between two commercial flakes made for cichlids, and between two commercial feeds for herbivorous fish, formulated from *Spirulina*. Tacon and DeSilva (1983) surveyed the mineral composition of 33 commercially available fish feeds in Europe and came to a similar conclusion, noting batch-to-batch variability as well. This conclusion emphasizes the need for experimentalists to measure the ionic composition of the foods they are using.

Even when freshwater trout were placed on a low daily ration (0.6%) of a specially formulated low [NaCl] pellet diet (D'Cruz and Wood 1998), they were still ingesting almost 20 μmol kg^{-1} h^{-1} of Na and Cl, and this increased to about 250 μmol kg^{-1} h^{-1} on a normal ration (3%) of regular commercial salmonid pellets (Bucking and Wood 2006), and 3,300 μmol kg^{-1} h^{-1} when a higher ration (4%) was spiked with NaCl (Kjoss et al. 2005). Relative to branchial net flux rates and urinary loss rates (both usually less than 50 μmol kg^{-1} h^{-1}), all these ingestion rates are significant, and it is remarkable that the fish were able to survive on such varied dietary salt loads. Dietary Ca^{2+} loads may be similarly variable. Trout coped with an eight-fold increase in daily Ca^{2+} load in the studies of Baldisserotto et al. (2004a,b). Taylor and Grosell (2006a) were able to vary the dietary Ca^{2+} load of marine toadfish more than 400-fold by simply switching from a pilchard (teleost) diet to a squid (invertebrate) diet. In a metal-contaminated lake, freshwater yellow perch were faced with a 95% depletion of Ca^{2+} from their natural diet of mixed invertebrates (Klinck et al. 2007).

The assessment of ration in wild fish is difficult and controversial (see, for example, Boisclair and Marchand 1993; Cortes 1997); natural ration is likely higher on a wet mass food to wet mass fish basis but lower on a dry mass food to wet mass fish basis than for fish fed dry pellets in aquaculture. Nevertheless, fish are capable of considerable compensation by altering stomach capacity so as to keep dry matter intake relatively constant in the face of large differences in the water content of the food (Ruohonen and Grove 1996). Ration is also greatly temperature-dependent, declining at temperatures above and below optimum, and size/age dependent, declining as fish grow. A general guideline for adult fish kept close to optimum temperature would be that 1% daily ration of commercial pellets would maintain body weight whereas 3% would ensure good growth.

An important additional consideration is the passage rate of the ingested meal through the gastrointestinal tract. Examining the ion levels or water content of a meal and ration size in a feeding study generates a good general idea about the amount of ion/water ingested. However, the rate of passage can affect how much of that ion or water is absorbed. Sibly (1981) generated a model that examined net energy obtained by an animal as a function of retention time. The model can be used to determine the theoretical optimal digestion time, based on the amount of energy invested in digestion by the animal and the amount of energy provided by the food. The model could potentially be modified to examine many different aspects of digestive efficiency—the efficiency of Na^+ assimilation, for example.

4. IONIC INTERACTIONS OF INTEREST IN THE DIET/CHYME

Regardless of the size or ionic composition of the meal, once ingested, its digestion begins. As a result, components of the diet are broken down and/or released into solution, solubility in general being a prerequisite for absorption. However, the contents of the gastrointestinal tract do not consist merely of partially dissolved chyme. In addition to the chyme, there also are multiple elements that are either secreted by the intestine on a continual basis (e.g. mucus) or secreted specifically in response to feeding (e.g. bile). These elements contain molecules that can bind to the dissolved components of the chyme, and/or alter their transport. Additionally, the undissolved portion of the chyme itself can also interact with ions and water from the solubilized fractions, reducing their bioavailability.

Mucus lines the gastrointestinal tract to protect the epithelium from exposure to digestive enzymes, mechanical abrasion from food stuffs, acidic gastric pH, etc. Mucus is a gel-like substance formed from glycoproteins (mucins) that contain negatively charged sugars (sialic acid or sulfosaccharides; Perez-Vilar and Hill 1999). These charged groups interact to form the protective gel (Strous and Dekker 1992). However, this also creates a negatively charged barrier that salts and water must cross to be absorbed. In fish, the mucins contained in the intestinal goblet cells appear to be species-specific. Some species secrete predominantly negatively charged mucins (e.g. Shi drum *Umbrina cirrosa*; Leknes 2009) and some secrete predominantly neutral mucins (e.g. common dentex *Dentex dentex*; Carrasson et al. 2006). There also appears to be heterogeneity in the type of mucins secreted (Leknes 2009). *In vitro* studies have shown ionic interactions between mammalian mucins and mono-, di-, and trivalent ions (Crowther and Marriott 1984). In fact, Crowther and Marriott (1984) found that mucins had a higher affinity for Ca^{2+} over Na^+, with different binding sites involved. Additionally, increasing the ionic strength of the solution surrounding an *in vitro* mucus layer resulted in increased ionization of the weak acid side groups (Healy and White 1978) and increasing the NaCl concentration, as well as the calcium concentration, of the solution resulted in a reversible expulsion of water from the mucous layer (Feldoto et al. 2008).

Another source for ionic interactions in the gastrointestinal tract is bile. Bile is necessary for solubilizing fats for transport across the intestinal epithelium and consists of bile acids dissolved in a salt-rich fluid. In fish, the principal bile acids are sulfated bile alcohols (cyprinids, cartilaginous fish, and ancient fish; Hofmann 2009) or sulfate/taurine/glycine conjugated bile acids (bony fishes; Haslewood 1967; Denton et al. 1974; Zhang et al. 2001). Conjugating the bile acid with taurine or glycine or even sulfate reduces their pKa's which increases their solubility (Fini and Roda 1987). Both conjugation and elevated NaCl concentrations increase the solubility of the Ca^{2+}-bile acid salts that form as a result of interactions with the negatively charged bile acids (Gu et al. 1992). Conjugated bile acids are unlikely to be heavily affected by the salt and calcium load presented by feeding, as the concentrations encountered in the intestine are similar to the concentration found in the bile itself (see Section 6). However, unabsorbed bile acids can become de-conjugated in the intestine (in mammals; Northfield and McColl 1973; Setchall et al. 1988) which increases the likelihood of precipitation with Ca^{2+} and Na^+ (Hofmann and Roda 1984), reducing their bioavailability. Bile acids are also suspected of affecting salt and water transport, reducing *in vitro* Na^+ and Cl^- transport (Frizzell and Schultz 1970; Binder and Rawlins 1973), as well as *in vivo* water and Na^+ transport (Mekhjian and Phillips 1970).

Finally, the food itself contains fibers and components that can bind water molecules and cations, especially recently developed commercial fish feeds. Commercial diets are traditionally formed from fish meal; however, dwindling fish stocks are forcing the incorporation of novel protein and energy bases for feeds, typically formed from land-derived vegetables. Many of these vegetable fibers contain components that bind cations. For example, many plants store phosphorous as phytic acid which can chelate cations (Duffus and Duffus 1991), making them unavailable for absorption. In fact, catfish had decreasing carcass Ca^{2+} levels that correlated with increasing phytic acid levels in the feed (Toko et al. 2008). Additionally, despite increasing levels of dietary Mg^{2+}, there was a decreasing carcass Mg^{2+} concentration with increasing levels of cottonseed incorporated into the feed of tilapia (Mbahinzireki et al. 2001). Notably, after addition of phytase to the feed to break down the phytic acid, tilapia were able to increase the amount of Ca^{2+} absorbed from the diet (Papatryphon et al. 1999). Cellulose contains hydroxyl groups that can trap cations, a characteristic that was used to explain reduced Na^+ absorption from the diet with increasing cellulose content in fish feeds (Storbakken et al. 1998; Sugiura et al. 1998a,b; Hansen and Storbakken 2007). Not only can cations bind to these negatively charged fibers, but water can as well (Boulos et al. 2000). In fact, feed pellets with increasing cellulose contents had increasing water-holding capacities although in practice this did not result in a significant increase in the water content of the feces in trout (Hansen and Storbakken 2007).

The gastrointestinal tract is a complex environment with multiple elements, all of which can interact. For example, binding of Ca^{2+} to mucus is affected by both the pH of the intestinal solution as well as the NaCl concentration (Crowther and Marriott 1984; Kuver and Lee 2004), both of which are in turn affected by bile secretions. Additionally, the oligosaccharide regions of mucins are well hydrated with water; however, absorbed bile salts may disrupt the hydrogen bonds with the carbohydrates and release water to the intestine (Wiedmann et al. 2004). Clearly there are many interacting facets of the gastrointestinal tract, and absorption of salts and water is more complicated than simply looking at the amount that is ingested. To date no studies have investigated the combined effects of all the fractions in the gastrointestinal tract on salt and water absorption.

5. IMPACT OF FEEDING ON INTERNAL IONS AND PLASMA COMPOSITION

Like humans, freshwater fish often prefer moderately salty food (e.g. Salman and Eddy 1988a; Niyogi et al. 2006) and it has been a long-standing practice in freshwater aquaculture to add NaCl to food in the belief that it improves voluntary food consumption and growth rates, perhaps by reducing ionoregulatory costs. This may explain the high Na^+ and Cl^- levels in several of the commercial feeds in Table 5.1. However, the evidence for growth improvement is at best equivocal (Shaw et al. 1975; Murray and Andrews 1979; Salman and Eddy 1988a; Gatlin et al. 1992; Fontainhas-Fernandes et al. 2000; Harpaz et al. 2005). Salt-feeding is also often used to pre-condition cultured salmonids for transfer to seawater, and here the evidence for beneficial effects with respect to both survival (e.g. Fig. 5.4A) and growth during the first few weeks in seawater is much stronger in both salmonids and tilapia (Zaugg and McLain 1969; Basulto 1976; Al-Amoudi 1987; Salman and Eddy 1990; Pelletier and Besner 1992; Fontainhas-Fernandes et al. 2001).

As early as 1944, Phillips demonstrated that a meal delivering 30,000 $\mu mol\ kg^{-1}$ NaCl would significantly increase plasma Cl^- levels in brook trout, whereas higher doses ($>46,000\ \mu mol\ kg^{-1}$) would produce edema. Smith et al. (1995) reported that the threshold dose for raising plasma Na^+ concentration in freshwater rainbow trout was about 18,000 $\mu mol\ kg^{-1}$, and several other studies have similarly reported that comparable or higher dietary doses will raise plasma (and/or tissue) Na^+ and/or Cl^- levels (Salman and Eddy 1988b; Pelletier and Besner 1992; Pyle et al. 2003; Kamunde et al. 2005; Kjoss et al. 2005; Niyogi et al. 2006). Even at doses around the threshold for plasma Na^+ and Cl^- elevation, long-term salt-feeding will cause chronic elevations in arterial blood pressure, just as in humans (Eddy et al. 1990; Chen et al. 2007; Olson and Hoagland 2008).

To our knowledge, there is no evidence that diets "deficient" in Na^+ and Cl^- affect internal ionic levels, because of the ability of the gills to compensate by increasing uptake from the water (but see Section 6). However, there is some evidence that dietary K^+ may be occasionally limiting; Shearer (1988) reported that a concentration of about 200 mmol kg^{-1} food in a hatchery diet was necessary to ensure maximum growth and whole body K^+ levels in juvenile Chinook salmon in freshwater, which seems surprisingly high relative to the dietary values in Table 5.1. Both Mg^{2+} and Ca^{2+} can be actively taken up from the water by gills and skin, but there is abundant evidence that dietary Mg^{2+} (especially, Dabrowska et al. 1991) and Ca^{2+} may be limiting, particularly when ambient levels are low, such as in soft freshwaters (reviewed by Lall 2002; Flik and Verbost 1993; Bijvelds et al. 1998). Muscle flaccidity and low plasma and bone Mg^{2+} content in response to low Mg^{2+}, poor mineralization of bone and scales in response to low Ca^{2+}, and reduced growth in response to both dietary deficits are the usual symptoms. To prevent these problems, minimum recommended dietary Mg^{2+} and Ca^{2+} concentrations in commercial diets for freshwater fish are about 20 mmol kg^{-1} food and 75 mmol kg^{-1} food, respectively. Meals spiked with high doses of Mg^{2+} (193 mmol kg^{-1} food; Oikari and Rankin 1985) and Ca^{2+} (1,450 mmol kg^{-1} food; Baldisserotto et al. 2004a) resulted in relatively short-lasting surges in plasma Mg^{2+} and Ca^{2+} concentrations. At least for the latter, these surges disappeared after the fish had been on the new diet for several days, suggesting an up-regulation of homeostatic mechanisms and/or a down-regulation of gut Ca^{2+} absorption.

One mineral for which the diet is overwhelmingly important in the aquatic environment is inorganic phosphate. Accordingly, to date the mechanisms of *in vitro* intestinal transport, as well as the influence of dietary phosphorus on intestinal transport and plasma phosphate concentrations, have been examined (eg. Avila et al. 1999, 2000; Coloso et al. 2003a). Recently, research has focused on phosphate transporter expression (e.g. Sugiura et al. 2003; Sugiura and Ferraris 2004a) and identifying candidate genes involved in phosphorus homeostasis (Sugiura and Ferraris 2004b; Kirchner et al. 2007). Additionally, there is an immense research effort in aquaculture to increase the digestibility and absorption efficiency of dietary phosphate (e.g. Vielma and Lall 1997; Avila et al. 1999; Coloso et al. 2001, 2003b; Sugiura et al. 2006). Inorganic phosphate, the end-product of the digestion of dietary phosphorus, is a major contaminant associated with aquaculture and efforts to reduce the environmental impact of the effluent are important in creating sustainable practices. However, diets with low phosphate availability result in growth limitation and markedly depressed levels of inorganic phosphate in blood plasma. One option with some success in meeting dietary requirements of phosphorus along with reducing the amount of phosphorus in the effluent is replacing animal-based feeds that contain bone phosphate (hydroxyapatite) with plant-based feeds supplemented with rock phosphorus (Ketola and Harland 1993), which is less soluble in water. Another option is incorporating vitamin D into the diets, as it is important in mammalian and avian phosphate homeostasis (DeLuca 2004) and is suspected to fill a similar role in fish (reviewed by Lock et al. 2010). While vitamin D incorporation had limited success in increasing the intestinal transport of phosphorus in fish, it did result in elevated plasma phosphorus levels possibly as a result of increased renal reabsorption (Avila et al. 1999; reviewed by Lock et al. 2010).

Do "normal" meals affect plasma ionic composition? With respect to natural diets the answer is generally no for major plasma ions. Taylor and Grosell (2006a) noted only insignificant variations in the plasma Na^+, K^+, Ca^{2+}, Mg^{2+} and Cl^- concentrations of gulf toadfish following large voluntary meals (9% ration) of two natural prey items with very different ionic compositions (see Table 15.1). Similarly, in European flounder consuming ragworms (5% ration; Taylor et al. 2007), and dogfish sharks consuming hake (5.5% ration; Wood et al. 2007b—see Table 5.1) postprandial fluctuations in these major plasma ions were negligible. However, for pelleted commercial feeds, rather variable results have been reported. Hille (1984) documented marked falls in plasma Na^+ and Cl^- concentrations of freshwater rainbow trout after 1.5–2.0% meals, and speculated that the decreases were due to increased gastrointestinal HCl and $NaHCO_3$ secretion. On the other hand, in the same species fed voluntary 3.1% meals in freshwater, Bucking and Wood (2006a,b, 2007) found increases in plasma Na^+ at 2 h through 8 h post-feeding, no change in plasma Cl^- and K^+, and minor fluctuations in Ca^{2+} and Mg^{2+}. However, in seawater-acclimated rainbow trout, there were no changes in any plasma ions following a comparable pellet meal (Bucking et al. 2010b). In freshwater-acclimated trout, Cooper and Wilson (2008) reported no changes in any of these plasma ions after voluntary 1% meals, but a marked decline in plasma Cl^- after 6 h and a persistent decline in plasma Mg^{2+} up to 72 h when the same ration was force-fed to trout as an aqueous slurry by an indwelling stomach tube. The gradients in ion concentrations, osmolality, and water content between chyme and plasma are very large for these pelleted diets (see Bucking and Wood 2006a,b, for example), so it is possible that variations in pellet composition and the amount of water ingested with the pellets (see Sections 3 and 5) explain this variability.

Chapter 6 discusses further changes in the plasma caused by digestion including increased ammonia levels and the alkaline tide.

6. FRESHWATER FISH: THE CRITICAL IMPORTANCE OF ION ACQUISITION FROM THE DIET

For most freshwater fish, the uptake of a significant portion of their major electrolytes from the diet is the normal situation, and in some circumstances it may become critical to survival. In the case of inorganic phosphate, the diet always likely predominates. One great advantage of uptake from the diet over uptake through the gills is that concentrations in the chyme are generally much higher (millimolar range) than in the external water (micromolar range). Thus gastrointestinal uptake is undoubtedly less costly than branchial uptake, especially since the cost of ion transport may be shared—i.e. many uptake pathways are coupled as Na-nutrient co-transport systems (see Chapter 2). For at least three groups which have been shown to lack active Cl^- uptake at the gills in freshwater (eels, Kirsch 1972; killifish, Patrick et al. 1997; bluegills, Tomasso and Grosell 2005), dietary Cl^- uptake is presumably essential (Scott et al. 2006, 2008).

There have been some estimates of the proportion of electrolytes taken from the food versus the water—e.g. Mg^{2+} ($> 50\%$; Bijvelds et al. 1998), Ca^{2+} (15%—Berg 1970; 30%—Flik and Verbost 1993), Na^+ (0–60%—Smith et al. 1989), inorganic phosphate ($> 50\%$—Lall 2002) but the true situation is likely so seasonal, water-specific, food-specific, and species-specific as to make these percentages not particularly useful. Certainly, the fact that there are established dietary minima (see Section 5) for K^+, Mg^{2+}, Ca^{2+}, and inorganic phosphate contents of commercial food used in aquaculture (Lall 2002) speaks to the importance of the diet. A more useful generalization may be that the lower the concentration of ions in natural water, the more important is uptake from the diet. This is particularly true in ion-poor softwater such as occurs on the Canadian Shield, Scandinavia, and the Amazon basin (e.g. Rodgers 1984; Gonzalez et al. 2005). Uptake from the diet also becomes critical to growth and survival when normal branchial transport mechanisms are blocked by environmental toxicants. For example, D'Cruz and Wood (1998) demonstrated that it was the NaCl content of the diet rather than its energy content which allowed rainbow trout to ionoregulate and survive chronic exposure to low pH (5.2) in softwater; given the opportunity, these acid-exposed fish ate more than the controls kept at circum-neutral pH (D'Cruz et al. 1998). Cu is another environmental toxicant that blocks Na^+ and Cl^- uptake at the gills; supplementation of the diet with NaCl improved survival and growth in rainbow trout subjected to chronic waterborne Cu exposure (Kamunde et al. 2005). However, when given a choice, Cu-exposed trout were unable to selectively choose the saltier food, perhaps because their chemosensory abilities were blocked by waterborne Cu (Niyogi et al. 2006).

Recently, Bucking and Wood (2006a,b, 2007, 2009) have quantitatively assessed the time-dependent contributions of different parts of the digestive tract of the freshwater rainbow trout to Na^+, K^+, Ca^{2+}, Mg^{2+}, Cl^- and water uptake or loss (Fig. 5.3) from a single meal of commercial trout pellets (3.1% ration; composition in Table 5.1). In Fig. 5.3, upward bars (positive values) represent addition of ions or water to the gastrointestinal tract section over the specified time span, while downward bars (negative values) represent removal of ions or water from the tract—i.e. absorption into the fish. Ballotini beads, quantified by X-radiography, were used as an inert marker (see Section 2), and care was taken to account for the hydration of the pellets during the short period in which they were in the water prior to ingestion. Overall absorption efficiencies were $Na^+ = -9\%$, $K^+ = +89\%$, $Ca^{2+} = +28\%$, $Mg^{2+} = +60\%$, and $Cl^- = +81\%$, whereas about 17 ml kg^{-1} of water was lost. The latter figure suggests that the high osmotic pressures generated in the chyme (1.3–2.7-fold higher than in blood plasma) by the relatively dry pellets are advantageous to osmoregulation by making the gut a route of net water loss, at about 10% of the rate of the kidney. However, this conclusion must be tempered by the fact that drinking was not measured, and there is some evidence that freshwater fish take in water orally during and after feeding (Tytler et al. 1990; Ruohonen et al. 1997; Kristiansen and Rankin 2001). Regardless, there was a very large addition of fluid to the chyme in the stomach initially, corresponding to the time of maximum osmotic gradient (Fig. 5.3A), and this continued through until 12 h after the meal. Kristiansen and Rankin (2001) provided similar evidence for large fluid addition in the stomach. In light of the simultaneous secretion of Cl^- (Fig. 5.3C) much of this was undoubtedly due to elevated gastric HCl secretion (Bucking and Wood 2009) at an estimated concentration of 240 mmol L^{-1} (i.e. pH=0.6). However, Cl^- was later taken back up in the stomach, followed by modest absorption in the other sections, in all cases against the concentration gradient. The generally similar trends in stomach K^+ fluxes (initial secretion Bucking and Wood 2009, later absorption; Fig. 5.3D) may reflect K^+ recycling associated with the K^+-dependent H^+-ATPase pump (Geibel and Wagner 2006). Fluid addition to the chyme continued in the anterior intestine, likely due to biliary and pancreatic secretions, whereas there was modest reabsorption in the distal parts of the intestine (Fig. 5.3A). In contrast to Cl^- (Fig. 5.3C), Na^+ was strongly absorbed in the stomach (against a concentration gradient) but secreted in the anterior intestine, likely in the biliary and pancreatic fluids, followed by modest absorption in the mid-intestine (Fig. 5.3B). Dabrowski et al. (1986) also provided evidence of Na^+ secretion in the anterior intestine. Ca^{2+} (Fig. 5.3E) and Mg^{2+} (Fig. 5.3F) were largely absorbed in the stomach, with small net secretory fluxes in the anterior intestine and variable fluxes in

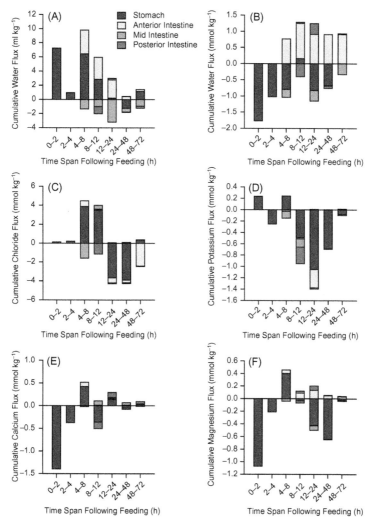

Fig. 5.3. Contribution of each section of the gastrointestinal tract to the total water (ml kg^{-1}) and ion (mmol kg^{-1}) fluxes during digestion in freshwater rainbow trout over a specified time span following ingestion of a meal (3.1% ration) of commercial pellets. (A) Water, (B) Na$^+$, (C) Cl$^-$, (D) K$^+$, (E) Ca^{2+}, and (F) Mg^{2+} fluxes. Upward bars (positive values) represent addition of ions or water to the gastrointestinal tract section over the specified time span, while downward bars (negative values) represent removal of ions or water from the tract—i.e. absorption into the fish. Feeding occurred immediately after time 0 h. Compiled from Bucking and Wood (2006a,b, 2007).

the lower tract. The net absorptions of K^+, Ca^{2+}, and Mg^{2+} all occurred in the direction of the chyme-to-plasma concentration gradients.

Surprising findings of these studies (Bucking and Wood 2006a,b, 2007) include the importance of the stomach as site of net uptake for most ions, of the anterior intestine as a site of net loss for most ions, the rather small net fluxes in the mid- and posterior intestine, and the lack of net Na^+ absorption. The overall absorption efficiencies documented for $K^+ = +89\%$, $Ca^{2+} = +28\%$, and $Mg^{2+} = +60\%$ were very typical of those ($K^+ = +96–99\%$; $Ca^{2+} = +13–60\%$; $Mg^{2+} = +61–73\%$) reported in several aquacultural studies on trout which employed the Cr_2O_3 or Y_2O_3 inert marker technique (Sugiura et al. 1998a,b, 1999; Overturf et al. 2003; see Section 2). The slightly negative value for Na^+ (-9%) was in the midrange of extremely variable Na^+ absorption efficiencies (-57% to $+71\%$) reported in these same studies. It will be interesting to see whether these same patterns occur in other freshwater teleosts.

While the diet may act as a source for certain ions for freshwater fish, digesting the meal itself may create ion and water losses, primarily through bile and pancreatic secretions. The pancreas in fish is diffuse, and isolating the secretions is difficult (see Chapter 7 for further information on pancreatic form and function). In contrast, there are several studies examining bile composition in fish. Extensive summaries of gallbladder bile composition from a variety of freshwater species have been presented by Diamond (1962) and Hunn (1972, 1976); typical concentrations are around Na^+ 200, Ca^{2+} 20, Mg^{2+} 8, K^+ 5, and Cl^- only 30 mmol L^{-1}, with the bile salts representing the predominant anions. However, the reader should be aware that these measurements were made on fasted fish, and so would represent the composition of bile released immediately on ingestion of a meal after a prolonged fast. The detailed study of Grosell et al. (2000) on rainbow trout demonstrated that there were substantial differences in concentration between hepatic bile (which is entering the gallbladder during fasting or passing directly into the intestine during prolonged digestion), and bile which had been sequestered in the gallbladder for some time. This is due to transport processes in the gallbladder epithelium which progressively modify the ionic composition of stored bile. Thus Na^+, Ca^{2+}, and Mg^{2+} concentration levels were approximately 50% lower, bile acid concentrations about 80% lower, and Cl^- concentrations five-fold higher in hepatic bile, which was produced at a constant rate of about 75 μl kg^{-1} h^{-1}. Regardless of the source, the secretion of bile into the anterior intestine may serve as a route of loss for many ions, as well as fluid, to the intestinal tract. For example, if the hepatic bile composition of the rainbow trout is taken as an example, these fish stand to lose roughly 12 μmol Na kg^{-1} h^{-1} during digestion (Grosell et al. 2000). This is after they have secreted potentially 600 μmol Na kg^{-1} from the gallbladder. If digestion occurs over a 48 hour period, rainbow trout could secrete roughly 1.2 mmol Na kg^{-1}, along with roughly 5.8 ml kg^{-1} water. As mentioned in Section 4, bile acids can reduce sodium and water absorption as well as contribute to additional water loss by dehydrating mucus, potentially aiding in osmo-regulation for freshwater fish (see Section 7).

7. FRESHWATER FISH: THE IMPACT OF FEEDING ON GILL AND KIDNEY FUNCTION

This is an extremely important area, but to date, responses to normal feeding have received scant attention, due to the preference of most investigators to work with fasted fish (see Section 1). In effect, most data in the literature on gill and kidney iono/osmoregulatory function have been collected under the abnormal, potentially stressful condition of fasting. Recent exceptions have mainly focused on the impacts of feeding on branchial ammonia excretion and its relationship to acid–base exchange.

7.1. Gills

Not surprisingly, branchial ammonia excretion increases dramatically after feeding in most teleosts, with as much as 50% of the N-content of the meal being lost over the next 12 h (reviewed by Wood 1995, 2001a). In one study on the rainbow trout, net base excretion to the water increased greatly at the same time as a compensation of a postprandial alkaline tide in the bloodstream (Bucking and Wood 2008), but this was not seen in another study on the trout although an internal alkaline tide again occurred (Cooper and Wilson 2008). It also did not occur in two species (European flounder—Taylor et al. 2007; common killifish—Wood et al. 2010) where an alkaline tide appeared to be absent. This area is explored in greater detail in Chapter 6.

Theoretically, we would expect perturbations in branchial acid–base balance to be associated with alterations in Na$^+$ and Cl$^-$ fluxes through Na$^+$/H$^+$ and Cl$^-$/HCO$_3^-$ linkages (Evans et al. 2005; Marshall and Grosell 2006) and perturbations in ammonia excretion to be associated with possible changes in Na$^+$/NH$_4^+$ exchange processes (Weihrauch et al. 2009; Wright and Wood 2009). On top of this, ions absorbed from the normal meal might be expected to alter gill Na$^+$, Cl$^-$, K$^+$, Ca^{2+}, Mg^{2+}, etc. fluxes. However, the sole measurements appear to be those of Taylor et al. (2007) on Na$^+$ and Cl$^-$ influx rates in the European flounder after a meal of ragworms; minor, only non-significant fluctuations occurred.

However, there has been much more attention paid to the branchial responses to ion-supplemented "abnormal" meals, and particularly to salt feeding. In rainbow trout fed diets containing 760 or 1,270 mmol kg^{-1} food

Na^+ (as NaCl, at a daily 3% ration), unidirectional Na^+ influx from the dilute, low Na^+ water was reduced by about 40% (Kamunde et al. 2003; Pyle et al. 2003). Unidirectional efflux rates were not affected, so net flux rates became negative—i.e. the fish lost Na^+ to the water. When the dietary Na^+ level was raised to 1,890 mmol kg^{-1} food, unidirectional Na^+ efflux rates were greatly increased, so net flux rates became more negative (Niyogi et al. 2006). Smith et al. (1995) worked with a still higher dietary salt load (2,100 mmol kg^{-1} food as NaCl) and allowed the trout to consume voluntary meals. Unidirectional Na^+ influx was reduced by about 70% regardless of meal size, whereas unidirectional Na^+ efflux increased only when more than 18,000 μmol kg^{-1} Na^+ was consumed in the single meal, which appears to be about the threshold dose for increasing plasma Na^+ and Cl^- concentrations (see Section 5). Overall, the inhibition of branchial influx appears to be more sensitive than the stimulation of efflux, but the latter is more important in clearing the Na^+ load at higher doses. Interestingly, the uptake of Cu at the gills, which is thought to occur in part via the Na^+ transport pathway (Grosell and Wood 2002), was also down-regulated in response to salt-feeding (Kamunde et al. 2003, 2005; Pyle et al. 2003; Niyogi et al. 2006).

Chronic feeding of freshwater rainbow trout with high NaCl diets (1,900 mmol kg^{-1} food; 1.5% daily ration for >4 weeks) induced changes in gill morphology, gene expression, and protein abundance reminiscent of those seen when salmonids enter seawater (Salman and Eddy 1987; Perry et al. 2006). These included increases in Na^+, K^+-ATPase activity (e.g. Fig. 5.4D, first shown by Zaugg and McLain (1969) in Pacific salmon), in expression of mRNAs for three transporters thought to be important for NaCl extrusion in seawater (CFTR; NKCC1, and Na^+,K^+-ATPase) and the appearance of seawater-type "chloride cells" (mitochondria-rich cells) together with accessory cells (Perry et al. 2006). The number of chloride cells (Fig. 5.4B) and overall Na^+, K^+-ATPase activity (Fig. 5.4D) increased more or less linearly with the NaCl content of the diet (Salman and Eddy 1987). This epithelial structure is thought to facilitate secondarily active Na^+ extrusion in seawater fish (Evans et al. 2005; Marshall and Grosell 2006), so it likely contributes to the reported increases in branchial Na^+ efflux rates in salt-fed fish. It also may explain how salt-feeding in aquaculture helps precondition salmonids for successful transfer to seawater (Fig. 5.4A) (Zaugg and McLain 1969; Basulto 1976; Zaugg et al. 1983; Salman and Eddy 1990; Pelletier and Besner 1992).

Branchial responses to high Ca^{2+} diets have also been examined recently. Perhaps not surprisingly, trout fed with commercial pellets enriched in Ca^{2+} as either $CaCO_3$ or $CaCl_2$ (approximately 1,500–1,600 mmol kg^{-1} food; 3% daily ration for >7 days) exhibited up to 70% decreases in unidirectional Ca^{2+} influx rates from the water (Baldisserotto et al. 2004a,b, 2005). Unidirectional Na^+ uptake rates were not affected; however, neither unidirectional Ca^{2+}

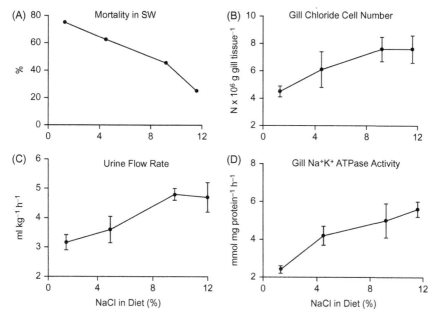

Fig. 5.4. The influence of feeding diets supplemented with NaCl on various physiological parameters in freshwater rainbow trout. (A) Mean % survival at 3 weeks after abrupt transfer to seawater (data from Salman and Eddy 1990); (B) gill chloride cell numbers (data from Salman and Eddy 1987); (C) urine flow rate (UFR) at 48 h after catheterization (data from Salman and Eddy 1988b); and (D) gill Na^+, K^+-ATPase activity (data from Salman and Eddy 1987). In all cases, the difference between the lowest and the highest [NaCl] diets was statistically significant.

efflux rates nor renal responses were measured. The high Ca^{2+} meals caused surges in plasma Ca^{2+} levels, so branchial Ca^{2+} uptake was probably decreased by the well-characterized stanniocalcin system which is keyed to homeostasis of plasma Ca^{2+} levels (Flik and Verbost 1993). Analogous to the reduction in unidirectional Cu uptake seen with high Na^+ diets, high Ca^{2+} diets reduced the unidirectional uptake of Cd (Baldisserotto et al. 2004a,b, 2005; Franklin et al. 2005) and Zn (Niyogi and Wood 2006), two metals which are known to be taken up at least in part through the branchial Ca^{2+} transport pathway (Verbost et al. 1987; Spry and Wood 1989; Wood et al. 2001b).

7.2. Kidney

Wood (1995) compared urinary parameters in rainbow trout which had been fed to satiation with a normal commercial pellet diet up until the time of bladder catheterization with those of trout which had been fasted for 7–10 days beforehand. Over the following 3 days, urine flow rate (UFR) and Na^+,

K^+, and Cl^- efflux rates all tended to be lower in the fed fish, whereas ammonia and acidic equivalent efflux rates were slightly higher. In a more detailed study, Bucking et al. (2010a) examined renal responses of trout (which had been fasted for a week beforehand) to a single meal of unamended commercial pellets (1.5% ration, administered as a slurry by gavage). The principal discrepancy between the two investigations was the finding by Bucking et al. (2010a) of increased metabolic base excretion through the kidney after feeding. Relative to a large branchial base excretion response, this accounted for only about 5% of the compensation of the postprandial alkaline tide. Possibly, the difference is due to the different protocols used in the two studies, because force-feeding seems to exacerbate the systemic alkaline tide (Cooper and Wilson 2008).

There was agreement that feeding results in lower UFR, and this was accompanied by a decrease in GFR (glomerular filtration rate; Bucking et al. 2010a). The reduction in water excretion amounted to about 18 ml kg^{-1}, almost identical to the 17 ml kg^{-1} lost from the digestive tract (Bucking and Wood 2006a), supporting the idea that dry commercial diets may reduce osmoregulatory costs in freshwater fish. In fact, an increase in growth and feed conversion efficiency was seen in Atlantic salmon when the water content of the diet was reduced from 30% to 10% (Hughes 1989) while the growth of rainbow trout was impaired at dietary water contents above 50% (Ruohonen et al. 1998). It appears that this effect might be species-specific as similar findings have not been found with different species (sea bass, Chou 1984; pink salmon, Higgs et al. 1985), although in some species a non-significant trend for increasing growth rate with decreasing water content was seen (turbot, Bromley, 1980; brown trout, Poston 1974).

Urinary K^+ excretion did not change after feeding (Bucking et al. 2010a), whereas there were slight increases in urinary Na^+, Cl^-, and Ca^{2+} effluxes, rather different from the pattern reported by Wood (1995). Nevertheless overall urinary ion fluxes in both studies were only a few percent of the dietary ion load, with the sole exception of Mg^{2+}. Clearance ratio analysis demonstrated that the kidney switched over from net Mg^{2+} reabsorption to net secretion, clearing 27% of the absorbed Mg^{2+} load over 48 h, in accord with an earlier study indicating the importance of the kidney in excreting this ion (Oikari and Rankin 1985).

When freshwater trout were fed satiation rations of salt-laden commercial diets (1,980 mmol kg^{-1} food) for two months prior to bladder catheterization, UFR and GFR increased by up to 50% (Fig. 5.4C), and urinary Na^+ and Cl^- excretion rates doubled, though they accounted for only a few percent of the daily NaCl load (Salman and Eddy 1988b). Clearance ratio analysis demonstrated that the increased UFR and NaCl excretion rates were due to both increased glomerular filtered load and reduced tubular reabsorption. The increase in GFR can be attributed to an increased arterial blood pressure (Eddy et al. 1990; Chen et al. 2007) associated with increases in cardiac output and blood volume (Olson and Hoagland 2008), undoubtedly associated with the frank edema first observed by Phillips (1944). Drinking rate probably also increased (Pyle et al. 2003).

8. SEAWATER FISH: THE CRITICAL IMPORTANCE OF DRINKING AND WATER CONSERVATION BY THE GASTROINTESTINAL TRACT

As detailed in Chapter 4, the basic iono- and osmoregulatory functions of the gastrointestinal tract in fasted marine teleosts were first characterized by the classic work of Smith (1930), but important new mechanistic information has been added in the past decade (reviewed by Grosell 2006 and Grosell et al. 2009b). In brief, seawater is continually imbibed, and undergoes considerable desalination (removal of Na^+ and Cl^-) as it passes down the esophagus, such that the resulting fluid is close to isotonic by the time it reaches the stomach. In the intestine, the progressive absorption of Na^+, Cl^- and K^+ drives the osmotic absorption of water, and more recent research has shown that this is further aided by Cl^-/HCO_3^- exchange processes that serve to precipitate calcium and magnesium carbonates, thereby lowering the osmolality of the intestinal fluid. The toxic divalent ions Mg^{2+} and SO_4^{2-} are largely excluded from absorption, while relative Ca^{2+} uptake is much lower than that of the monovalent ions which are more than 90% absorbed on a net basis. Key studies include those of Hickman (1967), Shehadeh and Gordon (1969), Fletcher (1978), Kirsch and Meister (1982), and Parmelee and Renfro (1983), while the more recent findings on the importance of Cl^-/HCO_3^- exchange have been summarized by Grosell (2006) and Grosell et al. (2009b).

There is only limited information on the same issues in fed marine teleosts. Given the abundance of most major electrolytes in seawater, marine fish are probably far less dependent upon ion uptake from the diet than are freshwater fish, because electrolytes can be acquired both via the gills and from seawater which is continually drunk as a critical part of their osmoregulatory strategy. Nevertheless, ions certainly are absorbed from the food, and indeed it becomes quite difficult to separate ionic uptake from the food versus ionic uptake from ingested seawater. However, what is likely of greatest importance is water uptake from natural food. For example, if a hyporegulating teleost eats another one or a hyporegulating invertebrate, then it will be less costly to "extract" water from this prey ($\sim$350 mOsmol kg^{-1}) than from the full-strength seawater ($\sim$1,050 mOsm kg^{-1}) which may be ingested along with the prey. To our knowledge, no study has directly addressed this issue. However, Taylor and Grosell (2006a) employed changes in the concentration of Mg^{2+} and SO_4^{2-} in the chyme as approximate "inert" markers for water fluxes, and concluded that marine toadfish clearly absorbed water from two natural diets, sardines and squid (see Table 5.1 for composition). Whether this absorption from natural food is greater than from seawater remains to be determined, but measurements of chyme composition suggest that the rate of intestinal Cl^-/HCO_3^- exchange is elevated after feeding in both the toadfish (Taylor and Grosell 2006a) and the European flounder (Taylor et al. 2007). This conclusion has been supported by *in vitro* experiments on gut sac preparations of the common killifish (Marshall et al. 2002; Wood et al. 2010) and the rainbow trout (Bucking et al. 2009), all showing enhanced rates of water absorption and either Cl^- uptake or HCO_3^- secretion after feeding, as well as Ussing chamber experiments on the toadfish demonstrating enhanced rates of intestinal HCO_3^- secretion at this time (Taylor and Grosell 2009).

With respect to "dry" commercial pellet diets, the low water contents and elevated ion concentrations (e.g. Table 5.1) may offer very different challenges from those of natural food, as first recognized by Dabrowski et al. (1986). Recently, Bucking et al. (2010b) have addressed these issues in seawater-acclimated rainbow trout of the steelhead strain (Fig. 5.5). Figure 5.5 uses the same format as Fig. 5.3: upward bars represent addition of ions or water to the gastrointestinal tract section while downward bars represent absorption into the fish. Methods (ballotini beads as an inert marker in the feed, 3.4% ration) were very similar to those employed with freshwater trout (Fig. 5.3), thereby allowing direct comparison of the two data sets, but included the additional measurement of drinking rate. In contrast to at least one previous study on seawater teleosts (Atlantic salmon; Usher et al. 1988) drinking rate remained unchanged following feeding and appeared to entirely account for water appearance in the stomach (71 ml kg^{-1} over 48 h). There was a significant secretion of fluid into the anterior intestine (7 ml kg^{-1}), likely due to postprandial biliary and pancreatic secretions, whereas there was absorption (52 ml kg^{-1}) in the distal parts of the intestine (Fig. 5.5A). Overall water absorption efficiency from the combined meal plus seawater drinking, over 48 h, was only +62%, rather lower than previously reported for unfed rainbow trout in seawater (+80–85%; Shehadeh and Gordon 1969; Wilson et al. 1996), as well as for marine southern flounder (+76%; Hickman 1967) and cod (> +68%; Fletcher 1978). Of course all these values are much higher than in fed

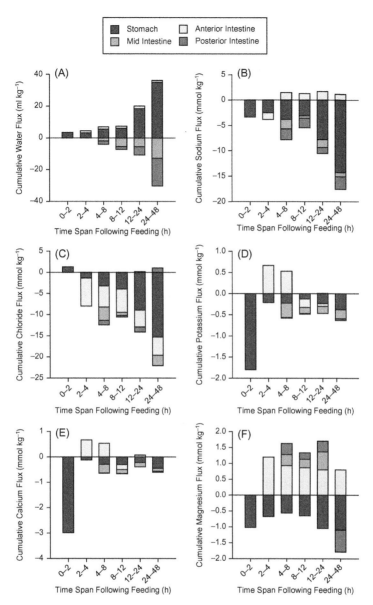

Fig. 5.5. Contribution of each section of the gastrointestinal tract to the total water (ml kg^{-1}) and ion (mmol kg^{-1}) fluxes during digestion in seawater rainbow trout over a specified time span following ingestion of a meal (3.1% ration) of commercial pellets. (A) Water, (B) Na^{+}, (C) Cl^{-}, (D) K^{+}, (E) Ca^{2+}, and (F) Mg^{2+} fluxes. Upward bars (positive values) represent addition of ions or water to the gastrointestinal tract section over the specified time span, while downward bars (negative values) represent removal of ions or water from the tract—i.e. absorption into the fish. Feeding occurred immediately after time 0 h. Note that ingested seawater provided additional ions for assimilation beyond those provided by the meal itself. Compiled from Bucking et al. (2010b).

freshwater trout (Fig. 5.3A) where a comparable pellet meal caused a net water loss (i.e. negative assimilation efficiency; Bucking and Wood 2006a). The lower water absorption efficiency in fed trout may be explained by the high osmotic pressures exerted by the solubilized pellets in the chyme, the water-trapping ability of the fibrous material in the pellets (see Section 4), and/or elevated biliary and pancreatic secretions.

Overall absorption efficiencies from the combined meal plus seawater drinking over 48 h were $Na^+ = +86\%$, $K^+ = +93\%$, $Ca^{2+} = +43\%$, $Mg^{2+} = -5\%$, and $Cl^- = +96\%$. This analysis confirms the traditional view that Mg^{2+} is largely excluded (Smith 1930), in contrast to the 60% absorption efficiency in freshwater trout, whereas for all other ions, the absorption efficiencies were greater than in fed freshwater trout (see Section 6, Fig. 5.3, and Bucking and Wood 2006b, 2007). However, Mg^{2+} absorption efficiencies in two other studies on fasted seawater trout were +48% (Shehadeh and Gordon 1969) and +41% (Wilson et al. 1996), and +16% in fasted seawater flounder (Hickman 1967), so it is possible that feeding actually inhibits Mg^{2+} absorption. Perhaps the increased HCO_3^- secretion increases magnesium carbonate precipitation in the tract. By way of comparison, values for other ions in studies on fasted marine teleosts were $Na^+ = +96–99\%$, $K^+ = +86–100\%$, $Ca^{2+} = +31–37\%$, and $Cl^- = +92–99\%$ (Hickman 1967; Shehadeh and Gordon 1969; Fletcher 1978; Wilson et al. 1996), very comparable to the values in fed trout. Given the high concentrations of all ions in seawater, these translate to absolute absorption rates which are much higher than in freshwater fish (compare Fig. 5.5 versus Fig. 5.3). For Na^+ and Cl^-, the vast majority of this absorption occurred from the imbibed seawater, which had Na^+ and Cl^- levels 2–3-fold higher than in the commercial pellets (Table 5.1), so the diet supplied at most 11–16%. However, for K^+ and Ca^{2+}, concentrations in the feed (Table 5.1) were 11-fold and 17-fold higher than in the seawater, so assimilation from the diet was 4–5-fold greater than from the seawater.

Regional analysis based on the ballotini beads indicated that there were large uptake fluxes of all ions (including even Mg^{2+}) in the stomach (Fig. 5.5), though it must be appreciated that this would include uptake across the esophagus, which is renowned for its NaCl uptake capacity (Parmelee and Renfro 1983). This swamped out any indication of HCl secretion by the stomach (Fig. 5.5C), though there was clear evidence of progressive gastric acidification (Bucking et al. 2010b). The anterior intestine was a site of net secretion for all ions (Fig. 5.5B, D, E, F) except Cl^- (Fig. 5.5C), presumably due to biliary and pancreatic secretions, whereas the mid- and posterior intestine were sites of further Na^+, K^+, and Cl^- absorption (Fig. 5.5B, C, D) but negligible Ca^{2+} or Mg^{2+} uptake (Fig. 5.5E, F). Again, surprising features and similarities to the situation in freshwater trout after feeding (compare Fig. 5.5 versus Fig. 5.3) were the importance of the stomach as a site of net uptake for most ions, of the anterior intestine as a site of net secretion for most ions, and the rather small net fluxes in the mid- and posterior intestine. The zero or low net uptakes of Mg^{2+} and Ca^{2+} clearly reflect recycling from the stomach to the anterior intestine, rather than impermeability of the entire tract.

9. SEAWATER FISH: THE IMPACT OF FEEDING ON GILL AND KIDNEY FUNCTION

With the exception of recent data on acid–base and ammonia fluxes, which are elaborated in greater detail in Chapter 6, almost nothing is known about the impact of feeding on gill or kidney function in marine teleosts, though there is some information on marine elasmobranchs (Section 10). The postprandial elevation in ammonia efflux appears to be somewhat faster in marine teleosts than in their freshwater counterparts fed similar rations (Taylor et al. 2007; Bucking et al. 2009 versus Bucking and Wood 2008; Wood et al. 2010), and most of this probably occurs at the gills, given the very low urine flow rates (Marshall and Grosell 2006) and negligible involvement of the kidney in acid–base balance (McDonald et al. 1982). Neither the European flounder (Taylor et al. 2007) nor the killifish (Wood et al. 2010) exhibited any disturbance in net acid–base flux to the water after feeding. However, in contrast to the elevated base efflux to the water seen after feeding in freshwater trout (Bucking and Wood 2008), there was actually an elevated base uptake from the external water during the period of the systemic alkaline tide in the seawater-adapted rainbow trout (Bucking et al. 2009). This may serve to fuel the increased rate of Cl^-/HCO_3^- exchange in the intestine during food processing (see Section 6), so a disturbance in branchial and/or renal Cl^- fluxes might be predicted at this time. Although urine flow rates are low, urinary Cl^- and Mg^{2+} concentrations are generally high in marine teleosts (Smith 1930; Hickman 1967; Shehadeh and Gordon 1969; Fletcher 1978; Wood et al. 2004).

The analysis of net gut fluxes in fed trout (Bucking et al. 2010b; see Section 8, Fig. 5.5) suggests that the other ions which may exhibit disturbed balance after a meal of commercial pellets are Mg^{2+}, Ca^{2+}, and K^+. If Mg^{2+} absorption is really reduced by feeding, while K^+ and Ca^{2+} absorption are increased 4–5-fold as the data seem to indicate, then decreased renal Mg^{2+} excretion (Beyenbach and Kirschner 1975; Oikari and Rankin 1985) and increased efflux of K^+ and Ca^{2+} would be expected. At least for the latter, the renal route seems to dominate (Bjornsson and Nilsson 1985), but it is likely that a significant portion of both may be put into the formation of muscle (K^+) and bone (Ca^{2+}) in actively growing fish. If gastrointestinal water absorption is increased from natural food (prey which are hypotonic to seawater) and decreased by pellet food (see Section 6), then water fluxes at both gills and kidney may also be altered by feeding. All these ideas remain to be tested experimentally.

10. MARINE ELASMOBRANCHS—A SPECIAL CASE

In addition to his classic work on marine teleosts (Smith 1930), Homer Smith (1931b, 1936) also elucidated the very different strategy of osmoregulation in marine elasmobranchs. These fish retain large amounts of "waste" urea, and to a lesser extent trimethylamine oxide to raise the osmolality of the body fluids approximately equal to that of seawater, thereby obviating the need for drinking. Theoretically, one might predict that this should also remove the need for intestinal Na^+ and Cl^- uptake and Cl^-/HCO_3^- exchange so as to drive water absorption, thereby reducing ionoregulatory costs at both gut and gills, relative to teleosts. Interestingly, a theoretical analysis has indicated that once the ATP expenditures for urea synthesis are taken into account, then the overall costs of the elasmobranch and teleost strategies are similar (Kirschner 1993). Furthermore, several recent studies have shown that small amounts of drinking and intestinal water uptake actually occur in fasted elasmobranchs, and these increase after hyperosmotic challenge (reviewed by Anderson et al. 2007). Indeed there is evidence in the bamboo shark (*Chiloscyllum plagiosum*) that intestinal Cl^-/HCO_3^- is turned on as a mechanism for enhancing water absorption in the latter circumstance (Taylor and Grosell 2006b).

What has been missing until recently has been any understanding of how feeding is integrated into this osmoregulatory strategy. Sharks tend to be opportunistic predators, eating large meals at irregular intervals, and digesting them much more slowly than teleosts (reviewed by Wood et al. 2007b, 2009). A series of studies on the Pacific spiny dogfish (*Squalus acanthias*) have now started to address these issues. The osmoregulatory benefits for a marine elasmobranch (internal osmolality ~950 mOsmol kg^{-1}) to eat a marine teleost (internal osmolality ~350 mOsmol kg^{-1}) are obvious. Sequential sampling of chyme from dogfish at various times after a large meal (5.5% ration of skate, *Merluccius productus*—see Table 5.1 for composition) indicated that most of the fluid content of the prey was absorbed on a net basis, though there was initial addition of fluid in the two sections of the stomach, likely associated with gastric HCl secretion (Wood et al. 2007b). Seawater drinking, though not measured directly, appeared

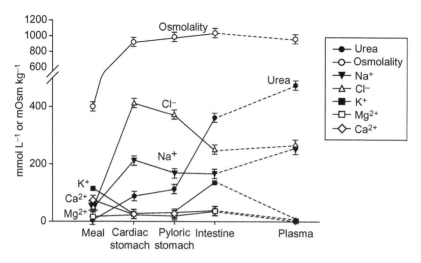

Fig. 5.6. Profiles of concentration of osmolality, urea, Na^+, Cl^-, Mg^{2+}, Ca^{2+}, and K^+ in the meal (hake), the chyme in the cardiac stomach, the chyme in the pyloric stomach, the chyme in the intestine, and the blood plasma of the dogfish shark (*Squalus acanthias*) at 20 h post-feeding. Data from Wood et al. (2007b).

to continue or increase slightly based on the analysis of chyme ions (Wood et al. 2007b, 2009). The volume of fluid in the intestine remained very constant during digestion, indicating efficient metering by the pyloric sphincter, and the colon was generally empty, indicating that most of the meal was absorbed. Plasma ion concentrations were largely unaffected (Wood et al. 2007b; Kajimura et al. 2008) despite Na^+, K^+, water, and Cl^- absorption in the intestine (Mg^{2+} and Ca^{2+} were largely excluded), a situation not dissimilar from that in marine teleosts (see Section 8).

Despite the low osmolality of the meal, the osmolality of the chyme had been brought into equilibrium with that of the plasma (a more than two-fold increase) by 6 h post-feeding, largely by the addition of Na^+ and especially Cl^- (HCl secretion) in the two stomach compartments, and by a very marked addition of urea in the intestine (Wood et al. 2007b). Figure 5.6 illustrates conditions in the chyme at 20 h post-feeding. Plasma urea concentrations and the activity of the ornithine–urea cycle (OUC) enzymes were significantly elevated, peaking at this time (Kajimura et al. 2006). In addition to the traditional sites of urea synthesis (liver, skeletal muscle), Kajimura et al. (2006) demonstrated that the intestine of the dogfish has a high level of the ammonia-trapping enzyme glutamine synthetase and a full complement of OUC enzymes. Anderson et al. (2010) have reported that there is mRNA expression for both urea (UT) and ammonia (Rh) transport proteins in the intestine of the little skate (*Leucoraja erinacea*), and that *in vitro* intestinal sac preparations from this species secrete urea at a high rate, while absorbing most ions. We speculate that a substantial portion of the secreted urea comes from endogenous production in the intestinal wall, fueled by high ammonia levels in the chyme (Wood et al. 2009). The almost complete absence of urea from colonic fluids (Anderson et al. 2009) and the fact that whole body urea excretion tends to decrease after feeding in the dogfish (Wood et al. 2007a) suggest that the great majority of this secreted urea is subsequently reabsorbed.

The impact of feeding on kidney function remains unknown, but there is clear evidence of large effects on gill function. In the dogfish, a marked systemic alkaline tide follows feeding (Wood et al. 2005, 2007b) and is due to enhanced gastric HCl secretion because it is inhibited by pretreatment with omeprazole (Wood et al. 2009). Blood pH is prevented from rising too high by a large secretion of basic equivalents across the gills (Wood et al. 2007a). Activation of a basolateral vacuolar proton ATPase creates an electro-chemical gradient driving apical Cl^-/HCO_3^- exchange, such that base is secreted to the environment and Cl^- is taken up (Tresguerres et al. 2007). This, plus increased intestinal Cl^- absorption (Wood et al. 2007b) (Fig. 5.6), may contribute to the slow postprandial rise in plasma Cl^- concentration seen in some studies (Wood et al. 2005; Kajimura et al. 2006). In contrast to teleosts, there is only a small increase in ammonia excretion, and either no change or a decrease in urea excretion after feeding, reflecting the importance of N-conservation for osmoregulation and growth in elasmo-branchs (Wood et al. 2005, 2007a; Kajimura et al. 2006, 2008). The branchial urea retention mechanisms described by Wood et al. (1995), Pärt et al. (1998) and Fines et al. (2001) may be activated at this time.

There is also clear evidence that feeding, possibly the alkaline tide itself (Wood et al. 2008), activates the rectal gland at enzymatic, proteomic, and structural levels (McKenzie et al. 2002; Walsh et al. 2006; Dowd et al. 2008, Matey et al. 2009). The rectal gland produces a solution which is almost pure 500 mmol L^{-1} NaCl (Wood et al. 2007c), thereby helping to prevent disturbances of extracellular Na^+, Cl^- and volume homeostasis caused by increased intake of salt and water with the meal.

Essentially nothing is known about these issues in freshwater elasmo-branchs, which tend to function like freshwater teleosts (Smith 1931a). However, one study of the stenohaline *Potomotrygon* sp. stingrays which live in the ion-poor waters of the Rio Negro, Brazil, concluded that they could not achieve Na^+ and Cl^- balance based on ion exchange with the external water alone, and that electrolyte uptake from feeding must be essential (Wood et al. 2002).

11. FUTURE DIRECTIONS

It should be clear from this review that feeding plays critical roles in salt and water homeostasis in both freshwater and marine fish, but that our overall understanding lags greatly behind our knowledge of the roles of gills and kidney in this regard. The following topics are important areas for future research.

1. At present, we really only understand the roles of the gills and kidney in the largely unnatural, potentially stressful condition of fasting. There is an urgent need to repeat much of this earlier work, but on fish which are allowed to feed naturally, using the approaches outlined in Section 2.
2. Most of the information which does exist comes from studies on fish eating commercial pellets. More work is needed on fish which are fed natural foods with much higher water contents and generally lower ion concentrations, such as the studies of Taylor and Grosell (2006a), Taylor et al. (2007), and Wood et al. (2007b).
3. To date, the pyloric ceca have not received sufficient attention. These structures may account for up to 70% of the total intestinal surface area in some fish such as salmonids, and they fill with chyme and empty at a comparable rate to the anterior intestine. They have long been known to be important sites of both nutrient absorption (Buddington and Diamond 1986, 1987) and Na^+-dependent water absorption (Bogé et al. 1988; Veillette et al. 2005), and more recent evidence indicates key roles in phosphate uptake (Sugiura and Ferrari 2004) and HCO_3^- secretion (Grosell et al. 2009a). In analyses (Figs. 5.3 and 5.5) such as those of Bucking and Wood (2006a,b, 2007) and Bucking et al. (2010b), their function is lumped into that of the anterior intestine and thereby may be "swamped" by the biliary, pancreatic, and intestinal mucosa secretions.
4. At present we know that gut blood flow is elevated following feeding (e.g. Grans et al. 2009), but almost nothing about how the ionic and acid–base composition of the plasma and lymph, and the transepithelial potential (TEP), may change at this time. PCO_2 levels in the chyme may also be greatly elevated during digestion, with beneficial consequences for O_2 unloading to working enterocytes (Wood et al. 2010). The hepatic portal vein technique has recently been exploited to reveal substantial increases in amino acid, ammonia, and urea levels in gut blood plasma following feeding in trout (Karlsson et al. 2006). This same approach could be used to investigate ion, acid–base, and TEP changes, and therefore the relevant electrochemical gradients for transport, during assimilation.

5. In the wild, fish choose the quantity and the quality of the food they eat. Does behavioral food selection play a role in iono/osmoregulation, or conversely, does the ionic content of the food influence behavior (e.g. salinity selection)? To our knowledge, these areas have so far been investigated only in the context of toxicology—i.e. whether metal exposures that are toxic to ionoregulation influence the selection of ion-rich food (e.g. Niyogi et al. 2006; Wood et al. 2006). This is an important area for future investigation in a purely physiological context.

ACKNOWLEDGMENTS

We thank Sunita Nadella for excellent bibliographic assistance. Susan Marsh-Rollo and Sigal Balshine kindly provided some of the feeds in Table 5.1. Our research program on feeding is supported by the NSERC (Canada) Discovery Grants Program. CMW is supported by the Canada Research Chair Program, and CB by an NSERC Postdoctoral Fellowship.

REFERENCES

Al-Amoudi, M. M. (1987). The effect of high salt diet on the direct transfer of *Oreochromis mossambicus*, *O. spirulus*/*O. niloticus* hybrids to seawater. *Aquaculture* **64**, 333–338.

Anderson, W. G., Taylor, J. R., Good, J. P., Hazon, N., and Grosell, M. (2007). Body volume regulation in elasmobranch fish. *Comp. Biochem. Physiol. A* **148**, 3–13.

Anderson, W. G., Dasiewicz, P. J., Liban, S., Ryan, C., Taylor, J. R., Grosell, M. and Weihrauch, D. (2010). Gastro-intestinal handling of water and solutes in three species of elasmobranch fish, the white-spotted bamboo shark, *Chiloscyllium plagiosum*, little skate, *Leucoraja erinacea* and the clear nose skate *Raja eglanteria*. *Comp. Biochem. Physiol. A.* **155**, 493–502.

Austreng, E. (1978). Digestibility determination in fish using chrome oxide marking and analysis of contents from different segments of the gastrointestinal tract. *Aquaculture* **13**, 266–272.

Avila, E. M., Basantes, S. P., and Ferraris, R. P. (1999). Cholecalciferol modulates plasma phosphate but not plasma vitamin D levels and intestinal phosphate absorption in rainbow trout (*Oncorhynchus mukiss*). *Gen. Comp. Endocrinol.* **111**, 460–469.

Avila, E. M., Tu, H., Basantes, S., and Ferraris, R. P. (2000). Dietary phosphorus regulates intestinal transport and plasma concentrations of phosphate in rainbow trout. *J. Comp. Physiol. B.* **170**, 201–209.

Baldisserotto, B., Kamunde, C., Matsuo, A., and Wood, C. M. (2004a). Acute waterborne cadmium uptake in rainbow trout is reduced by dietary calcium carbonate. *Comp. Biochem. Physiol. C* **137**, 363–372.

Baldisserotto, B., Kamunde, C., Matsuo, A., and Wood, C. M. (2004b). A protective effect of dietary calcium against acute waterborne cadmium uptake in rainbow trout. *Aquat. Toxicol.* **67**, 57–73.

Baldisserotto, B., Chowdhury, J. M., and Wood, C. M. (2005). Effects of dietary calcium and cadmium on cadmium accumulation, calcium and cadmium uptake from the water, and their interactions in juvenile rainbow trout. *Aquat. Toxicol.* **72**, 99–117.

Baldisserotto, B., Chowdhury, M. J., and Wood, C. M. (2006). Intestinal absorption of cadmium and calcium in juvenile rainbow trout fed with calcium and cadmium supplemented diets. *J. Fish Biol.* **69**, 658–667.

Basulto, S. (1976). Induced saltwater tolerance in connection with inorganic salts in the feeding of Atlantic salmon (*Salmo salar*). *Aquaculture* **8**, 45–55.

Beamish, F. W. H., and Thomas, E. (1984). Effects of dietary protein and lipid on nitrogen losses in rainbow trout, *Salmo gairdneri*. *Aquaculture* **41**, 359–371.

Berg, A. (1970). Studies on the metabolism of calcium and strontium in freshwater fish. II. Relative contribution of direct and intestinal absorption in growth conditions. *Mem. Ist. Ital. Idrobiol.* **26**, 241–255.

Bergman, A. N., Laurent, P., Otiang'a-Owiti, G., Bergman, H. L., Walsh, P. J., Wilson, P., and Wood, C. M. (2003). Physiological adaptations of the gut in the Lake Magadi tilapia, *Alcolapia grahami*, an alkaline- and saline-adapted fish. *Comp. Biochem. Physiol. A* **136**, 701–715.

Beyenbach, K. W., and Kirschner, L. B. (1975). Kidney and urinary bladder functions of the rainbow trout in Mg and Na excretion. *Am. J. Physiol.* **229**, 389–393.

Bijvelds, M. J. C., Van der velden, J. A., Kolar, Z., and Flik, G. (1998). Magnesium transport in freshwater teleosts. *J. Exp. Biol.* **201**, 1981–1990.

Binder, H. J., and Rawlins, C. I. (1973). Effect of conjugated dihydroxy bile salts on electrolyte transport in rat colon. *J. Clinical Invest.* **52**, 1460–1466.

Bjornsson, B. T., and Nilsson, S. (1985). Renal and extra-renal excretion of calcium in the marine teleost, *Gadus morhua*. *Am. J. Physiol.* **248**, R18–R22.

Bogé, G., Lopez, L., and Pérès, G. (1998). An *in vivo* study of the role of pyloric caeca in water absorption in rainbow trout (*Salmo gairdneri*). *Comp. Biochem. Physiol.* **91**, 9–13.

Boisclair, D. and Marchand, F. (1993). The guts to estimate fish daily ration. *Can. J. Fish. Aquat. Sci.* **50**, 1969–1975.

Boulos, N. N., Greenfield, H., and Wills, R. B. H. (2000). Water holding capacity of selected soluble and insoluble dietary fibre. *Int. J. Food Prop.* **3**, 217–231.

Bromley, P. J. (1980). The effect of dietary water content and feeding rate on the growth and food conversion efficiency of turbot (*Scophthalmus maximus* L). *Aquaculture* **20**, 91–99.

Bucking, C. and Wood, C. M. (2006a). Water dynamics in the digestive tract of freshwater rainbow trout during the processing of a single meal. *J. Exp. Biol.* **209**, 1883–1893.

Bucking, C., and Wood, C. M. (2006b). Gastrointestinal processing of monovalent ions (Na^+, Cl^-, K^+) during digestion: implications for homeostatic balance in freshwater rainbow trout. *Am. J. Physiol.* **291**, R1764–R1772.

Bucking, C. and Wood, C. M. (2007). Gastrointestinal transport of Ca^{2+} and Mg^{2+} during the digestion of a single meal in the freshwater rainbow trout. *J. Comp. Physiol. B* **177**, 349–360.

Bucking, C., and Wood, C. M. (2008). The alkaline tide and ammonia excretion after voluntary feeding in freshwater rainbow trout. *J. Exp. Biol.* **211**, 2533–2541.

Bucking, C. P. and Wood, C. M. (2009). The effect of postprandial changes in pH along the gastrointestinal tract on the distribution of ions between the solid and fluid phases of chyme in rainbow trout. *Aquacult. Nutr.* **15**, 282–296.

Bucking, C., Fitzpatrick, J. L., Nadella, S. R., and Wood, C. M. (2009). Post-prandial metabolic alkalosis in the seawater-acclimated trout: the alkaline tide comes in. *J. Exp. Biol.* **212**, 2159–2166.

Bucking, C., Landman, M. and Wood, C. M. (2010a). The role of the kidney in compensating the alkaline tide, electrolyte load, and fluid balance disturbance associated with feeding in the freshwater rainbow trout. *Comp. Biochem. Physiol. A*. In press.

Bucking, C., Fitzpatrick, J. L., Nadella, S. R., McGaw, I. J. and Wood, C. M. (2010b). Assimilation of water and dietary ions by the gastrointestinal tract during digestion in seawater-acclimated rainbow trout. Submitted.

Buddington, R. K., and Diamond, J. M. (1986). Aristotle revisited: the function of pyloric caeca in fish. *Proc. Natl. Acad. Sci.* **83**, 8012–8014.

Buddington, R. K., and Diamond, J. M. (1987). Pyloric ceca of fish: a "new" absorptive organ. *Am. J. Physiol.* **252**, G65–G76.

Carrasson, M., Grau, A., Dopazo, L. R., and Crespo, S. (2006). A histological, histochemical and ultrastructural study of the digestive tract of *Dentex dentex* (*Pisces, Sparidae*). *Histol. Histopathol.* **21**, 579–593.

Carter, C. G., McCarthy, I. D., Houlihan, D. F., Fonseca, M., Perera, W. M. K., and Sillah, A. B. S. (1993). The application of radiography to the study of fish nutrition. *J. Appl. Ichthyol.* **11**, 231–239.

Chen, X., Moon, T. W., Olson, K. R., Dombkowski, R. A., and Perry, S. F. (2007). The effects of salt-induced hypertension on α1-adrenoreceptor expression and cardiovascular physiology in the rainbow trout (*Oncorhynchus mykiss*). *Am. J. Physiol.* **293**, R1382–R1393.

Cho, C. Y., Slinger, S. J., and Bayley, H. S. (1982). Bioenergetics of salmonid fishes: energy intake, expenditure, and productivity. *Comp. Biochem. Physiol. B* **73**, 25–41.

Chou, R. (1984). The effect of dietary water content on the feed intake, food conversion efficiency and growth of young sea bass (*Lates calcarifer* Bloch). *Singapore J. Pri. Ind.* **12**, 120–127.

Clearwater, S. J., Baskin, S. J., Wood, C. M., and McDonald, D. G. (2000). Gastrointestinal uptake and distribution of copper in rainbow trout. *J. Exp. Biol.* **203**, 2455–2466.

Coloso, R. W., Basantes, S. P., King, K., Hendrix, M. A., Fletcher, J. W., Weis, P., and Ferraris, R. P. (2001). Effect of dietary phosphorus and vitamin D3 on phosphorus levels in effluent from the experimental culture of rainbow trout (*Oncorhynchus mykiss*). *Aquaculture* **202**, 145–161.

Coloso, R. M., King, K., Fletcher, J. W., Weis, P., Werner, A., and Ferraris, R. P. (2003a). Dietary P regulates phosphate transporter expression, phosphatase activity, and effluent P partitioning in trout culture. *J. Comp. Physiol.* **173**, 519–530.

Coloso, R. W., King, K., Fletcher, J. W., Hendrix, M. A., Subramanyam, M., Weis, P., and Ferraris, R. P. (2003b). Phosphorus utilization in rainbow trout (*Oncorhynchus mykiss*) fed practical diets and its consequences on effluent phosphorus levels. *Aquaculture* **220**, 801–820.

Conceicao, L. E. C., Morais, S., Dinis, M. T., and Ronnestad, I. (2008). Tracers studies in fish larvae. In: *Feeding and Digestive Functions of Fishes* (J. E. P. Cyrino, D. P. Bureau and B. G. Kapoor, eds), pp. 349–393. Science Publishers, New Hampshire.

Cooper, C. A., and Wilson, R. W. (2008). Post-prandial alkaline tide in freshwater rainbow trout: effects of meal anticipation on recovery from acid–base and ion regulatory disturbances. *J. Exp. Biol.* **211**, 2542–2550.

Cortes, E. (1997). A critical review of methods of studying fish feeding based on analysis of stomach contents: application to elasmobranch fishes. *Can. J. Fish. Aquat. Sci.* **54**, 726–738.

Crowther, R. S., and Marriott, C. (1984). Counter-ion binding to mucus glycoproteins. *J. Pharm. Pharmacol.* **36**, 21–26.

D'Cruz, L. M., and Wood, C. M. (1998). The influence of dietary salt and energy on the response to low pH in juvenile rainbow trout. *Physiol. Zool.* **71**, 642–657.

D'Cruz, L. M., Dockray, J. J., Morgan, I. J., and Wood, C. M. (1998). Physiological effects of sublethal acid exposure in juvenile rainbow trout on a limited or unlimited ration during a simulated global warming scenario. *Physiol. Zool.* **71**, 359–376.

Dabrowski, K., and Dabrowska, H. (1981). Digestion of protein by rainbow trout (*Salmo gairdneri* Rich.) and absorption of amino acids within the alimentary tract. *Comp. Biochem. Physiol. A* **69**, 99–111.

Dabrowski, K., Leray, C., Nonnotte, G., and Colin, D. A. (1986). Protein digestion and ion concentrations in rainbow trout (*Salmo gairdneri* Rich.) digestive tract in sea- and freshwater. *Comp. Biochem. Physiol. A* **83**, 27–39.

Dabrowska, H., Meyer-Burgdorff, K. H., and Gunther, K. D. (1991). Magnesium status in freshwater fish, common carp (*Cyprinus carpio*, L.) and the dietary protein–magnesium interaction. *Fish Physiol. Biochem.* **9**, 165–172.

DeLuca, H. F. (2004). Overview of general physiological features and functions of vitamin D. *Am. J. Nutr.* **80**, 1689S–1696S.

Denton, J. E., Yousel, M. K., Yousef, I. M., and Kuksis, A. (1974). Bile-acid composition of rainbow trout, *Salmo gairdneri*. *Lipids* **9**, 945–951.

DeSilva, S. S., Deng, D. F., and Rajendram, V. (1997). Digestibility in goldfish fed diets with and without chromic oxide and exposed to sublethal concentrations of cadmium. *Aquaculture Nutr.* **3**, 109–114.

Diamond, J. M. (1962). The reabsorptive function of the gall bladder. *J. Physiol. (London)* **161**, 442–473.

Dowd, W. W., Wood, C. M., Kajimura, M., Walsh, P. J., and Kültz, D. (2008). Natural feeding influences protein expression in the dogfish shark rectal gland: a proteomic analysis. *Comp. Biochem. Physiol. D* **3**, 118–127.

Duffus, C. M., and Duffus, J. H. (1991). *Toxic Substances in Crop Plants.* In: (F. J. P. D'Mello, C. M. Duffus and J. H. Duffus, eds), pp. 1–21. The Royal Society of Chemistry, Thomas Graham House, Science Park, Cambridge.

Eddy, F. B., Smith, N. F., Hazon, N., and Grierson, C. (1990). Circulatory and ionoregulatory effects of atrial natriuretic peptide on rainbow trout (*Salmo gairdneri* Richardson) fed normal or high levels of dietary salt. *Fish Physiol. Biochem.* **8**, 321–327.

Edwards, D. J. (1971). Effect of temperature on rate of passage of food through the alimentary canal of the plaice *Pleuronectes platessa* L. *J. Fish. Biol.* **3**, 433–439.

Edwards, D. J. (1973). The effects of drugs and nerve-section on the rate of passage of food through the gut of the plaice, *Pleuronectes platessa* (L.). *J. Fish. Biol.* **5**, 441–446.

Evans, D. H., Piermarini, P. M., and Choe, K. P. (2005). The multifunctional fish gill: dominant site of gas exchange, osmoregulation, acid–base regulation, and excretion of nitrogenous waste. *Physiol. Rev.* **85**, 97–177.

Feldoto, Z., Pettersson, T., and Dedinaite, A. (2008). Mucin–electrolyte interactions at the solid–liquid interface probed by QCM-D. *Langmuir* **24**, 3348–3357.

Fines, G. A., Ballantyne, J. S., and Wright, P. A. (2001). Active urea transport and an unusual basolateral membrane composition in the gills of a marine elasmobranch. *Am. J. Physiol.* **280**, R16–R24.

Fini, A., and Roda, A. (1987). Chemical properties of bile acids. IV. Acidity constants of glycine-conjugated bile acids. *J. Lipid Res.* **28**, 755–759.

Fletcher, C. R. (1978). Osmotic and ionic regulation in the cod (*Gadus callarias* L.). I. Water balance. *J. Comp. Physiol.* **124**, 149–155.

Flik, G., and Verbost, P. M. (1993). Calcium transport in fish gills and intestine. *J. Exp. Biol.* **184**, 17–29.

Fontainhas-Fernandes, A., Monteiro, M, Gomes, E., Reis-Henriques, M. A., and Coimbra, J. (2000). Effect of dietary sodium chloride acclimation on growth and plasma thyroid hormones in tilapia *Oreochromis niloticus* (L.) in relation to sex. *Aquacult. Res.* **31**, 507–517.

Fontainhas-Fernandes, A., Russell-Pinto, F., Gomes, E., Reis-Henriques, M., and Coimbra, J. (2001). The effect of dietary sodium chloride on some osmoregulatory parameters of the teleost *Oreochromis niloticus*, after transfer from freshwater to seawater. *Fish Physiol. Biochem.* **23**, 307–316.

Franklin, N. M., Glover, C. N., Nicol, J. A., and Wood, C. M. (2005). Calcium/cadmium interactions at uptake surfaces in rainbow trout: waterborne versus dietary routes of exposure. *Environ. Toxicol. Chem.* **24**, 2954–2964.

Frizzell, R. A., and Schultz, G. A. (1970). Effect of bile salts on transport across brush border of rabbit ileum. *Biochem. Biophys. Acta.* **211**, 589–592.

Gatlin, D. M., Mackenzie, D. S., Craig, S. R., and Neill, W. H. (1992). Effects of dietary sodium chloride on red drum juveniles in waters of various salinities. *Progress. Fish Cultur.* **54**, 220–227.

Geibel, J. P., and Wagner, C. (2006). An update on acid secretion. *Rev. Physiol. Biochem. Pharmacol.* **156**, 45–60.

Gonzalez, R. J., Wilson, R. W., and Wood, C. M. (2005). Ionoregulation in tropical fish from ion-poor, acidic blackwaters. In: *The Physiology of Tropical Fish, Fish Physiology* (A. L. Val, V. M. Almeida-Val and D. J. Randall, eds), Vol. 22, pp. 397–437. Academic Press, San Diego.

Grans, A., Albertsson, F., Axelsson, M., and Olsson, C. (2009). Postprandial changes in enteric electrical activity and gut blood flow in rainbow trout (*Oncorhynchus mykiss*) acclimated to different temperatures. *J. Exp. Biol.* **212**, 2550–2557.

Grosell, M. (2006). Intestinal anion exchange in marine fish osmoregulation. *J. Exp. Biol.* **209**, 2813–2827.

Grosell, M., and Wood, C. M. (2001). Branchial *versus* intestinal silver toxicity and uptake in the marine teleost (*Parophrys vetulus*). *J. Comp. Physiol.* **B171**, 585–596.

Grosell, M., and Wood, C. M. (2002). Copper uptake across rainbow trout gills: mechanisms of apical entry. *J. Exp. Biol.* **205**, 1179–1188.

Grosell, M., O'Donnell, M. J., and Wood, C. M. (2000). Hepatic *versus* gallbladder bile composition—*in vivo* transport physiology of the gallbladder in the rainbow trout. *Am. J. Physiol.* **278**, R1674–R1684.

Grosell, M., Genz, J., Taylor, J. R., Perry, S. F., and Gilmour, K. M. (2009a). The involvement of H^+-ATPase and carbonic anhydrase in intestinal HCO_3^- secretion in seawater-acclimated rainbow trout. *J. Exp. Biol.* **212**, 1940–1948.

Grosell, M., Mager, E. M., Williams, C., and Taylor, J. R. (2009b). High rates of HCO_3^- secretion and Cl^- absorption against adverse gradients in the marine teleost intestine: the involvement of an electrogenic anion exchanger and H^+ pump metabolon? *J. Exp. Biol.* **212**, 1684–1696.

Gu, J. J., Hofmann, A. F., Ton-Nu, H. T., Schteingart, C. D., and Mysels, K. J. (1992). Solubility of calcium salts of unconjugated and conjugated natural bile acids. *J. Lipid Res.* **33**, 635–646.

Hansen, J. M., Lazo., J. P., and Kling, L. J. (2009). A method to determine protein digestibility of microdiets for larval and early juvenile fish. *Aquacult. Nutr.* **15**, 615–626.

Hansen, J. O., and Storebakken, T. (2007). Effects of dietary cellulase level on pellet quality and nutrient digestibilities in rainbow trout (*Oncorhynchus mykiss*). *Aquaculture* **272**, 458–465.

Harpaz, S., Hakim, Y., Slosman, T., and Eroldogan, O. T. (2005). Effects of adding salt to the diet of Asian sea bass *Lates calcarifer* reared in fresh or salt water recirculating tanks, on growth and brush border enzyme activity. *Aquaculture* **248**, 315–324.

Haslewood, G. A. D. (1967). Bile salt evolution. *J. Lipid Res.* **8**, 535–550.

Healy, T. W., and White, L. R. (1978). Ionizable surface group models of aqueous interfaces. *Adv. Colloid Interface Sci.* **9**, 303–345.

Hickman, C. P., Jr. (1967). Ingestion, intestinal absorption, and elimination of seawater and salts in the southern flounder, *Paralichthys lethostigma*. *Can. J. Zool.* **46**, 457–466.

Higgs, D. A., Markert, J. R., Plotnikoff, M. D., McBride, J. R., and Dosanjh, B. S. (1985). Development of nutritional and environmental strategies for maximising the growth and survival of juvenile pink salmon (*Oncorhynchus gorbuscha*). *Aquaculture* **47**, 113–130.

Hille, S. (1984). The effect of environmental and endogenous factors on blood constituents of rainbow trout (*Salmo gairdneri*)—I. Food content of stomach and intestine. *Comp. Biochem. Physiol. A* **77**, 311–314.

Hofmann, A. F. (2009). Bile acids: trying to understand their chemistry and biology with the hope of helping patients. *Hepatol.* **49**, 1403–1418.

Hofmann, A. F., and Roda, A. (1984). Physicochemical properties of bile acids and their relationship to biological properties: an overview of the problem. *J. Lipid. Res.* **25**, 1477–1489.

Hughes, S. G. (1989). Effect of dietary moisture level on response to diet by Atlantic salmon. *Prog. Fish-Cult.* **51**, 20–23.

Hunn, J. B. (1972). Concentrations of some inorganic constituents in gallbladder bile from some freshwater teleosts. *Copeia* **1972**, 860–861.

Hunn, J. B. (1976). Inorganic composition of gallbladder bile from freshwater fishes. *Copeia* **1976**, 602–605.

Kajimura, M., Walsh, P. J., Mommsen, T. P., and Wood, C. M. (2006). The dogfish shark (*Squalus acanthias*) increases both hepatic and extra-hepatic ornithine–urea cycle enzyme activities for nitrogen conservation after feeding. *Physiol. Biochem. Zool.* **79**, 602–613.

Kajimura, K., Walsh, P. J., and Wood, C. M. (2008). The dogfish shark (*Squalus acanthias*) maintains its osmolytes during long term starvation. *J. Fish. Biol.* **72**, 656–670.

Kamunde, C. N., Pyle, G. G., McDonald, D. G., and Wood, C. M. (2003). Influence of dietary sodium and waterborne copper toxicity in rainbow trout, *Oncorhynchus mykiss*. *Environ. Toxicol. Chem.* **22**, 342–350.

Kamunde, C., Niyogi, S., and Wood, C. M. (2005). Interaction of dietary sodium chloride and waterborne copper in rainbow trout: sodium and chloride homeostasis, copper homeostasis, and chronic copper toxicity. *Can. J. Fish. Aquat. Sci.* **62**, 390–399.

Karasov, W. H., and Diamond, J. M. (1983). A simple method for measuring intestinal solute uptake in vitro. *J. Comp. Physiol.* **152**, 105–116.

Karlsson, A., Eliason, E. J., Mydland, L. T., Farrell, A. P., and Kiessling, A. (2006). Postprandial changes in plasma free amino acid levels obtained simultaneously from the hepatic portal vein and dorsal aorta in rainbow trout (*Onchorhynchus mykiss*). *J. Exp. Biol.* **209**, 4885–4894.

Ketola, H. G., and Harland, B. F. (1993). Influence of phosphorus in rainbow trout diets on phosphorus discharges in effluent water. *Trans. Am. Fish. Soc.* **122**, 1120–1126.

Kirchner, S., McDaniel, N. K., Sugiura, S. H., Soteropoulos, P., Tian, B., Fletcher, J. W., and Ferraris, R. P. (2007). Salmonid microarrays identify intestinal genes that reliably monitor P deficiency in rainbow trout aquaculture. *Anim. Gen.* **38**, 319–331.

Kirsch, R. (1972). The kinetics of peripheral exchanges of water and electrolytes in the silver eel (*Anguilla anguilla* L.) in fresh water and sea water. *J. Exp. Biol.* **57**, 489–512.

Kirsch, R., and Meister, M. F. (1982). Progressive processing of ingested water in the gut of seawater teleosts. *J. Exp. Biol.* **98**, 67–81.

Kirschner, L. B. (1993). The energetics of osmotic regulation in ureotelic and hypo-osmotic fishes. *J. Exp. Zool.* **267**, 19–26.

Kjoss, V. A., Kamunde, C. N., Niyogi, S., Grosell, M., and Wood, C. M. (2005). Dietary Na does not reduce dietary Cu uptake by juvenile rainbow trout. *J. Fish Biol.* **66**, 468–484.

Klinck, J. S., Green, W. W., Mirza, R., Nadella, S. R., Chowdhury, M. J., Wood, C. M., and Pyle, G. G. (2007). Branchial cadmium and copper binding and intestinal cadmium uptake in wild yellow perch (*Perca flavescens*) from clean and metal polluted lakes. *Aquat. Toxicol.* **84**, 198–207.

Klinck, J. S., Tania, Ng, T. Y.-T., and Wood, C. M. (2009). Cadmium accumulation and *in vitro* analysis of calcium and cadmium transport functions in the gastro-intestinal tract of trout following chronic dietary cadmium and calcium feeding. *Comp. Biochem. Physiol. C* **150**, 349–360.

Kristiansen, H. R., and Rankin, J. C. (2001). Discrimination between endogenous and exogenous water sources in juvenile rainbow trout fed extruded dry feed. *Aquat. Living Resources* **14**, 359–366.

Kuver, R., and Lee, S. P. (2004). Calcium binding to biliary mucins is dependent on sodium ion concentration: relevance to cystic fibrosis. *Biochim. Biophys. Res. Commun.* **314**, 330–334.

Lall, S. P. (2002). The minerals. In: *Fish Nutrition* (J. E. Halver and R. W. Hardy, eds), pp. 261–307.

Leknes, I. L. (2009). Histochemical study on the intestinal goblet cells in cichlid and poecilid species (Teleostei). *Tissue and Cell*. Academic Press, New York.

Leonard, E. M., Nadella, S. R., Bucking, C., and Wood, C. M. (2009). Characterization of dietary Ni uptake in the rainbow trout, *Oncorhynchus mykiss*. *Aquat. Toxicol.* **95**, 205–216.

Lock, E. J., Waagbø, R., Wendelaar Bonga, S., and Flik, G. (2010). The significance of vitamin D for fish: a review. *Aquacul. Nut.* **16**, 100–116.

Loretz, C. A. (1995). Electrophysiology of ion transport in teleost intestinal cells. In: *Cellular and Molecular Approaches to Fish Ionic Regulation. Fish Physiology* (C. M Wood and T. J Shuttleworth, eds), Vol. 14, pp. 25–56. Academic Press, San Diego, CA.

Marshall, W. S., and Grosell, M. (2006). Ion transport, osmoregulation, and acid–base balance. In: *The Physiology of Fishes* (D. H. Evans and J. B. Claiborne, eds), pp. 177–230. CRC Press, Boca Raton.

Marshall, W. S., Howard, J. A., Cozzi, R. R. F., and Lynch, E. M. (2002). NaCl and fluid secretion by the intestine of teleost *Fundulus heteroclitus*: involvement of CFTR. *J. Exp. Biol.* **205**, 745–758.

Matey, V., Wood, C. M., Dowd, W. W., Kültz, D., and Walsh, P. J. (2009). Morphology of the rectal gland of the dogfish shark (*Squalus acanthias*) in response to feeding. *Can. J. Zool.* **87**, 440–452.

Mbahinzireki, G. B., Dabrowski, K., Lee, K. J., El-Saidy, D., and Wisner, E. R. (2001). Growth, feed utilization and body composition of tilapia (*Oreochromis* sp.) fed with cottonseed meal-based diets in a recirculating system. *Aquacult. Nutr.* **7**, 189–200.

McCarthy, I. D., Houlihan, D. F., Carter, C. G., and Moutou, K. (1992). The effect of feeding hierarchy on individual variability in daily feeding of rainbow trout, *Oncorhynchus mykiss* (Walbaum). *J. Fish Biol.* **41**, 257–263.

McCarthy, I. D., Houlihan, D. F., Carter, C. G., and Moutou, K. (1993). Variation in individual food consumption rates of fish and its implications for the study of fish nutrition and physiology. *Proc. Nutr. Soc.* **52**, 427–436.

McDonald, D. G., Walker, R. L., Wilkes, P. R. H., and Wood, C. M. (1982). H$^+$ excretion in the marine teleost, *Parophrys vetulus*. *J. Exp. Biol.* **98**, 403–414.

McKenzie, S., Cutler, C. P., Hazon, N., and Cramb, G. (2002). The effects of dietary sodium loading on the activity and expression of Na$^+$, K$^+$-ATPase in the rectal gland of the European dogfish (*Scyliorhinus canicula*). *Comp. Biochem. Physiol. B.* **131**, 185–200.

Mekhjian, H. S., and Phillips, S. F. (1970). Perfusion of the canine colon with conjugated bile acids: effect on water and electrolyte transport. *Gastrentol.* **59**, 120–129.

Morais, S., and Conceicao, L. E. C. (2009). A new method for the study of essential fatty acid requirements in fish larvae. *Brit. J. Nutri.* **101**, 1564–1568.

Morais, S., Conceicao, L. E. C., Dinis, M. T., and Ronnestad, I. (2004). A method for radio labeling *Artemia* with applications in studies of food intake, digestibility, protein and amino acid metabolism in larval fish. *Aquaculture* **231**, 469–487.

Murray, M. W., and Andrews, J. W. (1979). Channel catfish: the absence of an effect of dietary salt on growth. *Prog. Fish-Cult.* **41**, 155–156.

Nadella, S. R., Bucking, C., Grosell, M., and Wood, C. M. (2006). Gastrointestinal assimilation of Cu during digestion of a single meal in the freshwater rainbow trout (*Oncorhynchus mykiss*). *Comp. Biochem. Physiol. C.* **143**, 394–401.

Niyogi, S., and Wood, C. M. (2006). Interaction between dietary calcium supplementation and chronic waterborne zinc exposure in juvenile rainbow trout, *Oncorhynchus mykiss*. *Comp. Biochem. Physiol. C* **143**, 94–102.

Niyogi, S., Kamunde, C. N., and Wood, C. M. (2006). Food selection, growth and physiology in relation to dietary sodium content in rainbow trout (*Oncorhynchus mykiss*) under chronic waterborne Cu exposure. *Aquat. Toxicol.* **77**, 210–221.

Northfield, M., and McColl, S. (1973). Postprandial concentrations of free and conjugated bile acids down the length of the normal human small intestine. *Gut* **14**, 513–518.

Oikari, A. O. J., and Rankin, J. C. (1985). Renal excretion of magnesium in a freshwater teleost, *Salmo gairdneri*. *J. Exp. Biol.* **117**, 319–333.

Olson, K. R., and Hoagland, T. M. (2008). Effects of freshwater and saltwater adaptation and dietary salt on fluid compartments, blood pressure, and venous capacitance in trout. *Am. J. Physiol.* **294**, R1061–R1067.

Overturf, K., Raboy, V., Cheng, Z. J., and Hardy, R. W. (2003). Mineral availability from barley low phytic acid grains in rainbow trout (*Oncorhynchus mykiss*) diets. *Aquacult. Nutr.* **9**, 239–246.

Papatryphon, E., Howell, R. A., and Soares, J. H. (1999). Growth and mineral absorption by striped bass *Monroe Saxatilis* fed a plant feed stuff based diet supplemented with phytase. *J. World Aquat. Soc.* **30**, 161–173.

Parmelee, J. T., and Renfro, L. J. (1983). Esophageal desalination of seawater in flounder: role of active sodium transport. *Am. J. Physiol.* **245**, R888–R893.

Pärt, P., Wright, P. A., and Wood, C. M. (1998). Urea and water permeability in dogfish (*Squalus acanthias*) gills. *Comp. Biochem. Physiol. A* **199**, 117–123.

Patrick, M. L., Pärt, P., Marshall, W. S., and Wood, C. M. (1997). Characterization of ion and acid–base transport in the fresh water adapted mummichog (*Fundulus heteroclitus*). *J. Exp. Zool.* **279**, 208–219.

Pelletier, D., and Besner, M. (1992). The effect of salty diets and gradual transfer to sea water on osmotic adaptation, gill Na^+, K^+-ATPase activation, and survival of brook charr, *Salvelinus fontinalis*, Mitchill. *J. Fish Biol.* **41**, 791–803.

Perez-Vilar, J., and Hill, R. (1999). The structure and assembly of secreted mucins. *J. Biol. Chem.* **274**, 31751–31754.

Perry, S. F., Rivero-Lopez, L., McNeill, B., and Wilson, J. (2006). Fooling a freshwater fish: how dietary salt transforms the rainbow trout gill into a seawater gill phenotype. *J. Exp. Biol.* **209**, 4591–4596.

Phillips, A. M., Jr. (1944). The physiological effect of sodium chloride upon brook trout. *Trans. Am. Fish. Soc.* **74**, 297–309.

Poston, H. A. (1974). Effect of feeding brown trout (*Salmo truttta*) a diet pelleted in dry and moist forms. *J. Fish. Res. Bd. Can.* **31**, 1824–1826.

Pyle, G. G., Kamunde, C. N., McDonald, D. G., and Wood, C. M. (2003). Dietary sodium inhibits aqueous copper uptake in rainbow trout (*Oncorhynchus mykiss*). *J. Exp. Biol.* **206**, 609–618.

Rodgers, D. W. (1984). Ambient pH and calcium concentration as modifiers of growth and calcium dynamics of brook trout, *Salvelinus fontinalis*. *Can. J. Fish. Aquat. Sci.* **41**, 1774–1780.

Ronnestad, I., Rojas-Garcia, C. R., Tonheim, S. K., and Conceicao, L. E. C. (2001). *In vivo* studies of digestion and nutrient assimilation in marine fish larvae. *Aquaculture* **201**, 161–175.

Ross, B., and Jauncey, K. (1981). A radiographic estimation of the effect of temperature on gastric emptying time in *Sarotherodon niloticus* (L) × *S. aurens* (Steindachner) hybrids. *J. Fish Biol.* **19**, 333–344.

Ruohonen, K., and Grove, D. J. (1996). Gastrointestinal responses of rainbow trout to dry pellet and low-fat herring diets. *J. Fish Biol.* **49**, 501–513.

Ruohonen, K., Grove, D. J., and McIlroy, J. T. (1997). The amount of food ingested in a single meal by rainbow trout offered chopped herring, dry and wet diets. *J. Fish Biol.* **51**, 93–105.

Ruohonen, K., Vielma, J., and Grove, D. J. (1998). High dietary inclusion level of fresh herring impairs growth of rainbow trout, *Oncorhynchus mykiss*. *Aquaculture* **163**, 263–273.

Salman, N. A., and Eddy, F. B. (1987). Response of chloride cell numbers and gill Na^+ K^+ ATPase activity of freshwater rainbow trout (*Salmo gairdneri* Richardson) to salt feeding. *Aquaculture* **61**, 41–48.

Salman, N. A., and Eddy, F. B. (1988a). Effect of dietary sodium chloride on growth, food intake and conversion efficiency in rainbow trout (*Salmo gairdneri* Richardson). *Aquaculture* **70**, 131–144.

Salman, N. A., and Eddy, F. B. (1988b). Kidney function in response to salt feeding in rainbow trout (*Salmo gairdneri* Richardson). *Comp. Biochem. Physiol.* **89**, 535–539.

Salman, N. A., and Eddy, F. B. (1990). Increased sea-water adaptability of non-smolting rainbow trout by salt feeding. *Aquaculture* **86**, 259–270.

Scott, G. R., Schulte, P. M., and Wood, C. M. (2006). Plasticity of osmoregulatory function in the killifish intestine: drinking rates, salt and water transport, and gene expression after freshwater transfer. *J. Exp. Biol.* **209**, 4040–4050.

Scott, G. R., Baker, D. W., Schulte, P. M., and Wood, C. M. (2008). Physiological and molecular mechanisms of osmoregulatory plasticity in killifish after seawater transfer. *J. Exp. Biol.* **211**, 2450–2459.

Secor, S. M. (2009). Specific dynamic action: a review of the postprandial metabolic response. *J. Comp. Physiol. B.* **179**, 1–56.

Setchall, K. D. R., Street, J. M., and Sjovall, J. (1988). Fecal bile acids. In: *The Bile Acids. Methods and Applications* (K. D. R. Setchall, D. Kritchevsky and P. P. Nair, eds), pp. 441–570. Plenum Press, New York.

Shaw, H. M., Saunders, R. L., Hall, H. C., and Henderson, E. B. (1975). Effect of dietary sodium chloride on growth of Atlantic salmon (*Salmo salar*). *J. Fish. Res. Bd. Can.* **32**, 1813–1819.

Shearer, K. D. (1988). Dietary potassium requirement of juvenile Chinook salmon. *Aquaculture* **73**, 119–129.

Shehadeh, Z. H., and Gordon, M. S. (1969). The role of the intestine in salinity adaptation of the rainbow trout, *Salmo gairdneri*. *Comp. Biochem. Physiol.* **30**, 397–418.

Shiau, S.-Y., and Chen, M.-J. (1993). Carbohydrate utilization by tilapia (*Oreochromis niloticus* × *O. aureus*) as influenced by different chromium sources. *J. Nutr.* **123**, 1747–1753.

Sibly, R. M. (1981). Strategies of digestion and defecation. In: *Physiological Ecology: An Evolutionary Approach to Resource Use* (C. R. Townsend and P. Calow, eds), pp. 109–139. Sinauer, Sunderland, MA.

Smith, H. W. (1930). The absorption and excretion of water and salts by marine teleosts. *Am. J. Physiol.* **93**, 480–505.

Smith, H. W. (1931a). The absorption and excretion of water and salts by the elasmobranch fishes. I. Fresh water elasmobranchs. *Am. J. Physiol.* **98**, 279–295.

Smith, H. W. (1931b). The absorption and excretion of water and salts by the elasmobranch fishes. I. Marine elasmobranchs. *Am. J. Physiol.* **98**, 296–310.

Smith, H. W. (1936). The retention and physiological role of urea in the elasmobranchii. *Biol. Rev.* **11**, 49–82.

Smith, N. F., Talbot, C., and Eddy, F. B. (1989). Dietary salt intake and its relevance to ionic regulation in freshwater salmonids. *J. Fish Biol.* **35**, 749–753.

Smith, N. F., Eddy, F. B., and Talbot, C. (1995). Effect of dietary salt load on transepithelial Na^+ exchange in freshwater rainbow trout (*Oncorhynchus mykiss*). *J. Exp. Biol.* **198**, 2359–2364.

Spry, D. J., and Wood, C. M. (1989). A kinetic method for the measurement of zinc influx in the rainbow trout and the effects of waterborne calcium. *J. Exp. Biol.* **142**, 425–446.

Storbakken, T., Shearer, K. D., and Roem, A. J. (1998). Availability of protein, phosphorus and other elements in fish meal, soy-protein concentrate and phytase-treated soy-protein-concentrate-based diets to Atlantic salmon, *Salmo salar*. *Aquaculture* **161**, 365–379.

Strous, G. J., and Dekker, J. (1992). Mucin-type glycoproteins. *Crit. Rev. Biochem. Mol. Biol.* **27**, 57–92.

Sugiura, S. H., and Ferraris, R. P. (2004a). Contributions of different NaPi cotransporter isoforms to dietary regulation of P transport in the pyloric caeca and intestine of rainbow trout. *J. Exp. Biol.* **207**, 2055–2064.

Sugiura, S. H., and Ferraris, R. P. (2004b). Dietary phosphorus-responsive genes in the intestine, pyloric ceca, and kidney of rainbow trout. *Am. J. Physiol.* **287**, R541–R550.

Sugiura, S. H., Dong, F. M., and Hardy, R. W. (1998a). Effects of dietary supplements on the availability of minerals in fish meal; preliminary observations. *Aquaculture* **160**, 283–303.

Sugiura, S. H., Dong, F. M., Rathbone, C. K., and Hardy, R. W. (1998b). Apparent protein digestibility and mineral availabilities in various feed ingredients for salmonid feeds. *Aquaculture* **159**, 177–202.

Sugiura, S. H., Raboy, V., Young, K. A., Dong, F. M., and Hardy, R. W. (1999). Availability of phosphorous and trace elements in low-phytate varieties of barley and corn for rainbow trout (*Oncorhynchus mykiss*). *Aquaculture* **170**, 285–296.

Sugiura, S. H., McDonald, N. K., and Ferraris, R. P. (2003). *In vivo* fractional P_i absorption and NaP_i-II mRNA expression in rainbow trout are upregulated by dietary P restriction. *Am. J. Physiol.* **285**, R770–R781.

Sugiura, S. H., Roy, P. K., and Ferraris, R. P. (2006). Dietary acidification enhances phosphorous digestibility but decreases H^+/K^+-ATPase expression in rainbow trout. *J. Exp. Biol.* **209**, 3719–3728.

Tacon, A. G. J., and DeSilva, S. S. (1983). Mineral composition of some commercial fish feeds available in Europe. *Aquaculture* **31**, 11–20.

Talbot, C., and Higgins, P. J. (1983). A radiographic method for feeding studies on fish using metallic iron powder as a marker. *J. Fish Biol.* **23**, 211–220.

Taylor, J. R., and Grosell, M. (2006a). Feeding and osmoregulation: dual function of the marine teleost intestine. *J. Exp. Biol.* **209**, 2939–2951.

Taylor, J. R., and Grosell, M. (2006b). Evolutionary aspects of intestinal bicarbonate secretion in fish. *Comp. Biochem. Physiol. A* **143**, 523–529.

Taylor, J. R. and Grosell, M. (2009). The intestinal response to feeding in seawater gulf toadfish, *Opsanus beta*, includes elevated base secretion and increased epithelial oxygen consumption. *J. Exp. Biol.* In press.

Taylor, J. R., Whittamore, J. M., Wilson, R. W., and Grosell, M. (2007). Postprandial acid–base balance and ion regulation in freshwater and seawater-acclimated European flounder, *Platichthys flesus*. *J. Comp. Physiol. B* **177**, 597–608.

Toko, I. I., Fiogbe, E. D., and Kestemont, P. (2008). Mineral status of African catfish (*Clarias gariepinus*) fed diets containing graded levels of soybean or cottonseed meals. *Aquaculture* **275**, 298–305.

Tomasso, J. R., Jr., and Grosell, M. (2005). Physiological basis for large differences in resistance to nitrite among freshwater and freshwater-acclimated euryhaline fishes. *Environ. Sci. Technol.* **39**, 98–102.

Tonheim, S. K., Espe, M., Raae, A. J., Darias, M. J., and Ronnestad, I. (2004). *In vivo* incorporation of [U]-14-C-amino acid: an alternative protein labeling procedure for use in examining larval digestive physiology. *Aquaculture* **235**, 553–567.

Tresguerres, M., Parks, S. K., Wood, C. M., and Goss, G. G. (2007). V-H$^+$-ATPase translocation during blood alkalosis in dogfish gills: interaction with carbonic anhydrase and involvement in the postfeeding alkaline tide. *Am. J. Physiol.* **292**, R2012–R2019.

Tytler, P., Tatner, M., and Findlay, C. (1990). The ontogeny of drinking in the rainbow trout, *Oncorhynchus mykiss* (Walbaum). *J. Fish Biol.* **36**, 867–875.

Usher, M. L., Talbot, C., and Eddy, F. B. (1988). Drinking in Atlantic salmon smolts transferred to seawater and the relationship between drinking and feeding. *Aquaculture* **73**, 237–246.

Veillette, P. A., White, R. J., Specker, J. L., and Young, G. (2005). Osmoregulatory physiology of pyloric ceca: regulated and adaptive changes in Chinook salmon. *J. Exp. Zool.* **303A**, 608–613.

Verbost, P. M., Flik, G., Lock, R. A. C., and Wendelaar Bonga, S. E. (1987). Cadmium inhibition of Ca^{2+} uptake in rainbow trout gills. *Am. J. Physiol.* **253**, R216–R221.

Vielma, J., and Lall, S. P. (1997). Dietary formic acid enhances apparent digestibility of minerals in rainbow trout, *Oncorhynchus mykiss* (Walbaum). *Aquaculture Nutr.* **3**, 265–268.

Walsh, P. J., Kajimura, M., Mommsen, T. P., and Wood, C. M. (2006). Metabolic organization and effects of feeding on enzyme activities of the dogfish shark (*Squalus acanthias*) rectal gland. *J. Exp. Biol.* **209**, 2929–2938.

Weihrauch, D., Wilkie, M. P., and Walsh, P. J. (2009). Ammonia and urea transporters in gills of fish and aquatic crustaceans. *J. Exp. Biol.* **207**, 4491–4504.

Wiedmann, T. S., Lang, W., and Herrington, H. (2004). Interaction of bile salts with gastrointestinal mucins. *Lipids* **39**, 51–58.

Wilson, R. W., and Grosell, M. (2003). Intestinal bicarbonate secretion in marine teleost fish-source of bicarbonate, pH sensitivity, and consequences for whole animal acid–base and calcium homeostasis. *Biochim. Biophys. Acta* **1618**, 163–174.

Wilson, R. W., Gilmour, K. M., Henry, R. P., and Wood, C. M. (1996). Intestinal base excretion in the seawater-adapted rainbow trout: a role in acid–base balance? *J. Exp. Biol.* **199**, 2331–2343.

Wilson, R. W., Wilson, J. M., and Grosell, M. (2002). Intestinal bicarbonate secretion by marine teleost fish—why and how? *Biochim. Biophys. Acta* **1566**, 182–193.

Wood, C. M., Wheatly, M. G., and Hobe, H. (1984). The mechanisms of acid–base and ionoregulation in the freshwater rainbow trout during environmental hyperoxia and subsequent normoxia III. Branchial exchanges. *Respir. Physiol.* **55**, 175–192.

Wood, C. M. (1995). Excretion. In: *Physiological Ecology of the Pacific Salmon* (C. Groot, L. Margolis and W. C. Clarke, eds), pp. 381–438. Government of Canada Special Publications Branch, UBC Press, Vancouver.

Wood, C. M., Pärt, P., and Wright, P. A. (1995). Ammonia and urea metabolism in relation to gill function and acid–base balance in a marine elasmobranch, the spiny dogfish (*Squalus acanthias*). *J. Exp. Biol.* **198**, 1545–1558.

Wood, C. M. (2001a). The influence of feeding, exercise, and temperature on nitrogen metabolism and excretion. In: *Fish Physiology* (P. A Anderson and P. A Wright, eds), Vol. 20, pp. 201–238. Academic Press, Orlando, FL.

Wood, C. M. (2001b). Toxic responses of the gill. In: *Target Organ Toxicity in Marine and Freshwater Teleosts, Volume 1—Organs* (D. W. Schlenk and W. H. Benson, eds), pp. 1–89. Taylor and Francis, Washington DC.

Wood, C. M., Matsuo, A. Y. O., Gonzalez, R. J., Wilson, R. W., Patrick, M. L., and Val, A. L. (2002). Mechanisms of ion transport in *Potamotrygon*, a stenohaline freshwater elasmobranch native to the ion-poor blackwaters of the Rio Negro. *J. Exp. Biol.* **205**, 3039–3054.

Wood, C. M., McDonald, M. D., Walker, P., Grosell, M., Barimo, J. F., Playle, R. C., and Walsh, P. J. (2004). Bioavailability of silver and its relationship to ionoregulation and silver speciation across a range of salinities in the gulf toadfish (*Opsanus beta*). *Aquat. Toxicol.* **70**, 137–157.

Wood, C. M., Kajimura, M., Mommsen, T. P., and Walsh, P. J. (2005). Alkaline tide and nitrogen conservation after feeding in an elasmobranch (*Squalus acanthias*). *J. Exp. Biol.* **208**, 2693–2705.

Wood, C. M., Franklin, N., and Niyogi, S. (2006). The protective role of dietary calcium against cadmium uptake and toxicity in freshwater fish: an important role for the stomach. *Environ. Chem.* **3**, 389–394.

Wood, C. M., Bucking, C. P., Fitzpatrick, J., and Nadella, S. R. (2007a). The alkaline tide goes out and the nitrogen stays in after feeding in the dogfish shark, *Squalus acanthias*. *Respir. Physiol. Neurobiol.* **159**, 163–170.

Wood, C. M., Kajimura, M., Bucking, C. P., and Walsh, P. J. (2007b). Osmoregulation, ionoregulation, and acid–base regulation by the gastrointestinal tract after feeding in the elasmobranch (*Squalus acanthias*). *J. Exp. Biol.* **210**, 1335–1349.

Wood, C. M., Munger, S. R., Thompson, J., and Shuttleworth, T. J. (2007c). Control of rectal gland secretion by blood acid–base status in the intact dogfish shark (*Squalus acanthias*). *Respir. Physiol. Neurobiol.* **156**, 220–228.

Wood, C. M., Kajimura, M., Mommsen, T. P., and Walsh, P. J. (2008). Is the alkaline tide a signal to activate metabolic or ionoregulatory enzymes in the dogfish shark (*Squalus acanthias*)? *Physiol. Biochem. Zool.* **81**, 278–287.

Wood, C. M., Schultz, A. G., Munger, R. S., and Walsh, P. J. (2009). Using omeprazole to link the components of the post-prandial alkaline tide in the spiny dogfish, *Squalus acanthias*. *J. Exp. Biol.* **212**, 684–692.

Wood, C. M., Bucking, C. and Grosell, M. (2010). Acid–base responses to feeding in freshwater and seawater acclimated killifish, *Fundulus heteroclitus*, an agastric euryhaline teleost. *J. Exp. Biol.* In Press.

Wright, P. A., and Wood, C. M. (2009). A new paradigm for ammonia excretion in aquatic animals: role of Rhesus (Rh) glycoproteins. *J. Exp. Biol.* **212**, 2303–2312.

Zaugg, W. S., and McLain, L. R. (1969). Inorganic salt effects on growth, salt water adaptation, and gill ATPase of Pacific salmon. In: *Fish in Research* (J. E. Halver and O. W. Neuhaus, eds), pp. 293–306. Academic Press, New York.

Zaugg, W. S., Roley, D. D., Prentice, E. F., Gores, K. X., and Waknitz, F. W. (1983). Increased seawater survival and contribution to the fishery of Chinook salmon (*Oncorhynchus tshawytscha*) by supplemental dietary salt. *Aquaculture* **32**, 183–188.

Zhang, C., Brown, S. B., and Hara, T. J. (2001). Biochemical and physiological evidence that bile acids produced and released by lake char (*Salvelinus namaycush*) function as chemical signals. *J. Comp. Physiol.* **171**, 161–171.

ADDITIONAL READING

MacLeod, M. G. (1978). Relationships between dietary sodium chloride, food intake, and food conversion in the rainbow trout. *J. Fish. Biol.* **13**, 73–78.

Sundell, K., and Bjornsson, B. T. (1988). Kinetics of calcium fluxes across the intestinal mucosa of the marine teleost, *Gadus morhua*, measured using an *in vitro* perfusion method. *J. Exp. Biol.* **140**, 171–186.

Vielma, J., Ruohonen, K., and Vogel, K. (2004). Top-spraying soybean meal-based diets with phytase improves protein and mineral digestibilities but not lysine utilization in rainbow trout, *Oncorhynchus mykiss* (Walbaum). *Aquaculture Res.* **35**, 955–964.

Wood, C. M., and Laurent, P. (2003). Na$^+$ versus Cl$^-$ transport in the intact killifish after rapid salinity transfers. *Biochim. Biophys. Acta.—Biomembranes* **1618**, 106–120.

6

IMPLICATIONS OF GI FUNCTION FOR GAS EXCHANGE, ACID–BASE BALANCE AND NITROGEN METABOLISM

J.R. TAYLOR

C.A. COOPER

T.P. MOMMSEN

The physiological trade-offs required for animals to maintain homeostasis under varying conditions have long been investigated by comparative physiologists. The gastrointestinal tract of fishes offers an abundance of examples in which gas exchange, acid–base balance, and nitrogen metabolism must be complementarily regulated in order to confer the smallest energetic burden to the animal. In the following we investigate the influence that, in particular, salinity acclimation and feeding have on the inter-regulated processes of gas exchange, acid–base balance, and nitrogen metabolism in the fish gut.

The Multifunctional Gut of Fish: Volume 30
FISH PHYSIOLOGY

1. INTRODUCTION

The overlapping gastrointestinal (GI) functions of nutrient absorption and salt-water balance necessitate a fine coordination of gas exchange, acid–base balance, and nitrogen metabolism in the fish gut. The role of the fish GI tract in homeostasis varies with environmental salinity (Chapters 4 and 5), but the coordination of physiological homeostatic processes involving the gut is particularly striking following feeding events. During the postprandial period, fish rapidly undergo a number of disturbances to steady state conditions. First, whole animal gas exchange is generally increased at least two-fold during a metabolic disturbance termed specific dynamic action (SDA) (Rubner 1902; Lusk 1931; Jobling 1981; McCue 2006; Secor 2009). Second, acid–base balance may be challenged by a systemic alkalosis or "alkaline tide" associated with gastric acid secretion (Rune 1965; Hersey and Sachs 1995; Niv and Fraser 2002; Wood et al. 2009). Finally, nitrogen excretion rates increase up to five- to six-fold in a matter of hours (Handy and Poxton 1993; Wood et al. 1995; Wood 2001), following intestinal processing of nitrogenous compounds associated with the meal. The extent of the compensatory responses required to maintain homeostasis during these postprandial disturbances can be expected to range from relatively minor in near-continual feeders ingesting modest rations, to quite remarkable physiological adjustments in more opportunistically feeding species. Pythons, as a case in point, ingest meals upwards of 20–30% body mass, and have become a classic model of extreme postprandial physiology (Secor and Diamond 1995, 1998; Andrade et al. 2004; Arvedsen et al. 2005; Lignot et al. 2005; Secor 2009). This example of August Krogh's principle that, "for such a large number of problems there will be some animal of choice, or a few such animals, on which it can be most conveniently studied" (Krogh 1929) can certainly be applied within the piscine taxa in which a wide range of feeding strategies are represented. Furthermore, the August Krogh principle can be applied at the individual level to study homeostatic mechanisms during the postprandial period, when they are particularly pronounced. An actively feeding individual is routinely subjected to physiological disequilibria resulting from meal ingestion and subsequent processing. The up-regulation of homeostatic mechanisms—including those of gas exchange, acid–base balance, and nitrogen metabolism—required to correct these disequilibria can often permit these mechanisms to be studied more clearly. Feeding has proven in recent years to be a useful means by which to study gastrointestinal mechanisms important in marine teleost osmoregulation. As opposed to freshwater teleosts in which the GI tract is largely inactive under steady state (fasting) conditions, marine teleosts drink the external medium at rates of 1–5 ml/kg/h.

This ingested seawater is processed using finely regulated gastrointestinal ion transport and subsequent branchial and renal ion excretion, allowing marine teleosts to maintain osmotic balance in a dehydrating environment (Smith 1930; Evans et al. 2005; Marshall and Grosell 2005).

Although the esophagus plays a role in the desalinization of ingested seawater (Stevens 1995; Loretz 2001; Ando et al. 2003), and the stomach is capable of some lipid absorption (Barrington 1942; Denstadli et al. 2004) and ion absorption associated with gastric acid secretion (Bucking and Wood 2006a,b), the great majority of active ion and nutrient absorption along the fish GI tract takes place in the post-gastric regions (Stevens 1995; Clements and Raubenheimer 2005). The intestinal tract expresses a wide range of nutrient and ion transport proteins, of which many have been characterized in molecular detail (Wright 1993; Loretz 2001; Grosell 2007). The uptake of carbohydrate and protein metabolites is often inextricably linked to ion transport, as in the classic example of Na^+-dependent glucose uptake (Collie 1985; Wright 1993). Furthermore, the processing of nitrogenous compounds is essential for maintenance of the oxidative substrates required for hydromineral transport to proceed. In addition to this overlap between nutrient absorption and salt-water balance, the intestinal transport of nutrients and ions can be linked, either directly or indirectly, to the transport of acid and base equivalents. In particular, movement of H^+ and HCO_3^- or their conjugates (including CO_2 and NH_4^+), between the intestinal lumen and the cellular (enterocyte) and/or extracellular fluids can have implications for systemic acid–base balance.

Intestinal HCO_3^- secretion by apical Cl^-/HCO_3^- exchange is the basis of a coordinated transport system that has become widely recognized for its role in marine fish osmoregulation (Walsh et al. 1991; Wilson et al. 1996, 2002, 2009; Grosell et al. 2005; Genz et al. 2008) and is described in detail by Grosell in Chapter 4. This system of intestinal HCO_3^- secretion has also received recent attention for its role in postprandial homeostasis in seawater and euryhaline teleosts (Bucking and Wood 2006a, 2008; Taylor and Grosell 2006b, 2009; Taylor et al. 2007; Bucking et al. 2009). Because of its relative novelty in the literature and its potential linkages to gas exchange, acid–base balance, and nitrogen metabolism, intestinal HCO_3^- secretion will be referenced as an exemplary mechanism by which the fish gut accomplishes multiple functions.

In the following sections, the central players are the enterocytes—polarized intestinal absorptive cells with villi at the apical surface exposed to the lumen, and basolateral portions that communicate with interstitial fluids and the plasma compartment. The physiology of these cells forms the basis for the function of the GI tissue and thereby its role in whole animal physiology, a link we strive to consistently make clear in the following sections. The intestine is a remarkably dynamic tissue, particularly in response to feeding/fasting, which not only exert massive effects on thickness of microvilli, number of enterocytes, thickness of muscular layer around the intestine and rate of material flow through the system, but can also affect the overall length of the gut (see Chapter 1).

2. GAS EXCHANGE

Gas exchange is accomplished primarily at the large surface area of the gill in most fishes examined, by mechanisms detailed in a number of recent reviews (Nikinmaa and Salama 1998; Tufts and Perry 1998; Evans et al. 2005; Perry et al. 2009). Tissue metabolic requirements thereby rely on the circulatory system for sufficient O_2 supply and CO_2 removal. The instrumental role of the GI tract in meal assimilation and thereby growth suggests a high demand for metabolic activity by this tissue. Metabolically active transport mechanisms may be in even greater demand and under more stringent regulation in marine teleosts, in which the GI tract has an additional vital function in osmoregulation. The GI tissue thereby is a considerable point of interest in the study of gas exchange in fish, particularly during salinity acclimation and feeding. However, methodological limitations combined with the interference of thermal and salinity acclimation, and biotic variables including species-specific differences, life stage and feeding strategy/history, beget a justifiably sparse and inconsistent body of literature describing gas exchange by GI tissue. However, recent years have brought about methodological advances that promise to aid the study of GI tissue metabolism considerably. In particular, isolated tissue respirometry chambers have recently allowed for the first (to our knowledge) quantification of oxygen consumption by isolated fish intestinal epithelia (Taylor and Grosell 2009). This approach holds promise for a number of under-represented inquiries including simple tissue-specific metabolic rate determination, fuel selection/metabolic substrate, and the influence of salinity challenge and feeding events on these parameters in fish GI tissue. By gaining a better understanding of metabolism at the cellular and isolated tissue levels, our inference of the contribution of the GI tract to gas exchange in fish (and the influence of variables that have led to inconsistency in the literature) will be more accurate. In the following we will survey what is known about the role of fish GI tissue in gas exchange, as inferred by (1) consumption of O_2 by this tissue as dictated by its metabolic requirements, and (2) excretion of metabolic CO_2 waste. Above all, we hope to inspire future investigations of the mechanisms by which the fish gut accomplishes a wide range of vital functions with striking metabolic efficiency.

2.1. Oxygen Consumption

Metabolic rates in fish are commonly measured indirectly by quantification of oxygen consumption (Cech 1990). Aerobic respiration requires the consumption of O_2 for the production of energy in the form of adenosine triphosphate (ATP). The complete oxidation of one glucose molecule to CO_2 and H_2O theoretically supplies up to 38 ATP molecules (Krebs et al. 1957; Lehninger 1970; Rich 2003), although estimates of actual yield are generally closer to 29–30 ATP per molecule of glucose (Rich 2003). The complex chemical interactions involved in aerobic respiration include glycolysis and subsequent oxidative phosphorylation via the citric acid (Krebs) cycle. These pathways can be simplified to the reaction given in Eqn (1).

$$C_6H_{12}O_6(aq) + 6O_2(g) \rightarrow 6CO_2(g) + 6H_2O(l) \tag{1}$$

The GI tract is a site of high O_2 use that amounts to 20–25% of the whole-body O_2 consumption in mammals, even in the fasting state (Britton and Krehbiel 1993; Duee et al. 1995). The metabolic activity of GI tissue is disproportionately greater than suggested by its mass (approx. 6% of body mass in these studies). While work by Taylor and Grosell (Taylor and Grosell 2009) investigating toadfish intestinal tissue O_2 consumption suggests baseline metabolic demand of this tissue to be in the range of a more modest 6% of whole animal O_2 consumption (still greater than that suggested by its mass of approx. 2% body mass), the demand of the GI tissue for O_2-consuming metabolic processes increases at least marginally in fish after feeding. This increased metabolic demand is linked in part to the synthesis and activation of enzymes (discussed below in Section 4) and transporters along the GI tract. Our current knowledge of transport processes in the fish gut indicates that baseline metabolic requirements of the intestinal tissue are closely correlated with the action of basolateral Na^+/K^+-ATPase (NKA). The electrochemical gradient generated by NKA favors the uptake of Na^+ by the enterocyte and is the driving force for a number of intestinal transporters important to both digestion and marine osmoregulation. Intestinal ion transporters fueled by NKA include apical NaCl and $Na^+/K^+/2Cl^-$ (NKCC) co-transporters, apical Cl^-/HCO_3^- and Na^+/H^+ (NHE) exchangers (Loretz 1995; Movileanu et al. 1998; Marshall and Grosell 2005; Grosell 2007), and basolateral Na^+/HCO_3^- (NBC) cotransport (Kurita et al. 2008; Taylor et al. 2010). An apical V-type H^+-ATPase has also been recently confirmed as an important player in rainbow trout seawater acclimation (Grosell et al. 2007, 2009b) and comprises a component of the multifaceted intestinal HCO_3^- secretion system, detailed by Grosell in Chapter 4. Further characterization of this H^+-ATPase will elucidate its potential contribution to the metabolic demand of the fish gut.

Carbonic anhydrase (CA) also plays an essential role in the intestinal HCO_3^- secretion system of marine teleosts. The action of CA is dependent on the alkalinization of its active site facilitated by mobile buffers including phosphate-containing compounds such as ATP (Henry and Heming 1998), making its activity susceptible to disturbances in gas exchange.

2.1.1. INFLUENCE OF SALINITY

While osmoregulatory costs have been suggested to comprise up to 20–68% of total energy expenditure in studies across a range of species, the most recent literature suggests osmoregulatory costs are closer to 10% of the total fish energy budget (Boeuf and Payan 2001; Tseng and Hwang 2008; Evans and Claiborne 2008). These studies are predominantly based on whole-fish O_2 consumption, however, and cannot distinguish the partitioning of energy costs to a particular osmoregulatory organ. Furthermore, the metabolic effects of salinity are subject to a number of complicated metabolic pathways occurring outside the principal osmoregulatory tissues. As recently reviewed by Tseng and Hwang (Tseng and Hwang 2008) and Evans and Claiborne (Evans and Claiborne 2008), methodologies for quantifying tissue-specific osmoregulatory costs have focused on the gills and are subject to a number of limitations begetting a wide range of results and subsequent conclusions. Febry and Lutz, for example, determined that osmoregulation in a euryhaline hybrid tilapia is more expensive in freshwater than in seawater, and that isosmotic conditions confer the least osmoregulatory cost (Febry and Lutz 1987). Further studies have shown that euryhaline Mozambique tilapia (*Oreochromis mossambicus*) have two- to three-fold higher growth rates in seawater than freshwater (Kuwaye et al. 1993; Ron et al. 1995; Shepherd et al. 1997; Sparks et al. 2003). Conversely, the "metabolic method" of comparing routine metabolic rates with the calculated cost of ATP utilization for ion transport by osmoregulatory organs including the gill, kidney, and intestine suggests osmoregulation is more costly in seawater than freshwater for trout and killifish. These discrepancies regarding osmoregulatory cost have been recently detailed by Evans and Claiborne (Evans and Claiborne 2008). Further complications arise when we consider the impact of dietary contribution to osmoregulation, particularly in freshwater where dietary intake can comprise an essential component of ion uptake as described by Wood and Bucking in Chapter 5.

Based on reports that NKA activity accounts for 20–40% of energy expenditure in mammals, a considerable proportion of the fasting metabolic rate of fishes can undoubtedly be attributed to ion transport via NKA (Jobling 1994). Owing to its importance in osmoregulation via active ion transport, the intestinal tissue of marine teleosts might be expected to have substantially higher metabolic demands and thereby gas exchange, than that of their freshwater or brackish water counterparts. However, the literature—and the methodology upon which it has relied—is lacking. Considering that osmoregulatory costs likely represent 10% of total energy expenditure, salinity-dependent differences in GI epithelial oxygen consumption are likely to be quantitatively small if detectable at all. Isolated tissue respirometry seems to be the most promising technique to accommodate investigations of salinity-dependent differences in GI (and branchial and renal) epithelial O_2 consumption, and would certainly allow for partitioning of the energetic costs associated with osmoregulation. During such isolated tissue experiments, pharmacological inhibition of NKA and H^+-ATPase using ouabain and bafilomycin, respectively, could also be employed to determine even more specifically the metabolic cost of GI epithelial ion transport in a variety of media. Recently, Sardella et al. (Sardella et al. 2008) used tissue microarray (TMA) and laser-scanning cytometry (LSC) to quantify NKA abundance per unit area and mitochondria-rich cell (MRC) number/size, respectively, in gills of freshwater- and saltwater-acclimated tilapia across a range of temperatures. The combined analyses showed that saltwater fish had larger but fewer MRCs that contained more NKA per unit area, and that changes to these parameters (in this case due to thermal acclimation) affected osmoregulatory ability. Likewise, comparative investigations of fish enterocyte MRC characteristics may offer a more exact means by which to investigate the energetics of this tissue.

2.1.2. INFLUENCE OF FEEDING

Similarly to its critical function in marine teleost osmoregulation, the role of the GI tract in meal processing inspires its consideration as a substantial contributor to the metabolic rate increases termed specific dynamic action (SDA) associated with feeding (Fig. 6.1). The SDA response of most fishes falls between the modest response seen in near-continually feeding mammals and the extreme response of opportunistically feeding reptiles. The primary literature investigating fish SDA has historically focused primarily on commercially important species. In the last quarter century, a particular interest in fish energetics by the aquaculture community promises to increase proportionately to our need for sustainable seafood in coming years. These studies expectedly focus on the influence of feed composition (most notably, protein content and type), holding conditions (particularly temperature, salinity, and hypoxia), and allometry, as they apply to cost–benefit relationships with respect to assimilation efficiency and from a business standpoint of maximizing biomass yield and

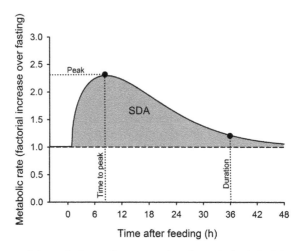

Fig. 6.1. Postprandial metabolic profile representing the typical whole fish specific dynamic action (SDA) response. The SDA response in fishes is characterized by peak metabolic rates occurring 3–12 h after feeding and averaging 2- to 3-fold over those during routine fasting (represented by dashed line). Elevated metabolic rates associated with SDA in fish generally persist in the 20–40 h range. Temperature is generally considered the strongest determinant in the shape of the SDA response profile. Figure adapted from Secor (2009).

quality for minimal resource input. These variables as they pertain to fish nutrition were detailed by Talbot following a symposium on "Fish and Nutrition" (Talbot 1993), and are presented alongside methodological considerations by Jobling (Jobling 1981, 1994, 1995) and Belal (Belal 2004).

A number of reviews of the SDA response have compiled results from the wide range of primary literature to suggest maximum postprandial metabolic rate increases averaging 2.36-fold in fishes, with elevated metabolic rates persisting in the range of 20–40 h after feeding (Jobling 1981, 1994; McCue 2006; Secor 2009). The abiotic and biotic variables influencing the SDA response and associated energy trade-offs offer myriad options for maximizing aquaculture efficiency, but lend an exceptional complexity to more fundamental investigations of postprandial gas exchange. The SDA response in all animals is dependent particularly on meal mass and composition, body temperature, and body size. A consequence of this variation in SDA is that the specific source(s) of the metabolic rate increase associated with feeding are difficult to discern. The SDA response corresponds to the summation of many energy-consuming processes, each of which is uniquely influenced by the described variables. For detailed analyses of the contribution of specific pre-absorptive, absorptive, and post-absorptive processes to SDA, we refer readers to recent comprehensive reviews by McCue (McCue 2006) and Secor (Secor 2009).

The post-absorptive oxidative processes associated with amino acid catabolism, and subsequent protein synthesis have become considered the principal components of SDA (Brown and Cameron 1991a,b). While post-absorptive amino acid catabolism occurs largely in the liver, the intestinal epithelium plays a gate-keeping role for amino acid absorption and as such gets first dibs at using these building blocks to meet its own needs. In the presence of sufficient oxygen availability and metabolic substrate, the enterocytes can be a site of considerable amino acid metabolism and protein synthesis as detailed in Section 4. Oxygen consumption has in fact been shown by Smith and Houlihan (Smith and Houlihan 1995) to be linearly correlated with protein synthesis in rainbow trout macrophages and scale cells. Accordingly, measurements of isolated GI tissue O_2 consumption may be a suitable proxy for the synthesis and activation of transport proteins and enzymes that could contribute considerably to the SDA response. Indeed, the isolated toadfish intestine is subject to a significant postprandial increase in O_2 consumption indicative of SDA in this tissue. Oxygen consumption peaked 6 h after feeding, at 187% of fasting rates, and accounts for a disproportionately large portion of whole animal SDA relative to its mass (Taylor and Grosell 2009).

2.2. Carbon Dioxide Excretion

Aerobic metabolism produces CO_2 in accordance with Eqn (1), at rates dictated by metabolic requirements/oxidative capacity, and substrate availability. In aqueous solutions, CO_2 acts as a weak acid with implications for acid–base balance that will be further discussed in Section 3. To avoid acid–base disequilibria to both the delicate intracellular environment and the whole animal, excretion of CO_2 must closely correspond to its production. In order to reach the gill for excretion, CO_2 is carried from tissues by the blood as physically dissolved CO_2, HCO_3^-, and carbamino CO_2, with HCO_3^- generally comprising $>90\%$ of the CO_2 within the blood compartment. Metabolism produces molecular CO_2 at rates far greater than the carrying capacity of the blood for physically dissolved CO_2, necessitating continual inter-conversion of CO_2 and HCO_3^- to meet the needs for CO_2 excretion (Tufts and Perry 1998). The enzyme carbonic anhydrase (CA) facilitates this inter-conversion via the equilibrium reaction.

$$CO_2 + H_2O \leftrightarrow H_2CO_3 \leftrightarrow H^+ + HCO_3^- \tag{2}$$

The hydration reaction that facilitates CO_2 excretion as HCO_3^- also generates H^+ (Eqn (2)) which must be removed from the reaction site in order to avoid acid–base disequilibria as HCO_3^- is carried away from tissues by the blood. Mechanisms by which cellular and systemic acid–base balance is accomplished are detailed in Section 3.

The transport and subsequent excretion of CO_2 from metabolically active tissues including those of the GI tract depends predominantly on its movement into the blood compartment. In the marine teleost intestinal epithelia, however, a considerable fraction of endogenously produced CO_2 is eliminated via the gut as HCO_3^-, secreted across the apical membrane through Cl^-/HCO_3^- exchange (Chapter 4). This relies on the facilitation of CO_2 hydration by intracellular CA and may also involve membrane-bound CA isoforms (McMurtrie et al. 2004; Becker and Deitmer 2007; Morgan et al. 2007). Transport of CO_2 across the basolateral membrane of the enterocyte has been inferred to occur via diffusion of molecular CO_2 and has more recently been shown also to occur as HCO_3^- via Na^+/HCO_3^- co-transport (NBC) (Kurita et al. 2008; Taylor et al. 2010). The reliance of these mechanisms of CO_2 transport and excretion on the inter-conversion of CO_2 and HCO_3^- indicates that CA plays an essential role in GI gas exchange that parallels its importance in the transport and excretion of CO_2 throughout the body. The importance of CA has been well substantiated in the operation of theoretical models for transport of the acid and base equivalents H^+ and HCO_3^- into or out of absorbing and secreting epithelial tissues (Parsons and Case 1982; Henry 1996; Tresguerres et al. 2006; Gilmour and Perry 2009). Furthermore, the critical role of CA in whole animal gas exchange has been detailed in a number of excellent reviews (Henry and Swenson 2000; Jensen 2004; Esbaugh and Tufts 2006a; Perry et al. 2009). The action of CA in the GI epithelia of fish is undeniably instrumental in the metabolic efficiency upon which GI function relies.

2.3. The Evolution of GI Metabolic Efficiency

The capacity of fish GI epithelia to increase ion/nutrient transport during both salinity challenge and digestion is disproportionately large compared to the modest metabolic cost to the tissue. The inherently dynamic nature of the GI tract in routinely feeding fishes likely at least to some extent underlies its efficiency in returning to equilibrium following such environmental challenges as salinity transfer. The activity of CA contributes appreciably to this efficiency by increasing the rate of inter-conversion between CO_2 and HCO_3^- up to three orders of magnitude (Henry and Heming 1998), facilitating multiple functions of the gut with minimal metabolic burden.

Ancestral fishes generally appear to use comparable GI ion and nutrient transport mechanisms to those characterized in teleosts inhabiting similar environments (Marshall and Grosell 2005; Taylor and Grosell 2006a; Grosell 2007). This suggests that the modest energetic cost of osmoregulation and digestion by GI epithelia is likely a universal characteristic across piscine taxa. Furthermore, data indicate that intestinal HCO_3^- secretion is a route of CO_2 excretion not only in marine teleosts but also in seawater-acclimated Siberian sturgeon, *Acipenser baerii,* salinity-challenged bamboo sharks, *Chiloscyllium plagiosum* (Taylor and Grosell 2006a), and sea lampey, *Petromyzon marinus* (J. R. Taylor, J. M. Wilson and M. Grosell unpublished). Furthermore, CA appears to have been selected as primarily an enzyme of facilitated CO_2 transport, and the evolution of its function continues to be one of the most fertile areas of study of this enzyme (Henry and Heming 1998). CA activity in the GI tissue has received relatively little attention from an evolutionary standpoint in comparison to that of red blood cells, for example, which play a more direct role in CO_2 excretion (Tufts et al. 2003; Esbaugh and Tufts 2006a). Scant studies of gut CA activity in ancestral fishes have shown an absence of membrane-bound CA isoforms in lamprey (Esbaugh and Tufts 2006b), and hagfish (Esbaugh et al. 2009), but have yet to investigate the activity of cytosolic CA isoforms in ancestral fish GI tissue.

3. ACID–BASE BALANCE

The regulation of acid–base status is of fundamental importance for life to persist. In fish, the production, uptake and excretion of acid–base relevant ions are achieved via a combination of processes primarily at the gill (McDonald and Wood 1981; Evans 1982; Goss et al. 1992) and to a lesser extent the kidney (Wood and Caldwell 1978; Perry et al. 1987; Goss and Wood 1990). However, the piscine GI tract also generates acid–base disturbances, and subsequent compensatory measures, that are unique among vertebrates (Wilson et al. 2002; Wilson and Grosell 2003; Grosell 2006; Grosell and Genz 2006; Grosell and Taylor 2007; Cooper and Wilson 2008; Grosell et al. 2009a,b).

3.1. Cellular Acid–Base Balance

In the term "acid–base balance," metabolic acidic equivalents are H^+ and NH_4^+, and basic equivalents are HCO_3^- and OH^-. CO_2 is a respiratory acid, and NH_3 is a respiratory base. Cellular respiration by enterocytes generates both acidic and basic compounds by production of CO_2 (Eqn (1)) and its subsequent hydration (Eqn (2)). Due to the gastrointestinal mechanisms by which homeostasis is maintained during both osmotic challenge and digestion, salinity and feeding status are of particular importance to cellular acid–base balance in the GI tract of fish. Overlap between gastrointestinal mechanisms of osmoregulation and nutrient absorption link these factors together, particularly in marine fish requiring continual secretion of HCO_3^- into the intestinal lumen as part of their osmoregulatory strategy. Accordingly, the requirement to actively maintain GI tract cellular acid–base balance varies between freshwater and marine fish, based on the range of osmoregulatory strategies (Chapter 4) represented across the piscine taxa.

3.1.1. INFLUENCE OF SALINITY

While a number of ancestral groups including hagfish and elasmobranchs are osmoconformers, the overwhelming majority of piscine vertebrates are osmoregulating bony fishes belonging to the taxonomic division Teleostei (Nelson 1994). In particular, marine teleosts osmoregulate in an environment having an osmolality approximately three-fold higher than that of their blood. In order to replace fluid lost to the environment, seawater ingested by marine teleosts must be processed by gastrointestinal mechanisms involving acid and base equivalents. These fish thereby face potential disturbances to acid–base balance (Grosell 2006). Freshwater teleosts, conversely, inhabit an environment hypo-osmotic to their bodies and thus gain water diffusively, eliminating the need for these gastrointestinal mechanisms under steady-state (unfed) conditions. Indeed, enterocytes of freshwater trout have been shown to be generally inactive during times of fasting (Bucking and Wood 2006a,b).

Osmoregulation in a hyperosmotic environment is facilitated by the production and secretion of intestinal HCO_3^-. This process liberates H^+ (Eqn (2)), which the enterocyte must then excrete. A number of recent studies on the marine teleost GI tract have helped to assemble an evolving model for acid–base balance at the cellular level which is summarized in Fig. 6.2 (Wilson et al. 2002; Wilson and Grosell 2003; Grosell 2006; Grosell and Genz 2006; Grosell and Taylor 2007; Grosell et al. 2009a,b). The intracellular hydration of CO_2 produces HCO_3^- and H^+ ions (Eqn (2)), a reaction that is catalyzed by CA (Gilmour et al. 2002; Perry and Gilmour 2006) (Fig. 6.2A). While secreting HCO_3^- into the lumen in exchange for Cl^- via apical anion exchange (Fig. 6.2B), the enterocyte must in parallel excrete the H^+ to maintain acid–base balance. Evidence suggests that the H^+ will be preferentially extruded from the intestinal cell basolaterally in exchange for Na^+ (Fig. 6.2C) (Grosell and Genz 2006) and also to some extent apically, via vacuolar H^+ ATPase (Grosell et al. 2009a,b) (Fig. 6.2D). In the apical boundary layer it is predicted that these H^+ will contribute to the dehydration of secreted HCO_3^-, producing CO_2 (Fig. 6.2E). This process may well involve an extracellular, membrane-bound isoform of CA, although this role of CA has yet to be established.

The HCO_3^- required for intestinal secretion is not always derived solely from endogenous CO_2 sources. During co-transport with Na^+ via NBC (Fig. 6.2F), HCO_3^- can also enter the enterocyte basolaterally from the

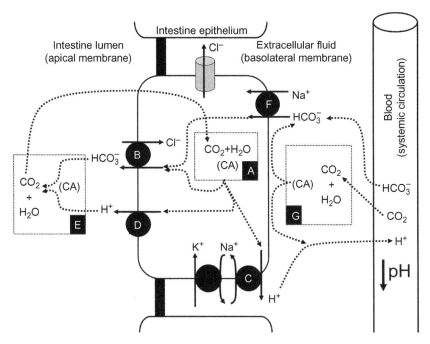

Fig. 6.2. Marine teleost osmoregulation involves finely coordinated inter-conversion and transport of acid–base equivalents, as detailed in the text.

extracellular fluid (Kurita et al. 2008; Taylor et al. 2010). Studies on gulf toadfish (*Opsanus beta*) have shown that the relative contributions of extracellular and endogenous CO_2 to supply intestinal HCO_3^- production are likely to vary in response to demand (Grosell 2006; Grosell and Genz 2006; Taylor and Grosell 2009; Taylor et al. 2010). The "demand" for intestinal HCO_3^- secretion may be determined not only by osmoregulatory needs, but also by the state of the extracellular compartment. The efficient inter-conversion of CO_2 and HCO_3^- by CA is apt to play a vital role in making blood CO_2 (Fig. 6.2G) and HCO_3^- available for intestinal secretion and subsequent excretion via the intestinal tract, although this role has yet to be characterized. Osmoregulation in a hyperosmotic environment creates cellular acid–base balance disturbances unique to marine teleost fish. As described, maintaining pH balance at the cellular level involves the inter-conversion and transportation of acid–base equivalent ions in and out of the enterocyte. Expectedly, these processes have the potential to impact whole animal acid–base status. Any such disequilibrium of systemic acid–base balance must be quickly compensated to maintain animal health.

It is generally perceived that the GI tract is not dynamically regulated to control whole animal acid–base homeostasis (Perry and Gilmour 2006). However, in circumstances such as feeding, the GI tract may act both to generate and compensate disturbances to acid–base balance. Two studies on rainbow trout have inferred that the intestine, in conjunction with the gills, is able to compensate for a postprandial alkaline tide (generated by the stomach) by excreting excess HCO_3^- into the intestinal lumen (Cooper and Wilson 2008; Bucking et al. 2009). Furthermore, *in vitro* work on toadfish intestine (Grosell and Genz 2006; Taylor et al. 2010) suggests the marine teleost intestine to be capable of a more general compensatory response to elevated serosal HCO_3^- (i.e. metabolic alkalosis). These examples are discussed in more detail below, and indicate that the GI tract of fishes can play a role in whole-body acid–base homeostasis at least during feeding.

3.1.2. INFLUENCE OF FEEDING

In addition to its essential role in marine teleost osmoregulation, the GI tract is also the obvious site for digestion in all fish species. Notably, some species of elasmobranchs have continuous acid-secreting stomachs regardless of the presence or absence of food (Menon and Kewalramani 1958; Williams et al. 1970; Papastamatiou and Lowe 2004; Wood et al. 2007c), and some species do not have stomachs at all (see Chapter 1). When food is present in the stomach, fish secrete isotonic HCl into the stomach lumen by parietal acid-secreting oxyntopeptic cells (Niv and Fraser 2002). The H^+ required for gastric acid secretion are produced when CA catalyzes the hydration of CO_2 (Fig. 6.3) shown in Eqn (2) (Perry and Gilmour 2006). It has been concluded in dogfish (*Squalus acanthias*) that, similarly to mammals and ectothermic terrestrial vertebrates, H^+ and Cl^- ions are extruded from the parietal cells into the lumen via apical H^+/K^+ ATPase pumps and Cl^- channels, respectively (Wood et al. 2009), as depicted in Fig. 6.3. To prevent intracellular alkalinization, an equimolar quantity of HCO_3^- is excreted basolaterally into the blood in exchange for Cl^-, thus fueling the net transcellular movement of Cl^- required for gastric HCl secretion (reviewed by Hersey and Sachs, 1995; Niv and Fraser, 2002). Furthermore, in mammals, ectothermic terrestrial vertebrates (Wang et al. 2001; Niv and Fraser 2002) and some species of fish (Wood et al. 2007b; Bucking and Wood 2008; Cooper and Wilson 2008; Bucking et al. 2009), the HCO_3^- loading into the blood has been shown to cause a metabolic alkalosis known as the postprandial "alkaline tide."

Following the gastric component of digestion, the acidic, partially digested food (chyme) enters the intestinal tract via the pyloric sphincter

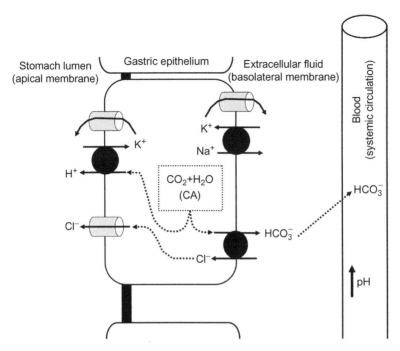

Fig. 6.3. Gastric acid secretion by piscine parietal cells results in HCl secretion into the stomach lumen and an equimolar secretion of HCO_3^- to the extracellular fluids. Figure based on Stevens (1995); Niv and Fraser (2002).

that commonly separates the distal stomach from the anterior intestine. Postprandial studies on freshwater (Shehadeh and Gordon 1969; Bucking and Wood 2006a,b) and marine (Taylor and Grosell 2006b, 2009; Taylor et al. 2007) teleost fish and an elasmobranch (Wood et al. 2007c) have shown that on entering the intestine, chyme pH is neutralized. However, the mechanism by which HCO_3^- enters the intestinal tract during digestion has not been well characterized in fish, due in part to the wide range of pancreatic structures represented among piscine species. Specifically, a number of fish including gulf toadfish (often used as a model marine teleost) have diffuse pancreatic tissue rather than a discrete organ secreting HCO_3^- via a pancreatic duct as is the case in mammals. Accordingly, in toadfish, Taylor and Grosell (Taylor and Grosell 2009) report postprandial increases in intestinal HCO_3^- secretion, in quantities sufficient to neutralize gastric acid secretion. This not only provides a link between HCl secretion and subsequent neutralization via HCO_3^- production and secretion, both of which originate from cells along the GI tract, but may also explain the lack of alkaline tide in toadfish (Taylor and Grosell 2006b) and flounder (Taylor et al. 2007).

3.2. Whole Animal Acid–Base Balance

The key parameters related to whole animal acid–base balance are plasma pH, bicarbonate (HCO_3^-) and the partial pressure of carbon dioxide (PCO_2) (Davenport 1958). The interactions between these three variables are related by the following equation:

$$pH = pK_{app} + \log \times ([HCO_3]/\alpha \times PCO_2) \tag{3}$$

where pK_{app} is the apparent dissociation of CO_2 (related to plasma pH and temperature) and α is the solubility of CO_2 (also related to temperature, see Boutilier et al. 1984). A diagram that plots these variables was first illustrated by Horace Davenport (Davenport 1958). Incidentally, when applying this model to fish, some considerations need to be made, as the [HCO_3^-] and the PCO_2 values of fish plasma are much lower than the corresponding values for air-breathing mammals (for reasons why, refer to Eddy 1976).

A deviation from steady-state plasma pH signifies a disturbance to acid–base balance. An increase in PCO_2 is termed a respiratory acidosis (Wood et al. 1977; Claiborne and Heisler 1986; McKenzie et al. 2002, 2003; Baker et al. 2009), whereas a decrease in PCO_2 signifies a respiratory alkalosis (Truchot and Forgue 1998; Chew et al. 2003; Leef et al. 2005; Wood et al. 2007a). Air-breathers typically compensate for these disturbances by hyperventilating to remove excess CO_2, or hypoventilating to retain CO_2. However, fish are essentially hyperventilated in terms of CO_2 (Eddy 1976; Gilmour 2001). Thus, to maintain acid–base balance they must compensate for changes in plasma pH by adjusting blood [HCO_3^-] via the gills (Eddy 1976; Claiborne and Heisler 1986; McKenzie et al. 2003; Baker et al. 2009) and the kidney (Curtis and Wood 1992; Wood et al. 1999). A change in plasma [HCO_3^-] and subsequently pH, while PCO_2 remains constant, is indicative of a metabolic disturbance to acid–base balance. An increase or decrease in plasma [HCO_3^-] results in a metabolic alkalosis or acidosis, respectively. The mechanisms by with the GI tract in fish carries out its functions can both cause and compensate metabolic alkalosis and acidosis.

3.2.1. IMPLICATIONS OF MARINE OSMOREGULATION

Maintaining osmoregulation in a hyperosmotic environment by continually imbibing seawater is a trait unique to marine teleost fish. If left uncompensated, the H^+ extruded into the blood compartment, combined with the removal of HCO_3^- and CO_2 for secretion into the intestinal lumen, would result in metabolic acidosis (Fig. 6.2). To compensate for this acidosis, fish up-regulate acid excretion via the gill. Indeed, Genz et al. (Genz et al. 2008) have shown that elevated salinity causes an increase in toadfish intestinal HCO_3^- secretion accompanied by a decrease in blood pH that is statistically significant but relatively modest due to compensatory branchial acid excretion.

Unlike marine teleost fish, most marine elasmobranchs have a completely different mechanism for preventing excessive fluid loss that eliminates the need to imbibe seawater. This negates the requirement for intestinal HCO_3^- secretion, and any consequential disturbances of systemic acid–base balance. However, several elasmobranch species have been shown to have a significant, albeit transient, drinking response when transferred to elevated salinity (Anderson et al. 2002; Taylor and Grosell 2006a). However, measurements of high intestinal fluid HCO_3^- concentrations have been shown only in bamboo shark (*Chiloscyllium plagiosum*) (Taylor and Grosell 2006a), suggesting that elasmobranchs may process ingested seawater by species-specific mechanisms (Anderson et al. 2010). For general reviews of osmoregulation in elasmobranchs and across piscine taxa, refer to Hazon et al. (Hazon et al. 2003) and Rankin (Rankin 2002), respectively, in addition to Chapter 4.

3.2.2. IMPLICATIONS OF FEEDING

Beyond marine teleost osmoregulation, the principal role of the GI tract in fishes is of course digestion. In order to maintain cellular acid–base balance during gastric acid secretion, systemic acid–base balance can be compromised. During gastric acid secretion, the equimolar secretion of base as HCO_3^- to the extracellular fluids (Fig. 6.3) can cause a metabolic alkalosis. This phenomenon has come to be known as the "alkaline tide." While recent years have seen a surge of interest in fish digestive physiology, evidence to date of a postprandial alkaline tide in fish has been limited to two species: a marine elasmobranch, the Pacific spiny dogfish (*Squalus acanthias*) (Wood et al. 2005), and a relatively euryhaline teleost fish, the rainbow trout (*Oncorhynchus mykiss*), acclimated to both freshwater (Bucking and Wood 2008; Cooper and Wilson 2008) and seawater (Bucking et al. 2009). In dogfish the HCO_3^- loading into the blood during gastric acid secretion resulted in a 0.20 pH unit rise in blood pH (Wood et al. 2005). Subsequent studies on rainbow trout have revealed that following a meal, blood pH rose between 0.15 pH units (Bucking and Wood 2008) and 0.05 pH units (Cooper and Wilson 2008). Interestingly, studies on toadfish (Taylor and Grosell 2006b), and euryhaline European flounder (*Platichthys flesus*) acclimated to both freshwater and seawater (Taylor et al. 2007), found no evidence of a postprandial alkaline tide. It was inferred that any acid–base balance disturbance caused by gastric acid secretion was being compensated for further along the GI tract of these fish. Because both species were adapted to, or in the case of freshwater flounder, able to quickly adapt to living in seawater, the enterocytes would have already been primed for HCO_3^- secretion. By exploiting these GI transport mechanisms, HCO_3^- loading to the blood during gastric acid secretion could be simultaneously matched by elevated intestinal HCO_3^- secretion, driving down blood pH and thereby negating an alkaline tide (Taylor and Grosell 2006b, 2009; Taylor et al. 2007). The only study so far to show evidence of a postprandial alkaline tide in a marine teleost was in seawater-acclimated rainbow trout (Bucking et al. 2009).

In addition to salinity acclimation, the quality and quantity of the meal influences the degree of postprandial disturbance to acid–base balance. According to Cooper and Wilson (Cooper and Wilson 2008), the amount of acid required to titrate commercial fish food pellets to pH 3.0 was 13 times greater than that required for the ragworm diet fed to flounder by Taylor et al. (Taylor et al. 2007). This difference is primarily due to the high calcium phosphate content (from skeletal material) present in the food pellets, whereas ragworms are soft-bodied invertebrates with no such skeletal material. The prediction follows then that the greater gastric acid secretion required to digest a calcium-rich meal would elicit a larger magnitude metabolic alkalosis. Interestingly, this was not the case for toadfish, as blood pH remained stable following ingestion of both high-calcium (pilchards, *Sardina pilchardus*) and low-calcium (common squid, *Loligo forbesi*) diets (Taylor and Grosell 2006b). Again, however, toadfish are osmoregulating marine teleosts that have continually operating intestinal HCO_3^- secretion mechanisms. Furthermore, the high-calcium diet did stimulate anion exchange (i.e. Cl^- uptake and HCO_3^- secretion), which would allow for compensation of increased HCO_3^- loading into the blood. Quantity of food may have a similar affect on gastric acid secretion as calcium content. For example, freshwater rainbow trout fed a 5% ration (Bucking and Wood 2008) exhibited a three-fold greater increase in blood pH compared to those fed a 1% ration (Cooper and Wilson 2008).

The question of "natural feeding" vs. "catheter feeding" has been another point of interest as the piscine alkaline tide has come under increasing scrutiny. It is presumed that a force-feeding situation short-circuits neural pathways involved in meal anticipation, perhaps reducing the ability of the GI tract to physiologically prepare and/or compensate for digestion. This idea is supported by results from Cooper and Wilson (Cooper and Wilson 2008) that show relatively greater systemic acid–base disturbance in catheter-fed freshwater trout than in voluntary-fed trout ingesting the same ration. Indeed, the magnitude of postprandial acid–base disturbance in catheter-fed trout was in line with that measured by Bucking and Wood (Bucking and Wood 2008) who used a five-fold greater ration in voluntary-fed trout.

Mammals and ectothermic terrestrial vertebrates compensate for a postprandial alkaline tide by the hypoventilatory retention of respiratory CO_2, which raises blood PCO_2 and restores blood pH (Wang et al. 1995, 2001; Overgaard et al. 1999; Andersen et al. 2003). In fish species studied thus far, blood PCO_2 does not change following feeding, indicating that rainbow trout (Bucking and Wood 2008; Cooper and Wilson 2008), like dogfish (Wood et al. 2005), do not adjust PCO_2 to compensate for a postprandial alkaline tide. This is not surprising considering fish gills in general are already hyperventilated with respect to CO_2. Wood et al. (Wood et al. 2007c) concluded that dogfish compensate for the alkaline tide by excreting the excess HCO_3^- at the gill. This was also evident in rainbow trout acclimated to freshwater (Bucking and Wood 2008). However, a similar study also on freshwater rainbow trout by Cooper and Wilson (Cooper and Wilson 2008) did not show an increase in net branchial base excretion and the authors inferred that instead the intestine was responsible for the excess base excretion. Indeed, further work on seawater-acclimated rainbow trout showed a significant increase in intestinal HCO_3^- secretion rates postprandially (Bucking et al. 2009). Bucking and co-workers also showed that in contrast to an earlier study on freshwater-acclimated trout (Bucking and Wood 2008), fish did not exhibit a compensating increase in net base excretion via the gill following a meal, but rather took up additional base from the external seawater. These studies indicate that for rainbow trout at least, both the gills and the intestine can contribute to the removal of excess HCO_3^- when the animal is faced with a metabolic alkalosis.

The gill is also responsible for excreting nitrogenous waste, both during feeding (Wood et al. 2005; Bucking and Wood 2008; Cooper and Wilson 2008; Bucking et al. 2009) and during osmoregulatory challenge (Genz et al. 2008). During digestion, excess branchial ammonia excretion is a result of amino acid deamination (Ballantyne 2001; Cooper and Wilson 2008). The inter-conversion of ammonia to ionic NH_4^+ means that its excretion can also contribute to whole-body acid–base balance in marine fish during osmoregulation. For example, Genz et al. (Genz et al. 2008) have shown that the H^+ loading into the blood during marine osmoregulation in toadfish is compensated by either branchial acquisition of HCO_3^- from the water and/or excretion of ammonia in the opposite direction.

It is evident that the problem of acid–base balance arising from the dual GI functions of osmoregulation and digestion still signifies a considerable gap in fish physiology literature to date. Evidence for species-specific differences, perhaps based on salinity tolerance and also feeding strategy, has laid the groundwork to direct future investigations of acid–base balance in the fish GI tract.

4. NITROGEN METABOLISM

Over the years, physiological and biochemical research on the fish intestine has focused on a number of aspects, but unfortunately the idea of metabolic support has usually entered the discussion only as a sideshow, or an afterthought, with little emphasis on the metabolic machinery itself. In this context, a synthesis of aspects of nitrogen metabolism must remain an incomplete quilt, devoid of defined overall structure and somewhat scattered considering the many potential nitrogenous inputs. These range from proteins through peptides, amino acids, nucleic acids, etc. to bacteria, with the only common theme that these substances contain nitrogen and can potentially be eliminated as common N-waste—namely ammonia, urea and uric acid. Little bits of information are usually buried in publications preoccupied with digestion, ion transport, acid–base regulation, expression of some favorite digestive enzyme, effects of feeding or production of one of the many fascinating intestinal hormones. Rather than following the flux of dietary nitrogen from the feeding process through to the voiding of N-waste, which only partially involves the digestive tract, we have focused on the processing of nitrogenous substrates involving the enterocytes, opening a different and necessarily somewhat biased view of nitrogen metabolism. Readers interested in production and elimination of nitrogenous end products are referred to recent reviews (Walsh et al. 1994; Wood 2001; Kajimura et al. 2004; Wright and Wood 2009).

4.1. The Multifunctional Fish Enterocyte

4.1.1. ENTEROCYTES HAVE A HIGH CAPACITY FOR OXIDATIVE SUPPORT

The range of GI functions presented in this volume can only be performed if the cells are metabolically active and have access to sufficient oxygen for oxidative support. To this end, the intestine has higher titers of citrate synthase, a key indicator of abundance of mitochondria, than the liver (two- to ten-fold higher, depending on the fish species), and titers in intestine exceed half of enzyme activity in heart, the undisputed leader in oxidative capacity in fishes. The idea presented in Section 2 that the intestine possesses a higher aerobic metabolism than suggested by its mass is corroborated by activity of glutamate dehydrogenase (GDH). GDH is another mitochondrial marker enzyme that also happens to play a key role in nitrogen metabolism, with intestinal activities in fish in the range of the liver—the top tissue for amino acid turnover. These observations on intestinal GDH are even more impressive when put into perspective with the mammalian situation, where rat intestine—although considered a metabolically very active tissue—has less than 10% of rat hepatic GDH activity. The piscine intestine is clearly maintained in a highly aerobic state, containing substantial amounts of mitochondria to support the many functions, independent of the actual oxidative substrates. This is supported by data on epithelial gas exchange presented in Section 2, and also by observations that intestinal blood flow increases postprandially (Grans et al. 2009). Of course, the relatively high mitochondrial abundance and metabolic rate of the intestinal cells come with the inherent drawbacks of mitochondrial proton leak, production of reactive oxygen species and therefore oxidative stress. Notably, in the distal regions of the intestine, the lumen may be anaerobic—favoring anaerobic bacteria (Mountfort et al. 2002)—which could draw oxygen from the enterocytes, or at least establish an interesting, yet unexplored, oxygen gradient across the cells.

4.1.2. THE ENTEROCYTE IS SUBJECT TO MULTIPLE STRESSORS

Oxidative stress is just one of the potential stressors affecting enterocytes. Apart from the mechanical stress of digestion, one has to add cell volume and perhaps even chemical stress to the equation. By the time ingested seawater reaches the intestine, the salinity will have somewhat decreased; however, enterocytes in seawater-adapted species may still experience hyperosmotic stress, which is reflected in cell shrinkage followed by regulatory volume increase (RVI). Osmotic stress is known to induce growth arrest and DNA damage and to initiate activation of DNA repair. The cell cycle restarts following DNA repair and cells resume proliferation; if DNA damage is severe, cells may enter apoptosis. Under other circumstances, cells may swell passively through uptake of water, which is usually followed by ATP-driven regulatory volume decrease (RVD); subsequently, cells regain their original cell volume by actively pumping internal osmolytes into the surrounding medium and associated with this transport, eliminate excess water. The osmolytes most closely associated with regulatory volume decrease in fish intestines are Cl, K, Ca, organic bases and amino acids. The ability to cope with or correct this cellular osmotic stress is important not only during digestion, but also underlies salinity tolerance in fishes. If that isn't enough, one has to add chemical stressors to the mix—the fish never really knows what type of assault may come down the gullet! Yet enterocytes have to find a careful balance between regulating all these stressors, and still perform their multiple functions. In addition to the meal itself, still other factors may come into play, regulating the enzymatic or hormonal activity of the intestine. This is likely to include diurnal or seasonal cycles as well as food anticipatory activity. While poorly researched in fish, there is some evidence that if fish are on a regular feeding cycle, some anticipatory activity is established. In the case of periodically fed goldfish, intestinal amylase, but not casein-protease, was entrained, while, as expected, the protease activity was increased following the meal (Vera et al. 2007).

4.1.3. The Role of Nitrogenous Compounds in the Enterocyte

In all these considerations, nitrogenous compounds, especially amino acids and their derivatives, play integral roles. While regulating acid–base, ion and water balance within the lumen, the tissue synthesizes hydrolytic enzymes, concurrently producing mucus to protect the intestinal tract from auto-digestion and potentially invading pathogens. In addition, enterocytes transport amino acids and small peptides at both apical and basolateral faces, hydrolyze small polypeptides and sort amino acids, either to funnel them into oxidative pathways, or to earmark them for endogenous protein synthesis (after all, the enterocytes proliferate actively) or destine them for export into the portal circulation, to be dealt with by the liver. Ammonia from the lumen or liberated in the enterocytes enters the plasma (Wright and Wood 2009) to be handled by liver, gill and other tissues. The intestine also establishes chemical feedback loops with other areas of the intestine and communicates through paracrine and endocrine mechanisms with other enterocytes, or tissues like the endocrine pancreas, liver or brain. An excellent example (borrowed from rodents) is the control of intestinal alanine absorption by alanine itself from a distant distal site in the intestine (Mourad et al. 2009). The role of the GI tract as an endocrine/neuroendocrine/paracrine organ is detailed by Takei and Loretz in Chapter 7.

Digestion of protein by proteases and peptidases results largely in a mixture of small peptides and only a small amount of free amino acids. Estimates for mammals put the proportion of small peptides in the range of 85% of the digesta. These small peptides can enter enterocytes either through co-transport with H^+ on an abundant high-capacity, low-specificity peptide transporter (PepT1), or through endocytosis. While amino acid transporters are fairly specific (see Chapter 2; Broer 2008; Hundal and Taylor 2009), PepT1 can transport any di- or tri-peptide potentially liberated from common proteins (Daniel 2004; Verri et al. 2009) and PepT1 apparently does so faster and more efficiently than transporters of individual amino acids. Enterocytes then hydrolyze the small peptides into free amino acids, using peptidases, although even in mammals the degree of hydrolysis is still under debate and it appears plausible, especially after identification of a distinct and unrelated basolateral peptide transporter (Shepherd et al. 2002) that di- and tri-peptides may enter into the circulation (see also Verri et al. 2009 for piscine models). Additionally, some peptides and proteins can cross into the mucosa and into the blood intact (McLean et al. 1999), bypassing the assault by assorted hydrolases through micropinocytosis or paracellular transport.

4.2. Metabolic Zonation along the Gut

Having established the high oxidative potential of enterocytes, it should be noted that large regional differences exist in many parameters along the intestine. Supporting ultrastructural data (Vernier 1990), the intestine shows clear and distinct zonation in function as diverse as the thickness of the mucosal layer, the concentrations and composition of luminal proteolytic enzymes, the activity of hydrolases anchored to the apical side of enterocytes, the PepT1 transporter (Bakke et al. 2010), detoxification enzymes (James 2005), distribution of MDR-like proteins (Hemmer et al. 1998; Kleinow et al. 2000), metabolic enzymes (Fig. 6.4), hormone responsiveness

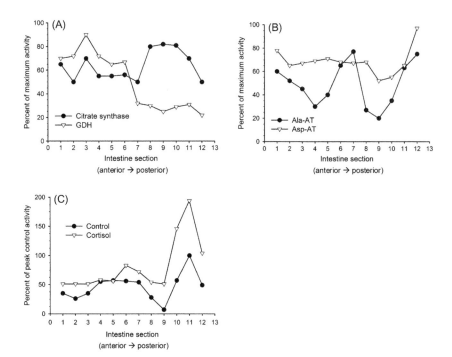

Fig. 6.4. Zonation of metabolic enzymes in fish small intestine. (A) Activities of citrate synthase and glutamate dehydrogenase in tilapia (redrawn from Mommsen et al. 2003b); (B) activities of alanine and aspartate aminotransferases in tilapia (redrawn from Mommsen et al. 2003b); (C) activity of glutamine synthetase in tilapia intestine and effects of cortisol (redrawn from Mommsen et al. 2003a).

(Fig. 6.4), nutrient absorption (Bakke-McKellep et al. 2000), membrane composition (Pelletier et al. 1986), protein endocytosis (Sire and Vernier 1992) and hormone production (Ostaszewska et al. 2010). Interestingly, the distributions for metabolic enzymes are distinct for three species with no obvious overriding patterns (Mommsen et al. 2003b), except for glutamine synthetase (GS), which invariably peaks in the distal portion of the intestine (Mommsen et al. 2003a); however, it is quite likely that additional patterns will become apparent as additional species are analyzed.

4.3. Practical Examples of Amino Acid Handling

In the following, we will single out three key amino acids, namely glutamine, glutamate and arginine, to illustrate important principles of nitrogen handling by the intestine:

1. Glutamine, because it is postulated to be the best oxidative substrate for enterocytes, while also assuming a central position in amino acids metabolism, as a sink for ammonia, as a nitrogen-donor required for synthetic pathways and as a source of ammonia and glutamate.
2. Glutamate, in turn, is known to be an excellent substrate for mitochondria and makes the link for both amino acids to various intermediary substrates in nitrogen metabolism and establishes functional links to the control of amino acid transport as well as intracellular redox.
3. Arginine, because of its unique biochemistry, abundance in diets and the multiple actions of its degradation products.

4.3.1. GLUTAMINE

Glutamine and glutamate tend to be abundant intracellularly and in plasma and occupy key points where carbon and nitrogen metabolism interface. Glutamine concentrations in plasma are about 10 times higher than those for glutamate in trout (Karlsson et al. 2006) and almost 30-fold higher in carp (Ogata and Arai 1985), while intracellularly, glutamate prevails and the ratio of glutamate to glutamine is about 3:1. Figure 6.5 provides a schematic overview of the interconnections between glutamine, glutamate and glutathione (which will be covered in more detail below). The figure also includes a list of physiological roles of the amino acids and some of their derivatives or intermediates. If luminal glutamine is to serve as an oxidative substrate for enterocytes, it will first have to cross the brush border membrane, then cross into the mitochondria where it is deaminated via mitochondrial glutaminase— an irreversible enzyme—into glutamate, followed by oxidative deamination by mitochondrial GDH into 2-oxo-glutarate. Dogma developed for feeding

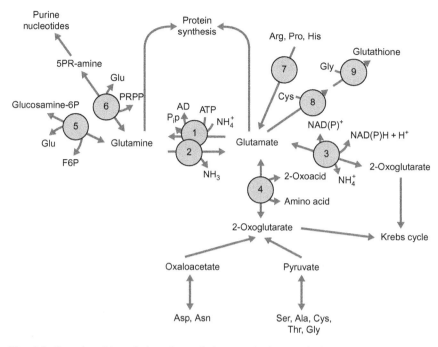

Fig. 6.5. Central position of glutamine and glutamate in fish metabolism, depicting the flow of nitrogen and carbon through intermediary products. Enzymes: 1. glutamine synthetase (EC: 6.3.1.2); 2. glutaminase (EC: 3.5.1.2); 3. glutamate dehydrogenase (EC: 1.4.1.3); 4. aminotransferases transfer the glutamate nitrogen to several 2-oxoacids to produce the corresponding amino acid, e.g. alanine aminotransferase (EC: 2.6.1.2), aspartate aminotransferase (EC: 2.6.1.1); 5. glutamine fructose-6-phosphate aminotransferase (EC: 2.6.1.16) leading into aminosugar metabolism and glycoproteins; 6. amidophosphoribosyltransferase (EC: 2.4.2.14), leading to purine biosynthesis; 7. path of arginine, proline and histidine from protein degradation, see also Fig. 6.6; 8. glutamate-cysteine ligase (EC: 6.3.2.2); 9. glutathione synthetase (EC: 6.3.2.3). Abbreviations: 2-OG: 2-oxoglutarate; 5-PR-amine: 5-phosphoribosyl-1-amine; PRPP: 5-phosphoribosyl-1-pyrophosphate; F6P: fructose 6-phosphate; glucosamine-6P: glucosamine-6-phosphate.

rats states that luminal glutamine, glutamate and alanine contribute 39% of respired CO_2 with arterial glutamine (38%) and luminal glucose (6%) accounting for most of the remainder (Windmueller and Spaeth 1980). Following this, it is usually assumed that glutamine is an excellent oxidative substrate also for the fish intestine and enterocytes. However, in the absence of direct data, indirect evidence clearly fails to support this idea. Compared with mammalian intestine (James et al. 1998), the activity of phosphate-dependent glutaminase in fish intestine seems relatively modest (Chamberlin et al. 1991). Further, the distribution and activity of glutamine synthetase differ substantially from mammals. While fish show substantial GS activity throughout their entire GI tract (Fig. 6.4), including the stomach, in mammals, the enzyme is prevalent in stomach only, and barely detectable in small intestine. Assuming that the ratio of these two enzymes partially controls the flux into and from glutamine, we must conclude that the fish intestine is much more inclined to produce glutamine than oxidize it, even though stationary concentrations of both amino acids are similar in rat (1–3 μMol/g wet tissue mass) and marbled goby (0.6–1.8 μMol/g), with ratios of 2.6 (rat) and 3.2 (goby), respectively, in favor of glutamate (Tng et al. 2008).

    Although glutamine can be largely dismissed as a key oxidative substrate, it is still important to the intestine, mainly as a buffer for glutamate, a potential sink for nitrogen and a key substrate for protein and nucleotide biosynthesis. Perhaps most importantly, an essential function of intestinal GS appears to be in preventing the high ammonia levels in the chyme from entering the bloodstream and poisoning the fish (Wood et al. 2005; Karlsson et al. 2006). Further evidence for specific transporters involved in ammonia and urea transport is mounting, particularly in the case of Rh ammonia transporters and UT urea transporters shown to be expressed in the gut of elasmobranchs (Anderson et al. 2010). Supporting the idea that glutamine formation is linked to removal and detoxification of ammonia is the observation that ammonia exposure increases abundance of GS transcripts in intestinal tissue (Anderson et al. 2002) and GS activity is increased by cortisol (Mommsen et al. 2003a), a steroid hormone usually associated with increased nitrogen turnover in fish and known to increase the availability of amino acids through activation of endogenous proteolysis (Mommsen et al. 1999). In goby intestine, GS activity is also positively correlated with feeding, increasing by 40% 12 hours after a meal (Tng et al. 2008). Interestingly, the stomach is the only tissue in the goby where GS expression is unaffected by ammonia, which is reminiscent of the situation in the rodent GI tract, where GS expression in stomach is controlled by completely different mechanisms than in intestine (Lie-Venema et al. 1998), although the fish stomach GS is subject to similar up-regulation as its intestinal counterpart (Mommsen et al. 2003a). Considering that a surge in ammonia production is unlikely to occur until digesta enter the intestine, the stomach is not really positioned to play an important role in ammonia removal. Unless stomach GS is oriented towards plasma substrates and not towards the lumen, the function of stomach GS remains somewhat enigmatic, but likely differs from its role in the intestine. While mammals possess only a single GS gene, the situation is more complex in fishes with at least two independent GS genes (Murray et al. 2002; Walsh et al. 2002), although at least in goby, the same transcripts occur in all tissues (Tng et al. 2008). Making things even a little more obscure is the observation that GS transcripts are evenly distributed along tilapia intestine (Mommsen et al.

2003a), even though actual activity shows distinct zonation (Fig. 6.4). The path from a few related mRNAs to active enzyme is not a simple one for glutamine synthetase. The enzyme is active mostly after assembly into 8-mers or 16-mers, with considerable post-translational modifications, in other words a bit of a regulatory nightmare, especially in pseudotetraploid species like rainbow trout that may differently express at least four GS subunit genes (Murray et al. 2002). Other aspects of GS presence should be considered, even though no data are as yet available for piscine models. Like most other enzymes, GS activity is likely controlled by concerted stability of its mRNA and protein. In mammalian models GS is rendered highly unstable in the presence of its product glutamine, which shortens the half-life of the protein by 90%, and by some covalent modifications that also increase turnover of active protein. For instance, during seawater adaptation in marbled goby, intestinal GS activity is decreased substantially (Chew et al. 2010), which is somewhat counterintuitive, since the enzyme is clearly subject to induction by cortisol which should be elevated at this time, but mechanisms of the surprising decrease and metabolic repercussion of this observation remain to be elucidated. Apart from these direct effects, indirect effects of GS catalysis cannot be ignored, for instance inhibition of GS-induced apoptosis in mammalian cells (Rotoli et al. 2005), again confirming a role of glutamine in regulation of cell proliferation.

While regulation of fish GI tract GS activity certainly deserves more attention, our current focus should remain with the overall role of glutamine. While all evidence points in the direction of glutamine as a nitrogen sink that secondarily regulates metabolism, under unusual physiological conditions, the flow of glutamine, carbon and nitrogen may be reversed by curtailing GS and activation of glutaminase. In this case, glutamine delivers glutamate into mitochondria for direct deamination and production of ammonia by glutamate dehydrogenase (GDH). Working with isolated fish (red muscle) mitochondria, Chamberlin et al. (Chamberlin et al. 1991) noticed that while mitochondria readily oxidized glutamine and glutamate, the pathways and end-products differed. Following deamination of glutamine into glutamate, ammonia was released and glutamate was oxidized by GDH and the resulting 2-oxoglutarate oxidized in the Krebs cycle. In contrast, glutamate added to mitochondria was preferentially transaminated with the glutamate nitrogen being transferred to other 2-oxo-acids to be released as amino acids—most likely aspartate and alanine. Murai and colleagues (Murai et al. 1987) showed that ammonia, proline and alanine increase in the hepatic portal vein postprandially in trout force-fed on casein and amino acids, and that these increases were largely independent of amino acid composition of the diet. These data imply that enterocytes actively produce these three compounds, where ammonia is likely to be derived from glutamate or asparagine, pointing at the involvement of GDH and asparaginase or any amino acid that can be transaminated with alpha-oxoglutarate as the nitrogen acceptor, and subsequent oxidative deamination in enterocyte mitochondria by GDH. Proline can only be derived from glutamate as precursor, while any carbons passing through pyruvate (serine, glycine, cysteine, threonine, glucose, etc.) could be sidetracked into alanine via the corresponding aminotransferase (ALT) (cf. Fig. 6.5). Intestinal ALT is indeed one of the enzymes up-regulated after a meal (Polakof et al. 2010).

In addition to its general use in protein synthesis, glutamine is also an essential starting point and nitrogen donor for nucleotide synthesis, specifically purine bases. Again, the fact that cod intestine contains about twice as much glutamine-dependent carbamoylphosphate synthetase (CPSase II), the committing enzyme in purine biosynthesis, as liver, supports the idea of high rates of cell proliferation in the intestine (Chadwick and Wright 1999). In fact, in cod, CPSase II accounts for all CPSase activity in the intestine, similar to the situation in trout CPSase II (Korte et al. 1997) while toadfish and largemouth bass intestine have both CPSase II and III (Kong et al. 1998, 2000). Independent of their actual metabolism, amino acids, especially leucine, glutamine and arginine, can regulate gene expression, mediated by the target of rapamycin (TOR) pathway, which is highly conserved in eukaryotes. While the actual amino acid sensing mechanisms are still under debate, TOR signaling positively targets genes involved in cell growth, proliferation and protein synthesis while curtailing genes that control autophagy. In developing zebrafish intestine, TOR, specifically complex 1 (TORC1), also governs the developmental program that directs epithelial morphogenesis and differentiation (Makky et al. 2007).

Apart from its direct actions, glutamine is much more than just a metabolic hub and joins the ranks of amino acids controlling numerous metabolic targets. Glutamine is part of an emerging group of amino acids that operate through nutrient sensors (Lindsley and Rutter 2004) and establish themselves as intracellular signaling molecules (Rhoads and Wu 2009), especially in the intestine. For instance, glutamine functions as a suppressor of apoptosis (Rotoli et al. 2005), thus directly opposing the intracellular effects of reactive oxygen species, one of the multiple stressors impacting fish enterocytes. Hitting the same target by a different route, glutamine stimulates formation of heat shock protein 70 by stabilizing its mRNA, influences redox potential of the cell by favoring production of glutathione, induces cellular anabolic effects by increasing cell volume, and activates mitogen-activated protein kinases (Rhoads et al. 2000). Glutamine also activates intestinal PPAR-gamma, a group of transcription factors regulating genes involved in cell differentiation, while also regulating barrier function easing cytokinase responses and ameliorating immune cell functions in the intestine (Rhoads and Wu 2009).

4.3.2. GLUTAMATE

Glutamate is at the hub of a complex arrangement of enzymes involved in transamination, deamination and synthetic pathways. To simplify the concept, glutamate is best treated as a 2-oxoglutarate with an alpha amino group, to indicate that these two components can be handled metabolically quite independent of each other. Ultimately, the amino group from most amino acids can be shuttled through glutamate, and similarly a number of key carbon compounds are closely related to 2-oxoglutarate. Figure 6.5 merely presents the most obvious metabolic interconnections and focuses largely on nitrogen flow.

When returning to potential oxidative substrates for fish intestine, we are left largely with glutamate as the key amino acid to generate ATP because of its central position. As pointed out, glutamate is the most abundant amino acid in the intestine and its concentration increases substantially following a meal: in a goby intestinal glutamate concentrations rise about 61% 12 hours after feeding and plasma glutamate almost doubles in the hepatic portal vein 6 hours after feeding in Atlantic salmon. However, it is impossible to calculate flux rates or predict nitrogen/carbon flow based on concentrations alone. After a meal, we can assume that the general flow of digestive products, including amino acids, will be unidirectional from the lumen to enterocytes, but it is unknown how much the enterocytes draw on plasma amino acids as oxidative substrates when the lumen is empty. Data for enzyme activities often provide a better idea about potential metabolic flux than stationary concentrations. As pointed out above, glutamate dehydrogenase is abundant in fish intestine, largely a reflection of the abundance of mitochondria in the tissue. While we can predict relatively high rates of glutamate oxidation based on values for GDH already, it is enzyme behavior after feeding that provides more detailed insight. The activity of GDH in intestine increases three- to five-fold postprandially in goby (Tng et al. 2008) and addition of amino acids is correlated with a doubling of GDH activity in isolated trout midgut pieces (Polakof et al. 2010), especially in the presence of insulin, which is usually increased postprandially since particularly basic amino acids such as arginine function as powerful insulinotropins in beta-cells of the endocrine pancreas (Ronner and Scarpa 1987). Increased flux through GDH is reflected in postprandial peaks in ammonia in the hepatic portal vein in Atlantic salmon (2.5-fold increase; Karlsson et al. 2006) and caudal vessels of a goby (10-fold increase; Tng et al. 2008). However, in goby tissue, concentration of ammonia was unaffected by feeding, either reflecting efficient removal of excess ammonia into glutamine via GS or shuttling of ammonia into the circulation by appropriate transporters (Wright and Wood 2009). A quick calculation, assuming an internal aqueous volume of 60% of cell mass, yields prefeeding ammonia concentrations of around 2.6 mM for enterocytes, which dwarfs the plasma value of 13 μM. Even at peak values postprandially, the tissue concentration for ammonia (12 h–2.8 mM) is still 21-fold higher than the plasma level (131 μM). While theoretically, the flow through GDH is reversible (Treberg et al. 2009), this is not the case in intestine (fish or mammalian), and the enzyme is unlikely to be a sink for ammonia to produce glutamate, but instead serves to deaminate amino acids. Once GDH gets involved, the carbons will be committed to oxidation, and therefore regulatory control likely lies with the access of the mitochondria to glutamate, i.e. the mitochondrial glutamate transporter(s) or feeder transaminases, like alanine and aspartate transaminases. With close to 70% of a meal retained as somatic growth (Tng et al. 2008), it becomes apparent that enterocytes control vast portions of the fate of ingested amino acids. Since it assumes such a pivotal role in committing amino acids to catabolic pathways, it does not come as a surprise that GDH is highly regulated. In addition to the usual allosteric activators ADP and leucine, the enzyme seems to be regulated through differential expression and covalent modification, including ADP-ribosylation. In the case that dietary proteins escape digestion and reach the distal intestine, these are likely broken down by bacteria. If glutamate is released into the lumen, it could constitute a precursor for butyrate, acetate and other volatiles. In mammalian colon, glutamate also serves as an excellent oxidative substrate, but it is thought to be provided from blood and not from the lumen.

Another, equally important, role for glutamate is as the starting point, together with cysteine and glycine, for synthesis of the tripeptide glutathione, γ-L-glutamyl-L-cysteinylglycine (GSH), which is present in all cells, predominantly in the thiol form. Figure 6.6 depicts some of the key pathways involving glutathione. All enzymes required for synthesis of glutathione are present in the fish intestine, as are a whole host of enzymes utilizing glutathione for diverse cellular functions. Because of its dual chemical nature as a reductant and as a nucleophile, glutathione is essential in maintaining cellular proteins, including enzymes, in a reduced state, and as a reactant with free radicals and xenobiotics. In addition, GSH regulates transcription of genes that control antioxidant defenses and cell cycle and functions as a storage and transport form of cysteine moieties. Finally, GSH is also instrumental in transmembrane transport of amino acids and glutathione. The importance of GSH to the vertebrate intestine cannot be understated and is probably best summarized by the title of a paper published 20 years ago by A. Meister, who had done most of the underlying research. It simply and unequivocally read "Glutathione is required for intestinal function" (Martensson et al. 1990). There is no reason to believe that the fish intestine differs substantially from

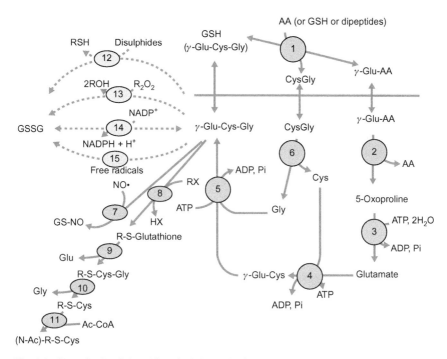

Fig. 6.6. Central role of glutathione in fish metabolism and glutamyl cycle for cross-membrane transport of amino acids, glutathione and dipeptides. Enzymes: 1. γ-glutamyltransferase (EC: 2.3.2.2); 2. γ-glutamyl cyclotransferase (EC: 2.3.2.4); 3. 5-oxoprolinase ATP hydrolyzing (EC: 3.5.2.9); 4. glutamate cysteine ligase (EC: 6.3.2.2); 5. glutathione synthetase (EC 6.3.2.3); 6. aminopeptidase (EC: 3.4.11.1; 3.4.11.5); 7. non-enzymatic reaction (Keszler et al. 2010); 8. glutathione S-transferase (EC: 2.5.1.18); 9. γ-glutamyltransferase (EC: 2.3.2.2); 10. aminopeptidase (See enzyme 6); 11. cysteine-S-conjugate N-acetyltransferase (EC: 2.3.1.80). The following reactions (indicated by stippled arrows) are not stoichiometric: 12. protein disulfide reductase (EC: 1.8.4.2); 13. peroxidase (includes EC: 1.11.1.9); 14. glutathione reductase (EC 1.8.1.7); 15. Direct reaction with free radicals (including glutathione oxidase (EC: 1.8.3.3)). Abbreviations: AA: amino acid; GSH: reduced glutathione; GSSG: oxidized glutathione; GS-NO: glutathione-nitric oxide adduct; NO: nitric oxide; AcCoA: acetyl-coenzyme A; (N-Ac)-R-S-Cys: N-acetyl-adduct of cycteine (mercapturate).

mammalian intestine, where GSH is synthesized in situ in large amounts and undergoes rapid turnover. For instance, application of buthionine sulfoximine, a specific inhibitor of gamma-glutamylcysteine synthetase, the rate-limiting step in GSH synthesis, leads to a drastic decrease in GSH within 30 min of application, and subsequently to substantial degeneration of epithelial cells of the intestine (Martensson et al. 1990). The intestinal glutathione system is affected by exposure of fish to toxicants. For instance, exposure of freshwater *Corydoras paleatus* to microcystin resulted in transient decreases in the activity of glutathione S-transferases and increases in the activity of glutathione peroxidase, likely affecting the concentration of glutathione in the tissue (Cazenave et al. 2006). African walking catfish exposed to dietary copper showed a two-fold increase in TBARS (thiobarbituric acid reactive substances), together with a doubling of total glutathione concentration in the intestinal tissue (Hoyle et al. 2007), from about 7 μMol/g to over 15 μMol/g over a 30-day experimental period, requiring substantial synthetic effort and resources to achieve this increase. Since intracellular glutathione is not attacked or broken down by cellular peptidases, GSH functions as a storage form of cysteine and thus thiol-residues.

GSH also provides a link to xenobiotics and the intestine is among the first tissues to be exposed. A group of proteins of the ATP-binding cassette (ABC) superfamily, also termed multidrug resistance protein (MRP)/cystic fibrosis transmembrane conductance regulator (CFTR) subfamily, are known as efflux proteins for both glutathione conjugates and glutathione. GSH itself may serve as a transport stimulant or be co-transported with the GSH-conjugate. In addition to being instrumental in the efflux of conjugates, these transporters can also function to prevent uptake of toxicants. Not surprising, then, is the observation that the luminal surface of guppy (*P. reticulata*) enterocytes stains positive for MDR-like proteins (Hemmer et al. 1998). Similarly, catfish (*I. punctatus*) and minnow (*Cyprinodon variegatus*) enterocytes are immuno-positive for MDR-like proteins both on their surface and in their cytoplasm (Hemmer et al. 1998; Kleinow et al. 2000). Using similar antibodies against (mammalian) P-groups, the epithelial cells of the posterior intestine of killifish (*F. heteroclitus*) appear to contain the piscine equivalents of these transporter proteins. As shown for catfish, these intestinal transporters are inducible by xenobiotics (Doi et al. 2001). The fish intestine is also involved in detoxification of xenobiotics and contains a full spectrum of enzyme machinery to process them. For instance, catfish intestine efficiently processes model xenobiotics at low concentrations, even though overall capacity of intestinal microsomes for glucuronidation was lower than in the liver (Sacco et al. 2008). Part of the detoxification machinery, of course, involves glutathione, both through its function as a redox regulator and the presence of multiple GSH-based enzyme systems, such as glutathione S-transferases, glutathione peroxidase and glutathione reductase. The mucosal GSH/GSSG ratio, with its highly reduced status, controls the elimination of peroxides in the intestinal lumen. The enzyme responsible, GSH peroxidase, is abundant in fish intestine as expected for a tissue with high intrinsic metabolic rates. Intestinal enzymes are important in the initial biotransformation of xenobiotics as a first line of defense—that is, before xenobiotics reach the systemic circulation. To this end, glutathione S-transferases and other detoxification enzymes (CYP1A, EROD and AHH) are highly active in the intestine (Gadagbui and James 2000) and respond to dietary exposure to toxicants.

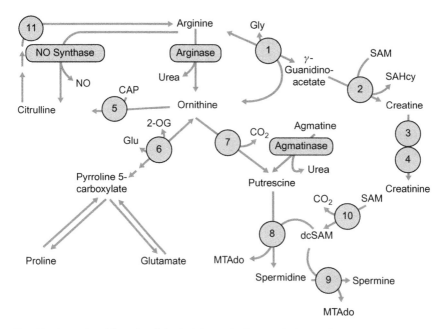

Fig. 6.7. Central position of arginine in fish metabolism, depicting the flow of arginine nitrogen into a multitude of intermediary and end products. Agmatine has been included as an additional source of polyamines; fish genomes contain agmatinase, an enzyme structurally similar to arginase. Only key cofactors/products of the main enzymatic reactions are included. Enzymes: arginase (EC 3.5.3.1); agmatinase (EC 3.5.3.11); NO synthase (EC 1.14.13.99); 1. γ-Glycine amidinotransferase (EC 2.1.4.1); 2. guanidinoacetate N-methyltransferase (2.1.1.2); 3. creatine phosphokinase (EC 2.7.3.2); 4. non-enzymatic decay of creatine phosphate; 5. ornithine carbamoyltransferase (EC 2.1.3.3); 6. ornithine aminotransferase (EC 2.6.1.13); 7. ornithine decarboxylase (EC 4.1.1.17); 8. spermidine synthase (EC 2.5.1.16); 9. spermine synthase (EC 2.5.1.22); 10. S-adenosylmethionine decarboxylase (EC 4.1.1.50); 11. urea cycle components (argininosuccinate synthetase and argininosuccinate lyase). Abbreviations: 2-OG: 2-oxoglutarate; CAP: carbamoylphosphate; dcSAM: decarboxylated SAM; Glu: glutamate; Gly: glycine; MTAdo: methylthioadenosine; NO: nitric oxide; SAHcy: S-adenosylhomocysteine; SAM: S-adenosylmethionine.

4.3.3. ARGININE

Arginine is abundant in the diet of carnivorous species, since it makes up about 7% (by mass) of fish muscle and other proteins; the reason for including arginine in our discussion on intestine is three-fold (Fig. 6.7). First, arginine as a substrate for arginase is a direct precursor of urea—a potentially key nitrogenous excretory product or a ready source of ammonia, assuming intestinal bacteria can handle urea. Second, arginine processing by NO synthase delivers nitric oxide, an important intracellular messenger, and citrulline. Third, through arginase, the amino acid is a source of intestinal ornithine, the pivotal amino acid in the synthesis of polyamines putrescine, spermine and spermidine. There is another key function for arginine, namely as precursor of muscle creatine, but that is unlikely to involve the intestine. Research indicates that dietary arginine is completely processed by intestine and liver, since no noticeable accumulation of hepatic arginine was noticed when trout were fed with relative high concentrations of this amino acid (Fauconneau et al. 1992). Even though arginase activity in the intestine—likely due to expression of two isoforms of arginase (Wright et al. 2004)—amounts to about one-tenth of that in liver (Lisowska-Myjak et al. 1978), rates are sufficient to account for urea increases in hepatic portal circulation after feeding (Karlsson et al. 2006).

Arginine is also an essential component in endogenous production of polyamines, which are positively linked to translation and RNA and DNA polymerase activity and therefore are essential for cell proliferation, differentiation and growth. Polyamines, either produced endogenously from arginine or obtained from the lumen, may in fact function as major regulators of enterocyte proliferation and growth. The pivotal position in the synthesis of polyamines is occupied by ornithine decarboxylase (ODC), an enzyme with a very short half-life—in the range of 1 to 2 min for human enterocyte ODC—which is highly regulated through a plethora of mechanisms, including a group of highly specific inhibitory antizymes (one of these also functions as a polyamine transporter and increases the rate of ODC degradation (Saito et al. 2000)), and antizyme inhibitors. While the existence of components of the ODC complex has been confirmed for fish (Hascilowicz et al. 2002; Tuziak et al. 2009) their tissue expression and physiological roles remain to be defined. As an aside, reprising the importance of osmotic stress in fish intestine, antizyme expression is up-regulated during high-salinity challenge in eel (Tseng and Hwang 2008), while ODC from a zebrafish cell line is activated more than 80-fold by short-term exposure to a hypo-osmotic medium (Hascilowicz et al. 2002). In addition to the indirect route involving arginine degradation, fish may be exposed to polyamines through their diets, reflected in the fact that polyamines elicit feeding behavior in goldfish (Rolen et al. 2003). Most cells contain polyamines and the compounds can also be used as indicators of bacterial spoilage in food and they can be contributed by intestinal bacterial flora, with a non-specific diamine transporter facilitating uptake of luminal polyamines. Seeing the ubiquity and powerful biological actions of polyamines, it seems to be a good strategy to use dietary polyamines as one route to regulate intestinal proliferation, with ODC providing an excellent branch point to adjust endogenous production of polyamines.

4.4. Other Considerations in Piscine Nitrogen Metabolism

Other compounds that may contribute to urinary and fecal nitrogen excretion, and are likely to have undergone some modifications within the intestine, include urea, uric acid, nitrite/nitrate, nucleic acids, biliverdin, taurine, methylamines, and muscle-specific dipeptides such as carnosine or anserine. Still other nitrogenous compounds, including creatinine and creatine, are by-products of intermediary metabolism. Another unusual amino acid relevant in the diet of carnivorous fish is D-alanine, which is abundant in crustaceans and mollusks. Fish have two methods to deal with D-amino acids; one is a fairly specific D-aspartate amino oxidase, and the other is a generic D-amino acid oxidase. Both occur, albeit not exclusively, in the intestine. Oral administration of modest amounts of D-alanine leads to large induction of D-amino acid oxidase activity together with increases in the relative amounts of its mRNA (Abe et al. 2005). The end products of this oxidation are pyruvate, ammonia, and hydrogen peroxide, which adds to the oxidative stress experienced by the tissue postprandially. An honorary mention should go to heme oxygenases (EC 1.14.99.3), an enzyme usually linked to heme metabolism. The enzyme catalyzes the rate limiting step in the degradation of heme to bilirubin, along the way generating three important biologically active compounds; namely, biliverdin (an antioxidant), iron (a regulator of gene expression, including expression of NO synthase), and carbon monoxide, which has recently joined the ranks of important signaling molecules. Recently, the enzyme has also been associated with suppression of hypoxia-induced cell death and has been implicated in oxygen sensing (Wang et al. 2008). The enzyme is expressed in intestine of *Dicentrarchus labrax* (Prevot-D'Alvise et al. 2008) and *Carassius auratus* (Wang et al. 2008), but presence of functional protein remains to be confirmed for any fish tissue.

Nitrite and nitrate, two compounds important in the global nitrogen cycle, are known contaminants of freshwater systems. While their impact on fish metabolism is likely through the gill, a different story emerges in marine fishes that drink continuously and thus are exposed to water contaminants. With few limits on NOx emissions, nitrates already contribute to freshwater acidification and little is known about its effects on marine systems. Two papers by Grosell and Jensen (Grosell and Jensen 1999, 2000) suggest the intestine is responsible for some 66% of whole-body nitrite uptake in exposed seawater-acclimated European flounder. The mechanism of intestinal NO_2^- uptake in these studies was suggested to be via the apical $Na^+/K^+/2Cl^-$ cotransporter rather than via a Cl^-/HCO_3^- exchanger, and may also include the Na^+/Cl^- cotransporter and conductive transport. It is thus foreseeable that ingested nitrite and nitrate will become important players in fish intestine, not least on the level of intestinal ion transporters, but also as potential sources of secondary products. For instance, formation of toxic and mutagenic N-nitrosamines from nitrite and secondary amines in the digestive tract—especially favored by acidic conditions in the stomach—has long been discussed in mammalian literature. In contrast, little attention has been devoted to this problem in fishes, although it has been documented that rainbow trout exposed to nitrite for extended periods will accumulate N-nitrosamines (De Flora and Arillo 1983). Similarly, carnivorous marine fish are likely to encounter substantial amounts of trimethylamine N-oxide (TMAO) in their (invertebrate) diets; TMAO is used by anaerobic bacteria in the gut to produce trimethylamines. Unfortunately, research on intestinal bacteria and their metabolism in fishes is in its infancy and although molecular techniques now make it possible to identify and classify many of these bacteria (Shiina et al. 2006), their contributions to intestinal metabolism and metabolic products remain to be elucidated. A notable exception is the bacteria-dependent hindgut fermentation described in a number of herbivorous fish (Mountfort et al. 2002; Skea et al. 2007).

As mentioned above, mucus plays an important part in the digestive tract, yet little information is available on gastrointestinal mucus production in fishes, apart from histological and histochemical evidence. At this point, a few gel-forming mucins have been identified in *Takifugu rubripes* with similarity to mammalian mucins (MUC2), containing anticipated repeats rich in S, T and P (Lang et al. 2004). In an intestinal mucin tentatively identified as MUC3A (XP_606380) in zebrafish, S, T and P make up almost 50 mol% of the residues, with glutamate accounting for 12% and glutamine for 1%. Most likely, these amino acids have to be drawn from the intracellular pool of gastrointestinal cells.

5. CONCLUDING REMARKS

As pointed out throughout this volume, the intestine is a highly diversified and multifunctional organ. At the same time as digesting and absorbing feedstuffs, the fish intestine regulates water, ion and acid–base balance while producing digestive enzymes and playing key roles in endocrine regulation of digestion, intermediary metabolism and endocrine activity in different tissues and in intestinal motility. The intestine also functions as a mechanical and chemical barrier and in the elimination of toxins, waste and other metabolic products. Other aspects that at times might impact the intestine as much as any of the previous components are the resident microbiological flora, the mucous layer protecting the enterocytes and not least the intestinal role in piscine immunity. We have identified three areas—intestinal gas exchange, acid–base balance, and nitrogen metabolism—in which the field of fish physiology has seen recent marked growth. However, the story of the interactions between these functions in the fish GI tract is far from complete and remains fertile ground for scientific discovery.

REFERENCES

Abe, H., Yoshikawa, N., Sarower, M. G., and Okada, S. (2005). Physiological function and metabolism of free D-alanine in aquatic animals. *Biol. Pharm. Bull.* **28**, 1571–1577.

Andersen, J. B., Andrade, D. V., and Wang, T. (2003). Effects of inhibition gastric acid secretion on arterial acid–base status during digestion in the toad. *Bufo marinus. Comp. Biochem. Physiol. A* **135**, 425–433.

Anderson, W. G., Takei, Y., and Hazon, N. (2002). Osmotic and voalemic effects on drinking rate in elasmobranch fish. *J. Exp. Biol.* **205**, 1115–1122.

Anderson, W. G., Dasiewicz, P. J., Liban, S., Ryan, C., Taylor, J. R., Grosell, M., and Weihrauch, D. (2010). Gastro-intestinal handling of water and solutes in three species of elasmobranch fish, the white-spotted bamboo shark, *Chiloscyllium plagiosum*, little skate, *Leucoraja erinacea* and the clear nose skate *Raja eglanteria. Comp. Biochem. Physiol. A* **155**, 493–502.

Ando, M., Mukuda, T., and Kozaka, T. (2003). Water metabolism in the eel acclimated to seawater: from mouth to intestine. *Comp. Biochem. Physiol. B* **136**, 621–633.

Andrade, D. V., De Toledo, L. F., Abe, A. S., and Wang, T. (2004). Ventilatory compensation of the alkaline tide during digestion in the snake *Boa constrictor. J. Exp. Biol.* **207**, 1379–1385.

Arvedsen, S. K., Andersen, J. B., Zaar, M., Andrade, D., Abe, A. U., and Wang, T. (2005). Arterial acid–base status during digestion and following vascular infusion of $NaHCO_3$ and HCl in the South American rattlesnake, *Crotalus durissus. Comp. Biochem. Physiol. A* **142**, 495–502.

Baker, D. W., Matey, V., Huynh, K. T., Wilson, J. M., Morgan, J. D., and Brauner, C. J. (2009). Complete intracellular pH protection during extracellular pH depression is associated with hypercarbia tolerance in white sturgeon, *Acipenser transmontanus. Am. J. Physiol. Regul. Integr. Comp. Physiol.* **296**, R1868–R1880.

Bakke, S., Olderbakk Jordal, A. E., Gomez-Requeni, P., Verri, T., Kousoulaki, K., Aksnes, A. and Ronnestad, I. (2010). Dietary protein hydrolysates and free amino acids affect the spatial expression of peptide transporter PepT1 in the digestive tract of Atlantic cod (*Gadus morhua*). *Comp. Biochem. Physiol. A*. In press.

Bakke-McKellep, A. M., Nordrum, S., Krogdahl, A., and Buddington, R. K. (2000). Absorption of glucose, amino acids, and dipeptides by the intestines of Atlantic salmon (*Salmo salar* L.). *Fish Physiol. Biochem.* **22**, 33–44.

Ballantyne, J. S. (2001). Amino acid metabolism. In: *Nitrogen Excretion* (P.A. Wright and P.M. Anderson, eds), pp. 77–108. Academic Press, New York.

Barrington, E. J. W. (1942). Gastric digestion in the lower vertebrates. *Biol. Rev.* **17**, 1–27.

Becker, H. M., and Deitmer, J. W. (2007). Carbonic anhydrase II increases the activity of the human electrogenic Na^+/HCO_3^- cotransporter. *J. Biol. Chem.* **282**, 13508–13521.

Belal, I. E. H. (2004). A review of some fish nutrition methodologies. *Bioresource Technology* **96**, 395–402.

Boeuf, G., and Payan, P. (2001). How should salinity influence fish growth? *Comp. Biochem. Physiol. C* **130**, 411–423.

Boutilier, R. G., Heming, T. A., and Iwama, G. K. (1984). Appendix: Physicochemical parameters for use in fish respiratory physiology. *Fish Physiology. Volume X Gills. Part A: Anatomy, Gas Transfer, and Acid–Base Regulation.*

Britton, R., and Krehbiel, C. (1993). Nutrient metabolism by gut tissues. *J. Dairy Sci.* **76**, 2125–2131.

Broer, S. (2008). Amino acid transport across mammalian intestinal and renal epithelia. *Physiol. Rev.* **88**, 249–286.

Brown, C. R., and Cameron, J. N. (1991a). The induction of specific dynamic action in channel catfish by infusion of essential amino acids. *Physiol. Zool.* **64**, 276–297.

Brown, C. R., and Cameron, J. N. (1991b). The relationship between specific dynamic action (SDA) and protein synthesis rates in the channel catfish. *Physiol. Zool.* **64**, 298–309.

Bucking, C., and Wood, C. M. (2006a). Gastrointestinal processing of Na^+, Cl^-, and K^+ during digestion: implications for homeostatic balance in freshwater rainbow trout. *Am. J. Physiol. Regul. Integr. Comp. Physiol.* **291**, R1764–R1772.

Bucking, C., and Wood, C. M. (2006b). Water dynamics in the digestive tract of the freshwater rainbow trout during the processing of a single meal. *J. Exp. Biol.* **209**, 1883–1893.

Bucking, C., and Wood, C. M. (2008). The alkaline tide and ammonia excretion after voluntary feeding in freshwater rainbow trout. *J. Exp. Biol.* **211**, 2533–2541.

Bucking, C., Fitzpatrick, J. L., Nadella, S. R., and Wood, C. M. (2009). Post-prandial metabolic alkalosis in the seawater-acclimated trout: the alkaline tide comes in. *J. Exp. Biol.* **212**, 2159–2166.

Cazenave, J., Bistoni Mde, L., Pesce, S. F., and Wunderlin, D. A. (2006). Differential detoxification and antioxidant response in diverse organs of *Corydoras paleatus* experimentally exposed to microcystin-RR. *Aquat. Tox.* **76**, 1–12.

Cech, J. J. (1990). Respirometry. In: *Methods for Fish Biology* (C.B. Schreck and P.B. Moyle, eds), pp. 335–363. American Fisheries Society, Bethesda, MD.

Chadwick, T. D., and Wright, P. A. (1999). Nitrogen excretion and expression of urea cycle enzymes in the Atlantic cod (*Gadus morhua* L.): a comparison of early life stages with adults. *J. Exp. Biol.* **202**, 2653–2662.

Chamberlin, M. E., Glemet, H. C., and Ballantyne, J. S. (1991). Glutamine metabolism in an holostean fish (*Amia calva*) and a teleost (*Salvelinus namaycush*). *Am. J. Physiol. Reg. Integr. Comp. Physiol.* **260**, R159–R166.

Chew, S. F., Hong, L. N., Wilson, J. M., Randall, D. J., and Ip, Y. K. (2003). Alkaline environmental pH has no effect on ammonia excretion in the mudskipper *Periophthalmodon schlosseri* but inhibits ammonia excretion in the related species *Boleophthalmus boddaerti*. *Physiol. Biochem. Zool.* **76**, 204–214.

Chew, S. F., Tng, Y. Y. M., Wee, N. L. J., Tok, C. Y. W. J. M. and Ip, Y. K. (2010). Intestinal osmoregulatory acclimation and nitrogen metabolism in juveniles of the freshwater marble goby exposed to seawater. *J. Comp. Physiol.* In press.

Claiborne, J. B., and Heisler, N. (1986). Acid–base regulation and ion transfers in the carp (*Cyprinus carpio*): pH compensation during graded long- and short-term environmental hypercapnia, and the effect of bicarbonate infusion. *J. Exp. Biol.* **126**, 41–61.

Clements, K. D., and Raubenheimer, D. (2005). Feeding and nutrition. In: *The Physiology of Fishes* (D.H. Evans and J.B. Claiborne, eds), pp. 47–83. CRC Press, Boca Raton.

Collie, N. L. (1985). Intestinal nutrient transport in coho salmon (*Onchorhynchus kisutch*) and the effects of development, starvation and seawater adaptation. *J. Comp. Physiol. B* **156**, 163–174.

Cooper, C. A., and Wilson, R. W. (2008). Post-prandial alkaline tide in freshwater rainbow trout: effects of meal anticipation on recovery from acid–base and ion regulatory disturbances. *J. Exp. Biol.* **211**, 2542–2550.

Curtis, B., and Wood, C. M. (1992). Kidney and urinary bladder responses of freshwater rainbow trout to isosmotic NaCl and NaHCO₃ infusion. *J. Exp. Biol.* **173**, 181–203.

Daniel, H. (2004). Molecular and integrative physiology of intestinal peptide transport. *Annu. Rev. Physiol.* **66**, 361–384.

Davenport, H. W. (1958). *The ABC of Acid–Base Chemistry*. 4th edition. University of Chicago Press, Chicago.

De Flora, S., and Arillo, A. (1983). Mutagenic and DNA damaging activity in muscle of trout exposed in vivo to nitrite. *Cancer Lett.* **20**, 147–155.

Denstadli, V., Vegusdal, A., Krogdahl, A., Bakke-McKellep, A. M., Berge, G. M., Holm, H., Hillestad, M., and Ruyter, B. (2004). Lipid absorption in different segments of the gastrointestinal tract of Atlantic salmon (*Salmo salar* L.). *Aquaculture* **240**, 385–398.

Doi, A. M., Holmes, E., and Kleinow, K. M. (2001). P-glycoprotein in the catfish intestine: inducibility by xenobiotics and functional properties. *Aquat. Tox.* **55**, 157–170.

Duee, P. H., Darcyvrillon, B., Blachier, F., and Morel, M. T. (1995). Fuel selection in intestinal-cells. *Proc. Nutr. Soc.* **54**, 83–94.

Eddy, F. B. (1976). Acid–base balance in rainbow trout (*Salmo gairdneri*) subjected to acid stresses. *J. Exp. Biol.* **64**, 159–171.

Esbaugh, A. J., and Tufts, B. L. (2006a). The structure and function of carbonic anhydrase isozymes in the respiratory system of vertebrates. *Resp. Physiol. Neurobiol.* **154**, 185–198.

Esbaugh, A. J., and Tufts, B. L. (2006b). Tribute to R. G. Boutilier: evidence of a high activity carbonic anhydrase isozyme in the red blood cells of an ancient vertebrate, the sea lamprey, *Petromyzon marinus*. *J. Exp. Biol.* **209**, 1169–1178.

Esbaugh, A. J., Gilmour, K. M., and Perry, S. F. (2009). Membrane-associated carbonic anhydrase in the respiratory system of the Pacific hagfish (*Eptatretus stouti*). *Resp. Physiol. Neurobiol.* **166**, 107–116.

Evans, D. H. (1982). Mechanisms of acid extrusion by two marine fishes: the teleost, *Opsanus beta*, and the elasmobranch, *Squalus acanthias*. *J. Exp. Biol.* **97**, 289–299.

Evans, D. H., and Claiborne, J. B. (2008). Osmotic and ionic regulation in fishes. In: *Osmotic and Ionic Regulation: Cells and Animals* (D.H. Evans, ed.), pp. 295–366. CRC Press, Boca Raton.

Evans, D. H., Piermarini, P. M., and Choe, K. P. (2005). The multifunctional fish gill: dominant site of gas exchange, osmoregulation, acid–base regulation, and excretion of nitrogenous waste. *Physiol. Rev.* **85**, 97–177.

Fauconneau, B., Basseres, A., and Kaushik, S. J. (1992). Oxidation of phenylalanine and threonine in response to dietary arginine supply in rainbow trout (*Salmo gairdneri* R.). *Comp. Biochem. Physiol. A.* **101**, 401.

Febry, R., and Lutz, P. (1987). Energy partitioning in fish: the activity related cost of osmoregulation in a euryhaline cichlid. *J. Exp. Biol.* **128**, 63–85.

Gadagbui, B. K., and James, M. O. (2000). The influence of diet on the regional distribution of glutathione S-transferase activity in channel catfish intestine. *J. Biochem. Mol. Toxicol.* **14**, 148–154.

Genz, J., Taylor, J. R., and Grosell, M. (2008). Effects of salinity on intestinal bicarbonate secretion and compensatory regulation of acid–base balance in *Opsanus beta*. *J. Exp. Biol.* **211**, 2327–2335.

Gilmour, K. M. (2001). The CO₂/pH ventilatory drive in fish. *Comp. Biochem. Physiol. A* **130**, 219–240.

Gilmour, K. M., and Perry, S. F. (2009). Carbonic anhydrase and acid–base regulation in fish. *J. Exp. Biol.* **212**, 1647–1661.

Gilmour, K. M., Shah, B., and Szebedinszky, C. (2002). An investigation of carbonic anhydrase activity in the gills and blood plasma of brown bullhead (*Ameiurus nebulosus*), longnose skate (*Raja rhina*), and spotted ratfish (*Hydrolagus colliei*). *J. Comp. Physiol. B* **172**, 77–86.

Goss, G. G., and Wood, C. M. (1990). Na$^+$ and Cl$^-$ uptake kinetics, diffusive effluxes and acidic equivalent fluxes across the gills of rainbow-trout. 1. Responses to environmental hyperoxia. *J. Exp. Biol.* **152**, 521–547.

Goss, G. G., Perry, S. F., Wood, C. M., and Laurent, P. (1992). Mechanisms of ion and acid–base regulation at the gills of fresh-water fish. *J. Exp. Zool.* **263**, 143–159.

Grans, A., Albertsson, F., Axelsson, M., and Olsson, C. (2009). Postprandial changes in enteric electrical activity and gut blood flow in rainbow trout (*Oncorhynchus mykiss*) acclimated to different temperatures. *J. Exp. Biol.* **212**, 2550–2557.

Grosell, M. (2006). Intestinal anion exchange in marine fish osmoregulation. *J. Exp. Biol.* **209**, 2813–2827.

Grosell, M. (2007). Intestinal transport processes in marine fish osmoregulation. In: *Fish Osmoregulation* (B. Baldisserotto, J. M. Mancera, B. G. Kapoor, eds.), pp. 332–357. Science Publishers Inc., Enfield, New Hampshire, USA.

Grosell, M., and Genz, J. (2006). Ouabain sensitive bicarbonate secretion and acid absorption by the marine fish intestine play a role in osmoregulation. *Am. J. Physiol.* **291**, R1145–R1156.

Grosell, M., and Jensen, F. B. (1999). NO$_2^-$ uptake and HCO$_3^-$ excretion in the intestine of the European flounder (*Platichthys flesus*). *J. Exp. Biol.* **202**, 2103–2110.

Grosell, M., and Jensen, F. B. (2000). Uptake and effects of nitrite in the marine teleost fish *Platichthys flesus*. *Aquat. Toxicol.* **50**, 97–107.

Grosell, M., and Taylor, J. R. (2007). Intestinal anion exchange in teleost water balance. *Comp. Biochem. Physiol. A* **148**, 14–22.

Grosell, M., Wood, C. M., Wilson, R. W., Bury, N. R., Hogstrand, C., Rankin, J. C., and Jensen, F. B. (2005). Bicarbonate secretion plays a role in chloride and water absorption of the European flounder intestine. *Am. J. Physiol.* **288**, R936–R946.

Grosell, M., Gilmour, K. M., and Perry, S. F. (2007). Intestinal carbonic anhydrase, bicarbonate, and proton carriers play a role in the acclimation of rainbow trout to seawater. *Am. J. Physiol. Regul. Integr. Comp. Physiol.* **293**, R2099–R2111.

Grosell, M., Genz, J., Taylor, J. R., Perry, S. F., and Gilmour, K. M. (2009a). The involvement of H+-ATPase and carbonic anhydrase in intestinal HCO$_3^-$ secretion in seawater acclimated rainbow trout. *J. Exp. Biol.* **212**, 1940–1948.

Grosell, M., Mager, E. M., Williams, C., and Taylor, J. R. (2009b). High rates of HCO$_3^-$ secretion and Cl$^-$ absorption against adverse gradients in the marine teleost intestine: the involvement of an electrogenic anion exchanger and H$^+$-pump metabolon? *J. Exp. Biol.* **212**, 1684–1696.

Handy, R. D., and Poxton, M. G. (1993). Nitrogen pollution in mariculture: toxicity and excretion of nitrogenous compounds by marine fish. *Reviews in Fish Biology and Fisheries* **3**, 205–241.

Hascilowicz, T., Murai, N., Matsufuji, S., and Murakami, Y. (2002). Regulation of ornithine decarboxylase by antizymes and antizyme inhibitor in zebrafish (*Danio rerio*). *Biochim. Biophys. Acta* **1578**, 21–28.

Hazon, N., Wells, A., Pillans, R. D., Good, J. P., Anderson, W. G., and Franklin, C. E. (2003). Urea based osmoregulation and endocrine control in elasmobranch fish with special reference to euryhalinity. *Comp. Biochem. Physiol. B* **136**, 685–700.

Hemmer, M. J., Courtney, L. A., and Benson, W. H. (1998). Comparison of three histological fixatives on the immunoreactivity of mammalian P-glycoprotein antibodies in the sheepshead minnow, *Cyprinodon variegatus*. *J. Exp. Zool.* **281**, 251–259.

Henry, R. P. (1996). Multiple roles of carbonic anhydrase in cellular transport and metabolism. *Annu. Rev. Physiol.* **58**, 523–538.

Henry, R. P., and Heming, T. A. (1998). Carbonic anhydrase and respiratory gas exchange. In: *Fish Respiration* (S.F. Perry and B.L. Tufts, eds), pp. 75–111. Academic Press, New York.

Henry, R. P., and Swenson, E. R. (2000). The distribution and physiological significance of carbonic anhydrase in vertebrate gas exchange organs. *Resp. Physiol.* **121**, 1–12.

Hersey, S. J., and Sachs, G. (1995). Gastric acid secretion. *Physiol. Rev.* **75**, 155–189.

Hoyle, I., Shaw, B. J., and Handy, R. D. (2007). Dietary copper exposure in the African walking catfish, *Clarias gariepinus*: transient osmoregulatory disturbances and oxidative stress. *Aquat. Tox.* **83**, 72.

Hundal, H. S., and Taylor, P. M. (2009). Amino acid transceptors: gate keepers of nutrient exchange and regulators of nutrient signaling. *Am. J. Physiol. Endocrinol. Metab.* **296**, E603–E613.

James, L. A., Lunn, P. G., and Elia, M. (1998). Glutamine metabolism in the gastrointestinal tract of the rat assessed by the relative activities of glutaminase (EC 3.5.1.2) and glutamine synthetase (EC 6.3.1.2). *Br. J. Nutr.* **79**, 365–372.

James, M. O., Lou, Z., Rowland-Faux, L., and Celander, M. C. (2005). Properties and regional expression of a CYP3A-like protein in channel catfish intestine. *Aquat. Toxicol.* **72**, 361–371.

Jensen, F. B. (2004). Red blood cell pH, the Bohr effect, and other oxygenation-linked phenomena in blood O_2 and CO_2 transport. *Acta Physiol. Scand.* **182**, 215–227.

Jobling, M. (1981). The influences of feeding on the metabolic-rate of fishes—a short review. *J. Fish Biol.* **18**, 385–400.

Jobling, M. (1994). *Fish Bioenergetics*. Springer.

Jobling, M. (1995). Environmental biology of fishes. *Fish and Fisheries* **16**, 211–249.

Kajimura, M., Croke, S. J., Glover, C. N., and Wood, C. M. (2004). Dogmas and controversies in the handling of nitrogenous wastes: the effect of feeding and fasting on the excretion of ammonia, urea and other nitrogenous waste products in rainbow trout. *J. Exp. Biol.* **207**, 1993–2002.

Karlsson, A., Eliason, E. J., Mydland, L. T., Farrell, A. P., and Kiessling, A. (2006). Postprandial changes in plasma free amino acid levels obtained simultaneously from the hepatic portal vein and the dorsal aorta in rainbow trout (*Oncorhynchus mykiss*). *J. Exp. Biol.* **209**, 4885–4894.

Keszler, A., Zhang, Y., and Hogg, N. (2010). Reaction between nitric oxide, glutathione, and oxygen in the presence and absence of protein: how are S-nitrosothiols formed? *Free Radic. Biol. Med.* **48**, 55–64.

Kleinow, K. M., Doi, A. M., and Smith, A. A. (2000). Distribution and inducibility of P-glycoprotein in the catfish: immunohistochemical detection using the mammalian C-219 monoclonal. *Mar. Env. Res.* **50**, 313–317.

Kong, H., Edberg, D. D., Korte, J. J., Salo, W. L., Wright, P. A., and Anderson, P. M. (1998). Nitrogen excretion and expression of carbamoyl-phosphate synthetase III activity and mRNA in extrahepatic tissues of largemouth bass (*Micropterus salmoides*). *Arch. Biochem. Biophys.* **350**, 157–168.

Kong, S. E., Hall, J. C., Cooper, D., and McCauley, R. D. (2000). Starvation alters the activity and mRNA level of glutaminase and glutamine synthetase in the rat intestine. *J. Nutr. Biochem.* **11**, 393–400.

Korte, J. J., Salo, W. L., Cabrera, V. M., Wright, P. A., Felskie, A. K., and Anderson, P. M. (1997). Expression of carbamoyl-phosphate synthetase III mRNA during the early stages of development and in muscle of adult rainbow trout (*Oncorhynchus mykiss*). *J. Biol. Chem.* **272**, 6270–6277.

Krebs, H. A., Kornberg, H. L. and Burton, K. (1957). A survey of the energy transformations in living matter. In: *Reviews of Physiology, Biochemistry and Pharmacology, Vol. 49*, pp. 212–298, Springer, Berlin Heidelberg.

Krogh, A. (1929). The progress of physiology. *Am. J. Physiol.* **90**, 243–251.

Kurita, Y., Nakada, T., Kato, A., Doi, H., Mistry, A. C., Chang, M. H., Romero, M. F., and Hirose, S. (2008). Identification of intestinal bicarbonate transporters involved in formation of carbonate precipitates to stimulate water absorption in marine teleost fish. *Am. J. Physiol. Regul. Integr. Comp. Physiol.* **294**, R1402–R1412.

Kuwaye, T. T., Okimoto, D. K., Shimoda, S. K., Howerton, R. D., Lin, H. R., Pang, P. K. T., and Grau, E. G. (1993). Effect of 17a-methyltestosterone on the growth of the euryhaline tilapia, *Oreochromis mossambicus*, in fresh water and in seawater. *Aquaculture* **113**, 137–152.

Lang, T., Alexandersson, M., Hansson, G. C., and Samuelsson, T. (2004). Bioinformatic identification of polymerizing and transmembrane mucins in the puffer fish, *Fugu rubripes*. *Glycobiology* **14**, 521–527.

Leef, M. J., Harris, J. O., and Powell, M. D. (2005). Respiratory pathogenesis of amoebic gill disease (AGD) in experimentally infected Atlantic salmon *Salmo salar*. *Dis. Aquat. Org.* **66**, 205–213.

Lehninger, A. L. (1970). *Biochemistry*. 2nd edition. Worth Publishers, Inc. New York.

Lie-Venema, H., Hakvoort, T. B., van Hemert, F. J., Moorman, A. F., and Lamers, W. H. (1998). Regulation of the spatiotemporal pattern of expression of the glutamine synthetase gene. *Prog. Nucleic Acid Res. Mol. Biol.* **61**, 243–308.

Lignot, J. H., Helmstetter, C., and Secor, S. M. (2005). Postprandial morphological response of the intestinal epithelium of the Burmese python (*Python, molurus*). *Comp. Biochem. Physiol. A* **141**, 280–291.

Lindsley, J. E., and Rutter, J. (2004). Nutrient sensing and metabolic decisions. *Comp. Biochem. Physiol. B* **139**, 143–159.

Lisowska-Myjak, B., Tomaszewski, L., and Hryckiewicz, L. (1978). Intestinal arginase in vertebrates and invertebrates. *Comp. Biochem. Physiol. B* **61**, 545–552.

Loretz, C. A. (1995). Electrophysiology of ion transport in the teleost intestinal cells. In: *Cellular and Molecular Approaches to Fish Ionic Regulation, Fish Physiology* (C. M. Wood and T. J. Shuttleworth, eds.), **14**, 25–56.

Loretz, C. A. (2001). Drinking and alimentary transport in teleost osmoregulation. *Proceedings of the 14th International Congress of Comparative Endocrinology* 723–732.

Lusk, G. (1931). The specific dynamic action. *J. Nutr.* **3**, 519–530.

Makky, K., Tekiela, J., and Mayer, A. N. (2007). Target of rapamycin (TOR) signaling controls epithelial morphogenesis in the vertebrate intestine. *Dev. Biol.* **303**, 501–513.

Marshall, W. S. and Grosell, M. (2005). Ion transport, osmoregulation and acid–base balance. In: (D. H. Evans and J. B. Claiborne, eds.). CRC Press, Boca Raton, FL.

Martensson, J., Jain, A., and Meister, A. (1990). Glutathione is required for intestinal function. *Proc. Natl. Acad. Sci.* **87**, 1715–1719.

McCue, M. D. (2006). Specific dynamic action: a century of investigation. *Comp. Biochem. Physiol. A* **144**, 381–394.

McDonald, D. G., and Wood, C. M. (1981). Branchial and renal acid and ion fluxes in the rainbow trout, *Salmo gairdneri*, at low environmental pH. *J. Exp. Biol.* **93**, 101–118.

McKenzie, D., Taylor, E., Dalla, V., Valle, A. D., and Steffensen, J. (2002). Tolerance of acute hypercapnic acidosis by the European eel (*Anguilla anguilla*). *J. Comp. Physiol. B* **172**, 339–346.

McKenzie, D. J., Piccolella, M., Valle, A. Z. D., Taylor, E. W., Bolis, C. L., and Steffensen, J. F. (2003). Tolerance of chronic hypercapnia by the European eel *Anguilla anguilla*. *J. Exp. Biol.* **206**, 1717–1726.

McLean, E., Ronsholdt, B., Sten, C., and Najamuddin (1999). Gastrointestinal delivery of peptide and protein drugs to aquacultured teleosts. *Aquaculture* **177**, 231–247.

McMurtrie, H. L., Cleary, H. J., Alvarez, B. V., Loiselle, F. B., Sterling, D., Morgan, P. E., Johnson, D. E., and Casey, J. R. (2004). The bicarbonate transport metabolon. *J. Enzyme Inhib. Med. Chem.* **19**, 231–236.

Menon, M., and Kewalramani, H. G. (1958). Studies on some physiological aspects of digestion in three species of elasmobranchs. *Proc. Plant Sci.* **50**, 26–39.

Mommsen, T. P., Vijayan, M. M., and Moon, T. W. (1999). Cortisol in teleosts: dynamics, mechanisms of action and metabolic regulation. *Rev. Fish Biol. Fish.* **9**, 211–268.

Mommsen, T. P., Busby, E. R., von Schalburg, K. R., Evans, J. C., Osachoff, H. L., and Elliott, M. E. (2003a). Glutamine synthetase in tilapia gastrointestinal tract: zonation, cDNA and induction by cortisol. *J. Comp. Physiol. B* **173**, 419–427.

Mommsen, T. P., Osachoff, H. L., and Elliott, M. E. (2003b). Metabolic zonation in teleost gastrointestinal tract—effects of fasting and cortisol in tilapia. *J. Comp. Physiol. B* **173**, 409–418.

Morgan, P. E., Pastorekova, S., Stuart-Tilley, A. K., Alper, S. L., and Casey, J. R. (2007). Interactions of transmembrane carbonic anhydrase, CAIX, with bicarbonate transporters. *Am. J. Physiol. Cell Physiol.* **293**, C738–C748.

Mountfort, D. O., Campbell, J., and Clements, K. D. (2002). Hindgut fermentation in three species of marine herbivorous fish. *Appl. Environ. Microbiol.* **68**, 1374–1380.

Mourad, F. H., Barada, K. A., Khoury, C., Hamdi, T., Saade, N. E., and Nassar, C. F. (2009). Amino acids in the rat intestinal lumen regulate their own absorption from a distant intestinal site. *Am. J. Physiol. Gastrointest. Liver Physiol.* **297**, G292–G298.

Movileanu, L., Flonta, M. L., Mihailescu, D., and Frangopol, P. T. (1998). Characteristics of ionic transport processes in fish intestinal epithelial cells. *BioSystems* **45**, 123–140.

Murai, T., Ogata, H., Hirasawa, Y., Akiyama, T., and Nose, T. (1987). Portal absorption and hepatic uptake of amino acids in rainbow trout force-fed complete diets containing casein or crystalline amino acids. *Nippon Suisan Gakkaishi* **53**, 1847–1859.

Murray, B. W., Busby, E. R., Mommsen, T. P., and Wright, P. A. (2002). Evolution of glutamine synthetase in vertebrates: multiple glutamine synthetase genes expressed in rainbow trout (*Oncorhynchus mykiss*). *J. Exp. Biol.* **206**, 1511–1521.

Nelson, J. S. (1994). *Fishes of the World*. 3rd edition. John Wiley & Sons, New York.

Nikinmaa, M., and Salama, A. (1998). Oxygen transport in fish. In: *Fish Respiration* (S.F. Perry and B.L. Tufts, eds), pp. 141–184. Academic Press, New York.

Niv, Y., and Fraser, G. M. (2002). The alkaline tide phenomenon. *J. Clin. Gastroenterol.* **35**.

Ogata, H., and Arai, S. (1985). Comparison of free amino acid contents in plasma, whole blood and erythrocytes of carp, coho salmon, rainbow trout, and channel catfish. *Bull. Jap. Soc. Sci. Fish* **51**, 1181–1186.

Ostaszewska, T., Kamaszewski, M., Grochowski, P., Dabrowski, K., Verri, T., Aksakal, E., Szatkowska, I., Nowak, Z., and Dobosz, S. (2010). The effect of peptide absorption on PepT1 gene expression and digestive system hormones in rainbow trout (*Oncorhynchus mykiss*). *Comp. Biochem. Physiol. A* **155**, 107–114.

Overgaard, J., Busk, M., Hicks, J. W., Jensen, F. B., and Wang, T. (1999). Respiratory consequences of feeding in the snake *Python molorus*. *Comp. Biochem. Physiol. A* **124**, 359–365.

Papastamatiou, Y. P., and Lowe, C. G. (2004). Postprandial response of gastric pH in leopard sharks (*Triakis semifasciata*) and its use to study foraging ecology. *J. Exp. Biol.* **207**, 225–232.

Parsons, D. S., and Case, R. M. (1982). Role of anions and carbonic anhydrase in epithelia [and discussion]. *Philosophical Transactions of the Royal Society of London. Series B, Biological Sciences* **299**, 369–381.

Pelletier, X., Duportail, G., and Leray, C. (1986). Isolation and characterization of brush-border membrane from trout intestine. Regional differences. *Biochim. Biophys. Acta* **856**, 267–273.

Perry, S. F., Malone, S., and Ewing, D. (1987). Hypercapnic acidosis in rainbow trout (*Salmo gairdneri*). II. Renal ionic fluxes. *J. Exp. Zool.* **65**, 896–902.

Perry, S. F., and Gilmour, K. M. (2006). Acid–base balance and CO_2 excretion in fish: unanswered questions and emerging models. *Resp. Physiol. Neurobiol.* **154**, 199–215.

Perry, S. F., Esbaugh, A., Braun, M., and Gilmour, K. M. (2009). Gas transport and gill function in water-breathing fish. In: *Cardio-Respiratory Control in Vertebrates* (M.L. Glass and S.C. Wood, eds), pp. 5–42. Springer, Berlin.

Polakof, S., Alvarez, R. and Soengas J. L. (2010). Gut glucose metabolism in rainbow trout: implications in glucose homeostasis and glucosensing capacity. *Am. J. Physiol.* Submitted.

Prevot-D'Alvise, N., Pierre, S., Gaillard, S., Gouze, E., Gouze, J. N., Aubert, J., Richard, S., and Grillasca, J. P. (2008). cDNA sequencing and expression analysis of *Dicentrarchus labrax* heme oxygenase-1. *Cell Mol. Biol.* **54 Suppl.**, OL1046–OL1054.

Rankin, J. C. (2002). In: *Osmoregulation and Drinking in Vertebrates* (N. Hazon and G. Flik, eds), pp. 1–17. BIOS Scientific Publishers Ltd., Oxford.

Rhoads, J. M., and Wu, G. (2009). Glutamine, arginine, and leucine signaling in the intestine. *Amino Acids* **37**, 111–122.

Rhoads, J. M., Argenzio, R. A., Chen, W., Graves, L. M., Licato, L. L., Blikslager, A. T., Smith, J., Gatzy, J., and Brenner, D. A. (2000). Glutamine metabolism stimulates intestinal cell MAPKs by a cAMP-inhibitable, Raf-independent mechanism. *Gastroenterology* **118**, 90–100.

Rich, P. R. (2003). The molecular machinery of Keilin's respiratory chain. *Biochem. Soc. Trans.* **31**, 1095–1105.

Rolen, S. H., Sorensen, P. W., Mattson, D., and Caprio, J. (2003). Polyamines as olfactory stimuli in the goldfish *Carassius auratus*. *J. Exp. Biol.* **206**, 1683–1696.

Ron, B., Shimoda, S. K., Iwama, G. K., and Grau, E. G. (1995). Relationships among ration, salinity, 17-methyltestosterone and growth in the euryhaline tilapia, *Oreochromis mossambicus*. *Aquaculture* **135**, 185–193.

Ronner, P., and Scarpa, A. (1987). Secretagogues for pancreatic hormone release in the channel catfish (*Ictalurus punctatus*). *Gen. Comp. Endocrinol.* **65**, 354–362.

Rotoli, B. M., Uggeri, J., Dall'Asta, V., Visigalli, R., Barilli, A., Gatti, R., Orlandini, G., Gazzola, G. C., and Bussolati, O. (2005). Inhibition of glutamine synthetase triggers apoptosis in asparaginase-resistant cells. *Cell Physiol. Biochem.* **15**, 281–292.

Rubner, M. (1902). *Die Gesetze des Energieverbrauchs bei der Ernährung*. Franz Deuticke, Lepizig.

Rune, S. J. (1965). The metabolic alkalosis following aspiration of gastric acid secretion. *Scand. J. Clin. Lab Invest.* **17**, 305–310.

Sacco, J. C., Lehmler, H. J., Robertson, L. W., Li, W., and James, M. O. (2008). Glucuronidation of polychlorinated biphenylols and UDP-glucuronic acid concentrations in channel catfish liver and intestine. *Drug Metab. Dispos.* **36**, 623–630.

Saito, T., Hascilowicz, T., Ohkido, I., Kikuchi, Y., Okamoto, H., Hayashi, S., Murakami, Y., and Matsufuji, S. (2000). Two zebrafish (*Danio rerio*) antizymes with different expression and activities. *Biochem. J.* **345**, 99–106.

Sardella, B., Kultz, D., Cech, J., and Brauner, C. (2008). Salinity-dependent changes in Na^+/K^+-ATPase content of mitochondria-rich cells contribute to differences in thermal tolerance of Mozambique tilapia. *J. Comp. Physiol. B* **178**, 249–256.

Secor, S. (2009). Specific dynamic action: a review of the postprandial metabolic response. *J. Comp. Physiol. B* **179**, 1–56.

Secor, S. M., and Diamond, J. (1995). Adaptive responses to feeding in Burmese pythons: pay before pumping. *J. Exp. Biol.* **198**, 1313–1325.

Secor, S. M., and Diamond, J. (1998). A vertebrate model of extreme physiological regulation. *Nature* **395**, 659–662.

Shehadeh, Z. H., and Gordon, M. S. (1969). The role of the intestine in salinity adaptation of the rainbow trout, *Salmo gairdneri*. *Comp. Biochem. Physiol.* **30**, 397–418.

Shepherd, B. S., Ron, B., Burch, A., Sparks, R., Richman, N. H., III, Shimoda, S. K., Stetson, M. H., Lim, C., and Grau, E. G. (1997). Effects of salinity, dietary level of protein and 17β-methyltestosterone on growth hormone (GH) and prolactin (tPRL177 and tPRL188) levels in the tilapia, *Oreochromis mossambicus*. *Fish Physiol. Biochem.* **17**, 279–288.

Shepherd, E. J., Lister, N., Affleck, J. A., Bronk, J. R., Kellett, G. L., Collier, I. D., Bailey, P. D., and Boyd, C. A. (2002). Identification of a candidate membrane protein for the basolateral peptide transporter of rat small intestine. *Biochem. Biophys. Res. Comm.* **296**, 918–922.

Shiina, A., Itoi, S., Washio, S., and Sugita, H. (2006). Molecular identification of intestinal microflora in *Takifugu niphobles*. *Comp. Biochem. Physiol. D.* **1**, 128–132.

Sire, M. F., and Vernier, J. M. (1992). Intestinal absorption of protein in teleost fish. *Comp. Biochem. Physiol. A* **103**, 771–781.

Skea, G. L., Mountfort, D. O., and Clements, K. D. (2007). Contrasting digestive strategies in four New Zealand herbivorous fishes as reflected by carbohydrase activity profiles. *Comp. Biochem. Physiol. A* **146**, 63–70.

Smith, H. W. (1930). The absorption and excretion of water and salts by marine teleosts. *Am. J. Physiol.* **93**, 480–505.

Smith, R. W., and Houlihan, D. F. (1995). Protein synthesis and oxygen consumption in fish cells. *J. Comp. Physiol. B* **165**, 93–101.

Sparks, R. T., Shepherd, B. S., Ron, B., Richman, N. H., Riley, L. G., Iwama, G. K., Hirano, T., and Grau, E. G. (2003). Effects of environmental salinity and 17β-methyltestosterone on growth and oxygen consumption in the tilapia, *Oreochromis mossambicus*. *Comp. Biochem. Physiol.* **136**, 657–665.

Stevens, C. E. (1995). *Comparative Physiology of the Vertebrate Digestive System.* 2nd edition. Cambridge University Press, Cambridge.

Talbot, C. (1993). Some aspects of the biology of feeding and growth in fish. *Proc. Nutr. Soc.* **52**, 403–416.

Taylor, J. R., and Grosell, M. (2006a). Evolutionary aspects of intestinal bicarbonate secretion in fish. *Comp. Biochem. Physiol. A.* **143**, 523–529.

Taylor, J. R., and Grosell, M. (2006b). Feeding and osmoregulation: dual function of the marine teleost intestine. *J. Exp. Biol.* **209**, 2939–2951.

Taylor, J. R., and Grosell, M. (2009). The intestinal response to feeding in seawater gulf toadfish, *Opsanus beta*, includes elevated base secretion and increased epithelial oxygen consumption. *J. Exp. Biol.* **212**, 3873–3881.

Taylor, J. R., Whittamore, J. M., Wilson, R. W., and Grosell, M. (2007). Postprandial acid–base balance in freshwater and seawater-acclimated European flounder, *Platichthys flesus*. *J. Comp. Physiol.* **177**, 597–608.

Taylor, J. R., Mager, E. M., and Grosell, M. (2010). Basolateral NBCe1 plays a rate-limiting role in transepithelial intestinal HCO_3^- secretion serving marine fish osmoregulation. *J. Exp. Biol.* **213**, 459–468.

Tng, Y. Y. M., Wee, N. L. J., Ip, Y. K., and Chew, S. F. (2008). Postprandial nitrogen metabolism and excretion in juvenile marble goby, *Oxyeleotris marmorata* (Bleeker, 1852). *Aquaculture* **284**, 260–267.

Treberg, J. R., Brosnan, M. E., Watford, M. and Brosnan, J. T. (2009). On the reversibility of glutamate dehydrogenase and the source of hyperammonemia in the hyperinsulinism/ hyperammonemia syndrome. *Adv. Enzyme Regul.* In press.

Tresguerres, M., Katoh, F., Orr, E., Parks, S. K., and Goss, G. G. (2006). Chloride uptake and base secretion in freshwater fish: a transepithelial ion-transport metabolon? *Physiol. Biochem. Zool.* **79**, 981–996.

Truchot, J. P., and Forgue, J. (1998). Effect of water alkalinity on gill CO_2 exchange and internal PCO_2 in aquatic animals. *Comp. Biochem. Physiol. A* **119**, 131–136.

Tseng, Y. C., and Hwang, P. P. (2008). Some insights into energy metabolism for osmoregulation in fish. *Comp. Biochem. Physiol. C* **148**, 419–429.

Tufts, B. L., and Perry, S. F. (1998). Carbon dioxide transport and excretion. In: *Fish Respiration* (S.F. Perry and B.L. Tufts, eds), pp. 220–282. Academic Press, New York.

Tufts, B. L., Esbaugh, A. J., and Lund, S. G. (2003). Comparative physiology and molecular evolution of carbonic anhydrase in the erythrocytes of early vertebrates. *Comp. Biochem. Physiol. A.* **136**, 259–269.

Tuziak, S. M., Moghadam, H. K., and Danzmann, R. G. (2009). Genetic mapping of the ornithine decarboxylase (odc) gene complex in rainbow trout (*Oncorhynchus mykiss*). *Cytogenet. Genome Res.* **125**, 279–285.

Vera, L. M., De Pedro, N., Gomez-Milan, E., Delgado, M. J., Sanchez-Muros, M. J., Madrid, J. A., and Sanchez-Vazquez, F. J. (2007). Feeding entrainment of locomotor activity rhythms, digestive enzymes and neuroendocrine factors in goldfish. *Physiol. Behavior* **90**, 518–524.

Vernier, J. M. (1990). Intestine ultrastructure in relation to lipid and protein absorption in teleost fish. In: *Animal Nutrition and Transport Processes. 1. Nutrition in Wild and Domestic Animals* (J. Mellinger, ed.), pp. 166–175. Karger, Basel.

Verri, T., Romano, A., Barca, A., Kottra, G., Daniel, H. and Storelli, C. (2009). Transport of di- and tripeptides in teleost fish intestine. *Aquaculture Research.* In press, DOI: 10.1111/ j.1365-2109.2009.02270.x.

Walsh, P. J., Blackwelder, P., Gill, K. A., Danulat, E., and Mommsen, T. P. (1991). Carbonate deposits in marine fish intestines: a new source of biomineralization. *Limnol. Oceanogr.* **36**, 1227–1232.

Walsh, P. J., Tucker, B. C., and Hopkins, T. E. (1994). Effects of confinement/crowding on ureogenesis in the Gulf toadfish *Opsanus-Beta. J. Exp. Biol.* **191**, 195–206.

Walsh, P. J., Mayer, G. D., Medina, M., Bernstein, M. L., Barimo, J. F., and Mommsen, T. P. (2002). A second glutamine synthetase gene with expression in the gills of the Gulf toadfish (*Opsanus beta*). *J. Exp. Biol.* **206**, 1523–1533.

Wang, D., Zhong, X. P., Qiao, Z. X., and Gui, J. F. (2008). Inductive transcription and protective role of fish heme oxygenase-1 under hypoxic stress. *J. Exp. Biol.* **211**, 2700–2706.

Wang, T., Burggren, W., and Nobrega, E. (1995). Metabolic, ventilatory, and acid–base disturbances associated with specific dynamic action in the toad *Bufo marinus. Physiol. Zool.* **68**, 192–205.

Wang, T., Busk, H., and Overgaard, J. (2001). The respiratory consequences of feeding in amphibians and reptiles. *Comp. Biochem. Physiol. A* **128**, 535–549.

Williams, H. H., McVicar, A. M., and Ralph, R. (1970). The alimentary canal of fish as an environment for helminth parasites. In: *Aspects of Fish Parasitology* (A. Taylor and R. Muller, eds). Blackwell Scientific Publications, Oxford.

Wilson, R. W., and Grosell, M. (2003). Intestinal bicarbonate secretion in marine teleost fish— source of bicarbonate, pH sensitivity, and consequence for whole animal acid–base and divalent cation homeostasis. *Biochim. Biophys. Acta* **1618**, 163–193.

Wilson, R. W., Gilmour, K., Henry, R., and Wood, C. (1996). Intestinal base excretion in the seawater-adapted rainbow trout: a role in acid–base balance? *J. Exp. Biol.* **199**, 2331–2343.

Wilson, R. W., Wilson, J. M., and Grosell, M. (2002). Intestinal bicarbonate secretion by marine teleost fish—why and how? *Biochim. Biophys. Acta* **1566**, 182–193.

Wilson, R. W., Millero, F. J., Taylor, J. R., Walsh, P. J., Christensen, V., Jennings, S., and Grosell, M. (2009). Contribution of fish to the marine inorganic carbon cycle. *Science* **323**, 359–362.

Windmueller, H. G., and Spaeth, A. E. (1980). Respiratory fuels and nitrogen metabolism in vivo in small intestine of fed rats. Quantitative importance of glutamine, glutamate, and aspartate. *J. Biol. Chem.* **255**, 107–112.

Wood, C. M. (2001). Influence of feeding, exercise, and temperature on nitrogen metabolism and excretion. In: *Nitrogen Excretion* (P. Wright and P. Anderson, eds), pp. 201–238. Academic Press, New York.

Wood, C. M., and Caldwell, F. H. (1978). Renal regulation of acid–base balance in a freshwater fish. *J. Exp. Zool.* **205**, 301–307.

Wood, C. M., Mcmahon, B. R., and McDonald, D. G. (1977). An analysis of changes in blood pH following exhausting activity in the starry flounder, *Platichthys stellatus*. *J. Exp. Biol.* **69**, 173–185.

Wood, C. M., Kajimura, M., Mommsen, T. P., and Walsh, P. J. (2005). Alkaline tide and nitrogen conservation after feeding in an elasmobranch (*Squalus acanthias*). *J. Exp. Biol.* **208**, 2693–2705.

Wood, C. M., Part, P., and Wright, P. A. (1995). Ammonia and urea metabolism in relation to gill function and acid–base-balance in a marine elasmobranch, the spiny dogfish (*Squalus-Acanthias*). *J. Exp. Biol.* **198**, 1545–1558.

Wood, C. M., Milligan, C. L., and Walsh, P. J. (1999). Renal responses of trout to chronic respiratory and metabolic acidoses and metabolic alkalosis. *Am. J. Physiol.* **277**, R482–R492.

Wood, C. M., Du, J., Rogers, J., Brauner, C. J., Richards, J. G., Semple, J. W., Murray, B. W., Chen, X. G., and Wang, Y. (2007a). Przewalski's naked carp (*Gymnocypris przewalskii*): an endangered species taking a metabolic holiday in Lake Qinghai, China. *Physiol. Biochem. Zool.* **80**, 59–77.

Wood, C. M., Bucking, C., Fitzpatrick, J., and Nadella, S. (2007b). The alkaline tide goes out and the nitrogen stays in after feeding in the dogfish shark, *Squalus acanthias*. *Resp. Physiol. Neurobiol.* **159**, 163–170.

Wood, C. M., Kajimura, M., Bucking, C., and Walsh, P. J. (2007c). Osmoregulation, ionoregulation and acid–base regulation by the gastrointestinal tract after feeding in the elasmobranch (*Squalus acanthias*). *J. Exp. Biol.* **210**, 1335–1349.

Wood, C. M., Schultz, A. G., Munger, R. S., and Walsh, P. J. (2009). Using omeprazole to link the components of the post-prandial alkaline tide in the spiny dogfish, *Squalus acanthias*. *J. Exp. Biol.* **212**, 684–692.

Wright, E. M. (1993). The intestinal Na$^+$/glucose cotransporter. *Annu. Rev. Physiol.* **55**, 575–589.

Wright, P. A., and Wood, C. M. (2009). A new paradigm for ammonia excretion in aquatic animals: role of Rhesus (Rh) glycoproteins. *J. Exp. Biol.* **212**, 2303–2312.

Wright, P. A., Campbell, A., Morgan, R. L., Rosenberger, A. G., and Murray, B. W. (2004). Dogmas and controversies in the handling of nitrogenous wastes: expression of arginase Type I and II genes in rainbow trout: influence of fasting on liver enzyme activity and mRNA levels in juveniles. *J. Exp. Biol.* **207**, 2033–2042.

7

THE GASTROINTESTINAL TRACT AS AN ENDOCRINE/NEUROENDOCRINE/PARACRINE ORGAN: ORGANIZATION, CHEMICAL MESSENGERS AND PHYSIOLOGICAL TARGETS

YOSHIO TAKEI
CHRISTOPHER A. LORETZ

The fish gut holds special status among internal organs since its lumen is continuous with the external medium, giving the gut direct contact in a technical sense with the outside world. Consequently, the multifunctionality

The Multifunctional Gut of Fish: Volume 30
FISH PHYSIOLOGY

of the gut extends to include luminal sensing by epithelial cells and signaling to physiological targets both near (paracrine/autocrine/luminocrine signaling) and far (traditional endocrine signaling by way of the blood vascular system). The chemical agents for this signaling are produced by cells of the gastroenteropancreatic series. Additionally, neuropeptides and neurotransmitters from axon terminals of the peripheral nervous system contribute to regulatory signaling at the gut. Combined, the humoral factors of these systems operate in parallel locally to direct digestive processing (secretion of digestive enzymes, and motility), to regulate nutrient and ion/water absorption, and to influence the secretion of other hormones, and more distantly at the brain to modulate feeding and drinking, and at various tissues and organs to regulate metabolism, growth and other essential physiological processes. This chapter summarizes the current state of knowledge regarding the gut as a hormone-secreting organ in fishes, and surveys key target actions in the gut and elsewhere of gastroenteropancreatic series chemical messengers.

1. INTRODUCTION

1.1. Chemical Signaling by the Gut

As with other functional elements of the alimentary tract of fishes, a system for integrative communication by way of chemical mediators is essential for efficient operation of the gut itself, and for the coordination of gut function with other body systems (Fig. 7.1). On a local scale, the management of digestive processing along, and absorption through, the tract following food ingestion and drinking, and the enhancement of vascular flow to and from the gut following nutrient and ion/water absorption are regulated by diffusely arranged hormone-secreting cells of the alimentary tract. On a global scale, signaling by gastrointestinal (GI) endocrine cells to the central nervous system and to peripheral tissues, including other endocrine glands, extends the functional reach of the gut and gut-derived organs through effects on absorbed nutrient processing and assimilation, and through influences on other physiological and behavioral homeostatic systems such as those relating to hydromineral balance, feeding, and drinking. Some aspects of functional coordination by and of enteric and enteric-derived tissues are covered elsewhere in this volume (Chapters 3 and 8), and in previous volumes of this series (Youson 2007; Holmgren and Olsson 2009). We have recently surveyed the endocrine system in fishes (Takei and Loretz 2006), and the reader is directed thereto for an overview of the topic. Therefore, this chapter will focus more

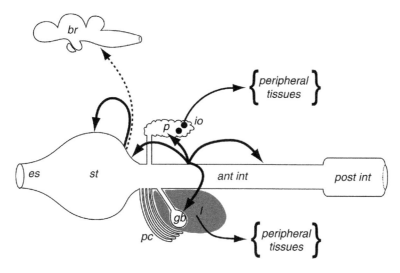

Fig. 7.1. Gastroenteropancreatic endocrine factors can coordinate gastrointestinal function through local or paracrine action among gut cells (heavy arrows), or they can coordinate metabolism and growth through blood-borne flow of factors from the gut or gut-derived tissues and organs (liver or pancreas, for example) to peripheral body tissues (light arrows), or they can signal gastrointestinal tract status to the brain where it may influence behavior (feeding or drinking, for example; broken arrow). Key to structures: *es*, esophagus; *st*, stomach; *ant int*, anterior intestine; *post int*, posterior intestine; *pc*, pyloric ceca; *l*, liver; *gb*, gall bladder; *p*, pancreas; *io*, islet organ(s); *br*, brain.

narrowly on the endocrine, neuroendocrine and paracrine system of the gut in several key aspects of integrative signaling in the fishes. Although the overall organization of chemical signaling in fishes resembles that in vertebrates generally, there are distinct, unique and sometimes novel expressions of signaling systems in fishes that reflect the substantial diversity of taxonomic groups within the fishes.

In the last decade, the availability of modern molecular biological techniques that have been applied to piscine endocrine systems, and of genomic resources for an expanding number of fish species have sharply changed our view of the endocrine and other chemical messenger systems of fishes. For example, the widely recognized intestinal hormone secretin, which stimulates pancreatic bicarbonate secretion to neutralize acidic chyme delivered from the stomach to the intestine, has yet to be found in fishes; it may be a tetrapod, or even mammalian, hormone with no ortholog in the fishes (Roch et al. 2009). Indeed, these deviations from the traditional endocrine story presented for the GI tract in general subject textbooks are not unique. Conversely, and importantly, studies of fish endocrine systems can fuel the discovery in other vertebrates of heretofore unknown chemical messengers and signaling systems (Takei et al. 2007).

1.2. The Gastroenteropancreatic "Space" of Fishes

More than that of tetrapods, the morphological diversity of piscine alimentary tracts, including those of agnathan groups, reveals the deep evolutionary developmental history of the gut endocrine system. Simply, the gastroenteropancreatic (GEP) system comprises the chemical messenger-secreting cells of the stomach, intestine and pancreas. In those fishes lacking a functional stomach, including the hagfish and lamprey, and some species and families of gnathostome fishes (Genten et al. 2009), a more restrictive enteropancreatic system terminology is more appropriate. General use of the GEP designation in this chapter should be understood to acknowledge the absence of a stomach in some species and groups. As elaborated below, the secretory cells of the GEP series are diffusely arranged in the gut mucosa and they may be present additionally as extramural aggregations of cells, such as pancreatic islet tissues, that resemble more typically endocrine glands of a classical type (Vigna 1986a). Regardless of their location, cells in both configurations secrete chemical messengers that support functional coupling between the GI tract and other parts of the organism. For example, hormones such as insulin and glucagon regulate intermediary metabolism in accordance with the nutritional status of the organism (Nelson and Sheridan 2006) that, in turn, is influenced by GEP factors such as ghrelin that stimulates feeding (Kaiya et al. 2008). Natriuretic peptides and guanylins from GEP cells influence drinking and other elements of hydromineral balance on an organismal scale (Loretz et al. 1997; Loretz and Pollina 2000; Takei and Yuge 2007). Individually or in combination, these cells and their near and distant influences on target tissues comprise the "GEP space" within the fish.

The diffusely arranged GEP cells within the gut mucosa commonly display a polarized structure that can span fully the epithelial layer, maintaining both basal anchorage to the basement membrane and apical exposure to the intestinal lumen; these are so-called "open"-type cells (Fig. 7.2). The term "paraneuron," coined by Fujita (Fujita and Kobayashi 1977; Fujita 1989; Day and Salzet 2002) for cells displaying "receptosecretory" function, seems particularly appropriate for many of these open-type GEP cells that sense, or "taste," the luminal fluid and respond appropriately with chemical messenger secretion at the basolateral pole. A well-known example of a paraneuron in tetrapods, representing one of the first enteroendocrine systems to be functionally described, is the secretin-secreting S cell of the duodenum that stimulates pancreatic ascinar cell

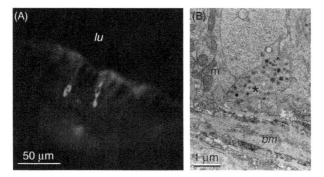

Fig. 7.2. Endocrine cells of the gastroenteropancreatic series are diffusely arranged within the gut mucosa. These polarized endocrine cells span the epithelium lining the gut, and maintain exposure to both the luminal compartment and the serosal (blood-side) face of the epithelium. The cells are commonly positioned to sense the luminal compartment at the apical cell pole, and to respond appropriately with chemical messenger secretion at the basolateral, or serosal, pole. (A) Immunofluorescence image of an atrial natriuretic peptide-secreting cell in the middle intestinal epithelium of a freshwater-adapted Japanese eel (*Anguilla japonica*). *lu*, intestinal lumen. (B) Electron microscopic image of the basolateral pole of an enteroendocrine cell in the middle intestine of a seawater-adapted goby (*Gillichthys mirabilis*). Secretory granules (*asterisk*) are visible at the basolateral cell pole. *m*, mitochondria; *bm*, basement membrane. See color plate section.

bicarbonate secretion in response to luminal acidification resulting from the delivery of acidic chyme from the stomach (Bayliss and Starling 1902). Cell polarization is also a feature of guanylin-secreting goblet cells (Yuge et al. 2003), and of defensive Paneth cells (Porter et al. 2002), although the secretory products of these cells are delivered instead to the intestinal lumen.

1.3. Evolutionary Developmental Biology of the Enteroendocrine System

Among the nearly 28,000 living species of fishes (of which the teleosts comprise about 27,000; Nelson 2006), there are many anatomical expressions of the alimentary tract that reflect in both striking and subtle ways hundreds of millions of years of evolutionary time in adaptive radiation. In addition to the stomachless nature of the tract in cyclostomes, holocephalans and some osteichthyans mentioned above, there are differences in the length and complexity of the intestine and in the number and size of pyloric ceca among carnivorous (fewer ceca) and herbivorous (more ceca) species (Buddington and Diamond 1987; Horn 1997), and segmental specializations such as the spiral valve of chondrichthyans and of some osteichthyans such as the bowfin and bichir (Hilton 1900; Abdel Magid 1975; Genten et al. 2009), and in the presence and elaboration of secretory tissues developed as evaginations of the posterior intestine in holocephalans (intramural glands) and in elasmobranchs and sarcopterygians (the rectal gland; Loretz 1987). Dietary differences among species are reflected in varied digestive system morphology, and in physiology through differentially regulated synthesis and secretion of digestive enzymes in carnivores and herbivores to match the biochemical nature of the food (Horn 1997; Clements and Raubenheimer 2006). The emergence in many fish species of osmoregulatory (as opposed to osmoconforming) strategies, and of euryhalinity from stenohalinity, during the colonization of the seas also brought the need for regulatory systems to coordinate ion and water transport processes across epithelial tissues, including the GI tract.

Underlying the remarkable adaptive radiations of the fishes are at least two rounds (1R and 2R) of whole-genome duplication (WGD) in the early vertebrate lineage during the Pre-Cambrian or Cambrian periods. Whereas there is general agreement that the 1R WGD occurred prior to the divergence of jawless and jawed vertebrates, there are conflicting opinions as to whether the 2R WGD occurred before or after the divergence of the jawed fishes about 520–550 mya (Panopoulou and Poustka 2009; Van de Peer et al. 2009). The third-round (3R) WGD, occurring in the ray-finned fish lineage during the Permian-Triassic (226–316 mya), generated again more new genetic material (Christoffels et al. 2006; Panopoulou and Poustka 2009; Roch et al. 2009; Van de Peer et al. 2009). The evidence of WGD (and of additional segmental or small-scale duplications that might have occurred) is seen in (super)families of hormones and their receptors (Sherwood et al. 2000; Conlon and Larhammer 2005; Sundstrom et al. 2008). Whereas one fate of a duplicated copy of a gene is loss or silencing (non-functionalization), alternate fates of neofunctionalization or subfunctionalization can support diversification of endocrine signaling systems and the persistence of functional paralogues in descendent lineages (Conant and Wolfe 2008). As a consequence of this extraordinary diversity in morphology, physiology and biochemistry, and of the modest number of species that have been examined in reality, the scope of our knowledge on the topic of GEP endocrine systems in fishes is still incomplete and is necessarily limited. Nevertheless, some trends are evident within the scope of available data, and many interesting questions remain to be answered.

The completion of the genome database and advances in the knowledge of developmental biology and genetic models in mice have revealed how enteroendocrine cells develop and differentiate from a common precursor cell localized in the intestinal crypts (Schonhoff et al. 2004). It has been suggested that the fate of the stem cell to differentiate into four types of intestinal epithelial cells, absorptive enterocytes and three secretory cells (enteroendocrine cells, goblet cells, and Paneth cells) is principally directed by the Notch signaling pathway. The differentiation of secretory-type cells

Table 7.1

Gastroenteropancreatic (GEP) hormone families of the fish gut. Family groupings of GEP hormones reflect sequence and structural similarities, and infer shared evolutionary ancestry through paralogy

(Super)family/Members	Source	Target/Function
Ghrelin		
ghrelin (GLN)	stomach	stimulate food intake and gastric motility
motilin	intestine	stimulate gastric motility
Gastrin		
cholecystokinin (CCK)	intestine	inhibit food intake; stimulate pancreatic enzyme secretion and gall bladder contraction
gastrin	stomach G-cells	stimulate gastric gland secretion and gastric motility
Insulin		
insulin (INS)	B-cells of pancreas, intestine	inhibit food intake; hyperglycemia
insulin-like growth factor-I (IGF-I)	liver	stimulate cartilage growth
insulin-like growth factor-II (IGF-II)	liver	unknown
Secretin		
pituitary adenylate cyclase- activating polypeptide (PACAP)	GI tract	inhibit food intake
vasoactive intestinal polypeptide (VIP)	stomach, intestine	alter intestinal and gall bladder motility; inhibit food intake
glucagon (GLUC)	A-cells of pancreas, intestine	stimulate glycogenolysis and gluconeogenesis (hyperglycemia); stimulate lipolysis (hyperlipidemia)
glucagon-like peptide (GLP-1)	A-cells of pancreas, intestine	glycogenolysis and gluconeogenesis (hyperglycemia); lipolysis
glucagon-like peptide (GLP-2)	intestine	unknown
growth hormone-releasing hormone (GHRH)		rectal gland secretion?
glucose-dependent insulinotropic peptide (GIP)	K-cell of GI tract	stimulate INS secretion
Somatostatin (SST)		
somatostatin-14 (SST-14)	D-cells of pancreas, intestine	modulate insulin, glucagon and GLP secretion; promote lipid mobilization (hyperlipidemia) and glycogenolysis (hyperglycemia)

(Continued)

Table 7.1 (*continued*)

(Super)family/Members	Source	Target/Function
"large somatostatins" (SST-22, -25, -26, -27, -28, -34)	D-cells of pancreas, intestine	modulate insulin, glucagon and GLP secretion; promote lipid mobilization (hyperlipidemia) and glycogenolysis (hyperglycemia)
Neuropeptide Y		
neuropeptide Y (NPY)	nervous system	stimulate food intake
peptide tyrosine-tyrosine (PYY)	endocrine pancreas	stimulate food intake?
pancreatic polypeptide (PP)	pancreatic islet F-cells	unknown
peptide Y (PY)	pancreatic islet, intestine, brain	unknown
Guanylin		
guanylin (GN)	goblet cells of intestine	stimulate Cl^- secretion into lumen
uroguanylin (UGN)	enterochromaffin cells	stimulate Cl^- secretion
renoguanylin (RGN)	intestine and kidney	stimulate Cl^- secretion
Natriuretic peptide		
atrial natriuretic peptide (ANP)	intestine	inhibit NaCl absorption
ventricular natriuretic peptide (VNP)	intestine	inhibit NaCl absorption
C-type natriuretic peptide (CNP)	intestine	stimulate rectal gland secretion
Calcitonin gene-related peptide		
Calcitonin gene-related peptide (CGRP)	intestine	inhibit food intake
amylin (AMY)	pancreatic B-cell	stimulate INS secretion; inhibit food intake
adrenomedullin/adrenomedullin 1 (AM)	liver	inhibit food intake
adrenomedullin 2/intermedin (AM2)	intestine	stimulate drinking
adrenomedullin 5 (AM5)	intestine	stimulate drinking
Leptin	mesenteric adipose tissue	inhibit food intake
Galanin		
galanin	stomach, intestine	stimulate food intake
galanin-like peptide	intestine	stimulate food intake
Tachykinins		
substance P and neurokinins	nerve terminals in GI tracts	possible appetite regulation
Neuromedin U (NMU)	GI tract	inhibit food intake
Gastrin-releasing peptide (GRP)	GI tract	inhibit food intake

is further directed by a distal factor and enteroendocrine cells are further differentiated into more than ten different cell types. Moreover, the similarity of enteroendocrine cells to neurosecretory cells as mentioned above is signaled by the role of Notch signaling pathways in differentiation of nervous system (Lathia et al. 2008), although enteroendocrine cells are derived from the endoderm, not from neuroectoderm (Andrew et al. 1998). A similar pathway may be involved in the differentiation of enteroendocrine cells of fishes.

2. HORMONES SECRETED FROM GASTROINTESTINAL TRACTS AND ACCESSORY ORGANS AND TISSUES

2.1. Hormone Inventory

Table 7.1 lists by family groupings the principal, and some minor, less well-characterized hormones that are known from GEP tissues in the fishes. The family groupings of structurally related hormones in the table reflect an evolutionary history of ancestral WGD and segmental duplications, and subsequent evolutionary divergence on a grand scale. Information on the membership of these hormone families has been revealed by the comparative genomic analyses through data mining of an increasingly available number of fish genome databases (Oleksiak and Crawford 2006; Takei et al. 2007). Further, enzymatic processing of prohormone forms expands the membership in familial categories even more. Consequently, much more information exists on the membership of the hormone families, relative to the demonstrated functional activities of a smaller number of these hormones. And the tabular accounting does not include the many chemical neurotransmitters and other signaling molecules that are synthesized and released by the enteric nervous system. For a full discussion of the enteric nervous system influence on GI processes such as gut motility (including peristalsis and gall bladder contraction), secretion and digestion and splanchnic blood flow, the reader is directed to Chapter 8 and to recent surveys and reviews (Bosi et al. 2004; Holmgren and Olsson 2009).

Despite the various arrangements of GEP cells among diverse fish anatomies, the reader will recognize similar complements of chemical messengers to those that are typically present in vertebrates. Importantly, hormones uncovered in studies of fish systems might signal yet-to-be-discovered homologues in tetrapod vertebrates (Takei et al. 2007).

2.1.1. GHRELIN

Ghrelin (GRLN), discovered as an endogenous ligand of the growth hormone secretagogue receptor (GHS-R), is a single-chain polypeptide molecule with a characteristic acylation at position Ser^3 (Kojima et al. 1999; Uniappan and Peter 2005; Kaiya et al. 2008). Processed from preprohormone (>100 amino acid (aa) residues in length), fish GRLNs, at 19–25 aa residues, are shorter than tetrapod GRLNs of 25–28 aa residues (Uniappan et al. 2002; Kaiya et al. 2003a,b,c). Teleost GRLNs are distinguished additionally by an amidated C-terminus (Uniappan et al. 2002; Kaiya et al. 2003a,b,c, 2008; Uniappan and Peter, 2005). GRLN gene expression measured by RT-PCR is highest in the stomach, with lower expression in some other tissues (Uniappan et al. 2002; Kaiya et al. 2003a,b,c). The GRLN receptor (GHS-R) is a member of the Class A G protein-coupled receptor (GPCR) superfamily with its characteristic heptahelical transmembrane domain (Uniappan and Peter 2005; Kaiya et al. 2008).

2.1.2. CHOLECYSTOKININ/GASTRIN

Cholecystokinin (CCK) and gastrin comprise a small hormone family. Both hormones are synthesized as larger preprohormones (>100 aa residues) that are enzymatically processed from the N-terminus into multiple shorter forms with the C-terminus of these molecules containing the 7- to 8-aa-long active core for biological activity (Vigna 1986b, 2000). Within the bioactive core, the C-terminal pentapeptide with an amidated C-terminus is invariant between the two hormones and is preceded by a sulfated tyrosine residue. It is proposed that gastrin arose by gene duplication from a CCK-like ancestral hormone molecule, as evidenced by the presence of just a single CCK-like molecule (cionin) in *Ciona*, a tunicate protochordate (Johnsen 1998; Vigna 2000). The discovery of both CCK and gastrin in sharks places the gene duplication event early in vertebrate evolution, and marks that gastrin's appearance in the vertebrate lineage leading to chondrichthyans may have coincided with the evolutionary development of an acid-secreting stomach, consistent with gastrin's present function (Johnsen et al. 1997; Johnsen 1998). The actions of both hormones are mediated by two CCK receptor subtypes (CCK-AR and CCK-BR) belonging to the Class A GPCR superfamily (Wank 1998).

2.1.3. INSULIN AND INSULIN-LIKE GROWTH FACTOR

Insulin (INS) sequences are known from several taxonomic groups of fishes, including actinopterygians, sarcopterygians, chondrichthyans and agnathans (Hobart et al. 1980b; Plisetskaya et al. 1988; Chan et al. 1990; Conlon et al. 1997, 1999; Chan and Steiner 2000; Conlon 2000, 2001; Anderson et al. 2002b; Irwin 2004). The structures of fish INS are highly conserved relative to INSs of other vertebrates. As in vertebrates generally, fish INS is composed of separate A- (about 21 aa residues) and B-chains (about 30 aa residues) following removal of the intervening C-segment, and there is an intramolecular disulfide bond within the A-chain, and two interchain (A-B) disulfide bonds. The six cysteine residues of INS are fully conserved. Extensions to both N- and C-termini of the two chains are observed in fishes (Conlon 2001). The C-terminal extension to the B-chain appears not to alter biological activity, based on receptor binding and functional assay in the holocephalan ratfish (*Hydrolagus colliei*), where multiple forms of INS with differing lengths of amino acid extension to the C-terminus of the B-chain result from alternative splicing of a prohormone (Conlon et al. 1989). Examination of the genomes of fugu (*Takifugu rubripes*) and zebrafish (*Danio rerio*) revealed two copies of the INS gene (*INS1* and *INS2*), both of which contain full, but non-identical, INS-coding sequences (Irwin 2004). Probably resulting from the 3R WGD in the ancestral teleost lineage, the retention of both gene copies and the phylogenetic grouping of the INS1 sequence with other fish INSs suggest the possibility of sub- or neofunctionalization, with perhaps *INS1* being the predominantly expressed form in islet organs based on sequencing of cDNA from pancreas (Irwin 2004). INS is secreted by B-cells of the pancreatic islets as well as by the diffusely arranged cells of the intestine.

IGFs are strongly conserved among vertebrates, and are structurally similar to proinsulin. Fish IGFs are single-chain polypeptides of about 70 aa length, depending on species, and contain three intramolecular disulfide bonds (Reinecke et al. 1997; Upton et al. 1998). The evolutionary divergence of the INS and IGF genes from an ancient metazoan INS-like gene likely occurred after the origin of protochordates from invertebrate stock. Since only a single IGF that is more similar to IGF-I has been identified in agnathans, the divergence of IGF-I and -II occurred after the origin of gnathostome fishes (Chan et al. 1990; Nagamatsu et al. 1991; Chan and Steiner, 2000; Wood et al. 2005). An important part of the IGF signaling system is IGF binding protein (IGFBP). IGFBPs in the circulation can inhibit or potentiate the biological activity of IGFs through their interactions with the hormone (Zhou et al. 2008).

The INS receptor (INSR) and the IGF receptor (with the predominant form being the type 1 IGF receptor, IGF-1R) are members of the heterotetrameric receptor tyrosine kinase superfamily (Navarro et al. 1999; Planas et al. 2000; Wood et al. 2005). Consistent with the observed glucose intolerance of fishes generally (see Section 5.1.1), insulin receptor expression is lower in fishes than that in mammals (Párrizas et al. 1994; Planas et al. 2000). Further, within teleosts, receptor expression, measured as receptor tyrosine kinase activity, is dependent on diet; specifically, receptor expression in skeletal muscle of carnivorous species is lower than in omnivorous and herbivorous species (Párrizas et al. 1994; Planas et al. 2000). Receptor pairs of INSR, IGF-1R (as IGF-1R_a and IGF-1R_b) and IGFBPs are known from several species of teleosts, with the substantial sequence divergence of these paralogues supporting an early duplication event (Chan et al. 1997; Wood et al. 2005; Zhou et al. 2008).

2.1.4. GLUCAGON AND GLUCAGON-LIKE PEPTIDES

Glucagon (GLUC, a 29-aa peptide) and glucagon-like peptide-1 (GLP-1, 31-32 aa) and -2 (GLP-2, 32-33 aa) are members of the secretin superfamily of hormones (Roch et al. 2009). This superfamily also includes pituitary adenylate cyclase-activating polypeptide (PACAP), vasoactive intestinal polypeptide (VIP), peptide histidine-isoleucine/methionine (PHI/PHM), growth hormone-releasing hormone (GHRH), secretin, and glucose-dependent insulinotropic polypeptide (GIP) (Roch et al. 2009). The proglucagon gene in vertebrates encodes the glucagon and glucagon-like peptide hormones. The encoding order of the proglucagon gene, in the 5'-to-3' direction, is GLUC—GLP-1—GLP-2, but alternative splicing of mRNAs may result in tissue-specific expression patterns in some species (Irwin 2001). For example, GLUC and GLP-1, but not GLP-2, are encoded in mRNA from pancreas (in anglerfish, *Lophius americanus*; Lund et al. 1982), whereas GLP-2 is encoded in mRNA from intestine (in rainbow trout, *Oncorhynchus mykiss*; Irwin and Wong 1995). GLUC and GLP-1 are produced by the A-cells of the pancreatic islets, and by the L-cells of the intestine (Nozaki et al. 1988; Duguay and Mommsen 1994; Alexander et al. 2006).

Insight into the evolution of proglucagon and of the receptors to GLUC and GLPs has been improved with the aid of modern genomic and molecular phylogenetic tools. The GLUC gene evolved by internal duplication of an ancestral GLUC gene to yield GLUC and a proto-GLP, followed by another internal duplication to produce GLP-1 and -2 before the agnathan–gnathostome split (Irwin et al. 1999). Fishes have multiple copies of the GLUC gene. Again, this reflects 3R WGD in the teleost lineage, or subsequent tetraploidization in some species such as salmonids, or gene duplication in lampreys (Irwin et al. 1999; Irwin 2001; Roch et al. 2009). Important questions related to tissue-specific patterns of hormone and receptor expression and of relative functional activities of paralogous proteins remain unanswered at present. The duplicated status of GLUC genes in the osteichthyans may underlie the greater rate of evolution of proglucagon-derived peptides relative to other vertebrates, including the chondrichthyans (Irwin 2001).

The GLUC receptor (GLUCR) and GLP-1 receptor (GLP-1R) are Class B members of the GPCR superfamily (Navarro et al. 1999; Chow et al. 2004; Cardoso et al. 2005; Roch et al. 2009). Functional data for the goldfish (*Carassius auratus*) GLUCR and GLP-1R are reported (Chow et al. 2004). Phylogenetic and genomic analyses are consistent with 3R WGD of the teleost GLUCR gene, and of the secretin receptor gene family (Chow et al. 2004; Cardoso et al. 2005; Roch et al. 2009).

2.1.5. SOMATOSTATIN

Somatostatin (SST) is produced and secreted by the islet organs and by diffusely arranged enteroendocrine cells in all fishes, including agnathans. Teleost fishes possess at least two SST genes, each of which encodes a prosomatostatin (PSS-I or -II) that after translation is proteolytically processed into products of several lengths (Hobart et al. 1980a; Lin and Peter 2001; Nelson and Sheridan 2005). SST-14 is identical in all vertebrates, but some species may possess additional variant forms of SST-14; the identical SST-14 form is contained within PSS-I at the C-terminus, and the variant form at the C-terminus of PSS-II (Lin and Peter 2001). "Large SSTs" represent N-terminal extended forms of SST-14; lengths are variable, including -22, -25, -26, -27, -28, -34 and -37 forms, according to species and PSS type (Sheridan et al. 2000; Lin and Peter 2001; Youson 2007). SSTs are produced in the D-cells of the pancreatic islets and by diffusely arranged cells in the intestine (Youson and Al-Mahrouki 1999; Youson, 2007). Within the pancreatic islets, there are two populations of D-cells according to their SST products; centrally located D1-cells produce SST-14, whereas peripheral D2-cells produce larger SSTs (Youson et al. 2006; Alexander et al. 2006; Youson 2007).

SST receptors (sst) are members of the Class A GPCR superfamily. There are at least five subtypes of SST receptor that are widely distributed and have overlapping patterns of expression (Sheridan et al. 2000; Gong et al. 2004; Nelson and Sheridan 2005; Tulipano and Schulz 2007). Coupled with the multiplicity of SST forms and with overlapping ligand-binding ability, there is the potential for great diversity in physiological response to SST (Tulipano and Schulz 2007; Schonbrunn 2008).

2.1.6. NEUROPEPTIDE Y

The neuropeptide Y (NPY) family of hormones, which arose from an NPY ancestor through WGD events including the 3R WGD of teleosts, has been difficult to resolve. Early studies based on analyses of evolutionarily conserved mature NPY family peptide sequences apparently incorrectly assigned NPY subfamily identities to fish peptides. By application of synteny and of phylogenetic analysis of genomic sequences to discover orthology and paralogy, NPY family and subfamily membership has been largely resolved (Larhammer 1996; Cerdá-Reverter and Larhammer 2000; Cerdá-Reverter et al. 2000; Kurokawa and Suzuki 2002; Sundström et al. 2008). Consequently, it seems that only the NPY and PYY subfamilies are present in fishes; and that in teleosts, the 3R WGD has generated duplicate genes for both that are now identified as NPY_a and NPY_b, and PYY_a and PYY_b, and subsume the earlier-recognized fish PY designation into PYY_b (Sundström et al. 2008). Extrapolating from RT-PCR mRNA expression and in situ hybridization findings on Japanese flounder (*Paralichthys olivaceus*) and fugu, NPY family expression is limited in fishes to brain and kidney, whereas PYY (PY) expression is broader and includes strong expression in intestine and pancreas (F-cells), but not in liver (Kurokawa and Suzuki 2002; Sundström et al. 2008). The receptors are Class A GPCRs.

2.1.7. VASOACTIVE INTESTINAL POLYPEPTIDE AND PEPTIDE HISTIDINE-ISOLEUCINE

Vasoactive intestinal polypeptide (VIP) is another member of the secretin superfamily of hormones (see Section 2.1.4). Described for actinopterygians and chondrichthyans, it is a highly conserved 28-aa residue single-chain polypeptide with demonstrated expression in the stomach and intestine (Wang and Conlon 1995; Nam et al. 2009; Roch et al. 2009). The VIP gene from Japanese flounder has been cloned and sequenced, revealing that in addition to encoding VIP, it encodes peptide histidine-isoleucine (PHI), a 27-aa residue peptide that is also a member of the secretin superfamily; VIP, PHI, and so-called cryptic peptides are all encoded within a single open reading frame of the VIP gene (Nam et al. 2009). Duplicate VIP genes are reported for fugu and for rainbow trout (Roch et al. 2009). VIP and PHI, together with another member of the secretin superfamily (PACAP), act through three Class B GPCRs ($VPAC_1R$, $VPAC_2R$/PHIR, and PAC_1R) (Cardoso et al. 2005; Dickson and Finlayson 2009; Roch et al. 2009).

2.1.8. GUANYLINS AND NATRIURETIC PEPTIDES

The guanylins (GNs) are a family of small cysteine-rich peptide hormones. Typically 15 aa residues in length for GN, C-terminal extensions of up to several aa residues are seen in uroguanylin (UGN) and renoguanylin (RGN). The GN molecules are stabilized by two intramolecular disulfide bonds. Despite the names GN, UGN and RGN, and their connotations of tissue-specific biosynthetic origin, all three hormones are produced by the intestine and the kidney of eels, where they have best been studied in the fishes (Yuge et al. 2003; Cramb et al. 2005; Kalujnaia et al. 2009).

The natriuretic peptides (NPs) are a family of hormones, 21–35 aa residues in length. Each NP family hormone possesses a 17-aa ring structure formed by a single intramolecular disulfide bond. In teleostean and chondrostean bony fishes, seven NP subtypes are described (ANP, BNP, VNP and four CNPs, CNP1–4) that vary in the length of the N- and C-terminal tails and in composition of the central highly conserved ring structure. All NP-subtypes are derived from an ancestral NP before the 3R WGD (Kawakoshi et al. 2004) by 1R and 2R WGD and tandem segmental duplications as determined by synteny and linkage analysis (Inoue et al. 2003a, 2005). The ancestral NP was likely a CNP-like molecule (probably CNP4), since hagfish and lamprey possess only CNP4 and elasmobranch and holocephalan possess only CNP3 (Kawakoshi et al. 2006).

The GNs and NPs exert their effects through guanylyl cyclase (GC)-coupled receptors, but different GC-coupled receptors are engaged for the two families of ligand (GC-A and GC-B for NPs; GC-C for GNs) (Loretz and Pollina 2000; Hirose et al. 2001; Nakauchi and Suzuki 2005). As a consequence, therefore, the cellular mechanisms of action of GNs and NPs may intersect to some extent.

2.1.9. UNDISCOVERED AND ABSENT HORMONES IN FISHES

Interestingly, some very familiar tetrapod hormones are now thought to be absent from fishes. Prior to the development and application of molecular biological and genomic techniques with sufficiently strict specificity, antibodies with imperfect specificity might have detected in immunological assays some closely related, but different, molecules through cross-reaction. Or, in the absence of clear evidence to their existence in fishes (so-called "negative data"), perhaps commonly accepted and functionally characterized tetrapod hormones were assumed incorrectly to exist in fishes. Based on examination of five fish genomes, secretin and the secretin receptor are both absent from teleosts (Roch et al. 2009). And pancreatic polypeptide, a member of the NPY superfamily of hormones, is also missing from teleosts, apparently having arisen through a segmental duplication of the PYY gene in the tetrapod lineage (Sundström et al. 2008). Clearly, additional work is needed on selected taxa to complete the inventory of gut hormones in the fishes, and to discover the deeper evolutionary history of some GI hormones.

2.2. The Endocrine Pancreas

2.2.1. EVOLUTIONARY ORIGIN OF THE PANCREAS

For some hormones, particularly those involved in regulation of digestive function, or in signaling to the brain regarding GI tract status, the advantage of messenger secretion from cells positioned in the intestinal epithelium is easily appreciated, since digestion is a highly sequenced process and serves critical functions in nutrition and osmoregulation. Examination of the comparative morphologies and the cell/tissue/organ source of GEP series hormones among protochordates and several groups of fishes exposes the most likely evolutionary origin of the endocrine pancreas as the escape of hormone-secreting cells from the gut epithelium (or the epithelium of the common bile duct) and their subsequent establishment in aggregate outside the epithelium. As vascularized aggregates of endocrine cells outside the epithelium, an islet organ (that is, a collection of islets) may exist separately from exocrine pancreatic tissue as follicles within the gut wall in proximity to the common (extrahepatic) bile duct where it enters the anterior intestine (in hagfish and larval lampreys, for example), or as a large single cranial islet (Southern Hemisphere (SH) lamprey in Fig. 7.3) or as large cranial and caudal islets (Northern Hemisphere (NH) lamprey in Fig. 7.3) positioned close to the gut wall or within the liver. Or, it may exist within a morphologically compact gland as aggregates of islet tissue either associated with or embedded within exocrine tissue in chondrichthyans and lungfishes (with several principal islets) and in actinopterygian fishes (Epple and Brinn

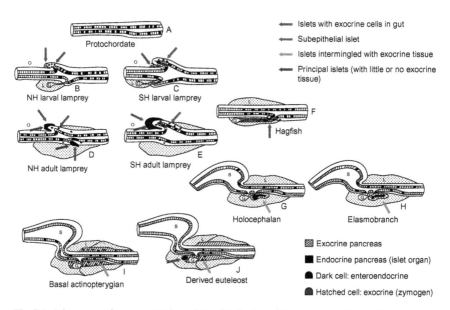

Fig. 7.3. Diagrammatic representation of the distribution of the endocrine tissue (islet organ) in fishes relative to the exocrine pancreatic tissue, esophagus (O), stomach (S), intestine (unlabeled portion of the gut), gall bladder (G), and the bile duct (tube leading from the gall bladder). The epithelial cells of the digestive tract are enlarged relative to the cells of the other structure in order to show endocrine (dark) and exocrine (hatched) cells; the latter are seen in (B)–(F), inclusive. The protochordate has endocrine cells restricted to the gut with the islet tissue and exocrine tissue relationships indicated by the key and the accompanying arrows for the various fish groups. See the text for further detail for each fish group. Reprinted from *Primitive Fishes (Fish Physiology Series)*, Vol. 26 (D. J. McKenzie, A. P. Farrell and C. J. Brauner, eds.), Youson, J. H., Peripheral endocrine glands. I. The gastroenteropancreatic endocrine system and the thyroid gland, pp. 381–455, 2000, San Diego: Academic Press, with permission from Elsevier Academic Press. See color plate section.

1986; Youson and Al-Mahrouki 1999; Youson 2000, 2007). Within the actinopterygians, there is great diversity in islet organ structure. Islets can be diffusely arranged (chondrosteans and the bowfin), or embedded within hepatic tissue (bichir and garpikes), or organized as a large aggregate of islet tissue with an investing layer of exocrine tissue (the so-called Brockmann body of teleosts) (Epple and Brinn 1986).

The evolutionary migration of enteroendocrine cells from the gut epithelium appears to have occurred gradually since the number of hormone-secreting cell types comprising the islet tissue increases progressively in morphological series from hagfish and larval lampreys (largely INS-secreting B-cells in the pancreatic islets, with an intestinal mucosa rich in hormone-secreting cells of the GEP series) to adult lampreys (B-cells and SST-secreting D-cells in the islets) to gnathostome fishes (B-cells and D-cells, and GLUC-secreting A-cells and PP-secreting F-cells in the islets, with a corresponding reduction in the gut epithelium of cells secreting these hormones) (Epple and Lewis 1973; Youson and Al-Mahrouki 1999; Youson 2000, 2007; Sherwood et al. 2005; Youson et al. 2006). This expansion and relocation to the pancreas may better position some of these endocrine cells for sensing and responding to systemic conditions without the influences of local GI activity. For example, compared with a location in the intestinal mucosa or even in the gut wall, INS-secreting cells in pancreatic islet tissue would experience less exposure to a local hyperglycemia and hyperaminoacidemia resulting from the glucose- and amino acid-absorbing activities of the gut. By examination of a morphological series of alimentary tracts in living chordates and vertebrates, the possible intermediate steps in the evolution of the vertebrate endocrine pancreas can be assembled. The morphological series is simply illustrated in Fig. 7.3. For detailed discussions of the evolutionary emergence of the vertebrate endocrine pancreas and for compilations of supporting data from the fishes, the reader is directed to earlier reviews (Epple and Lewis 1973; Youson and Al-Mahrouki 1999; Youson 2000, 2007; Takei and Loretz 2006; Youson et al. 2006).

2.2.2. CELL TYPE COMPOSITION OF THE ISLET ORGAN

The composition by cell type of the teleost pancreatic islets has been determined by histological and immunohistochemical methods. Generally, the same complement of cell types comprising the mammalian islets is seen in fish islets (Epple and Lewis 1973; Duguay and Mommsen 1994). In the Nile tilapia (*Oreochromis niloticus*), as an example of the teleost condition, four main histologically distinguishable cell types are present as in mammals: A-cells (secreting GLUC and GLPs), B-cells (INS), D-cells (SST), and F-cells (NPY) (Yang et al. 1999; Morrison et al. 2003, 2004; Alexander et al. 2006). In addition, there are two subpopulations of D-cells that produce different molecular isoforms of SST; the centrally located D1-cells produce SST-14 that is common to vertebrates, whereas the peripheral D2-cells produce the longer SST-28 with its processed hormonal products (Yang et al. 1999; Morrison et al. 2003).

2.3. The Liver

Derived as it is in other vertebrate classes from the anterior segment of embryonic intestine, the liver maintains a direct functional connection to the alimentary tract through its production of bile. The gall bladder with its bile duct that drains into the intestine sometimes maintains a morphological connection with the exocrine pancreas, which may share the duct (Clements and Raubenheimer 2006), and with the endocrine pancreas since islet organs in some species are derived from the duct and retain a nearby position (Youson and Al-Mahrouki 1999; Youson 2007). A functional connection between the intestine and gall bladder is also evident since the GEP hormone CCK stimulates gall bladder contraction. The liver expresses many receptors for circulating hormones, and it is a source of chemical messengers and precursors to messengers that predominantly have their effects outside the GI tract.

The liver is the major source of IGF-I and -II, but extrahepatic IGF production also occurs where the hormone may act in paracrine or autocrine fashion (Wood et al. 2005). The liver produces IGFs in response to GH. The liver synthesizes precursors to other hormone systems as well. These secreted products include angiotensinogen, the substrate for the renin-angiotensin system (Kobayashi and Takei 1996), and high-molecular-weight kininogen and Factor XII, which are components of the kallikrein-kinin system (Conlon 1999).

3. DIGESTION

3.1. Regulation of Digestion

With delivery of food from the mouth and esophagus into the stomach and the intestine, the chemical and enzymatic steps of digestion begin. Generally, diffusely arranged GEP cells along the length of the alimentary tract from stomach to the posterior intestine coordinate digestion. The progression of chyme through the digestive tract sequentially stimulates hormone-secreting paraneuronal cells in the intestinal mucosa, with the secreted hormonal signaling products from these cells, in turn, terminating or moderating earlier digestive steps such as gastric acid secretion, and triggering subsequent steps in digestion through paracrine signaling and via local vascular flow to nearby targets. Extrapolation from the mammalian and other tetrapod vertebrate classes is widespread in the literature, but herein we will limit remarks to some demonstrated endocrine signaling pathways in the fishes that exemplify regulatory pathways. Digestive physiology as a topic is considered elsewhere in this volume (see Chapter 2).

3.1.1. ENZYME AND ACID SECRETION

Appropriately phased gastric and pancreatic secretion of acid and enzymes is critical to efficient digestion. Stomach G-cells secrete gastrin and enterochromaffin cells secrete histamine in response to protein food (polypeptides and amino acids) delivered to the stomach; gastrin stimulates combined hydrochloric acid and pepsinogen/pepsin secretion by oxyntico-peptic cells of the stomach and increases motility (Buddington and Krogdahl 2004; Holmgren and Olsson 2009). Bombesin, histamine and tachykinins stimulate, and SST and VIP inhibit, gastric acid and pepsinogen secretion from oxynticopeptic cells in Atlantic cod, *Gadus morhua* (Holstein and Cederberg 1988; Holmgren and Olsson 2009). There is some evidence that the secretion of acid and pepsinogen may be separately regulated according to the balance among endocrine and neural messengers (Buddington and Krogdahl 2004; Holmgren and Olsson 2009). G-cells are present in the intestine of fishes lacking a stomach, suggesting functions other than stimulation of gastric acid and pepsin secretion (Buddington and Krogdahl 2004).

GLUC and GLP immunoreactivity in pyloric cacae and intestine of salmonids suggests that a local signaling pathway to inhibit gastric secretion after the digesta has entered the intestine may exist in fishes, as it does in other vertebrates (Nozaki et al. 1988).

As demonstrated in several fish species, CCK stimulates and PY antagonistically inhibits the synthesis and secretion of pancreatic lipase, trypsin and amylase; and CCK stimulates gall bladder contraction to release bile into the intestine (Buddington and Krogdahl 2004; Murashita et al. 2007, 2008a; Rønnestad et al. 2007). Consistent with their antagonistic effects on exocrine pancreatic secretion, the enteric delivery of fat and protein nutrients stimulates the expression of CCK and inhibits the expression of PY (Murashita et al. 2007, 2008a). With some species variability, CCK expression in the midgut begins early in the larval life of fishes, and mirrors the high functional capability of this segment of the larval digestive tract (Rønnestad et al. 2007). Moreover, the distributions of CCK-positive cells within the midgut correspond to species-specific patterns of enzymatic food processing in the anterior or posterior midgut segments.

3.1.2. MOTILITY

Gut motility, through peristaltic contractions of the smooth musculature in the alimentary tract wall, mixes and transports the digesta through the tract. Quiescence or relaxation of the musculature increases retention and processing times. Gastric receptive relaxation is an enteric reflex that couples food intake into the stomach with gastric smooth muscle relaxation to increase the capacity of the stomach (Olsson and Holmgren 2001). Intestinal motility is regulated by chemical mediators from both GEP cells and the enteric nervous system, with the patterns of regulation perhaps being different from those described for mammals (Buddington and Krogdahl 2004). In the fishes, generally, both excitatory (acetylcholine, histamine, serotonin, and tachykinins) and inhibitory (NO, VIP and PACAP) actions on peristaltic activity are reported; and the reported effects of CCK vary by species (Olsson and Holmgren 2001; Holmgren and Olsson 2009; and see Chapter 8).

Similar to its effect in other vertebrates, GRLN, from the stomach, has a stimulatory effect on intestinal motility in zebrafish (Kaiya et al. 2008; Olsson et al. 2008). Motilin also stimulates motility in zebrafish, but the status of the motilin gene and protein in fishes is uncertain (Olsson et al. 2008). Delivery of chyme from the stomach to the intestine is modulated by CCK, which reduces gastric emptying (Olsson and Holmgren 2001; Holmgren and Olsson 2009).

The observation of GLUC and GLP immunoreactivity in pyloric ceca and intestine of salmonids suggests that an intestinal brake mechanism may exist in fishes as in other vertebrates (Nozaki et al. 1988). GLP-1, however, failed to produce any effect on smooth muscle contraction in Chinook salmon (*Oncorhynchus tshawytscha*) GI ring contraction; since the GLP-1 tested was the non-native mammalian hormone, the possibility of GLP inhibition of motility should not be simply dismissed (Forgan and Forster 2007).

3.2. Signaling of Gastrointestinal Status

As discussed in Section 4, and elsewhere (Takei and Loretz 2006; Volkoff 2006), a number of GEP hormones signal GI status to the central nervous system. These hormones influence both feeding (orexogenic and anorexo-genic effects) and drinking (dipsogenic and anti-dipsogenic) behaviors in fishes.

4. APPETITE REGULATION

4.1. General Scheme of Endocrine Regulation of Appetite

Appetite control is a major target of recent research in endocrinology in relation to the metabolic syndrome and obesity in developed countries. This interest in appetite regulation appears to have been initiated by the discovery of leptin, the product of the *obese* gene, the disruption of which results in persistent appetite and obese phenotype (Zhang et al. 1994). The subsequent explosion of research has revealed the elaborate interactions of orexigenic and anorectic hormones, and of peripheral hormones and central neuropeptides for appetite regulation in mammals. To summarize the overall regulatory system (Fig. 7.4), peripheral hormones, anorectic leptin from the adipose tissue, orexigenic GRLN from the stomach, anorectic INS from the pancreas, and anorectic CCK/gastrin from the GI tract act on the hypothalamic neurons in the arcuate nucleus (ARC) that express orexigenic NPY/agouti-related protein (AgRP) or anorectic melanocyte-stimulating hormone (MSH)/cocaine- and amphetamine-regulated transcript (CART)

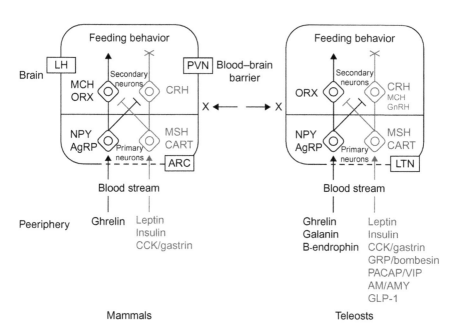

Fig. 7.4. Comparison of hormonal regulation of appetite between teleosts and mammals. Peripheral orexigenic (black letters) and anorectic hormones (red letters) act on the lateral tuberal nucleus (LTN) or arcuate nucleus (ARC) where a blood–brain barrier is insufficient. The primary orexigenic neurons containing neuropeptide Y (NPY)/agouti-related peptide (AgRP) and anorectic neurons containing α-melanocyte stimulating hormone (MSH)/cocaine- and amphetamine-regulated transcript (CART) send their axons to the secondary neurons in the lateral hypothalamic area (LH) and paraventricular nucleus (PVN), respectively. The secondary neurons are protected by the blood–brain barrier. The location of the secondary neurons in teleost brain is yet unknown. For other abbreviations, see Table 7.1 and text. ORX, orexin. See color plate section.

(see Schwartz et al. 2000; Currie et al. 2005). From the blood stream, the peripheral hormones can reach the neurons since the ARC has an incomplete blood–brain barrier. These primary neurons send their axons to the secondary neurons in the lateral hypothalamus that express anorectic orexins (orexin A and B) or melanin-concentrating hormone (MCH) and those in the paraventricular nucleus that express anorectic corticotrophin releasing hormone (CRH), resulting in appetite regulation (Fig. 7.4). In addition to the major regulatory system, there are many minor regulatory hormones and neuropeptides that are involved in the fine-tuning of appetite regulation; most of these are brain-gut hormones such as galanin, galanin-like peptide, GLP-1, PYY, amylin (AMY), neuromedin U (NMU), etc. (Funahashi et al. 2003; Valassi et al. 2008).

In fishes, appetite regulation is an important topic of research in relation to aquaculture. Thus, the regulatory mechanisms have been actively investigated in various teleost species, and particularly in detail in the goldfish (Lin et al. 2000). These studies revealed a basic scheme of regulation that is similar between fishes and mammals. Interested readers can refer to reviews that summarize recent progress in appetite regulation in teleost fish, including the cerebral mechanisms (Volkoff et al. 2005, 2009; Gorissen et al. 2006). Therefore, the focus here will be on the role of GEP-derived hormones in appetite regulation in the fishes.

4.2. Appetite-inducing (Orexigenic) Hormones

Most of those that have been defined as orexigenic hormones in mammals also induce food intake in fishes when administered peripherally and/or into the cerebral ventricle (i.c.v.). These include GRLN, galanin, NPY, AgRP, orexins, and β-endorphin (β-END) (Fig. 7.4). An exception is MCH, which is a potent orexigenic neuropeptide in mammals, but its i.c.v. injection suppressed food intake in the goldfish (Matsuda et al. 2006). However, barfin flounder (*Verasper moseri*) reared under white background had enhanced MCH production in the brain and increased food intake and growth rate compared with those kept under black background (Amiya et al. 2005; Sunuma et al. 2009), indicating a species difference. β-END and α-MSH are produced by alternative splicing of the same precursor, pro-opiomelanocortin (POMC), but the former is orexigenic and the latter anorectic through its MC4 receptor. AgRP exerts its orexigenic effect through its action as an endogenous receptor antagonist of α-MSH (Cerdá-Reverter and Peter 2003). Most orexigenic hormones listed above act on the brain as neuropeptides synthesized locally in the brain in fishes, but some are also synthesized in the GEP tissues and act on the brain through bloodstream delivery (Table 7.1; see Volkoff et al. 2009). The i.c.v. injection of orexigenic neuropeptides induced food intake and their receptor antagonists depressed normal appetite. In addition, food deprivation increased the expression of their genes and the amount of immunoreactive peptides in the brain. Some of the peptides such as NPY and AgRP are expressed in the lateral tuberal nucleus that is thought to be a teleost homologue of mammalian arcuate nucleus (Cerdá-Reverter and Peter 2003).

Among the orexigenic hormones, GRLN and galanin are produced in the stomach and intestine, respectively, and act on a brain area not protected by the blood–brain barrier, probably the lateral tuberal nucleus (Gorissen et al. 2006; Volkoff et al. 2009), although they are also synthesized locally in various brain areas including the appetite-regulating sites (Unniappan et al. 2002; Kaiya et al. 2008). The GRLN receptor is expressed abundantly in the brain of few teleost species (see Kaiya et al. 2008). NPY is a neuropeptide, but its fish paralogue PY (PYY) (Table 7.1) is a GEP hormone identified in some teleost species (Cerdá-Reverter et al. 2000). Since NPY and these gut hormones share a receptor and since the PYY gene expression was suppressed after feeding in the yellowtail, *Seriola quinqueradata* (Murashita et al. 2007), it is possible that PYY acts as a peripheral orexigenic hormone to mimic the NPY action in the brain.

4.3. Appetite-suppressing (Anorectic) Hormones

Many anorectic hormones and neuropeptides have been identified in fishes including leptin, CCK/gastrin, INS, gastrin-releasing peptide/bombesin, PACAP/VIP, CGRP/adrenomedullin (AM)/AMY, tachykinins, NMU, serotonin, α-MSH, CART, CRH, GnRH, and RFamides (Table 7.1; see Volkoff et al. 2009) (Fig. 7.4). Most of the hormones are synthesized briskly in the GEP system and can be designated as "brain-gut hormones" except for α-MSH, CART, CRH, and RFamides that seem to be genuine neuropeptides. Leptin is synthesized almost exclusively in the adipose tissue in mammals, but its gene is expressed most abundantly in the liver of carp (Huising et al. 2006) and rainbow trout (Murashita et al. 2008b). Initial studies using heterologous mammalian leptin in teleost fish are controversial probably because of low sequence identity between mammalian and teleost leptins (14–25%). However, more recent studies using homologous peptide showed that peripheral leptin injection suppressed food intake in the rainbow trout with concomitant decreases in the hypothalamic NPY gene expression and increases in the POMC (α-MSH) gene expression (Murashita et al. 2008b). In the carp, two leptin genes were identified, and fasting did not alter the leptin gene expression but the expression of both genes was up-regulated after re-feeding (Huising et al. 2006).

CCK and gastrin are paralogous GI peptides with a common C-terminal sequence in several teleosts and elasmobranchs (Johnsen et al. 1997; Peyon et al. 1998; Kurokawa et al. 2003). CCK is also expressed in the teleost brain and functions as a neuropeptide. Both central and peripheral injection of sulfated octapeptide CCK-8S, an active form of CCK, inhibit food intake, and i.c.v. injection of CCK antagonist enhanced food intake in a few teleost species (see Volkoff et al. 2009). Furthermore, fasting resulted in increased CCK gene expression in the brain and the GI tract. Interaction of CCK with other hormones has been suggested for appetite regulation. It seems that CCK is a major GI hormone that inhibits food intake in fishes.

Members of the secretin superfamily are principal GI peptides that are also identified in the brain, and thus underlie the origin of the term "brain-gut peptides." Among the superfamily members, GLUC and two GLPs (GLP-1 and -2) are localized on a single gene in teleost fish and expressed in the pancreas and alimentary tract (Navarro et al. 1999). The effect of GLUC on feeding has not been examined yet in fish, but GLP-1 was shown to have a potent anorectic effect in channel catfish, with i.c.v. injection being much more potent than intraperitoneal (i.p.) injection (Silverstein et al. 2001). However, i.c.v. injection of GLP-1 antiserum was without effect on food intake. VIP and PACAP are kin molecules that belong to the secretin superfamily (Sherwood et al. 2000). They are also known to be brain-gut peptides and influence brain function locally and through bloodstream delivery. Both i.p. and i.c.v. injection of VIP or PACAP suppressed food intake in the goldfish (Matsuda et al. 2005). As with GLP-1, higher doses are required for the i.p. effect than for the i.c.v. effect.

As will be discussed below, INS is a regulatory factor in glucose metabolism and is deeply involved in appetite regulation directly and via glucose metabolism in mammals and other tetrapods (Schwartz et al. 2000). However, teleost fish, particularly carnivorous fish, are glucose intolerant (see Section 5.1.1), and amino acids rather than glucose may be a major regulator of INS secretion in the fish such as barfin flounder (Andoh 2007). The effect of INS on food intake is controversial; both i.c.v. and i.p. injection inhibited long-term food intake in the rainbow trout (Soengas and Aldegunde 2004), but INS was without effect when injected i.c.v. in the channel catfish, *Ictalurus punctatus* (Silverstein and Plisetskaya 2000). IGF-1 and -2, members of the INS family, secreted from the liver increase somatic growth, which is accompanied by enhanced food intake. However, there is no evidence showing that IGFs act directly on the brain to modulate feeding in the fishes.

Gastrin-releasing peptide (GRP) is another important endogenous gut peptide that has high sequence identity with bombesin, a peptide isolated from skin of the frog, *Bombina bombina*. Bombesin is anorectic after i.c.v.

injection in the goldfish (Himick and Peter 1994), but plasma GRP concentrations are unaltered by feeding in the rainbow trout (Jonsson et al. 2006). The CGRP family consists of CGRPs, AMY and AMs in teleost fish (Table 7.1; Takei et al. 2010). AMs are multifunctional peptides synthesized in various tissues including the GEP system of fishes (Nobata et al. 2008). The i.c.v. injection of AM2, but not AM1 that is orthologous to mammalian AM, dose-dependently suppressed fasting-induced food intake in the goldfish (Martinez-Alvarez et al. 2009). AMY is synthesized and secreted from the pancreatic B-cells in response to a meal. AMY, injected i.p. or i.c.v. alone or in combination with CCK-8S, inhibits food intake in the goldfish with synergic interaction of the two peptides (Thavanathan and Volkoff 2006). Inhibition of food intake by endogenous AMY was suggested as both i.p. and i.c.v. AMY receptor blocker increased food intake. NMU was first identified in the porcine uterus (Minamino et al. 1985) and comprises a group of hormones together with neuromedin S. In the goldfish, the NMU gene was expressed as four splice variants and their transcripts are most abundant in the GI tract followed by the brain (Maruyama et al. 2008). Synthetic NMU-21 injected i.c.v. inhibited food intake and locomotor activity in a dose-dependent manner in the goldfish. It is not known whether endogenous NMU secreted from the gut exerts an inhibitory effect.

Other hormones that are involved in feeding regulation are RFamides that are divided roughly into five groups, prolactin-releasing peptide (PrRP) group, kisspeptin group, QFRP (26RFa) group, LPXRFa group, and PQRFa (NPFF) group based on the peptides isolated from mammals (Bechtold and Luckman 2007). The interest in this peptide family is focused on its central actions, so that their production sites have been examined only in the brain. Thus it is not known whether these peptide genes are expressed in the GEP system. In the goldfish, PrRP caused dose-dependent decreases in food intake by both i.c.v. and i.p. injections (Kelly and Peter 2006). The tachykinins comprise more than 40 peptides isolated from vertebrates and invertebrates including substance P, neurokinin A, neuropeptide K and neuropeptide γ which are coded by a single gene. The tachykinin gene expression and the peptides are detected in the CNS, skin, and GI tracts (mostly in the nerve terminals), and novel types, named carassin and scyliorhinin, were identified in fishes (Conlon and Larhammer 2005). Apparent postprandial increases in tachykinin gene expression have been reported in the hypothalamus of goldfish, suggesting a possible involvement in feeding regulation (Peyon et al. 2000), but the effect of tachykinin injection has not yet been examined. Another hormone involved in feeding regulation is serotonin that is secreted locally in the brain or from the enterochromaffin cells of the intestine (Leibowitz and Alexander 1998). In the goldfish, i.c.v. injection of serotonin was anorectic for the initial 2 h in a dose-dependent fashion, but i.p. injection was without effect (De Pedro et al. 1998). The inhibitory effect of central serotonin was partially mediated by inhibition of the hypothalamic CRH system. The anorectic action of hyperglycemia and i.p. INS is correlated with an increase in hypothalamic serotonin level in the rainbow trout (Ruibal et al. 2002).

5. METABOLISM AND GROWTH

5.1. Metabolism

GEP series hormones probably have far more effects on energy metabolism than have been described to date. Hormonal effects on metabolism may be direct on nutrient transport by cells and on metabolically important enzymes. Or they may be indirect by way of their actions on the synthesis and secretion of other chemical messengers with metabolic effects, or by effects on feeding to trigger nutrient-responsive chemical signaling (Buddington and Krogdahl 2004).

5.1.1. GLUCOSE HOMEOSTASIS

Whereas a similar panel of regulatory hormones is recognized among fishes and mammals, there are some clear differences in glucose homeostatic patterns. A combination of dietary, inherent metabolic and endocrine regulatory differences may underlie the relative glucose intolerance that is reported in teleosts (Moon 2001, 2004). In fishes, glucose may play a smaller metabolic role than it does in other vertebrates; and this may strongly be the case in carnivorous fishes that display a tendency for hyperglycemia following delivery of a glucose load. Teleosts generally exhibit lesser control over blood glucose concentrations, and are more tolerant to bouts of hypoglycemia (Wright et al. 2000). Teleosts are glucose intolerant compared with mammals, even though carbohydrate and amino acid nutrients stimulate insulin secretion (Mommsen and Plisetskaya 1991) and, more specifically, pancreatic islet B-cells appropriately secrete INS in response to elevations in blood glucose (Wright et al. 2000; Moon 2001). Since glucose intolerance appears not to result from INS deficiency, lower peripheral utilization relative to other vertebrates may be the causal factor. Earlier studies on several different teleost species indicated that peripheral glucose uptake and utilization by skeletal muscle was not supported in fishes by the expression of a mammalian GLUT4-type INS-sensitive glucose transporter (Wright et al. 1998, 2000; Legate et al. 2001). The negative findings of these studies may have resulted from the application of heterologous mammalian oligonucleotide probes and antisera to fish systems (Alexander et al. 2006).

More recently, using fish-specific molecular biological reagents, GLUT4 homologues have been cloned and sequenced from brown trout (*Salmo trutta*) and coho salmon (*Oncorhynchus kisutch*), and have been demonstrated to express glucose transporting functional activity (Capilla et al. 2002, 2004; Alexander et al. 2006). Moreover, in brown trout, GLUT4 expression in red muscle was stimulated by INS, whereas expression in white muscle was not (Capilla et al. 2002), and INS stimulated both GLUT4 translocation to the plasma membrane and glucose uptake by cells (Capilla et al. 2004). The biological actions of INS on fish liver are incompletely known and deserve additional study (Moon 2001, 2004).

The physiological activity of GLP-1 in fishes is hyperglycemic, in contrast to its activity in mammals (Mojsov 2000; Mommsen 2000). In fishes, GLUC and GLP-1 both oppose the actions of INS, with GLP-1 being more potent than GLUC; at the liver GLUC and GLP-1 stimulate glycogenolysis, gluconeogenesis and lipolysis (Moon 1998, 2004; Mojsov 2000; Mommsen 2000). In enterocytes, GLP-1 and GLUC stimulate glucose uptake into the blood from the intestinal lumen and reduce glucose oxidation (Mommsen 2000). Basic amino acids stimulate, and glucose inhibits, the secretion of GLUC (Moon 2004).

SST is hyperglycemic; it stimulates glycogenolysis in lamprey and teleosts (Sheridan et al. 2000; Sheridan and Kittilson 2004; Nelson and Sheridan 2006). In addition to direct effects on cell metabolism, SST has indirect effects through inhibitory influences on the secretion of other hormones, notably pancreatic INS secretion (Sheridan and Kittilson 2004).

5.1.2. LIPID MOBILIZATION

INS promotes fat storage, especially in liver and adipose tissue, through both lipid synthesis-stimulating and anti-lipolytic actions (Nelson and Sheridan 2006). Opposing the effects of INS, SST stimulates lipid mobilization (Sheridan et al. 2000; Sheridan and Kittilson 2004).

5.2. Growth

Growth in fishes depends on adequate nutrient supply, so it is not surprising that the control of growth processes in fishes involves endocrine signaling from the gut. The principal regulator of somatic growth in fishes is the growth hormone (GH)/IGF-I axis; specifically, IGF-I levels are directly correlated with growth rate (Dickhoff et al. 1997; Boeuf and Payan 2001; Company et al. 2001; Wood et al. 2005). IGF-I promotes growth through its stimulatory action on cartilage proliferation (Duan 1997; Datuin et al. 2001). Responsiveness of the liver to GH can be altered by nutritional status and other factors (Wood et al. 2005). In salmonids and Mozambique tilapia

(*Oreochromis mossambicus*), for example, nutritional restriction reduces the responsiveness of the liver to GH, despite steady or elevated circulating levels of GH (Dickhoff et al. 1997; Datuin et al. 2001; Uchida et al. 2003).

6. OSMOREGULATION

As the gills are the site where extracellular fluids most closely contact environmental water of varying salinities, this organ has been recognized as a primary osmoregulatory site for aquatic fishes. However, the gut is also a critical site for osmoregulation in teleost fishes, particularly for seawater-adapted (SW) fishes. The role of the gut in osmoregulation is detailed in Chapter 4 (Grosell 2010).

6.1. Drinking

As oral intake of SW into the alimentary tract is the first step for marine teleosts to counter dehydration in the hypertonic media, regulation of drinking is important for body fluid homeostasis in fishes, particularly in euryhaline species that change drinking rate according to the environmental salinities. Hormones play important roles for regulation of drinking in teleosts (Takei 2002) and elasmobranchs (Anderson et al. 2001). Generally, vasodepressor substances are dipsogenic and vasopressor substances are anti-dipsogenic (Hirano and Hasegawa 1981; Kozaka et al. 2003), but there are several exceptions such as dipsogenic angiotensin II (ANG II) and anti-dipsogenic ANP. In spite of the importance of drinking, inhibitory hormones are generally more predominant for regulation in fishes in contrast to the dominance of dipsogenic hormones in mammals (Takei 2002). This may be due to the differences in habitat; aquatic fishes can drink environmental water anytime and thus are always subject to the crisis of excess drinking, while mammals need to seek water, motivated by the sensation of thirst before oral drinking. Therefore, thirst is ensured by multiple mechanisms including hormones in mammals (Fitzsimons 1979). There are other interesting differences between fishes and tetrapods in the mechanisms regulating drinking: (1) volemic stimulus (hypovolemia) is a primary regulator of drinking in fishes, while osmotic stimulus (hyper-osmoremia) is primary in tetrapods; and (2) drinking is regulated at the hindbrain probably via reflex swallowing in fishes, while it is regulated by the thirst center in the hypothalamus (forebrain) in tetrapods. These and other regulatory mechanisms of drinking in fishes, particularly involving neuroendocrine control, have been described in detail in a previous volume of this series (Takei and Balment 2008).

6.1.1. Hormones that Stimulate Drinking

ANG II is known as a potent dipsogen in all vertebrate species except cyclostomes, hagfishes and lampreys, in which no appreciable enhancement of drinking was observed after i.p. injections (Kobayashi et al. 1979, 1983; Rankin 2002). The dipsogenic effect is more potent by i.c.v. injection than peripheral injections, suggesting that ANG II action is on the brain (Kozaka et al. 2003). The site of action of ANG II may be in the area postrema (AP), a circumventricular structure in the medulla oblongata that lacks a functional blood–brain barrier. This idea originated from data showing in eel that following removal of the entire forebrain and midbrain ("decerebrated eel") peripheral injection of ANG II still stimulated drinking (Takei et al. 1979). Similarly, i.c.v. injection of ANG II into the fourth ventricle near the AP is highly effective in eliciting drinking (Kozaka et al. 2003). After peripheral injection of ANG II, a transient burst of drinking is followed by a prolonged inhibition of drinking for ~ 15 min, but such delayed inhibition was not observed after i.c.v. injection. The inhibition may be secondary to the increased arterial pressure (stimulation of sympathetic system) or by some other hormone(s) whose secretion is stimulated by ANG II. The minimum effective dose for the dipsogenic effect of ANG II is generally much higher in fishes (~ 100-fold higher) than in mammals and birds, reflecting the dominance of anti-dipsogenic mechanisms in fishes (Takei 2002).

Another dipsogenic hormone in teleost fishes is AM (Ogoshi et al. 2006). Among AMs, AM2 and AM5 increased drinking rate in the eel when injected peripherally, but AM1, an orthologue of mammalian AM, was without effect on drinking (Ogoshi et al. 2008). The i.c.v. injections into the third or fourth ventricle failed to induce drinking, though ANG II was effective in the same fish. As AMs are much larger peptides than ANG II, they may not be able to cross the cerebrospinal fluid (CSF)–brain barrier to act on the neurons in the parenchyma. In mammals, however, AM and AM2 are weakly anti-dipsogenic in the rat following i.c.v. delivery (Taylor et al. 2005). The dipsogenic effect is not caused by the induced hypotension because slow infusion of non-vasodepressor doses of AM2 and AM5 into the circulation increased drinking rate in a dose-dependent manner. The minimum effective dose was comparable to that of ANG II and the delayed inhibition of drinking observed after ANG II treatment did not occur after AM administration. It was reported that substance P (SP) and CCK are weakly dipsogenic when injected peripherally in the eel (Kozaka et al. 2003).

6.1.2. Hormones that Inhibit Drinking

ANP is the most potent anti-dipsogenic hormone thus far known when injected into the circulation in the eel (Tsuchida and Takei 1998). The inhibitory effect was dose-dependent and >100-fold more potent in eel than in the rat. Endogenous ANP appears to be involved in the regulation of drinking in the eel since removal of circulating ANP by anti-serum infusion increased drinking and increased plasma osmolality in SW eel (Tsukada and Takei 2006). Among other cardiac NPs, VNP was as potent as ANP, but BNP was less potent and CNPs were without appreciable effect on drinking in the eel (H. Miyanishi et al. unpublished data). ANP in blood may act on the AP since ANP receptors (NPR-A) exist there, and electric and chemical lesioning of AP neurons abolished the anti-dipsogenic effect of ANP in eels (Tsukada et al. 2007).

GRLN secreted from the stomach is a potent anti-dipsogenic hormone when injected both centrally and peripherally in the eel (Kozaka et al. 2003). A detailed analysis showed that GRLN is as potent as ANP for inhibition of drinking, and it also acts on the AP to exert its effect (H. Kaiya et al. unpublished data). Recently, GRLN was found to be a potent anti-dipsogen also in mammals even though food intake was increased and prandial drinking was stimulated (Hashimoto et al. 2007). Although it is dipsogenic in mammals, bradykinin inhibited drinking in the eel when infused into the circulation at non-hemodynamic doses (Takei et al. 2001). Other inhibitory hormones include arginine vasotocin, VIP, SST and prolactin (PRL) in the eel (Kozaka et al. 2003).

6.2. Ion and Water Absorption

Two possible mechanisms are in operation for hormonal regulation of ion and water absorption in the alimentary tracts. One involves *de novo* synthesis of transporters/channels and cell adhesion molecules by protein or steroid hormones (long-term regulation), and the other involves immediate regulation of existing transporting molecules by oligopeptide hormones (short-term regulation) after changes in environmental salinity. Detection of salinity (or osmotic pressure) changes initiates signaling that leads to changes in transporter activity or expression of new transporter genes. Possible osmoreceptors have been suggested such as members of the transient receptor potential vanilloid family of ion channels (TRPV1, 2 and 4) (Sharif-Naeini et al. 2008) and calcium-sensing receptor (CaSR) (Nearing et al. 2002; Hebert 2004; Loretz 2008) as well as osmoresponsive transcription factors such as osmotic stress transcription factor (Ostf1) (Fiol and Kurtz 2007) and tonicity-responsive enhancer binding protein (TonEBP) (Jeon et al. 2006). These transcription factors may regulate the expression of transporter genes differentially in response to osmotic changes. However, information transfer from the sensor to transcription factors and to gene expression has not been fully clarified yet. In addition, detection in fishes by volume or stretch receptors may be involved since ion and water movements are essentially inseparable in an aquatic environment. Volume sensing and regulation of extracellular fluid volume have been extensively reviewed by Olson (1992).

6.2.1. LONG-TERM REGULATION

The ability of the intestine to absorb water increases during parr-smolt transformation in salmonid fishes (Collie and Bern 1982; Usher et al. 1991). In the Atlantic salmon, smolting fish underwent a decrease of fluid transport in the middle intestine and an increase in the posterior intestine. The increase in the posterior intestine was twice greater in SW smolts than in freshwater-adapted (FW) ones (Veillette et al. 1993), although transfer of the juveniles from FW to SW did not alter fluid transport (Usher et al. 1991). The thyroid hormones T3 and T4 induce changes in body color and shape, but smoltification is a complex process that involves other physiological and histological changes. Supporting enhanced salinity tolerance, GH/IGF-I and cortisol play an important role (McCormick 2001; Bjornsson and Bradley 2007). These osmoregulatory hormones may change the function and morphology of the intestine during smoltification.

GH is known as a "seawater-adapting" hormone in salmonids, and it also plays a part in the intestinal function through its direct actions on the synthesis of transporters and intercellular matrices (Sakamoto et al. 1993; Bjornsson 1997). GH secreted from the pituitary acts directly on the pyloric regions of intestine (Koppang et al. 1998) through GH receptors in these tissues (Sakamoto and Hirano 1991) or through increased IGF-I secretion from the liver. However, IGF-I given alone or in combination with GH decreased Na^+,K^+-ATPase activity in the pyloric ceca and middle intestine of brown trout, although both GH and IGF enhanced hyposmoregulatory performance (Seidelin et al. 1999). GH increases the length of intestine, microvillus height and surface area of its epithelia, and amino acid absorption in several teleost species. For instance, GH increased intestinal proline transport in pyloric ceca in the yearing coho salmon (Collie and Stevens 1985), leucine absorption through Na^+-coupled and Na^+-uncoupled transport in the striped bass (genus *Morone*) hybrid (Sun and Farmanfarmaian 1992), and leucine absorption after long oral administration in the goldfish (Walker et al. 2004). It is important to determine which proteins and genes are up-regulated by GH/IGF-I for intestinal ion and water transport.

Cortisol is another important seawater-adapting hormone in teleost fish. The first report showed that water and Na^+ transport in isolated intestine of the eel is dose-dependently augmented by ACTH and by cortisol (Hirano and Utida 1968). In salmonids, cortisol improved the ability to regulate plasma osmolality with increased Na^+,K^+-ATPase activity in the gills, but it failed to increase the enzyme activity in the intestine of Atlantic salmon (Bisbal and Specker 1991). However, cortisol increased water transport in the posterior intestine of parr and postsmolt Atlantic salmon juveniles when plasma cortisol level is low, and the inhibitor RU486 decreased water transport in the smolt when plasma cortisol is high (Veillette et al. 1995). Glucocorticoid receptor and its gene expression have been documented in the intestine of Mozambique tilapia (Takahashi et al. 2006b) and sea bass (Vazzana et al. 2008).

As described in Chapter 4, water absorption is achieved by transcellular water movement through water channels (aquaporins), which is driven by ion fluxes principally through apical NKCC2 transporter and basolateral Na^+, K^+-ATPase, and by paracellular water flow through intercellular matrices such as occludin (Grosell 2010). NKCC2 is critically important for water absorption and inhibition of this transporter compromises the absorptive process (Loretz 1995). Sundell et al. (2003) showed that intestinal Na^+,K^+-ATPase activity increased and paracellular permeability decreased after seawater transfer of juvenile Atlantic salmon with concomitant increases in plasma cortisol level. Thus, cortisol seems to increase transcellular water movement by enzyme activation while decreasing paracellular movement. Cortisol addition to medium maintained the Na^+,K^+-ATPase activity in the explants of pyloric ceca and posterior intestine above the level in control explants in sockeye salmon (*Oncorhynchus nerka*) as observed *in vivo* after cortisol implantation (Veillette and Young 2005). In the eel, three aquaporins (AQP1, AQP3 and an aquaglyceroporin, AQPe) were identified in the intestine, of which the AQP1 gene expression was profoundly up-regulated after SW transfer (Aoki et al. 2002; Lignot et al. 2002; Martinez et al. 2005). AQP1 proteins are localized at the apical brush border membrane of intestinal epithelial cells, while AQP3 was not detected on the absorptive epithelial cells but rather on the mucus-secreting goblet cells and macrophage-like cells. Since cellular localization of AQPe has not been reported, it is not yet known which AQP is responsible for water transport at the basolateral membrane of intestinal epithelial cells. Cortisol increases the AQP1 gene expression in the eel intestine but not AQPe (Martinez et al. 2005). In the gilthead seabream (*Sparus aurata*), two types of AQP1 are identified and one type is localized at both apical and lateral membrane of epithelial cells of duodenum and hindgut (Raldua et al. 2008).

Another candidate hormone that may change transporter gene expression and intestinal water and ion transport is PRL, which is considered as a "freshwater-adapting" hormone (Marshall and Grosell 2005). In fact, PRL receptors are identified in the intestine of a few teleost species (Sandra et al. 2001; Huang et al. 2007; Pierce et al. 2007; An et al. 2008), and the receptor's gene expression is regulated by cortisol and osmotic challenge. In the euryhaline Japanese flounder (*Paralichthys olivaceus*), PRL gene expression in the pituitary paralleled PRL receptor gene expression in the intestine (An et al. 2008). However, how PRL influences transport processes is not known yet.

The "long-acting" hormones such as cortisol, GH/IGF-I and PRL may also be involved in the morphogenesis of the alimentary tract. Transfer of euryhaline teleost fish from freshwater to seawater induces marked changes in the intestinal morphology in rainbow trout (MacLeod 1978) and in glass eels (*Anguilla anguilla*) (Ciccotti et al. 1993); decreases in mucus-secreting goblet cells and increases in cross-sectional area and in villar height of absorptive epithelia were reported. Dramatic changes were also observed in Japanese eel esophagus (Yamamoto and Hirano 1978; Meister et al. 1983) where absorptive activity of Na^+ and Cl^- was profoundly augmented after seawater adaptation (Hirano and Mayer-Gostan 1976). During the course of seawater adaptation, stratified epithelium was changed to simple form and blood vessels developed just beneath the epithelium, which corresponds well to increased absorptive activity. Similar changes in esophageal morphology are reported in tilapia (Cataldi et al. 1988). PRL and its receptor appear to be involved in active cell proliferation of esophageal epithelia of Mozambique tilapia (Takahashi et al. 2007), while cortisol and its receptor may be involved in apoptosis and cell proliferation of epithelial cells (Takahashi et al. 2006b). Similar actions of PRL and cortisol have been reported in the anterior intestine including esophagus in the euryhaline mudskipper, *Periophthalmus modestus* (Takahashi et al. 2006a). It is reasonable to speculate that cortisol remodeled the multilayered epithelium of esophagus into a monolayer by apoptosis and expanded the luminal surface by cell proliferation after seawater transfer. 11-Deoxycorticosterone, the inferred ligand of the mineralcorticoid receptor, was without effect.

6.2.2. SHORT-TERM REGULATION

Oligopeptide hormones and amines are known to regulate through phosphorylation and dephosphorylation the activity of transport proteins in the intestine immediately after changes in environmental salinity. These peptide hormones are secreted locally from osmosensitive endocrine cells in the alimentary tract and act in a paracrine fashion on neighboring epithelial cells. Alternatively, chemical messengers can be secreted from nerve terminals innervating the epithelial cells, with the osmosensing function located in the central nervous system. They are also secreted from tissue outside the alimentary tract and act on the epithelial cells in a traditional endocrine fashion. As mentioned above, various transporters exist on the apical and basolateral membrane of enteric epithelial cells of teleost fishes.

Several hormones stimulate, or more appropriately release the inhibition of, ion and water absorption in the SW teleost intestine (Ando et al. 2003). These include SSTs (Uesaka et al. 1994), NPY (Uesaka et al. 1996), urotensin II (Mainoya and Bern 1984) and adrenaline (epinephrine) and other catecholamines (Ando and Hara 1994). On the other hand, hormones that inhibit ion and water absorption include NPs (O'Grady et al. 1985; Ando et al. 1992; Loretz and Takei 1997), GNs (Yuge and Takei 2007), VIP (Mainoya and Bern 1984; O'Grady 1989), urotensin I (Loretz et al. 1981), acetylcholine, serotonin and histamine (Mori and Ando 1991). These peptide hormones or transmitters are secreted soon after environmental changes and disappear quickly because of short half life. Therefore, they are a form of emergency hormone for euryhaline teleosts to cope with abrupt changes in environmental salinities by countering a sudden increase in plasma osmolality and NaCl concentration.

O'Grady et al. (1985) first reported that ANP and its intracellular messenger, cGMP, potently inhibited transepithelial fluxes of Na^+ and Cl^- in the intestinal mucosa of winter flounder (*Pseudopleuronectes americanus*) using newly isolated truncated forms of rat ANP (atriopeptins; Geller et al. 1984). The inhibition of NaCl transport is most likely through NKCC2 because ANP also inhibited unidirectional Rb^+ (a substitute for K^+) influx. Ando et al. (1992) further showed, using homologous eel peptide, that ANP is the most potent and efficacious inhibitor for ion and water absorption thus far known in the eel intestine. In fact, the ANP effect was two to three orders of magnitude more potent than other inhibitory substances such as serotonin, acetylcholine and histamine. Eel ANP was >100-fold more potent than rat ANP that has 60% sequence identity. The effect of ANP was comparable in the FW and SW eel intestine (Ando and Hara 1994), which is consistent with the hypothesis that ANP is secreted upon migration of FW eels to seawater and inhibits NaCl absorption from ingested seawater to ameliorate excess increase in plasma NaCl concentration (Kaiya and Takei 1996). In addition to circulating ANP secreted from the heart, ANP is synthesized locally in the intestinal epithelial cells of FW and SW eels (Loretz et al. 1997), and may act to inhibit NaCl absorption in a paracrine fashion following seawater drinking and/or food intake. It is suggested that ANP inhibits Na^+ absorption and accelerates Na^+-linked nutrient absorption (Loretz and Takei 1997).

VIP is another hormone that inhibits NaCl absorption by acting on the basolateral (blood) side of the epithelial cells (Mainoya and Bern 1984; O'Grady 1989). As mentioned above, VIP uses cAMP as a second messenger to express its function (Laburthe et al. 2007). The important roles of cAMP and cGMP in intestinal ion and water absorption have been known for a long time in the flounder (Rao et al. 1984) and the eel (Trischitta et al. 1996);

cGMP is more potent and efficacious for the inhibition than cAMP but the effects are not additive. Consistently, the effect of VIP is less potent and less efficacious than ANP (Trischitta et al. 1996; Ando et al. 2003). The mechanisms of action of cAMP and cGMP appear to differ, because cAMP, but not cGMP, abolished Na^+ selectivity of the paracellular transport of teleost intestine and increased Cl^- permeability (Rao et al. 1984).

GN is a unique intestinal hormone and was first identified as an endogenous ligand of GC-C that had been known as a receptor for heat-stable enterotoxin secreted from enteric bacteria (Schulz et al. 1990). Therefore, the GC-C is located on the apical (luminal) membrane, and after binding of the toxin increases intracellular cGMP, which activates CFTR Cl^- channel and strongly stimulates Cl^- secretion into the lumen, resulting in severe diarrhea. Therefore, GN is secreted into the intestinal lumen and acts to cause water secretion in mammals, which is certainly disadvantageous for seawater adaptation by fishes. However, expression of guanylin family genes (GN, UGN and RGN) and their receptors (GC-C1 and GC-C2) are consistently up-regulated in SW eels compared with FW eels (Yuge et al. 2003, 2006). GN and GC-C gene expressions were much higher in the anterior intestine than in the posterior intestine and particularly increased in the anterior part after seawater adaptation, suggesting its local action on the anterior segment of eel intestine. Immunoreactive GN was localized in the goblet cells, from which GN may be secreted into the lumen with mucus (Yuge et al. 2003). GNs profoundly inhibited ion absorption *in vitro* when applied only on the apical side of the intestinal epithelium (Yuge and Takei 2007). The effect of GN was more potent in the anterior intestine than in the posterior intestine, further supporting its action on the anterior segment. The inhibitory effect of GN was abolished by a CFTR inhibitor, suggesting similar mechanisms of action occur in the eel as in mammals. In fact, luminal Cl^- concentration does not decrease but Na^+ concentration is profoundly decreased along the length of the intestine; active functioning of NKCC should decrease Cl^- more than Na^+, suggesting the presence of a Cl^- secretory mechanism (Tsukada and Takei 2006). Our recent data showed that Cl^- secretion actually occurs into the lumen of the intestinal sac after GN administration, and that HCO_3^- secretion also occurs into the luminal fluid as determined using the pH-stat method (Ando, M. and Takei, Y. unpublished data).

Based on these data, the following hypothesis was suggested (Fig. 7.5). Since Na^+ and Cl^- are present in similar concentrations in SW, active transport of ions via NKCC2 causes a deficit of Cl^- ions in the posterior intestine where active water absorption occurs. GNs are synthesized and secreted abundantly in the anterior intestine and act on the receptors there which exhibit higher affinity than those in the posterior intestine.

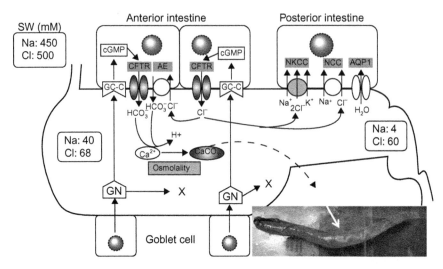

Fig. 7.5. Possible mechanisms of guanylin (GN) action on water absorption in eel intestine. GN is mostly secreted in the anterior intestine to stimulate CFTR and to increase Cl^-, and possibly HCO_3^-, secretion into the lumen. The secreted Cl^- was used to facilitate Na-K-2Cl cotransporter (NKCC2) and Na-Cl cotransporter (NCC), resulting in water absorption via aquaporin 1 (AQP1). Luminal HCO_3^- secreted by anion exchanger (AE) and CFTR was used for $CaCO_3$ precipitation, which decreases luminal fluid osmolality and further facilitates water absorption. Inset shows SW eel intestine with developed vasculature and white precipitates of $CaCO_3$ in the lumen that are visible because of thin intestinal epithelium of SW-adapted eels. Na^+ and Cl^- concentrations of luminal fluid at each segment of intestine are shown. For more details and other abbreviations, see text. See color plate section.

The resulting secretion of Cl^- into the lumen delivers the ion to support NKCC2 function in the posterior intestine (Fig. 7.5). GNs also increase HCO_3^- secretion, which may facilitate precipitation of concentrated Ca^{2+} as $CaCO_3$. Since a major route for HCO_3^- secretion appears to be an anion exchanger in marine teleosts (Grosell 2006), it remains to be determined whether the increased HCO_3^- secretion is via CFTR or facilitated Cl^-/HCO_3^- exchange. HCO_3^- secretion via anion channels is thermodynamically challenged by the high luminal HCO_3^- concentration (80 mM) compared with the intracellular concentration of HCO_3^- in epithelial cells (1–2 mM), but the luminal concentration may be lowered in the unstirred layer at the apical surface because of active H^+ secretion by V-ATPase (Grosell et al. 2009). The precipitation decreases luminal osmotic pressure and further facilitates water absorption in the posterior segment (Fig. 7.5). In fact, white precipitates are observed in the SW eel intestine. Secreted GNs may be digested in the lumen before reaching the posterior part where GN and GC-C syntheses are much reduced and GC-Cs have lower affinity compared with the anterior intestine. In this way, GNs, which cause water secretion in mammals, may increase water absorption and promote seawater adaptation in the eel, although the mechanism of action (luminal Cl^- secretion) is the same for mammals and teleost fish.

NPs and GNs utilize the same intracellular messenger, cGMP, to exert their actions in the intestine, but NPs act on the basolateral side and GNs on the apical side through GC-A and GC-B for NPs and through GC-C for GNs. Since the intestinal lumen is technically an external space, luminocrine secretion, as a form of exocrine secretion, may represent a primitive form of chemical signal transmission compared with internal endocrine signaling. Reflecting the difference in mode of action, the potency of ANP applied on the basolateral side for inhibition of ion transport is >100-fold greater than that of GN applied on the apical side of intestinal epithelium of SW eels (Loretz and Takei 1997; Yuge and Takei 2007). Interestingly, the affinity of GNs to eel GC-Cs is 2–3 orders of magnitude greater than to rat GC-C, and eel GC-Cs have little affinity to enterotoxin although rat GC-C has much higher affinity to enterotoxin than to GNs (Yuge et al. 2006). This may suggest that GNs play more important roles for intestinal osmoregulation in teleost fish compared with that in mammals. In contrast to the inhibitory hormones, the mechanisms of action of facilitative hormones, SST and NPY, on ion and water transport have not yet been investigated.

In addition to monovalent ions such as Na^+ and Cl^-, divalent ions such as Ca^{2+}, Mg^{2+} and SO_4^{2-} are under deliberate control in fishes. However, hormonal control has been investigated only for Ca^{2+} regulation. In teleost fish, calcitonin does not seem to be a major hypocalcemic hormone but stanniocalcin (STC) synthesized in the unique fish organ, the corpuscles of Stannius, takes its role. Salmon STC decreased Ca^{2+} influx in the isolated, perfused intestine of cod, *Gadus morhua* (Sundell et al. 1992). A second STC, named STC 2, was later identified in teleosts and mammals. Since fishes lack a parathyroid gland, parathyroid hormone (PTH) and its paralogue, parathyroid hormone-related peptide (PTHrP), were thought to be absent in fishes. However, five paralogous peptides, PTHA and B, PTHrPA and B, and PTH-like peptide (PTHL), were identified and were shown to form a family in teleosts (Guerreiro et al. 2007). In teleost fish, PTHrP and PTHL have hypocalcemic effects but PTH is without effect, which differs from mammals in which PTH and PTHrP increase plasma Ca^{2+} levels. In seabream, about 60% of Ca^{2+} is taken up from the intestine, and PTHrP increases whole-body Ca^{2+} content (Guerreiro et al. 2001). An *in vitro* study using intestinal epithelia of seabream further showed that $^{45}Ca^{2+}$ influx was enhanced by serosal addition of PTHrP (Fuentes et al. 2006).

6.3. Rectal Gland Secretion

The rectal gland is a small glandular organ formed as a diverticulum of the distal intestine (Silva et al. 1990b). Although the rectal gland shares its origin with the intestinal tissue, its epithelial cells have secretory function as opposed to the absorptive function of the intestine (Loretz 1987). The secreted solution is largely concentrated Na^+ and Cl^- (~ 500 mM), which is two- to three-fold higher than the concentration in plasma, while urea concentration (18 mM) is about 20-fold less than that in plasma because of much reduced permeability to urea (Zeidel et al. 2005). To concentrate Na^+ and Cl^- in the lumen of the gland, the epithelial cells possess a suite of transporters that are identified in another ion-secretory cell type, the mitochondria-rich cells of the gills; Na^+,K^+-ATPase and secretory-type Na-K-2Cl cotransporter (NKCC1) on the basolateral membrane, and CFTR Cl^- channel on the apical membrane (Marshall and Grosell 2005). In the shark rectal gland, a member of the FXYD domain-containing proteins (FXYD10) binds to the α-subunit of Na^+,K^+-ATPase and probably regulates enzyme activity (Mahmmoud et al. 2005). The CaSR is suggested to act in the rectal gland as a salinity sensor to regulate its secretion (Nearing et al. 2002). Several humoral factors have been implicated in the regulation of rectal gland function. These include stimulatory VIP, scyliorhinin II (one of the tachykinins), NPs, prostaglandins, adenosine, inhibitory NPY, bombesin, SST (Silva et al. 1990a), endothelin, and norepinephrine (Evans 2001). The mode of action of hormones on rectal gland secretion is diverse; acting directly on the epithelium to regulate transporter activities, acting on the arteries supplying blood to the gland (rectal gland artery or posterior mesenteric artery) to regulate blood perfusion of the epithelial cells (Anderson et al. 2002a), and/or acting on the smooth muscle that circumscribes the sub-capsular region to induce contraction of the rectal gland itself for excretion of the luminal fluid (Evans and Peirmarini 2001). Endocrine control of elasmobranch osmoregulation including rectal gland function has been reviewed by Hazon et al. (2003).

VIP was the first hormone shown to be a rectal gland secretagogue in the dogfish shark (*Squalus acanthias*) (Stoff et al. 1979). VIP was immunohis-tochemically detected in nerve terminals innervating the gland, suggesting neural regulation (Stoff et al. 1988). VIP acts directly on the epithelial cells of the gland via intracellular accumulation of cAMP, Ca^{2+} and/or IP3. Since it increased Cl^- conductance in primary monolayer culture of rectal gland cells, the effect is directly on the epithelium (Valentich and Forrest 1991). More recently, a VIP receptor has been cloned from the rectal gland of dogfish, which exhibits affinity to other members of the secretin family of peptides in the order VIP > GHRH=PHI > PACAP > SCT, indicating that other members also have secretagogue action (Bewley et al. 2006). Administration of VIP to *Xenopus* oocytes co-expressing the cloned VIP receptor and CFTR activates Cl^- conductance. It was shown that VIP, acting via cAMP, stimulates insertion of the channel protein in the intracellular vesicles to the apical membrane of rectal gland epithelial cells (Lehrich et al. 1998), but its potent vasodilatory action may also contribute to the increased Cl^- secretion. VIP has no stimulatory effect on Cl^- secretion of the rectal gland in another dogfish species, *Scyliorhinus canicula*, which led to an isolation of new endogenous secretagogue from the rectal gland, named scyliorhinin II (Shuttleworth and Thorndyke 1984; Anderson et al. 1995). The cyclo-oxygenase that is responsible for prostaglandin production exists in the rectal gland (Yang et al. 2002), and an enzyme blocker attenuates VIP action on Cl^- secretion, probably because the vasodilatory action of VIP is mediated in part by prostaglandin (Evans 2001).

Of the CNP group of hormones, elasmobranchs express only CNP3 in the heart and brain, and significant amounts of that peptide are stored in the different segments of digestive tract (Suzuki et al. 1994; Inoue et al. 2003a). The CNP receptor (natriuretic peptide receptor B, NPR-B) was cloned from the dogfish rectal gland; it is stimulated by CNP to increase Cl^- secretion from *Xenopus* oocytes co-expressing CFTR (Aller et al. 1999). Pioneering studies showed that shark heart extract and mammalian ANP stimulate Cl^- secretion from the rectal gland *in vivo* with concomitant hypotension in dogfish (Solomon et al. 1985a). Both are effective stimulators *in vitro* of Cl^- secretion and vasodilatation in the perfused gland. Later, the ANP action was suggested to be mediated by VIP release from the nerve terminal, because blockade of VIP release impaired the ANP action and cGMP failed to stimulate Cl^- secretion in the perfused rectal gland (Silva et al. 1987). However, this notion was challenged by data showing that ANP was effective on Cl^- secretion in monolayer culture of rectal gland cells free from innervation (Karnaky et al. 1991). Therefore, two different mechanisms, direct action on the epithelial cells via cGMP and indirect action via VIP and cAMP, may underlie CNP stimulation of Cl^- secretion in vivo (Silva et al. 1999). Electrophysiological studies on the isolated rectal gland tubule showed that CNP administered on the basolateral side increased short circuit current with concomitant decreases in transepithelial resistance, suggesting facilitated Cl^- conductance by CFTR (Greger et al. 1999). VIP, adenosine, and cAMP had similar effects, but mammalian GN had no effect when applied to either apical or basolateral sides.

It is reported that the primary factor regulating rectal gland secretion is not an osmotic stimulus but rather a volemic one (Solomon et al. 1985b), which relates well to the fact that the primary stimulus for ANP release from atria is the stretch of atrial myocytes induced by increased blood volume in mammals (Ruskoaho 1992) and rainbow trout (Cousins and Farrell 1996). Thus, it is reasonable to speculate that increased blood volume triggers CNP secretion from the heart and stimulates rectal gland secretion. In fact, rectal glands isolated from volume-expanded *S. canicula* in 70% seawater had higher secretion rates compared with those from control seawater-acclimated fish even though plasma osmolality was decreased in 70% SW fish (Anderson et al. 2002a). Rectal gland secretion from fish in 120% SW was suppressed compared with controls. Since CNP has potent secretagogue activity in this dogfish, plasma CNP may be increased in the volume-expanded fish. However, transfer of euryhaline bullshark (*Carcharhinus leucas*) from FW to SW did not alter plasma CNP concentration for a week, although it is higher in SW fish than in FW fish (Anderson et al. 2006).

A few humoral factors that inhibit rectal gland secretion have been reported. Bombesin-like peptide is identified in the nerve terminal of dogfish rectal gland, and bombesin inhibited VIP-induced Cl$^-$ secretion (Silva et al. 1990a). The bombesin inhibition is mediated by SST released from the nerve terminals as blockade of neurotransmitter release abolished the bombesin effect. NPY also inhibits Cl$^-$ secretion augmented by VIP and/or cAMP through direct action on the gland (Silva et al. 1993). Endothelin 1 potently inhibited the luminal transport of xenobiotics via multidrug resistance-associated protein 2 through ET_B receptors in the shark rectal gland (Miller et al. 2002). The inhibitory effect of endothelin on secretion may be due to its potent vasoconstrictive action on the shark vessels (Evans 2001). In another well-studied salt gland, avian supraorbital nasal gland, ANG II inhibits Cl$^-$ secretion probably through the central nervous system (Gerstberger et al. 1984) and through vascular constriction (Heinz and Gray 2001), but ANG II did not change secretion from the perfused rectal gland of *S. canicula* (Anderson et al. 2002a). Corticosteroids also inhibit NaCl secretion from the nasal salt gland of birds and reptiles (Bradshaw et al. 1984), but the effect of 1α-hydroxycorticosterone, a likely mineralocorticoid in elasmobranchs, and removal of its secretory organ, the interrenal tissues, had variable effects on rectal gland secretion (see Hazon et al. 2003).

7. CONCLUSIONS AND FUTURE PERSPECTIVES

Since the lumen of the alimentary tract is continuous with the external environment, epithelial cells lining the tract of fishes are exposed sometimes to abrupt changes in bathing fluid caused by oral intake of various food stuffs and environmental fluids. Food and fluid intake are regulated by ingestive behaviors motivated by appetite or thirst caused by various peripheral hormones acting on those brain areas lacking an effective blood–brain barrier, or by neuropeptides synthesized in the brain and acting in paracrine fashion. Food introduced into the alimentary tract is digested by enzymes secreted into the lumen, with the secretion of digestive juices regulated by hormones acting in long-distance endocrine and short-distance paracrine fashion. The food and digestive juices are mixed properly by gut motility, which is under dual hormonal and neural control. FW fish drink when they eat, and nutrient absorption is regulated by hormones such as GH. For SW fish, on the other hand, drinking is essential for survival to maintain body fluid homeostasis. Multiple hormones are involved in drinking regulation in fishes, and the central mechanisms have been recently unraveled. In marine fish, ingested seawater is modified during passage along the alimentary tract for maximal absorption at the distal intestine. Various ion and water channels/transporters and intercellular matrices are involved in the absorptive processes and the activity of these proteins is most likely regulated by GI hormones. Studies on the hormonal regulation of transport proteins are still in their infancy. Similarly, little is known about how changes in the luminal composition are sensed by the detectors or receptors in the alimentary tract and how that information is processed to result in altered synthesis and secretion of the GI hormones. Future studies will expose these important functional linkages.

ACKNOWLEDGMENT

The authors are grateful to Drs. Masaaki Ando and Jillian Healy for helpful discussion and critical reading of this manuscript.

REFERENCES

Abdel Magid, A. M. (1975). The epithelium of the gastro-intestinal tract of *Polypterus senegalus* (Pisces: Brachiopterygii). *J. Morphol.* **146**, 445–447.

Alexander, E. L. R., Dooley, K. C., Pohajdak, B., Xu, B.-Y., and Wright, J. R., Jr. (2006). Things we have learned from tilapia islet xenotransplantation. *Gen. Comp. Endocrinol.* **148**, 125–131.

Aller, S. G., Lombardo, I. D., Bhanot, S., and Forrest, J. N., Jr. (1999). Cloning, characterization, and functional expression of a CNP receptor regulating CFTR in the shark rectal gland. *Am. J. Physiol.* **276**, C442–C449.

Amiya, N., Amano, M., Takahashi, A., Yamanome, T., Kawauchi, H., and Yamamori, K. (2005). Effects of tank color on melanin-concentrating hormone levels in the brain, pituitary gland and plasma of the barfin flounder as revealed by a newly developed time-resolved fluoroimmunoassay. *Gen. Comp. Endocrinol.* **143**, 251–256.

Anderson, W. G., Conlon, J. M., and Hazon, N. (1995). Characterization of the endogenous intestinal peptide that stimulates the rectal gland of *Scyliorhinus canicula*. *Am. J. Physiol.* **268**, R1359–R1364.

Anderson, W. G., Good, J. P., and Hazon, N. (2002a). Changes in chloride secretion rate and vascular perfusion in the rectal gland of the Enropean lesser-spotted dogfish in response to environmental and hormonal stimuli. *J. Fish Biol.* **60**, 1580–1590.

Anderson, W. G., Ali, M. F., Einarsdóttir, I. E., Schäffer, L., Hazon, N., and Conlon, J. M. (2002b). Purification, characterization, and biological activity of insulins from the spotted dogfish, *Scyliorhinus canicula*, and the hammerhead shark, *Sphyrna lewini*. *Gen. Comp. Endocrinol.* **126**, 113–122.

Anderson, W. G., Pillans, R. D., Hyoso, S., Tsukada, T., Good, J. P., Takei, Y., Franklin, C. E., and Hazon, N. (2006). The effects of freshwater to seawater transfer on circulating levels of angiotensin II, C-type natriuretic peptide and afginine vasotocin in the euryhaline elasmobranch, *Carcharhinus leucas*. *Gen. Comp. Endocrinol.* **147**, 39–46.

Anderson, W. G., Takei, Y., and Hazon, N. (2001). The dipsogenic effect of the renin-angiotensin system in elasmobranch fish. *Gen. Comp. Endocrinol.* **125**, 300–307.

An, K. W., Shin, H. S., An, M. I., Jo, P. G., Choi, Y. K., and Choi, C. Y. (2008). Physiological responses and expression of arginine vasotocin receptor, prolactin and prolactin receptor mRNA in olive flounder *Paralichthys olovaceus* during osmotic stress. *Mar. Fresh. Behav. Physiol.* **41**, 191–203.

Ando, M., and Hara, I. (1994). Alteration of sensitivity to various regulators in the intestine of the eel following seawater acclimation. *Comp. Biochem. Physiol.* **109A**, 447–453.

Ando, M., Kondo, K., and Takei, Y. (1992). Effects of eel atrial natriuretic peptide on NaCl and water transport across the intestine of the seawater eel. *J. Comp. Physiol. B* **162**, 436–439.

Ando, M., Mukuda, M., and Kozaka, T. (2003). Water metabolism in the eel acclimated to sea water: from mouth to intestine. *Comp. Biochem. Physiol.* **136B**, 621–633.

Andoh, T. (2007). Amino acids are more important insulinotropins than glucose in a teleost fish, barfin flounder (*Verasper moseri*). *Gen. Comp. Endocrinol.* **151**, 308–317.

Andrew, A., Kramer, B., and Rawdon, B. B. (1998). The origin of gut and pancreatic neuroendocrine (APUD) cells—the last word? *J. Pathol.* **186**, 117–118.

Bayliss, W. M., and Starling, E. H. (1902). The mechanism of pancreatic secretion. *J. Physiol.* **28**, 325–353.

Bechtold, D. A., and Luckman, S. M. (2007). The role of RFamide peptides in feeding. *J. Endocrinol.* **192**, 3–15.

Bewley, M. S., Pena, J. T. G., Plesch, F. N., Decker, S. E., Weber, G. J., and Forrest, J. N., Jr. (2006). Shark rectal gland vasoactive intestinal peptide receptor: cloning, functional expression, and regulation of CFTR chloride channels. *Am. J. Physiol.* **291**, R1157–R1164.

Bisbal, G. A., and Specker, J. L. (1991). Cortisol stimulates hypo-osmoregulatory ability in Atlantic salmon, *Salmo salar* L. *J. Fish Biol.* **39**, 421–432.

Bjornsson, B. T. (1997). The biology of salmon growth hormone: from daylight to dominance. *Fish Physiol. Biochem.* **17**, 9–24.

Bjornsson, B. T., and Bradley, T. M. (2007). Epilogue: past successes, present misconceptions and future milestones in salmon smoltification research. *Aquaculture* **273**, 384–391.

Bosi, G., Di Giancamillo, A., Arrighi., S., and Domeneghini, C. (2004). An immunohisto-chemical study on the neuroendocrine system in the alimentary canal of the brown trout, *Salmo trutta* L. *Gen. Comp. Endocrinol.* **138**, 166–181.

Bœuf, G., and Payan, P. (2001). How should salinity influence fish growth? *Comp. Biochem. Physiol. C* **130**, 411–423.

Bradshaw, S. D., Lemire, M., Vernet, R., and Grenot, C. J. (1984). Aldosterone and the control of secretion by the nasal salt gland of the North African desert lizard, *Uromstix acanthinurus*. *Gen. Comp. Endocrino.* **54**, 314–323.

Buddington, R. K., and Diamond, J. M. (1987). Pyloric caca of fish: a "new" absorptive organ. *Am. J. Physiol.* **252**, G65–G76.

Buddington, R. K., and Krogdahl, A. (2004). Hormonal regulation of the fish gastrointestinal tract. *Comp. Biochem. Physiol. A* **139**, 261–271.

Capilla, E., Díaz, M., Gutiérrez, J., and Planas, J. V. (2002). Physiological regulation of the expression of a GLUT4 homolog in fish skeletal muscle. *Am. J. Physiol.* **283**, E44–E49.

Capilla, E., Díaz, M., Albalat, A., Navarro, I., Pessin, J. E., Keller, K., and Planas, J. V. (2004). Functional characterization of an insulin-responsive glucose transporter (GLUT4) from fish adipose tissue. *Am. J. Endocrinol.* **287**, E348–E357.

Cardoso, J. C. R., Clark, M. S., Viera, F. A., Bridge, P. D., Gilles, A., and Power, D. M. (2005). The secretin G-protein-coupled receptor family: teleost receptors. *J. Mol. Endocrinol.* **34**, 753–756.

Cataldi, E., Crosetti, D., Conte, G., D'Ovidio, D., and Cataudella, S. (1988). Morphological changes in the esophageal epithelium during adaptation to salinities in *Oreochromis mossambicus, O. niloticus* and their hybrid. *J. Fish Biol.* **32**, 191–196.

Cerdá-Reverter, J. M., and Larhammer, D. (2000). Neuropeptide Y family of peptides: structure, anatomical expression, function, and molecular evolution. *Biochem. Cell. Biol.* **78**, 371–392.

Cerdá-Reverter, J. M., and Peter, R. E. (2003). Endogenous melanocortin antagonist in fish: structure, brain mapping, and regulation by fasting of the goldfish agouti-related protein gene. *Endocrinology* **144**, 4552–4561.

Cerdá-Reverter, J. M., Martínez-Rodríguez, G., Zanuy, S., Carrillo, M., and Larhammer, D. (2000). Molecular evolution of the neuropeptide Y (NPY) family of peptides: cloning of three NPY-related peptides from the sea bass (*Dicentrarchus labrax*). *Regul. Peptides* **95**, 25–34.

Chan, S. J., and Steiner, D. F. (2000). Insulin through the ages: phylogeny of a growth-promoting and metabolic regulatory hormone. *Amer. Zool.* **40**, 213–222.

Chan, S. J., Cao, Q.-P., and Steiner, D. F. (1990). Evolution of the insulin superfamily: cloning of a hybrid insulin/insulin-like growth factor cDNA from amphioxus. *Proc. Natl. Acad. Sci. USA* **87**, 9319–9323.

Chan, S. J., Plisetskaya, E. M., Urbinati, E., Jin, Y., and Steiner, D. F. (1997). Expression of multiple insulin and insulin-like growth factor receptor genes in salmon gill cartilage. *Proc. Natl. Acad. Sci. USA* **94**, 12446–12451.

Chow, B. K. C., Moon, T. W., Hoo, R. L. C., Yeung, C.-M., Müller, M., Christos, P. J., and Mojsov, S. (2004). Identification and characterization of a glucagon receptor from the goldfish *Carassius auratus*: implications for the evolution of the ligand specificity of glucagon receptors in vertebrates. *Endocrinology* **145**, 3273–3288.

Christoffels, A., Brenner, S., and Venkatesh, B. (2006). Tetraodon genome analysis provides further evidence for whole-genome duplication in the ray-finned fish lineage. *Comp. Biochem. Physiol. D* **1**, 13–19.

Ciccotti, E., Macchi, E., Rosser, A., Cataldi, E., and Cataudella, S. (1993). Glass eel (*Anguilla anguilla*) acclimation to freshwater and seawater: morphological changes of the digestive tracts. *J. Appl. Ichthyol.* **9**, 7–81.

Clements, K. D., and Raubenheimer, D. (2006). Endocrinology. In: *The Physiology of Fishes (CRC Marine Science Series)* (D. H. Evans and J. B. Claiborne, eds), 3d edition, pp. 47–82. CRC Press, Boca Raton.

Collie, N. L., and Bern, H. A. (1982). Changes in intestinal fluid transport associated with smoltification and seawater adaptation in coho salmon, *Oncorhynchus kisutch* (Walbaum). *J. Fish Biol.* **21**, 337–348.

Collie, N. L., and Stvens, J. J. (1985). Hormonal effects on L-proline transport in coho salmon (*Oncorhynchus kisutch*) intestine. *Gen. Comp. Endocrinol.* **59**, 399–409.

Company, R., Astola, A., Pedón, C., Valdivia, M. M., and Pérez-Sánchez, J. (2001). Somatotropic regulation of fish growth and adiposity: growth hormone (GH) and somatolactin (SL) relationship. *Comp. Biochem. Physiol.* C **130**, 435–445.

Conant, G. C., and Wolfe, K. H. (2008). Turning a hobby into a job: how duplicated genes find new functions. *Nature Rev. Genet.* **9**, 938–950.

Conlon, J. M. (1999). Bradykinin and its receptors in non-mammalian vertebrates. *Regul. Pept.* **79**, 71–81.

Conlon, J. M. (2000). Molecular evolution of insulin in non-mammalian vertebrates. *Amer. Zool.* **40**, 200–212.

Conlon, J. M. (2001). Evolution of the insulin molecule: insights into structure-activity and phylogenetic relationships. *Peptides* **22**, 1183–1193.

Conlon, J. M., and Larhammar, D. (2005). The evolution of neuroendocrine peptides. *Gen. Comp. Endocrinol.* **142**, 53–59.

Conlon, J. M., Göke, R., Andrews, P. C., and Thim, L. (1989). Multiple molecular forms of insulin and glucagon-like peptide from the pacific ratfish (*Hydrolagus colliei*). *Gen. Comp. Endocrinol.* **73**, 136–145.

Conlon, J. M., Platz, J. E., Nielsen, P. E., Vaudry, H., and Vallarino, M. (1997). Primary structure of insulin from the African lungfish, *Protopterus annectens. Gen. Comp. Endocrinol.* **107**, 421–427.

Conlon, J. M., Basir, Y., and Joss, J. M. P. (1999). Purification and characterization of insulin from the Australian lungfish, *Neoceratodus forsteri* (Dipnoi). *Gen. Comp. Endocrinol.* **116**, 1–9.

Cousins, K. L., and Farrell, A. P. (1996). Stretch-induced release of atrial natriuretic factor from the heart of rainbow trout (*Oncorhynchus mykiss*). *Can. J. Zool.* **74**, 380–387.

Cramb, G., Martinez, A.-S., McWilliam, I. S., and Wilson, G. D. (2005). Cloning and expression of guanylin-like peptides in teleost fish. *Ann. NY Acad. Sci.* **1040**, 277–280.

Currie, P. J., Mirza, A., Fuld, R., Park, D., and Vasselli, J. R. (2005). Ghrelin is an orexigenic and metabolic signaling peptide in the arcuate and paraventricular nucleus. *Am. J. Physiol.* **289**, R353–R358.

Datuin, J. P., Ng, K. P., Hayes, T. B., and Bern, H. A. (2001). Effects of glucocorticoids on cartilage growth and response to IGF-I in the tilapia (*Oreochromis mossambicus*). *Gen. Comp. Endocrinol.* **121**, 289–294.

Day, R., and Salzet, M. (2002). The neuroendocrine phenotype, cellular plasticity, and the search for genetic switches: redefining the diffuse neuroendocrine system. *Neuroendocrinol. Lett.* **23**, 447–451.

De Pedro, N., Pinillos, M. L., Valenciano, I., Alonso-Bedate, M., and Delgado, M. J. (1998). Inhibitory effect of serotonin on feeding behavior in goldfish: involvement of CRF. *Peptides* **19**, 505–511.

Dickhoff, W. W., Beckman, B. R., Larsen, D. A., Duan, C., and Moriyama, S. (1997). The role of growth in endocrine regulation of salmon smoltification. *Fish Physiol. Biochem.* **17**, 231–236.

Dickson, L., and Finlayson, K. (2009). VPAC and PAC receptors: from ligands to function. *Pharmacol. Therapeutics* **121**, 294–316.

Duan, C. (1997). The insulin-like growth factor system and its biological actions in fish. *Amer. Zool.* **37**, 491–503.

Duguay, S. J., and Mommsen, T. P. (1994). Molecular aspects of pancreatic peptides. In: *Fish Physiology. Vol. XIII, Molecular Endocrinology of Fish* (N. M Sherwood and C. L Hew, eds), pp. 225–271. Academic Press, San Diego.

Epple, A., and Brinn, J. E. (1986). Pancreatic islets. In: *Vertebrate Endocrinology: Fundamentals and Biomedical Implications. Vol. 1, Morphological Considerations* (P. K. T Pang and M. P Schreibman, eds), pp. 279–317. Academic Press, San Diego.

Epple, A., and Lewis, T. L. (1973). Comparative histophysiology of the pancreatic islets. *Am. Zool.* **13**, 567–590.

Evans, D. H. (2001). Vasoactive receptors in abdominal blood vessels of the dogfish shark, *Squalus acanthias. Physiol. Biochem. Zool.* **74**, 120–126.

Evans, D. H., and Piermarini, P. M. (2001). Contractile properties of the elasmobranch rectal gland. *J. Exp. Biol.* **204**, 59–67.

Fiol, D. F., and Kurtz, D. (2007). Osmotic stress sensing and signaling in fishes. *FEBS J.* **274**, 5790–5798.

Fitzsimons, J. T. (1979). *The Physiology of Thirst and Sodium Appetite.* Cambridge University Press, Cambridge.

Forgan, L. G., and Forster, M. E. (2007). Effects of potential mediators of an intestinal brake mechanism on gut motility in Chinook salmon (*Oncorhynchus tshawytscha*). *Comp. Biochem. Physiol. C* **145**, 343–347.

Fuentes, J., Figueiredo, J., Power, D. M., and Canario, A. V. M. (2006). Parathyroid hormone-related protein regulates intestinal calcium transport in sea bream (*Sparus auratus*). *Am. J. Physiol.* (**291**), R1499–R1506.

Fujita, T. (1989). Present status of paraneuron concept. *Arch. Histol. Cytol.* (**52**), S1–S8.

Fujita, T., and Kobayashi, S. (1977). Structure and function of gut endocrine cells. *Intl. Rev. Cytol. Suppl.* **6**, 187–233.

Funahashi, H., Takenoya, F., Guan, J.-L., Kageyama, H., Yada, T., and Shioda, S. (2003). Hypothalamic neuronal networks and feeding-related peptides involved in the regulation of feeding. *Anat. Sci. Int.* **78**, 123–138.

Geller, D. M., Currie, M. G., Wakitani, K., Cole, B. R., Adams, S. P., Fok, K. F., Siegel, N. R., Eubanks, S. R., Galluppi, G. R., and Needleman, P. (1984). Atriopeptins: A family of potent biologically active peptides derived from mammalian atria. *Biochem. Biophys. Res. Commun.* **120**, 333–338.

Genten, F., Terwinghe, E., and Danguy, A. (2009). *Atlas of Fish Histology*, 215 pp. Science Publishers, Enfield.

Gerstberger, R., Gray, D. A., and Simon, E. (1984). Circulatory and osmoregulatory effects of angiotensin II perfusion of the third ventricle in a bird with salt gland. *J. Physiol.* **349**, 167–182.

Gong, J.-Y., Kittelson, J. D., Slagter, B. J., and Sheridan, M. A. (2004). The two subtype 1 somatostatin receptors of rainbow trout, Tsst$_{1A}$ and Tsst$_{1B}$, possess both distinct and overlapping ligand binding and agonist-induced regulation features. *Comp. Biochem. Physiol. B* **138**, 295–303.

Gorissen, M. H. A. G., Flik, G., and Huising, M. O. (2006). Peptides and proteins regulating food intake: a comparative view. *Animal Biol.* **56**, 447–473.

Greger, R., Bleich, M., Warth, R., Thiele, I., and Forrest, J. N. (1999). The cellular mechanisms of Cl^- secretion induced by C-type natriuretic peptide (CNP). Experiments in isolated in vitro perfused rectal gland tubules of *Squalus acanthias. Pflugers Arch.* **438**, 15–22.

Grosell, M. (2006). Intestinal anion exchange in marine fish osmoregulation. *J. Exp. Biol.* **209**, 2813–2827.

Grosell, M., Mager, E. M., Williams, C., and Taylor, J. R. (2009). High rates of HCO_3^- secretion and Cl^- absorption against adverse gradients in the marine teleost intestine: the involvement of an electrogenic anion exchanger and H^+-pump metalolon? *J. Exp. Biol.* **212**, 1684–1696.

Grosell, M. (2010). The role of the gastrointestinal tract in salt and water balance. In: *Multifunctional Gut of Fish* (M. Grosell, A. P. Farrell and C. J. Brauner, eds), pp. 136–165. Academic Press, San Diego.

Guerreiro, P. M., Fuentes, J., Power, D. M., Ingleton, P. M., Flik, G., and Canario, A. V. (2001). Parathyroid hormone-related protein: a calcium regulatory factor in sea bream (*Sparus autarta* L.) larvae. *Am. J. Physiol.* **281**, R855–R860.

Guerreiro, P. M., Renfro, J. L., Power, D. M., and Canario, A. V. M. (2007). The parathyroid hormone family of peptides: structure, tissue distribution, regulation, and potential functional roles in calcium and phosphate balance in fish. *Am. J. Physiol.* **292**, R679–R696.

Hashimoto, H., Fujihara, H., Kawasaki, M., Saito, T., Shibata, M., Takei, Y., and Ueta, Y. (2007). Centrally and peripherally administered ghrelin potently inhibits water intake in rats. *Endocrinology* **148**, 1638–1647.

Hazon, N., Wells, A., Pillans, R. D., Good, J. P., Anderson, G. W., and Franklin, C. E. (2003). Urea based osmoregulation and endocrine control in elasmobranch fish with special reference to euryhalinity. *Comp. Biochem. Physiol.* **136B**, 685–700.

Heinz, M. K., and Gray, D. A. (2001). Role of plasma ANG II in the excretion of acute sodium load in a bird with salt glands (*Anas platychynchos*). *Am. J. Physiol.* **281**, R346–R351.

Herbert, S. C. (2004). Calcium and salinity sensing by the thick ascending limb: a journey from mammals to fish and back again. *Kid. Internatl.* **91**, S28–S33.

Hilton, W. A. (1900). On the intestine of *Amia calva*. *Amer. Naturalist* **34**, 717–735.

Himick, B. A., and Peter, R. E. (1994). Bombesin acts to suppress feeding behavior and alter serum growth hormone in goldfish. *Physiol. Behav.* **55**, 65–72.

Hirano, T., and Mayer-Gostan, N. (1976). Eel esophagus as an osmoregulatory organ. *Proc. Natl. Acad. Sci. USA* **73**, 1348–1350.

Hirano, T., and Utida, S. (1968). Effects of ACTH and cortisol on water movement in isolated intestine of the eel, *Anguilla japonica*. *Gen. Comp. Endocrinol.* **11**, 373–380.

Hirose, S., Hagiwara, H., and Takei, Y. (2001). Comparative molecular biology of natriuretic peptide receptors. *Can. J. Physiol. Pharmacol.* **79**, 665–672.

Hobart, P., Crawford, R., Shen, L., Pictet, R., and Rutter, W. J. (1980a). Cloning and sequence analysis of cDNAs encoding two distinct somatostatin precursors found in the endocrine pancreas of anglerfish. *Nature* **288**, 137–141.

Hobart, P., Shen, L., Crawford, R., Pictet, R., and Rutter, W. J. (1980b). Comparison of the nucleic acid sequence of anglerfish and mammalian insulin mRNA's from cloned cDNA's. *Science* **210**, 1360–1363.

Holmgren, S., and Olsson, C. (2009). The neuroendocrine regulation of gut function. In: *Fish Neuroendocrinology (Fish Physiology Series)* (N. J Bernier, G Van der Kraak, A. P Farrell and C. J Brauner, eds), *Vol. 28*, pp. 467–512. Academic Press, San Diego.

Holstein, B., and Cederberg, C. (1988). Effect of somatostatin on basal and stimulated gastric secretion in the cod, *Gadus morhua*. *Am. J. Physiol.* **254**, G183–G188.

Horn, M. H. (1997). Feeding and digestion. In: *The Physiology of Fishes* (D. H. Evans, ed.), 2nd edition, pp. 43–63. CRC Press, Boca Raton.

Huang, X., Jiao, B., Fung, C. K., Zhang, Y., Ho, W. K. K., Chan, C. B., Lin, H., Wang, D., and Cheng, C. H. K. (2007). The presence of two distinct prolactin receptors in seabream with different tissue distribution patterns, signal transduction pathways and regulation of gene expression by steroid hormones. *J. Endocrinol.* **194**, 373–392.

Huising, M. O, Geven, E. J. W., Kruiswijk, C. P., Nabuurs, S. B., Stolte, E. H., Spanings, F. A. T., Verburg-van Kemennade, B. M. L., and Flik, G. (2006). Increased leptin expression in common carp (*Cyprinus carpio*) after food intake but not after fasting or feeding to satiation. *Endocrinology* **147**, 5786–5797.

Inoue, K., Naruse, K., Yamagami, S., Mitani, H., Suzuki, N., and Takei, Y. (2003a). Four functionally distinct C-type natriuretic peptides found in fish reveal evolutionary history of the natriuretic peptide system. *Proc. Natl. Acad. Sci. USA* **100**, 10079–10084.

Inoue, K., Sakamoto, T., Yuge, S., Iwatani, H., Yamagami, S., Tsutsumi, H., Hori, H., Cerra, M. C., Tota, B., Suzuki, N., Okamoto, N., and Takei, T. (2005). Structural and functional evolution of three cardiac natriuretic peptides. *Mol. Biol. Evol.* **22**, 2428–2434.

Irwin, D. M. (2001). Molecular evolution of proglucagon. *Regul. Pept.* **98**, 1–12.

Irwin, D. M. (2004). A second insulin gene in fish genomes. *Gen. Comp. Endocrinol.* **135**, 150–158.

Irwin, D. M., and Wong, J. (1995). Trout and chicken proglucagon: alternative splicing generates mRNA transcripts encoding glucagon-like peptide 2. *Mol. Endocrinol.* **9**, 267–277.

Irwin, D. M., Huner, O., and Youson, J. H. (1999). Lamprey glucagon and the origin of glucagon-like peptides. *Mol. Biol. Evol.* **16**, 1548–1557.

Jeon, U. S., Kim, J.-A., Sheen, M. R., and Kwon, H. M. (2006). How tonicity regulates genes: story of TonEBP transcriptional activator. *Acta Physiol.* **187**, 241–247.

Johnsen, A. H. (1998). Phylogeny of the cholecystokinin/gastrin family. *Frontiers Neuroendocrinol.* **19**, 73–99.

Johnsen, A. H., Jønson, L., Rourke, I. J., and Rehfeld, J. F. (1997). Elasmobranchs express separate cholecystokinin and gastrin genes. *Proc. Natl. Acad. Sci. USA* **94**, 10221–10226.

Jonsson, E., Forsman, A., Einarsdottir, I. E., Egner, B., Ruohonen, K., and Bjornsson, B. T. (2006). Circulating levels of cholecystokinin and gastrin-releasing peptide in rainbow trout fed different diets. *Gen. Comp. Endocrinol.* **148**, 187–194.

Kaiya, H., Kojima, M., Hosoda, H., Moriyama, S., Takahashi, A., Kawauchi, H., and Kangawa, K. (2003a). Peptide purification, complementary deoxyribonucleic acid (DNA) and genomic DNA cloning, and functional characterization of ghrelin in rainbow trout. *Endocrinology* **144**, 5215–5226.

Kaiya, H., Kojima, M., Hosoda, H., Riley, L. G., Hirano, T., Grau, E. G., and Kangawa, K. (2003b). Identification of tilapia ghrelin and its effects on growth hormone and prolactin release in the tilapia, *Oreochromis mossambicus. Comp. Biochem. Physiol. B* **135**, 421–429.

Kaiya, H., Kojima, M., Hosoda, H., Riley, L. G., Hirano, T., Grau, E. G., and Kangawa, K. (2003c). Amidated fish ghrelin: purification, cDNA cloning in the Japanese eel and its biological activity. *J. Endocrinol.* **176**, 415–423.

Kaiya, H., Miyazato, M., Kangawa, K., Peter, R. A., and Unniappan, S. (2008). Ghrelin: a multifunctional hormone in non-mammalian vertebrates. *Comp. Biochem. Physiol. A* **149**, 109–128.

Kalujnaia, S., Wilson, G. D., Feilen, A. L., and Cramb, G. (2009). Guanylin-like peptides, guanylate cyclase and osmoregulation in the European eel (*Anguilla anguilla*). *Gen. Comp. Endocrinol.* **161**, 103–114.

Karnaky, K. J., Valentich, J. D., and Currie, M. G. (1991). Atriopeptin stimulates chloride secretion in cultured shark rectal gland cells. *Am. J. Physiol.* **260**, C1125–C1130.

Kawakoshi, A., Hyodo, S., Inoue, K., Kobayashi, Y., and Takei, Y. (2004). Four natriuretic peptides (ANP, BNP, VNP and CNP) coexist in the sturgeon: identification of BNP in fish lineage. *J. Mol. Endocrinol.* **32**, 547–555.

Kawakoshi, A., Hyodo, S., Nozaki, M., and Takei, Y. (2006). Identification of a natriuretic peptide (NP) in cyclostomes (lamprey and hagfish): CNP-4 is the ancestral gene of the NP family. *Gen. Comp. Endocrinol.* **148**, 41–47.

Kelly, S. P., and Peter, R. E. (2006). Prolactin-releasing peptide, food intake, and hydromineral balance in goldfish. *Am. J. Physiol.* **291**, R1474–R1481.

Kobayashi, H., and Takei, Y. (1996). *The Renin-Angiotensin System: Comparative Aspects. Zoophysiology Vol. 35.* Springer, Heidelberg.

Kobayashi, H., Uemura, H., Wada, M., and Takei, Y. (1979). Ecological adaptation of angiotensin II-induced thirst mechanism in tetrapods. *Gen. Comp. Endocrinil.* **38**, 93–104.

Kobayashi, H., Uemura, H., Takei, Y., Itazu, N., Ozawa, M., and Ichinohe, K. (1983). Drinking induced by angiotensin II in fishes. *Gen. Comp. Endocrinol.* **49**, 295–306.

Kojima, M., Hosoda, H., Date, Y., Nakazato, M., Matsuo, H., and Kangawa, K. (1999). Ghrelin is a growth-hormone-releasing acetylated peptide from stomach. *Nature* **402**, 656–660.

Koppang, E. O., Tomas, G. A., Ronningen, K., and Press., C. L. (1998). Expression of insulin-like growth factor-I in the gastrointestinal tract of Atlantic salmon (*Salmo salar* L.). *Fish Physiol. Biochem.* **18**, 167–175.

Kurokawa, T., and Suzuki, T. (2002). Development of neuropeptide Y-related peptides in the digestive organs during the larval stage of Japanese flounder, *Paralichthys olivaceus*. *Gen. Comp. Endocrinol.* **126**, 30–38.

Kurokawa, T., Suzuki, T., and Hashimoto, H. (2003). Identification of gastrin and multiple cholecystokinin genes in teleost. *Peptides* **24**, 227–235.

Laburthe, M., Ouvineau, A., and Tan, V. (2007). Class II G protein-coupled receptors for VIP and PACAP: structure, models for activation and pharmacology. *Peptides* **28**, 1631–1639.

Larhammer, D. (1996). Evolution of neuropeptide Y, peptide YY and pancreatic polypeptide. *Regul. Pept.* **62**, 1–11.

Lathia, J. D., Mattson, M. P., and Cheng, A. (2008). Notch: from neural development to neurological disorders. *J. Neurochem.* **107**, 1471–1481.

Legate, N. J., Bonen, A., and Moon, T. W. (2001). Glucose tolerance and peripheral glucose utilization in rainbow trout (*Oncorhynchus mykiss*), American eel (*Anguilla rostrata*), and black bullhead catfish (*Ameiurus melas*). *Gen. Comp. Endocrinol.* **122**, 48–59.

Lehrich, R. W., Aller, S. G., Webster, P., Marino, C. R., and Forrest, J. N., Jr. (1998). Vasoactive intestinal peptide, forskolin, and genistein increase apical CFTR trafficking in the rectal gland of the spiny dogfish, *Squalus acanthias*. Acute regulation of CFTR trafficking in an intact epithelium. *J. Clin. Invest.* **101**, 737–745.

Leibowitz, S. F., and Alexander, J. T. (1998). Hypothalamic serotonin in control of eating behavior, meal size, and body weight. *Biol. Psychiatry* **44**, 851–864.

Lignot, J. H., Cutler, C. P., Hazon, N., and Cramb, G. (2002). Immunolocalisation of aquaporin 3 in the gill and the gastrointestinal tract of the European eel *Anguilla antuilla* L. *J. Exp. Biol.* **205**, 2653–2663.

Lin, X., and Peter, R. E. (2001). Somatostatins and their receptors in fish. *Comp. Biochem. Physiol. B* **129**, 543–550.

Lin, X., Volkoff, H., Narnaware, Y., Bernier, N. J., Peyon, P., and Peter, R. E. (2000). Brain regulation of feeding behavior and food intake in fish. *Comp. Biochem. Physiol.* **126A**, 415–434.

Loretz, C. A. (1987). Rectal glands and crypts of Lieberkühn: is there a phylogenetic basis for functional similarity? *Zool. Sci.* **4**, 933–944.

Loretz, C. A. (1995). Electrophysiology of ion transport in teleost intestinal cells. In: *Cellular and Molecular Approaches to Fish Ionic Regulation. Fish Physiology Vol. 14* (C. M. Wood and T. J. Shuttleworth, eds), pp. 25–56. Academic Press, San Diego.

Loretz, C. A. (2008). Extracellular calcium-sensing receptors in fishes. *Comp. Biochem. Physiol. A: Molec. Integr. Physiol* **149**, 225–245.

Loretz, C. A., and Pollina, C. (2000). Natriuretic peptides in fish physiology. *Comp. Biochem. Physiol. A* **125**, 169–187.

Loretz, C. A., and Takei, Y. (1997). Natriuretic peptide inhibition of intestinal salt absorption in the Japanese eel: physiological significance. *Fish Physiol. Biochem.* **17**, 319–324.

Loretz, C. A., Bern, H. A., Foskett, J. K., and Mainoya, J. R. (1981). The caudal neurosecretory system and osmoregulation in fish. In: *Neurosecretion: Molecules, Cells, Systems* (D. S. Farner and K. Lederis, eds), pp. 319–328. Plenum, New York.

Loretz, C. A., Pollina, C., Kaiya, H., Sakaguchi, H., and Takei, Y. (1997). Local synthesis of natriuretic peptides in the eel intestine. *Biochem. Biophys. Res. Comm.* **238**, 817–822.

Lund, P. K., Goodman, R. H., Dee, P. C., and Habener, J. F. (1982). Pancreatic preproglucagon cDNA contains two glucagon-related coding sequences arranged in tandem. *Proc. Natl. Acad. Sci. USA* **79**, 345–349.

MacLeod, M. G. (1978). Effects of salinity and starvation on the alimentary canal anatomy of the rainbow trout *Salmo gairdneri* Richardson. *J. Fish Biol.* **12**, 71–79.

McCormick, S. D. (2001). Endocrine control of osmoregulation in teleost fish. *Amer. Zoologist* **41**, 781–794.

Mainoya, J. R., and Bern, H. A. (1984). Influence of vasoactive intestinal peptide and urotensin II on the absorption of water and NaCl by the anterior intestine of the tilapia. *Sarotherondon mossambicus. Zool. Sci.* **1**, 100–105.

Marshall, W. S., and Grosell, M. (2005). Ion transport, osmoregulation, and acid base balance. In: *Physiology of Fishes* (D. H. Evans and J. B. Claiborne, eds), Third Edition, pp. 177–230. CRC Press, Boca Raton.

Martinez, A. S., Cutler, C. P., Wilson, G. D., Phillips, C., Hazon, N., and Cramb, G. (2005). Regulation of expression of two aquaporin homologs in the intestine of the European eel: effects of seawater acclimation and cortisol treatment. *Am. J. Physiol.* **288**, R1733–R1743.

Martinez-Alvarez, R. M., Volkoff, H., Munoz-Cueto, J. A., and Delgado, M. J. (2009). Effect of calcitonin gene-related peptide (CGRP), adrenomedullin and adrenomedullin-2/intermedin on food intake in goldfish (*Carassius auratus*). *Peptides* **30**, 803–807.

Maruyama, K., Konno, N., Ishiguro, K., Wakasugi, T., Uchiyama, M., Shioda, S., and Matsuda, K. (2008). Isolation and characterisation of four cDNAs encoding neuromedin U (NMU) from the brain and gut of goldfish, and the inhibitory effect of a deduced NMU on food intake and locomotor activity. *J. Neuroendocrinol.* **20**, 71–78.

Matsuda, K., Maruyama, K., Nakamachi, T., Miura, T., Uchiyama, M., and Shioda, S. (2005). Inhibitory effects of pituitary adenylate cyclase-activating polypeptide (PACAP) and vasoactive intestinal peptide (VIP) on food intake in the goldfish, *Carassium auratus. Peptides* **26**, 1611–1616.

Matsuda, K., Shimakura, S., Murayama, K., Miura, T., Uchiyama, M., Kawauchi, H., Shioda, S., and Takahashi, A. (2006). Central administration of melanin-concentrating hormone (MCH) suppresses food intake, but not locomotor activity, in the goldfish, *Carassius auratus. Neurosci. Lett.* **399**, 259–263.

Meister, M. F., Humbert, W., Kirsch, R., and Vivien-Roels, B. (1983). Structure and ultrastructure of the oesophagus in sea-water and fresh-water teleosts (*Pisces*). *Zoomorphology* **102**, 33–51.

Minamino, N., Kangawa, K., and Matsuo, H. (1985). Neuromedin-U-8 and neuromedin-U-25 – Novel uterus stimulating and hypertensive peptides identified in procine spinal cord. *Biochem. Biophys. Res. Commun.* **130**, 1078–1085.

Miller, D. S., Masereeuw, R., and Karnaky, K. J., Jr. (2002). Regulation of MRP2-mediated transport in shark rectal salt gland tubules. *Am. J. Physiol.* **282**, R774–R781.

Mojsov, S. (2000). Glucagon-like peptide-1 (GLP-1) and the control of glucose metabolism in mammals and teleost fish. *Amer. Zool.* **40**, 246–258.

Mommsen, T. P. (2000). Glucagon-like peptide-1 in fishes: the liver and beyond. *Amer. Zool.* **40**, 259–268.

Mommsen, T. P., and Plisetskaya, E. M. (1991). Insulin in fishes and agnathans: history, structure, and metabolic regulation. *Rev. Aquat. Sci.* **4**, 225–259.

Moon, T. W. (2001). Glucose intolerance in teleost fish: fact or fiction? *Comp. Biochem. Physiol.* B **129**, 243–249.

Moon, T. W. (2004). Hormones and fish hepatocyte metabolism: "the good, the bad and the ugly!". *Comp. Biochem. Physiol.* B **139**, 335–345.

Mori, Y., and Ando, M. (1991). Regulation of ion and water transport across the eel intestine: effects of acetylcholine and serotonin. *J. Comp. Phsyiol.* B **161**, 387–392.

Morrison, C. M., Yang, H., Al-Jazaeri, A., Tam, J., Plisetskaya, E. M., and Wright, J. R., Jr. (2003). Xenogeneic milieu markedly remodels endocrine cell populations after transplantation of fish islets into streptozotocin-diabetic nude mice. *Xenotransplantation* **10**, 60–65.

Morrison, C. M., Pohajdak, B., Tam, J., and Wright, J. M., Jr. (2004). Development of the islets, exocrine pancreas and related ducts in the Nile tilapia, *Oreochromis niloticus* (Pisces: Cichlidae). *J. Morphol.* **261**, 377–389.

Murashita, K., Fukada, H., Hosokawa, H., and Masumoto, T. (2007). Changes in cholecystokinin and peptide Y gene expression with feeding in yellowtail (*Seriola quinqueradiata*): relation to pancreatic exocrine regulation. *Comp. Biochem. Physiol.* B **146**, 318–325.

Murashita, K., Fukada, H., Rønnestad, I., Kurokawa, T., and Masumoto, T. (2008a). Nutrient control of release of pancreatic enzymes in yellowtail (*Seriola quinqueradiata*): involvement of CCK and PY in the regulatory loop. *Comp. Biochem. Physiol.* A **150**, 438–443.

Murashita, K., Uji, S., Yamamoto, T., Ronnestad, I., and Kurokawa, T. (2008b). Production of recombinant leptin and its effects on food intake in rainbow trout (*Oncorhynchus mykiss*). *Comp. Biochem. Physiol.* **150B**, 377–384.

Nagamatsu, S., Chan, S. J., Falkmer, S., and Steiner, D. F. (1991). Evolution of the insulin gene superfamily: sequence of a preproinsulin-like growth factor cDNA from the Atlantic hagfish. *J. Biol. Chem.* **266**, 2397–2402.

Nakauchi, M., and Suzuki, N. (2005). Enterotoxin/guanylin receptor type guanylyl cyclases in non-mammalian vertebrates. *Zool. Sci.* **22**, 501–509.

Nam, B.-H., Kim, Y.-O., Hee, J. K., Kim, W.-J., Lee, S. J., and Choi, T.-J. (2009). Identification and characterization of the prepro-vasoactive intestinal peptide gene from the teleost *Paralichthys olivaceus*. *Vet. Immunol. Immunopathol.* **127**, 249–258.

Navarro, I., Leibush, B., Moon, T. W., Plisetskaya, E. M., Banos, N., Méndez, E., Planas, J. V., and Gutiérrez, J. (1999). Insulin, insulin-like growth factor-I (IGF-I) and glucagon: the evolution of their receptors. *Comp. Biochem. Physiol.* B **122**, 137–153.

Nearing, J., Betka, M., Quinn, S., Hentchel, H., Elger, M., Baum, M., Chattopadyhay, N., Brown, E. M., Hebert, S. C., and Harris, H. W. (2002). Polyvalent cation receptor proteins (CaRs) are salinity sensors in fish. *Proc. Natl. Acad. Sci. USA* **99**, 9231–9236.

Nelson, J. S. (2006). *Fishes of the World* (4th edition). Wiley, Hoboken, 601 pp.

Nelson, L. E., and Sheridan, M. A. (2005). Regulation of somatostatins and their receptors in fish. *Gen. Comp. Endocrinol.* **142**, 117–133.

Nelson, L. E., and Sheridan, M. A. (2006). Gastroenteropancreatic hormones and metabolism in fish. *Gen. Comp. Endocrinol.* **148**, 116–124.

Nobata, S., Ogoshi, M., and Takei, Y. (2008). Potent cardiovascular actions of homologous adrenomedullins in eels. *Am. J. Physiol.* **294**, R1544–R1553.

Nozaki, M., Miyata, K., Oota, Y., Gorbman, A., and Plisetskaya, E. M. (1988). Colocalization of glucagon-like peptide and glucagon immunoreactivities in pancreatic islets and intestine of salmonids. *Cell Tissue Res.* **253**, 371–375.

O'Grady, S. M. (1989). Cyclic nucleotide-mediated effects of ANF and VIP on flounder intestinal ion transport. *Am. J. Physiol.* **256**, C142–C146.

O'Grady, S. M., Field, M., Nash, N. T., and Rao, M. C. (1985). Atrial natriuretic factor inhibits Na-K-Cl cotransport in teleost intestine. *Am. J. Physiol.* **249**, C531–C534.

Ogoshi, M., Inoue, K., Naruse, K., and Takei, Y. (2006). Evolutionary history of the calcitonin gene-related peptide family in vertebrates revealed by comparative genomic analyses. *Peptides* **27**, 3154–3164.

Ogoshi, M., Nobata, S., and Takei, Y. (2008). Potent osmoregulatory actions of homologous adrenomedullins administered peripherally and centrally in eels. *Am. J. Physiol.* **295**, R2075–R2083.

Oleksiak, M. F., and Crawford, D. L. (2006). Functional genomics in fishes: insights into physiological complexity. In: *The Physiology of Fishes (CRC Marine Science Series)* (D. H. Evans and J. B. Claiborne, eds), 3rd edition, pp. 523–549. CRC Press, Boca Raton.

Olson, K. R. (1992). Blood and extracellular fluid volume regulation: role of the renin-angiotensin, kallikrein-kinin systems and atrial natriuretic peptides. In: *Fish Physiology Vol. 12B* (W. S. Hoar, D. J. Randall and A. P. Farrell, eds), pp. 135–254. Academic Press, San Diego.

Olsson, C., and Holmgren, S. (2001). The control of gut motility. *Comp. Biochem. Physiol. A* **128**, 479–501.

Olsson, C., Holbrook, J. D., Bompadre, G., Jönsson, E., Hoyle, C. H., Sanger, G. J., Holmgren, S., and Andrews, P. L. (2008). Identification of genes for the ghrelin and motilin receptors and a novel related gene in fish, and stimulation of intestinal motility in zebrafish (*Danio rerio*) by ghrelin and motilin. *Gen. Comp. Endocrinol.* **155**, 217–226.

Párrizas, M., Planas, J., Plisetskaya, E. M., and Gutiérrez, J. (1994). Insulin binding and receptor tyrosine kinase activity in skeletal muscle of carnivorous and omnivorous fish. *Am. J. Physiol.* **266**, 1944–1950.

Peyon, P., Lin, X., Himick, B. A., and Peter, R. E. (1998). Molecular cloning and expression of cDNA encoding precholecystokinin in goldfish. *Peptides* **19**, 199–210.

Peyon, P., Saied, H., Lin, X., and Peter, R. E. (2000). Preprotachykinin gene expression in goldfish brain: sexual, seasonal, and postprandial variations. *Peptides* **21**, 225–231.

Pierce, A. L., Fox, B. K., Davis, L. K., Visitacion, N., Kitahashi, T., Hirano, T., and Grau, E. G. (2007). Prolactin receptor, growth hormone receptor, and putative somatolactin receptor in Mozambique tilapia: tissue specific expression and differential regulation by salinity and fasting. *Gen. Comp. Endocrinol.* **154**, 31–40.

Planas, J. V., Méndez, E., Baños, N., Capilla, E., Castillo, J., Navarro, I., and Gutiérrez, J. (2000). Fish insulin, IGF-I and IGF-II receptors: a phylogenetic approach. *Amer. Zool.* **40**, 223–233.

Plisetskaya, E. M., Pollock, H. G., Elliott, W. M., Youson, J. H., and Andrews, P. C. (1988). Isolation and structure of lamprey (*Petromyzon marinus*) insulin. *Gen. Comp. Endocrinol.* **69**, 46–55.

Porter, E. M., Bevins, C. L., Ghosh, D., and Ganz, T. (2002). The multifaceted Paneth cell. *Cell. Mol. Life Sci.* **59**, 156–170.

Raldua, D., Otero, D., Fabra, M., and Cerda, J. (2008). Differential localization and regulation of two aquaporin 1 homologs in the intestinal epithelia of the marine teleost, *Sparus aurata. Am. J. Physiol.* **294**, R993–R1003.

Rankin, J. C. (2002). Drinking in hagfishes and lampreys. In: *Osmoregulation and Drinking in Vertebrates* (N. Hazon and G. Flik, eds), pp. 1–18. BIOS Scientific Publishers Ltd, Oxford.

Rao, M. C., Nash, N. T., and Field, M. (1984). Differing effects of cGMP and cAMP on ion transport across flounder intestine. *Am. J. Physiol.* **246**, C167–C171.

Reinecke, M., Schmid, A., Ermatinger, R., and Loffing-Cueni, D. (1997). Insulin-like growth factor I in the teleost *Oreochromis mossambicus*, the tilapia: gene sequence, tissue expression, and cellular localization. *Endocrinology* **138**, 3613–3619.

Roch, G. J., Wu, S., and Sherwood, N. M. (2009). Hormones and receptors in fish: do duplicates matter? *Gen. Comp. Endocrinol.* **161**, 3–12.

Rønnestad, I., Kamisaki, Y., Conceição, L. E. C., Morais, S., and Tonheim, S. K. (2007). Digestive physiology of marine fish larvae: hormonal control and processing capacity for proteins, peptides, and amino acids. *Aquaculture* **268**, 82–97.

Ruibal, C., Soengas, J. L., and Aldegunde, M. (2002). Brain serotonin and the control of food intake in rainbow trout (*Oncorhynchus mykiss*): effects of changes in plasma glucose levels. *J. Comp. Physiol. A* **188**, 479–484.

Ruskoaho, H. (1992). Atrial natriuretic peptide: synthesis, release, and metabolism. *Pharmacol. Rev.* **44**, 479–602.

Sakamoto, T., and Hirano, T. (1991). Growth hormone receptors in the liver and osmoregulatory organs of rainbow trout: characterization and dynamics during adaptation to seawater. *J. Endocrinol.* **130**, 425–433.

Sakamoto, T., McCormick, S. D., and Hirano, T. (1993). Osmoregulatory actions of growth hormone and its mode of action in salmonids: a review. *Fish Physiol, Biochem.* **11**, 155–164.

Sandra, O., Rouzic, P. L., Rentier-Delrue, F., and Prunet, P. (2001). Transfer of tilapia (*Oreochromis niloticus*) to a hyperosmotic environment is associated with sustained expression of prolactin receptor in intestine, gill, and kidney. *Gen. Comp. Endocrinol.* **123**, 295–307.

Schonbrunn, A. (2008). Selective agonism in somatostatin receptor signaling and regulation. *Mol. Cell. Endocrinol.* **286**, 35–39.

Schonhoff, S. E., Giel-Maloney, M., and Leiter, A. B. (2004). Minireview: development and differentiation of gut endocrine cells. *Endocrinology* **145**, 2639–2644.

Schulz, S., Green, C. K., Yuen, P. S. T., and Garbers, D. L. (1990). Guanylyl cycloase is a heat-stable enterotoxin receptor. *Cell* **63**, 941–948.

Schwartz, M. W., Woods, S. C., Porte, D., Jr., Seeley, R. J., and Baskin, D. G. (2000). Central nervous system control of food intake. *Nature* **404**, 661–671.

Seidelin, M., Madsen, S. S., Byrialsen, A., and Kristiansen, K. (1999). Effects of insulin-like growth factor-I and cortisol on Na^+, K^+-ATPase expression in osmoregulatory tissues of brown trout (*Salmo trutta*). *Gen. Comp. Endocrinol.* **113**, 331–342.

Sharif-Naeini, R., Ciuts, S., Zhang, Z., and Bourque, C. W. (2008). Contribution of TRPV channels to osmosensory transduction, thirst, and vasopressin release. *Kid. Internatl.* **73**, 811–815.

Sheridan, M. A., and Kittilson, J. D. (2004). The role of somatostatins in the regulation of metabolism in fish. *Comp. Biochem. Physiol. B* **138**, 323–330.

Sheridan, M. A., Kittilson, J. D., and Slagter, B. J. (2000). Structure-function relationships of the signaling system for the somatostatin peptide hormone family. *Amer. Zool.* **40**, 269–286.

Sherwood, N. M., Krueckl, S. L., and McRory, J. E. (2000). The origin and function of the pituitary adenylate cyclase-activating peptide (PACAP)/glucagon superfamily. *Endocrine Rev.* **21**, 619–670.

Sherwood, N. M., Adams, B. A., and Tello, J. A. (2005). Endocrinology of protochordates. *Can. J. Zool.* **83**, 225–255.

Silva, P., Stoff, J. S., Solomon, R. J., Lear, S., Kniaz, D., Greger, R., and Epstein, F. H. (1987). Atrial natriuretic peptide stimulates salt secretion by shark rectal gland by releasing VIP. *Am. J. Physiol.* **252**, F99–F103.

Silva, P., Lear, S., Reichlin, S., and Epstein, F. H. (1990a). Somatostatin mediates bombesin inhibition of chloride secretion in the shark rectal gland. *Am. J. Physiol.* **258**, R1459–R1463.

Silva, P., Solomon, R. J., and Epstein, F. H. (1990b). Shark rectal gland. In: *Methods in Enzymology* (S. Fleischer and B. Fleischer, eds.), *Vol. 192*, pp. 754–766. Academic Press, San Diego, CA, USA.

Silva, P., Epstein, F. H., Karnaky, K. J., Jr., Reichlin, S., and Forrest, J. N., Jr. (1993). Neuropeptide Y inhibits chloride secretion in the shark rectal gland. *Am. J. Physiol.* **265**, R439–R466.

Silva, P., Solomon, R. J., and Epstein, F. H. (1999). Mode of activation of salt secretion by C-type natriuretic peptide in the shark rectal gland. *Am. J. Physiol.* **277**, R1725–R1732.

Silverstein, J. T., and Plisetskaya, E. M. (2000). The effects of NPY and insulin on food intake regulation in fish. *Amer. Zool.* **40**, 296–308.

Silverstein, J. T., Bondareva, V. M., Leonard, J. B. K., and Plisetsukaya, E. M. (2001). Neuropeptide regulation of feeding in catfish, *Ictalurus punctatus*: a role for glucagon-like peptide-1 (GLP-1)? *Comp. Biochem. Physiol.* **129**, 623–631.

Soengas, J. L., and Aldegunde, M. (2004). Brain glucose and insulin: effects on food intake and brain biogenic amines of rainbow trout. *J. Comp. Physiol. A* **190**, 641–649.

Solomon, R., Taylor, M., Dorsey, D., Silva., P., and Epstein, F. H. (1985a). Atriopeptin stimulation of rectal gland function in *Squalus acanthias*. *Am. J. Physiol.* **249**, R348–R354.

Solomon, R., Taylor, M., Sheth, S., Silva, P., and Epstein, F. H. (1985b). Primary role of volume expansion in stimulation of rectal gland function. *Am. J. Physiol.* **248**, R638–R640.

Stoff, J. S., Rosa, R., Hallac, R., Silva, P., and Epstein, F. H. (1979). Hormonal regulation of active chloride transport in the dogfish rectal gland. *Am. J. Physiol.* **237**, F138–F144.

Stoff, J. S., Silva, P., Lechan, R., Solomon, R., and Epstein, F. H. (1988). Neural control of shark rectal gland. *Am. J. Physiol.* **255**, R212–R216.

Sun, L. Z., and Farmanfarmaian, A. (1992). Biphasic action of growth hormone on intestinal amino acid absorption in striped bass hybrids. *Comp. Biochem. Physiol.* **103A**, 381–390.

Sundell, K., Bjornssson, B. T., Itoh, H., and Kawauchi, H. (1992). Chum salmon (*Onchorhynchus keta*) stanniocalcin inhibits in vitro intestinal calcium uptake in Atlantic cod (*Gadus morhua*). *J. Comp. Physiol. B* **162**, 489–495.

Sundell, K., Jutfelt, F., Agustsson, T., Olsen, R. E., Sandblom, E., Hansen, T., and Bjorsson, B. T. (2003). Intestinal transport mechanisms and plasma cortisol levels during normal and out-of-season parr-smolt transformation of Atlantic salmon, *Salmo salar*. *Aquaculture* **222**, 265–285.

Sundström, G., Larrson, T. A., Brenner, S., Venkatesh, B., and Larhammar, D. (2008). Evolution of the neuropeptide Y family: new genes by chromosome duplications in early vertebrates and teleost fishes. *Gen. Comp. Endocrinol.* **155**, 705–716.

Sunuma, T., Yamanome, T., Amano, M., Takahashi, A., and Yamamori, K. (2009). White background stimulates the food intake of a pleuronectiform fish the barfin flounder, *Verasper moseri* (Jordan and Gilbert). *Aquaculture Res.* **40**, 748–751.

Suzuki, R., Togashi, K., Ando, K., and Takei, Y. (1994). Distribution and molecular forms of C-type natriuretic peptide in plasma and tissue of a dogfish, *Triakis scyllia*. *Gen. Comp. Endocrinol.* **96**, 378–384.

Takahashi, H., Takahashi, A., and Sakamoto, T. (2006a). In vivo effects of thyroid hormone, corticosteroids and prolactin on cell proliferation and apoptosis in the anterior intestine of the euryhaline mudskipper (*Periophthalmus modestus*). *Life Sci.* **79**, 1873–1880.

Takahashi, H., Sakamoto, T., Hyodo, S., Shepherd, B. S., Kaneko, T., and Grau, E. G. (2006b). Expression of glucocorticoid receptor in the intestine of a euryhaline teleost, the Mozambique tilapia (*Oreochromis mossambicus*): effect of seawater exposure and cortisol treatment. *Life Sci.* **78**, 2329–2335.

Takahashi, H., Prunet, P., Kitahashi, T., Kajimura, S., Hirano, T., Grau, E. G., and Sakamoto, T. (2007). Prolactin receptor and proliferating/apoptotic cells in esophagus of the

Mozambique tilapia (*Oreochromis mossambicus*) in fresh water and in seawater. *Gen. Comp. Endocrinol.* **152**, 326–331.

Takei, Y. (2002). Hormonal control of drinking in eels: an evolutionary approach. In: *Osmoregulation and Drinking in Vertebrates* (N. Hazon and G. Flik, eds), pp. 61–82. BIOS Scientific Publishers Ltd., Oxford.

Takei, Y., and Balment, R. J. (2008). The neuroendocrine regulation of fluid intake and fluid balance. In: *Fish Neuroendocrinology* (N. J. Bernier, G. Van Der Kraak, A. P. Farrell and C. J. Brauner, eds), pp. 365–419. Academic Press, San Diego.

Takei, Y., and Loretz, C. A. (2006). Endocrinology. In: *The Physiology of Fishes* (D. H. Evans and J. B. Claiborne, eds), 3rd edition, pp. 271–318. CRC Press, Boca Raton.

Takei, Y., and Yuge, S. (2007). The intestinal guanylin system and seawater adaptation in eels. *Gen. Comp. Endocrinol.* **152**, 339–351.

Takei, Y., Hirano, T., and Kobayashi, H. (1979). Angiotensin and water intake in the Japanese eel, *Anguilla japonica. Gen. Comp. Endocrinol.* **38**, 446–475.

Takei, Y., Tsuchida, T., Li, Z., and Conlon, J. M. (2001). Antidipsogenic effect of eel bradykinin in the eel, *Anguilla japonica. Am. J. Physiol.* **281**, R1090–R1096.

Takei, Y., Ogoshi, M., and Inoue, K. (2007). A "reverse" phylogenetic approach for identification of novel osmoregulatory and cardiovascular hormones in vertebrates. *Frontiers Neuroendocrinol.* **28**, 143–160.

Takei, Y., Ogoshi, M., Wong, M. K. S., and Nobata, S. (2010). Molecular and functional evolution of the adrenomedullin family in vertebrates: what do fish studies tell us? In: *The Calcitonin Gene-Related Peptide Family. Form, Function and Future Perspectives* (D. L. Hay and I. M. Dickerson, eds), pp. 1–21. Springer, Dordrecht.

Taylor, M. M., Bagley, S. L., and Samson, W. K. (2005). Intermedin/adrenomedullin-2 acts within central nervous system to elevate blood pressure and inhibit food and water intake. *Am. J. Physiol.* **288**, R919–R927.

Thavanathan, R., and Volkoff, H. (2006). Effects of amylin on feeding of goldfish: interactions with CCK. *Regul. Pept.* **133**, 90–96.

Trischitta, F., Denaro, M. G., Faggio, C., Mandolfino, M., and Schettino, T. (1996). Different effects of cGMP and cAMP in the intestine of the European eel, *Anguilla anguilla. J. Comp. Physiol. B* **166**, 30–36.

Tsuchida, T., and Takei, Y. (1998). Effects of homologous atrial natriuretic peptide on drinking and plasma angiotensin II level in eels. *Am. J. Physiol.* **275**, R1605–R1610.

Tsukada, T., and Takei, Y. (2006). Integrative approach to osmoregulatory action of atrial natriuretic peptide in seawater eels. *Gen. Comp. Endocrinol.* **147**, 31–38.

Tsukada, T., Nobata, S., Hyodo, S., and Takei, Y. (2007). Area postrema, a brain circumventricular organ, is the site of antidipsogenic action of circulating atrial natriuretic peptide in eels. *J. Exp. Biol.* **210**, 3970–3978.

Tulipano, G., and Schulz, S. (2007). Novel insights in somatostatin receptor physiology. *Eur. J. Endocrinol.* **156**, S3–S11.

Uchida, K., Kajimura, S., Riley, L. G., Hirano, T., Aida, K., and Grau, E. G. (2003). Effects of fasting on growth hormone/insulin-like growth factor I axis in the tilapia, *Oreochromis mossambicus. Comp. Biochem. Physiol. A* **134**, 429–439.

Uesaka, T., Yano, K., Yamaguchi, M., Nagashima, K., and Ando, M. (1994). Somatostatin-related peptides isolated from the eel gut: effects on ion and water absorption across the intestine of the seawater eel. *J. Exp. Biol.* **188**, 205–216.

Uesaka, T., Yano, K., Sugimoto, S., and Ando, M. (1996). Effects of neuropeptide Y on ion and water transport across the seawater eel intestine. *Zool. Sci.* **13**, 341–346.

Unniappan, S., Lin, X., Cervini, L., Rivier, J., Kaiya, H., Kangawa, K., and Peter, R. H. (2002). Goldfish ghrelin: molecular characterization of the complementary deoxyribonucleic acid, partial gene structure and evidence for its stimulatory role in food intake. *Endocrinology* **143**, 4143–4146.

Uniappan, S., and Peter, R. E. (2005). Structure, distribution and physiological functions of ghrelin in fish. *Comp. Biochem. Physiol. A* **140**, 396–408.

Upton, Z., Yandell, C. A., Degger, B. G., Chan, S. J., Moriyama, S., Francis, G. L., and Ballard, F. J. (1998). Evolution of insulin-like growth factor-I (IGF-I) action: in vitro characterization of vertebrate IGF-I proteins. *Comp. Biochem. Physiol. B* **121**, 35–41.

Usher, M. L., Talbot, C., and Eddy, F. B. (1991). Intestinal water transport in juvenile Atlantic salmon (*Salmo salar* L.) during smolting and following transfer to seawater. *Comp. Biochem. Physiol.* **100A**, 813–818.

Valassi, E., Scacchi, M., and Cavagnini, F. (2008). Neuroendocrine control of food intake. *Nutr. Metab. Cardiovasc. Dis.* **18**, 158–168.

Valentich, J. D., and Forrest, J. N., Jr. (1991). Cl$^-$ secretion by cultured shark rectal gland cells I. Transepithalial transport. *Am. J. Physiol.* **260**, C813–C823.

Van de Peer, Y., Maere, S., and Meyer, A. (2009). The evolutionary significance of ancient genome duplications. *Nature Rev. Gen.* **10**, 725–732.

Vazana, M., Vizzini, A., Salerno, G., Di Bella, M. L., Celi, M., and Parrinello, N. (2008). Expression of a glucocorticoid receptor (DIGR1) in several tissues of the teleost fish *Dicentrarchus labrax*. *Tissue Cell* **40**, 89–94.

Veillette, P. A., and Young, G. (2005). Tissue culture of sockeye salmon intestine: functional response of Na$^+$,K$^+$-ATPase to cortisol. *Am. J. Physiol.* **288**, R1598–R1605.

Veillette, P. A., White, R. J., and Specker, J. L. (1993). Changes in intestinal fluid transport in Atlantic salmon (*Salmo salar* L.) during parr-smolt transformation. *Fish Physiol. Biochem.* **12**, 193–202.

Veilette, P. A., Sundell, K., and Specker, J. L. (1995). Cortisol mediates the increase in intestinal fluid absorption in Atlantic salmon during parr-smolt transformation. *Gen. Comp. Endocrinol.* **97**, 250–258.

Vigna, S. R. (1986a). Gastrointestinal tract. In: *Vertebrate Endocrinology: Fundamentals and Biomedical Implications. Vol. 1: Morphological Considerations* (P. K. T Pang and M. P Schreibman, eds), pp. 261–278. Academic Press, San Diego.

Vigna, S. R. (1986b). Evolution of hormone and receptor diversity: cholecystokinin and gastrin. *Amer. Zool.* **26**, 1033–1040.

Vigna, S. R. (2000). Evolution of the cholecystokinin and gastrin peptides and receptors. *Amer. Zool.* **40**, 287–295.

Volkoff, H. (2006). The role of neuropeptide Y, orexins, cocaine and amphetamine-related transcript, cholecystokinin, amylin and leptin in the regulation of feeding in fish. *Comp. Biochem. Physiol. A* **144**, 325–331.

Volkoff, H., Canosa, L. F., Unniappan, S., Cerda-Reverter, J. M., Bernier, N. J., Kelley, S. P., and Peter, R. E. (2005). Neuropeptides and the control of food intake in fish. *Gen. Comp. Endocrinol.* **142**, 3–19.

Volkoff, H., Uniappan, S., and Kelly, S. P. (2009). The endocrine regulation of food intake. In: *Fish Neuroendocrinology (Fish Physiology Series)* (N. J Bernier, G Van der Kraak, A. P Farrell and C. J Brauner, eds), Vol. 28, pp. 421–465. Academic Press, San Diego.

Walker, R. L., Buret, A. G., Jackson, C. L., Scott, K. G.-E., Bajwa, R., and Habibi, H. R. (2004). Effects of growth hormone on leucine absorption, intestinal morphology, and ultrastructure of the goldfish intestine. *Can. J. Physiol. Pharmacol.* **82**, 951–959.

Wang, Y., and Conlon, J. M. (1995). Purification and structural characterization of vasoactive intestinal polypeptide from the trout and bowfin. *Gen. Comp. Endocrinol.* **98**, 94–101.

Wank, S. A. (1998). G protein-coupled receptors in gastrointestinal physiology I. CCK receptors: an exemplary family. *Am. J. Physiol.* **274**, G607–G613.

Wood, A. W., Duan, C., and Bern, H. A. (2005). Insulin-like growth factor signaling in fish. *Int. Rev. Cytol.* **243**, 215–285.

Wright, J. R., Jr., O'Hali, W., Yang, H., Han, X.-X., and Bonen, A. (1998). GLUT-4 deficiency and severe peripheral resistance to insulin in the teleost fish tilapia. *Gen. Comp. Endocrinol.* **111**, 20–27.

Wright, J. R., Jr., Bonen, A., Conlon, J. M., and Pohajdak, B. (2000). Glucose homoeostasis in the teleost fish tilapia: insights from Brockmann body xenotransplantation studies. *Amer. Zool.* **40**, 234–245.

Yamamoto, M., and Hirano, T. (1978). Morphological changes in the esophageal epithelium of the eel, *Anguilla japonica*, during adaptation to seawater. *Cell Tissue Res.* **192**, 25–38.

Yang, H., Morrison, C. M., Conlon, J. M., Laybolt, K., and Wright, J. R., Jr. (1999). Immunocytochemical characterization of the pancreatic islet cells of the Nile tilapia (*Oreochromis niloticus*). *Gen. Comp. Endocrinol.* **114**, 47–56.

Yang, T., Forrest, S. J., Stine, N., Endo, Y., Pasumarthy, A., Castrop, H., Aller, S., Forrest, J. N., Jr., Schnermann, J., and Briggs, J. (2002). Cyclooxygenase cloning in dogfish shark, *Squalus acanthias*, and its role in rectal gland Cl secretion. *Am. J. Physiol.* **283**, R631–R637.

Youson, J. (2000). The agnathan enteropancreatic endocrine system: phylogenetic and ontogenetic histories, structure, and function. *Amer. Zool.* **40**, 179–199.

Youson, J. H. (2007). Peripheral endocrine glands. I. The gastroenteropancreatic endocrine system and the thyroid gland. In: *Primitive Fishes (Fish Physiology Series)* (D. J McKenzie, A. P Farrell and C. J Brauner, eds), Vol. 26, pp. 381–455. Academic Press, San Diego.

Youson, J. H., and Al-Mahrouki, A. A. (1999). Ontogenetic and phylogenetic development of the endocrine pancreas (islet organ) in fishes. *Gen. Comp. Endocrinol.* **116**, 303–335.

Youson, J. H., Al-Mahrouki, A. A., Amemiya, Y., Graham, L. C., Montpetit, C. J., and Irwin, D. M. (2006). The fish endocrine pancreas: review, new data, and future research directions in ontogeny and phylogeny. *Gen. Comp. Endocrinol.* **148**, 105–115.

Yuge, S., and Takei, Y. (2007). Regulation of ion transport in eel intestine by the homologous guanylin family of peptides. *Zool. Sci.* **24**, 1222–1230.

Yuge, S., Inoue, K., Hyodo, S., and Takei, Y. (2003). A novel guanylin family (guanylin, uroguanylin and renoguanylin) in eels: possible osmoregulatory hormones in intestine and kidney. *J. Biol. Chem.* **278**, 22726–22733.

Yuge, S., Yamagami, S., Inoue, K., Suzuki, N., and Takei, Y. (2006). Identification of two functional guanylin receptors in eel: multiple hormone-receptor system for osmoregulation in fish intestine and kidney. *Gen. Comp. Endocrinol.* **149**, 10–20.

Zeidel, J. D., Mathai, J. C., Campbell, J. D., Ruiz, W. G., Apodaca, G. L., Riordan, J., and Zeidel, M. L. (2005). Selective permeability to urea in shark rectal gland. *Am. J. Physiol.* **289**, F83–F89.

Zhang, Y., Proenca, R., Maffei, M., Barone, M., Leopold, L., and Friedman, J. M. (1994). Positional cloning of the mouse obese gene and its human homologue. *Nature* **372**, 425–432.

Zhou, J., Li, W., Kamei, H., and Duan, C. (2008). Duplication of the IGFBP-2 gene in teleost fish: protein structure and functionality conservation and gene expression divergence. *PLoS ONE* **3**(12), e3926.

ADDITIONAL READING

Aoki, M., Kaneko, T., Katoh, F., Hasegawa, S., Tsutsui, N., and Aida, K. (2003). Intestinal water absorption through aquaporin 1 expressed in the apical membrane of mucosal epithelial cells in seawater-acclimated Japanese eel. *J. Exp. Biol.* **206**, 3495–3505.

Chan, S. J., and Steiner, D. F. (1994). Structure and expression of insulin like growth factor genes in fish. In: *Fish Physiology, Vol. XIII, Molecular Endocrinology of Fish* (N. M. Sherwood and C. L. Hew, eds), pp. 213–224. Academic Press, San Diego.

Hirano, T., and Hasegawa, S. (1984). Effects of angiotensins and other vasoactive substances on drinking in the eel, *Anguilla japonica. Zool. Sci.* **1**, 106–113.

Inoue, K., Russel, M. J., Olson, K. R., and Takei, Y. (2003b). C-type natriuretic peptide of rainbow trout (*Oncorhynchus mykiss*): primary structure and vasorelaxant activities. *Gen. Comp. Endocrinol.* **130**, 185–192.

Kozaka, T., and Ando, M. (2003). Cholinergic innervations to the upper esophageal sphincter muscle in the eel, with special reference to drinking behavior. *J. Comp. Physiol. B* **173**, 135–140.

Lundin, K., and Holmgren, S. (1984). Vasoactive intestinal polypeptide-like immunoreactivity and effects of VIP in the swimbladder of the cod, *Gadus morhua. J. Comp. Physiol. B* **154**, 627–633.

Matsuda, K., Shimakura, S., Miura, T., Murayama, K., Uchiyama, M., Kawauchi, H., Shioda, S., and Takahashi, A. (2007). Feeding-induced changes of melanin-concentrating hormone (MCH)-like immunoreactivity in goldfish brain. *Cell Tissue Res.* **328**, 375–382.

Moon, T. W. (1998). Glucagon: from hepatic binding to metabolism in teleost fish. *Comp. Biochem. Physiol. B* **121**, 27–34.

Murayama, K., Konno, N., Ishiguro, K., Wakasugi, T., Uchiyama, M., Shioda, S., and Matsuda, K. (2008). Isolation and characterization of four cDNAs encoding neuromedin U (NMU) from the brain and gut of goldfish, and the inhibitory effect of a deduced NMU on food intake and locomotor activity. *J. Neuroendocrinol.* **20**, 71–78.

Panopoulou, G., and Poustka, A. J. (2005). Timing and mechanism of ancient vertebrate genome duplications—the adventure of a hypothesis. *Trends Genet.* **21**, 559–567.

Shuttleworth, T. J., and Thorndyke, M. C. (1986). An endogenous peptide stimulates secretory activity in the elasmobranch rectal gland. *Science* **225**, 319–321.

Sower, S. A., Suzuki, K., and Reed, K. L. (2000). Perspective: research activity of enteropancreatic and brain/central nervous system hormones across invertebrates and vertebrates. *Amer. Zool.* **40**, 165–178.

Veillette, P. A., and Young, G. (2004). Temporal changes in intestinal Na^+,K^+-ATPase activity and in vitro responsiveness to cortisol in juvenile Chinook salmon. *Comp. Biochem. Physiol.* **138A**, 297–303.

8

THE ENTERIC NERVOUS SYSTEM

CATHARINA OLSSON

In most fish species investigated, the gastrointestinal tract is a highly innervated organ. Like in other vertebrates, the autonomic nerves control a variety of gut functions, from breakdown and transport of food to blood flow, osmoregulation and barrier functions. The autonomic innervation includes both local intrinsic nerves (the enteric nervous system) and extrinsic sympathetic and parasympathetic nerves. The autonomic nerves take part in involuntary reflex signaling. Hence, sensory extrinsic neurons, although strictly speaking not autonomic, are often considered in the same context as the autonomic motor neurons. They respond to various stimuli, many of which can be related to the presence of food in the gastrointestinal tract, and

The Multifunctional Gut of Fish: Volume 30
FISH PHYSIOLOGY

make up elaborate reflex pathways. The final effectors include smooth muscle cells in the gut wall and blood vessels, as well as glandular cells. They may also include other cell types that are involved in the signaling pathways like endocrine cells, interstitial cells of Cajal (ICCs) and glial cells. In this chapter, the layout of the enteric innervation in fish is presented, focusing on transmitter distribution and effect on gut motility. Furthermore, areas where knowledge of fish is still substantially lacking, like functional cell types and reflexes, are discussed.

1. ANATOMICAL OVERVIEW OF THE ENTERIC NERVOUS SYSTEM

The enteric nervous system follows a similar anatomical plan in most vertebrates including fish (Furness and Costa 1987; Holmgren and Olsson 2010). In most fish species it contains a high number of nerve cell bodies situated within the gut wall, with the majority found in the myenteric plexus between the two layers of smooth muscle (Fig. 8.1) (Monti 1895; Burnstock 1959; Olsson 2009). The number of neurons may differ among species and regions of the gut, but the neuronal density is usually similar to what is seen in mammals of a comparable size (Li and Furness 1993; Olsson and Karila 1995).

The vertebrate enteric nervous system also contains glial cells. In birds and mammals, the glial cells outnumber neurons by up to four to one (Gabella 1981; Ruhl et al. 2004). Glial cells play several roles in the mammalian gut, including support and protection for the neurons, but they can also take part in neurotransmission. So far little is known about the distribution and function of enteric glia in fish. Recent studies have shown putative glial cells immunoreactive to glial fibrillary acidic protein (GFAP) in the gut of some teleost species (Kelsh and Eisen 2000; Hagström and Olsson 2010).

Another cell type that may be involved in neurotransmission is the interstitial cells of Cajal (ICCs). While commonly found in close proximity to enteric neurons throughout the mammalian gut (e.g. Sanders et al. 2006), there are only few reports of ICCs in fish. Using methylene blue staining, Kirtisinghe (1940) described cells similar to mammalian ICCs in both a teleost (*Ophiocephalus striatus*) and an elasmobranch (*Scyliorhinus canicula*). Beyond this, their distribution and characteristics have been little investigated. One reason is the lack of suitable tools. The most common marker for ICCs in mammals is Kit, a tyrosine kinase receptor involved in the development of ICCs (Sanders et al. 2006). Kit isoforms are present in fish but seem to differ from the mammalian forms and most (but not all) attempts to label Kit-expressing ICCs in fish have failed (Parichy et al. 1999; Mellgren and Johnson 2005; Wallace et al. 2005; Rich et al. 2007).

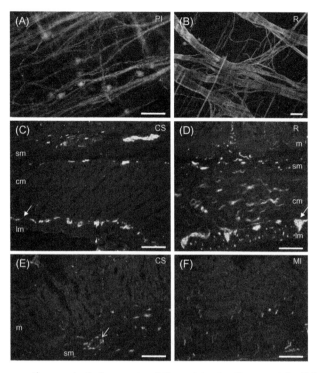

Fig. 8.1. Cross-sections and whole mounts of the gut in shorthorn sculpin (*Myoxocephalus scorpius*) (A, B), Arctic sculpin (*Myoxocephalus scorpioides*) (C, D, F) and Arctic staghorn sculpin (*Gymnocanthus tricuspis*) (E). Nerves are labeled with a combination of antibodies against acetylated tubulin and human neuronal protein (Hu) C/D. Whole mount of the myenteric plexus in the proximal intestine (A) and rectum (B) showing nerve cell bodies and bundles of nerve fibers. C. Cross-section of the cardiac stomach showing dense innervation of the myenteric plexus (arrow) and submucosa. D. Cross-section of the rectum showing innervation of all layers of the gut wall. E. Cross-section of the cardiac stomach, note innervation of blood vessel (open arrow) in the submucosa. F. Cross-section of the mid-intestine with nerve fibers in the submucosa and mucosa. CS, cardiac stomach; MI, mid-intestine; PI, proximal intestine; R, rectum; cm, circular muscle; lm, longitudinal muscle; m, mucosa; sm, submucosa. Scale bars: 100 μm. See color plate section

1.1. Myenteric Neurons

In most fish species, the majority of enteric neurons have their cell bodies within the myenteric plexus. Occasionally, cell bodies are also present in a submucous plexus, situated between the circular muscle layer and the submucosa. The myenteric nerve cells represent the three major functional classes necessary to make up multicellular reflexes, that is sensory, inter- and motor neurons (Furness 2006).

The appearance of the myenteric plexus varies among fish species and regions of the gut, but the cells are generally more evenly distributed than in mammals (Fig. 8.1). This is true for cyclostomes, as well as for elasmobranchs and teleosts (see Olsson 2009). In mammals, the myenteric neurons form ganglia, clusters of cells containing up to approximately 200 cell bodies (in, e.g., guinea pig; Furness 2006). If present in fish, myenteric ganglia generally contain only a few cells.

In fish as in other vertebrates, the myenteric neurons send axons and innervate the other layers of the gut wall, including both muscle layers (circular and longitudinal) as well as the submucosa and mucosa (Fig. 8.1). In most species, nerve processes run together in smaller or larger nerve bundles, making up the neuronal network (plexus). It is commonly observed that the main bundles run in the longitudinal direction with circumferentially oriented bundles being thinner. The cell bodies are often situated where nerve bundles branch.

The size of the nerve cell bodies varies among species, often reflecting the size of the fish. While in some fish the neuronal population is quite homogeneous in size (e.g. spiny dogfish, *Squalus acanthias*), there is a large variation in others like Atlantic cod (*Gadus morhua*) (Olsson and Karila 1995). Similarly, the morphology of the cells also differs among species, with some being almost exclusively unipolar and having a smooth cell soma, while others are multipolar or more star-like, with several processes protruding from the cell body (Monti 1895; Kirtisinghe 1940; Burnstock 1959; Olsson and Karila 1995; see also Olsson 2009).

2. EXTRINSIC INNERVATION OF THE GUT

When looking at histological sections of the gut it is usually impossible to tell the intrinsic enteric innervation from the extrinsic innervation. Fibers of either origin may innervate all layers of the gut wall, and they may also contain the same transmitters (see below). There have been few attempts to trace the origins of any nerve fibers in the fish gut and most of our knowledge of the extrinsic innervation comes from general morphological observations. It is suggested that the extrinsic innervation of the fish gastrointestinal tract, like in, e.g., mammals, includes efferent autonomic motor pathways as well as a high number of afferent sensory fibers.

2.1. Efferent (Motor) Innervation

The extrinsic efferent innervation comprises the autonomic parasympathetic and sympathetic pathways. In fish in general, it includes cranial and spinal nerve trunks (Nilsson 1983; Olsson 2009; Holmgren and Olsson 2010) (Fig. 8.2). In contrast to most mammals, little extrinsic innervation originates in the sacral (pelvic) region of the spinal cord in most fish.

The cranial innervation of fish is mainly via a pair of vagal nerves and corresponds to the parasympathetic innervation of the upper part of the gastrointestinal tract of most mammals (Nicol 1952; Burnstock 1969; see also Nilsson 1983). The preganglionic nerves leave the central nervous system in the brainstem, while the postganglionic cells are found in the lower part of the vagus or in the gut itself (i.e. the enteric nerves). In Atlantic hagfish (*Myxine glutinosa*), the left and right vagi join together before innervating the gut (Young 1931; Nicol 1952). In most other teleost and elasmobranch species, the vagal nerves innervate mainly the anterior part of the gut, including the esophagus, stomach and proximal intestine (Nicol 1952; Burnstock 1959, 1969; Nilsson 1983) (Fig. 8.2). In stomachless teleosts like zebrafish (*Danio rerio*) and tench (*Tinca tinca*) as well as in hagfish, the vagus innervates the intestine down to the distal regions (see Nilsson 1983; Olsson et al. 2008).

The spinal innervation involves one or several splanchnic nerves (Fig. 8.2). In cyclostomes, these spinal nerves mainly innervate the hindgut (Nicol 1952). In most elasmobranchs, a pair of anterior splanchnic nerves innervates the posterior stomach and anterior intestine, while mid and posterior splanchnic nerves innervate the spiral intestine and rectum (Nicol

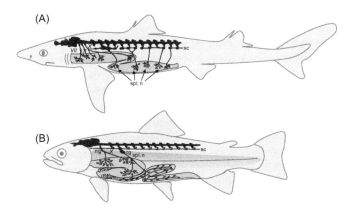

Fig. 8.2. Schematic drawing of the extrinsic autonomic innervation of the gastrointestinal tract in an elasmobranch (spiny dogfish) (A) and a teleost (rainbow trout) (B). Cranial outflow (parasympathetic nerves) is mainly via the vagus (cranial nerves X). Spinal pathways run in the splanchnic nerve(s) to the gut. IX, glossopharyngeal nerve; VII, facial nerve; X, vagal nerve; cg, celiac ganglion; ng, nodose ganglion; sc, sympathetic chain; spl. n, splanchnic nerve(s). (Courtesy of Albin Gräns, modified from Holmgren and Olsson 2010.)

1952). In teleosts, branches of the anterior splanchnic nerve innervate the posterior stomach (if present) and most of the intestine (Burnstock 1959; Nilsson 1983). The splanchnic nerves contain postganglionic sympathetic fibers that originate in the sympathetic chain ganglia. In teleosts, elasmobranchs and cyclostomes, anterior spinal fibers also enter the gut via the vagus (i.e. a vago-sympathetic trunk) (Nilsson 1983).

2.2. Afferent (Sensory) Innervation

The extrinsic sensory (afferent) innervation sends information about the state of the gut to higher centra in the central nervous system as well as being involved in the long autonomic reflexes. In mammals, these sensory nerves constitute the majority of fibers in both cranial and spinal nerves (Grundy and Schemann 2002). The proportion of sensory fibers in the fish extrinsic nerves is so far little investigated, to a large extent because of the lack of reliable markers for sensory neurons (see more under Section 4.1).

Afferent vagal nerves have their cell bodies in the nodose ganglia while spinal sensory nerves have their cell bodies in the dorsal root ganglia (see Furness 2006) (Figs 8.2, 8.4). Retrograde tracing of nerve fibers in Atlantic cod has shown that a proportion of the cells in the nodose ganglia have their peripheral endings in the stomach (Karila 1997) but the extent to which cells in the dorsal root ganglia innervate the fish gut remains to be investigated.

3. TRANSMITTER CONTENT AND DISTRIBUTION

Enteric nerves express a wide variety of transmitters; most transmitters found in the central nervous system are found in identical or similar forms in the gut. Furthermore, most transmitters found in the mammalian gut are also found in most fish species. Nonetheless, distribution and abundance show species and regional differences (Table 8.1). Most of our knowledge about distribution comes from immunohistochemical studies, often using commercially available antisera raised against mammalian sequences. Generally, antisera against neuropeptides still recognize the corresponding sequence in fish species, while antisera against larger molecules, e.g. enzymes, transporters or receptors, may have less affinity. For a comprehensive summary of the distribution of various transmitters in fish see Holmgren and Olsson (2010).

3.1. Acetylcholine

One of the most important transmitters in the gut, controlling, e.g., motility and acid secretion, is probably acetylcholine (see, e.g., Tobin et al.

Table 8.1

The distribution of some common transmitters in the gastrointestinal tract of selected fish species. For references, see Holmgren and Olsson, 2010

	Ach	CA	5-HT	NOS	BM/GRP	CGRP	Gal	G/CCK	NPY	NT	PACAP	Som	Tach	VIP
Cephalaspidomorphi														
European river lamprey *Lampetra fluviatilis*		nf, nc	nf, nc		nf, nc	nf, nc	nf, nc							
Chondrichthyes														
Spiny dogfish *Squalus acanthias*		nf	nf	nf, nc	nf, nc			nf, nc				nf	nf, nc	nf, nc
Actinopterygii														
Zebrafish *Danio rerio*			nf, nc	nf, nc		nf, nc					nf		nf	nf
Rainbow trout *Oncorhynchus mykiss*		nf, nc	nf, nc	nf, nc	nf				nf	nf	nf, nc		nf	nf, nc
Atlantic cod *Gadus morhua*	nf, nc	nf	nf, nc	nf, nc	nf	nf, nc	nf, nc	nf	nf		nf, nc		nf, nc	nf, nc
Sarcopterygii														
South American lungfish *Lepidosiren paradoxa*						nf	nf	nf		nf		nf	nf	nf

nf, nerve fiber; nc, nerve cell body.

2009). Commonly used markers for cholinergic neurons are the synthesizing enzyme choline acetyl transferase (ChAT) and the vesicular acetylcholine transporter (VAChT), responsible for transferring acetylcholine into vesicles so it can be released from nerve terminals (Schemann et al. 1993; Arvidsson et al. 1997). However, lack of fish-specific VAChT and ChAT antisera has hampered the knowledge about the distribution of cholinergic neurons in fish species.

In the few studies on fish where the distribution of cholinergic nerves has been demonstrated, cell bodies are seen in the myenteric plexus (e.g. Karila et al. 1998). In Atlantic cod, immunohistochemistry, myotomies and pharmacological experiments together suggest that ChAT- and VAChT-immunoreactive nerve cell bodies include both motor and interneurons (Karila and Holmgren 1995; Karila et al. 1998). Pharmacological data have also shown extrinsic (vagal) cholinergic innervation in teleost (Nilsson 1983).

Since, at least in mammals, more than 80% of the enteric nerve population is cholinergic, the cells must co-express other signaling substances (Furness 2006). This most commonly includes neuropeptides and nitric oxide, see below.

3.2. Neuropeptides

The neuropeptides comprise a large group of signaling molecules, often released together with other transmitters (Holmgren and Jensen 2001). They commonly vary from a few up to around 40 amino acids, with peptides of similar origin grouped in families. The best-studied neuropeptides regarding distribution in the fish gut are the tachykinins and the vasoactive intestinal polypeptide (VIP) family, but other neuropeptides are also common.

Usually, the amino acid sequence of a peptide differs to some extent between species, and the degree of similarity is often used to assess the relationship among species. For details on peptide sequences and evolutionary aspects see Hoyle (1998) and Holmgren and Jensen (2001).

Tachykinins constitute a fairly large peptide family of which substance P and neurokinin A (NKA) are the most studied in the vertebrate gut (Holmgren and Jensen 2001; Severini et al. 2002). Several related peptides have been isolated from various species, including scyliorhinin I and II expressed in, e.g., lesser spotted dogfish (*Scyliorhinus canicula*) (Conlon et al. 1986).

Tachykinins are found throughout the gastrointestinal tract in most elasmobranchs and teleosts investigated. Commonly, numerous immuno-reactive nerve fibers are found in the myenteric plexus but fibers also innervate other gut layers (Jensen and Holmgren 1985, 1991; Jensen et al. 1993b; Karila et al. 1998) (Fig. 8.3). Generally, the available antisera do not

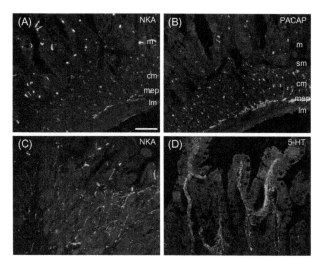

Fig. 8.3. Examples of transmitters found in the enteric nervous system of shorthorn sculpin (*Myoxocephalus scorpius*). (A) Dense innervation of NKA (neurokinin A)-immunoreactive nerve fibers in, e.g., the circular muscle and the myenteric plexus. (B) Dense innervation of PACAP (pituitary adenylate cyclase-activating polypeptide)-immunoreactive nerve fibers in, e.g., the circular muscle and the myenteric plexus. (C) NKA-immunoreactive nerve fibers in the mucosa and submucosa. (D) Serotonin (5-hydroxytryptamine, 5-HT)-immunoreactive nerve fibers in the mucosa. cm, circular muscle; lm, longitudinal muscle; m, mucosa; mep, myenteric plexus; sm, submucosa. Scale bar: 100 μm. See color plate section.

distinguish between substance P and NKA in fish; hence the staining is referred to as tachykinin immunoreactivity. Tachykinin-immunoreactive nerve cell bodies are common also in the myenteric plexus, indicating an enteric origin for many fibers (Karila et al. 1998). This does not rule out the possibility that some tachykinin-immunoreactive fibers may also be extrinsic. No neuronal tachykinin immunoreactivity has been detected in the cyclostome gut (Jensen and Holmgren 1991).

Tachykinin immunoreactivity, using antisera raised against fish tachy-kinins, is found in nerve fibers innervating both the celiac and mesenteric arteries in rainbow trout (*Onchorhynchus mykiss*) and spiny dogfish (Kågström et al. 1996; Kågström and Holmgren 1998).

VIP and *pituitary adenylate cyclase-activating polypeptide* (*PACAP*) are two closely related members of the secretin-glucagon superfamily and share 67% sequence identity in mammals (Campbell and Scanes 1992).

VIP was demonstrated in the fish enteric nerves by Langer et al. (1979). Since then it has been found in nerves in most species examined, except in cyclostomes where it is only present in endocrine cells (Reinecke et al. 1981;

Rombout and Reinecke 1984; Bjenning and Holmgren 1988; Van Noorden 1990). PACAP has also been demonstrated in nerves in the gut in several teleost species (Olsson and Holmgren 1994; Holmberg et al. 2004).

VIP and PACAP fibers innervate all tissue layers of the gut wall as well as its blood vessels (Fig. 8.3). Although most studies have reported few labeled nerve cell bodies, it has been argued that the innervation is most likely of intrinsic origin. In Atlantic cod, rainbow trout and spiny dogfish VIP and PACAP co-localize to 100% in enteric neurons (Olsson and Holmgren 1994; C. Olsson unpublished).

Calcitonin gene-related peptide (*CGRP*) belongs to the calcitonin gene peptide superfamily. CGRP-immunoreactive nerve fibers are found in both the stomach and intestine of Atlantic cod (Karila 1997; Shahbazi et al. 1998). Since nerve cell bodies are seen only in the intestine, it is believed the gastric CGRP-related innervation is mainly extrinsic. In zebrafish, CGRP-immunoreactive nerve fibers run in the vagus with additional intrinsic nerves in the distal part (Olsson et al. 2008). CGRP is also found in gut nerves in the European eel (*Anguilla anguilla*) (Domeneghini et al. 2000), Arctic lamprey (*Lampetra japonica*) (Yui et al. 1988) and Australian lungfish (*Neoceratodus forsteri*) (Holmgren et al. 1994).

CGRP co-localizes with substance P in certain nerves in some, but not all, species (Holmgren et al. 1994; Shahbazi et al. 1998). In the Arctic lamprey, CGRP coexists with serotonin in mucosal nerve cells and fibers (Yui et al. 1988).

Other more or less common neuropeptides identified in the fish gut include galanin, gastrin releasing peptide (GRP)/bombesin, peptide YY (PYY), cholecystokinin (CCK), opioids and somatostatin. *Galanin*-immunoreactive nerve fibers are most prominent in the mucosa and submucosa indicating a role in control of secretion (Karila et al. 1993; Holmgren et al. 1994; Preston et al. 1995). Fibers are also present in the muscle layers and may innervate blood vessels. The density of fibers is often higher in the stomach compared to the intestine. Most of the galanin-positive neurons also show immuno-reactivity to ChAT and some express VIP (Karila 1997; Bosi et al. 2007).

GRP/bombesin-like immunoreactive nerves are present in the gut of lampreys, and various elasmobranch and teleost species (Cimini et al. 1985; Bjenning and Holmgren 1988; Tagliafierro et al. 1988b; Yui et al. 1988; Bjenning et al. 1991). The presence of nerve cell bodies indicates an intrinsic origin of the innervation (Holmgren and Nilsson 1983; Bjenning and Holmgren 1988). While few fibers were found in West African lungfish (*Protopterus annectens*) and bowfin (*Amia calva*), the dogfish rectum is densely innervated (Lundin et al. 1984; Tagliafierro et al. 1988b; Rajjo et al. 1989). In Arctic lamprey, GRP/bombesin coexists with serotonin in nerves in the muscular layer (Yui et al. 1988).

PYY is considered the most common member of the NPY (neuropeptide Y) family in enteric nerves, although the exact nature of the peptide content is seldom evaluated (see Sundström et al. 2008). PYY/NPY-like material is present in nerves in the myenteric plexus and the muscle layer, as well as surrounding blood vessels in elasmobranchs and teleosts (Bjenning and Holmgren 1988; Bjenning et al. 1989; Burkhardt-Holm and Holmgren 1989). *Opioid* (mainly enkephalin)-immunoreactive nerve fibers and nerve cell bodies are also found in the gut of several teleost species (e.g. Langer et al. 1979; Bjenning and Holmgren 1988) while *somatostatin* is found in nerve fibers in European eel, rosy barb (*Puntius conchonius*) and spiny dogfish (Holmgren and Nilsson 1983; Rombout and Reinecke 1984; Domeneghini et al. 2000).

3.3. Nitric Oxide

Nitric oxide first emerged as an important neurotransmitter in the gut in the early 1990s (see Lincoln et al. 1997). It is synthesized by the enzyme nitric oxide synthase (NOS), and the presence or activity of this enzyme is commonly used to demonstrate nitrergic innervation. The first report of NOS immunoreactivity in the fish gut was in rainbow trout (Li and Furness 1993), showing that nitrergic nerve cell bodies were common in the myenteric plexus in all regions of the gut. A subsequent study demonstrated that nitrergic cells comprised on average close to or just over half the enteric population in Atlantic cod and spiny dogfish (Olsson and Karila 1995). Nitrergic neurons have since then been confirmed in several more teleost species including zebrafish, goldfish (*Carassius auratus*), sea bass (*Dicentrarchus labrax*) and shorthorn sculpin (*Myoxocephalus scorpius*) (Brüning et al. 1996; Holmberg et al. 2006; Pederzoli et al. 2007; C. Olsson unpublished). No nitrergic innervation was found in the intestine of hagfish (Olsson and Karila 1995). In general, nitrergic fibers innervate the myenteric plexus and the muscle layers.

In both Atlantic cod and spiny dogfish, around 50% of the NOS-immunoreactive neurons also express VIP/PACAP; the proportion of the VIP/PACAP-immunoreactive population expressing NOS is similar (Olsson and Karila 1995).

3.4. Amines

The most common amines found in autonomic nerves in the fish gut are catecholamines (noradrenaline and adrenaline) and serotonin (5-hydroxytryptamine, 5-HT).

3.4.1. CATECHOLAMINE

Over the years different methods have been applied to localize aminergic neurons. Early studies involving the Falk-Hillarp catecholamine fluorescence method (Falk and Owman 1965) showed that the gut of most classes of fish is densely innervated by adrenergic nerves (e.g. Baumgarten et al. 1973; Anderson 1983; Holmgren and Nilsson 1983). More recently, catecholaminergic neurons have been identified by immunohistochemical demonstration of one of the synthesizing enzymes, e.g. tyrosine hydroxylase (TH), dopamine-beta-hydroxylase (DBH) or phenylethanolamine-N-methyl transferase (PNMT).

The adrenergic fibers generally surround myenteric nerve cell bodies and innervate blood vessels in the submucosa as well as major vessels to the teleost gut (e.g. Anderson 1983). The lack of fluorescent nerve cell bodies in, e.g., goldfish and shortfin eel (*Anguilla australis*) suggests that the origin of these nerve fibers is mainly extrinsic. However, TH-positive cell bodies are reported in the teleost gut (Domeneghini et al. 2000; Olsson to be published). Since TH catalyzes the first step in catecholamine synthesis, TH-positive cells may include noradrenergic, adrenergic and dopaminergic cells. The presence of dopaminergic nerves in the fish gut is poorly investigated. Early reports on lampreys (*Lampetra*) suggest that there are both dopaminergic and adrenergic nerve cell bodies (Baumgarten et al. 1973; Sakharov and Salimova 1980). It has also been suggested that some catecholamine fluorescent nerves in the teleost intestine contain dopamine rather than noradrenaline or adrenaline (Nilsson 1983).

3.4.2. SEROTONIN

Serotonin-immunoreactive nerves are present to a varying degree in most teleosts (Fig. 8.3). The interspecies variations are often correlated with the presence or absence of serotonin in muscosal endocrine cells (Anderson 1983). In shortfin eel, both nerves and endocrine cells contain serotonin (Anderson and Campbell 1988), while no serotonergic nerves are found in the European eel (Domeneghini et al. 2000). In cyprinids like goldfish and zebrafish as well as in cyclostomes, serotonin is present only in nerves (Baumgarten et al. 1973; Goodrich et al. 1980; Sakharov and Salimova 1980; Pederzoli et al. 1996; Olsson et al. 2008). In sturgeons (*Chondrostei*) both endocrine and nerve cells contain serotonin (Salimova and Fehér 1982), while serotonergic nerves are sparse in elasmobranchs (Holmgren and Nilsson 1983; Cimini et al. 1985).

3.5. Other Transmitters in the Gut

Other transmitters in the innervation of the gut include amino acids like γ-amino butyric acid (GABA) and glutamate. There are so far no reports of glutamate and few reports only of GABA from fish gut (Gábriel et al. 1990; C. Olsson unpublished). Purinergic nerves may use purine derivates like adenosine and ATP as transmitters when released in high enough concentrations (for recent review see Gourine et al. 2009).

4. FUNCTIONAL CELL TYPES

To make up functional reflex pathways, the myenteric neurons must comprise different neuronal classes (see below). These include sensory, inter- and motor neurons. To further improve functionality of the nervous signaling, there should be both inhibitory and excitatory neurons. Different nerve types also have different targets. So far there are few studies looking at the functions of individual nerve cells in fish. For example, there are no electrophysiological studies that distinguish between the properties of different neuronal classes.

In mammals, sensory and motor/interneurons have distinct electrical properties that are commonly used to discriminate between cell types. Sensory neurons (AH neurons) are generally characterized by a prominent slow after-hyperpolarization taking place after an action potential; this is not seen in inter- and motorneurons (S neurons) (e.g. Furness 2006).

Furthermore, in most mammals AH neurons can also be distinguished from S neurons by their morphology. While AH neurons have a smooth cell body with several long processes, S neurons have flattened dendrites and a single axon. These morphological distinctions are often less apparent in fish. Although some studies report the presence of both morphologies, the cells seem more uniform in other species, with few cells resembling the mammalian AH cells (see above and Olsson 2009). The significance of these differences remains unclear. There are so far no studies showing a correlation between morphology and function in fish neurons and little can be said of the possible prevalence of sensory neurons in the fish gut.

4.1. Chemical Coding

An additional way to approach the lack in our understanding of the enteric nervous system in fish (and many other non-mammalian species) is to expand the number of co-localization studies. In mammals, all myenteric neurons identified so far contain more than one transmitter (see, e.g., Costa et al. 1996). Plus they often contain other substances that alone or in combination with the transmitters characterize a certain neuronal class. Such markers include calcium-binding proteins like calbindin and calretinin (Costa et al. 1996). Cells can also be identified by the selective binding of isolectin B4 (IB4) (Furness 2006). Combining immunohistochemistry with morphological traits (e.g. shape and projection) and electrophysiological studies in mammals have given quite a good picture of the chemical code representing different functional cell types (Costa et al. 1996; Furness 2006).

Double or triple labeling of nerve cells in fish is limited and has so far been performed in a somewhat random fashion, not least because of lack of specific antisera. Some of these results are already mentioned, e.g. the coexistence of VIP, PACAP and nitric oxide in most fish species. More recent studies combine the presence of calbindin with the presence or absence of different neurotransmitters (Olsson 2010).

Chemical coding also relates to the extrinsic innervation. In mammals, co-localization of CGRP and substance P is a common marker for extrinsic spinal sensory pathways (Green and Dockray 1988). Similarly, CGRP is present in about 20% of the substance P-immunoreactive nerve fibers in the gut of the Australian lungfish (Holmgren et al. 1994), indicating that these may be sensory although it has not been correlated to any physiological data. In contrast, only few fibers co-express the two peptides in the Atlantic cod (Shahbazi et al. 1998) and there are no reports of coexistence between tachykinins and CGRP in elasmobranchs. If we assume that fish, like mammals, have extrinsic sensory pathways, these data suggest that the chemical coding is different. However, until we have electrophysiological data to supplement the histochemical and morphological studies, we can only speculate about how to interpret the code in fish.

It must also be noted that the combination of markers characterizing one cell type in one species may not necessarily play the same role in other even closely related species. For example, calbindin is believed to be selectively expressed in intrinsic sensory neurons in some mammals but have a broader distribution in others. There may even be regional differences in the same species.

So far, much of the knowledge on functional nerve cell types in fish comes from pharmacological studies, combining, e.g., application of transmitters with nerve blockers like tetrodotoxin (TTX) (see below). In a few cases, these results are correlated to the projection of the neurons.

5. DEVELOPMENT OF THE ENTERIC NERVOUS SYSTEM

The enteric nervous system develops from neural crest-derived precursors, migrating along predefined pathways to the developing gut (Lamers et al. 1981; Sadaghiani and Vielkind 1990; Raible et al. 1992). The neural crest is an embryonic transient structure, formed from the ectoderm along the dorsal margins of the developing neural tube and different regions of the crest give rise to different cell populations. The anterior (vagal) crest region is suggested to give rise to most enteric neurons in fish (Sadaghiani and Vielkind 1990; Bisgrove et al. 1997). Whether the sacral region contributes (as it does in mammals) is unknown.

Zebrafish are a model for early development (Raible et al. 1992; Shepherd et al. 2001; Holmberg et al. 2003, 2004, 2006, 2007), although additional data exist for other species, e.g. turbot (*Psetta maxima*) and green swordtail (*Xiphophorus helleri*) (Sadaghiani and Vielkind 1990; Reinecke et al. 1997). In zebrafish, presumptive enteric neuron precursors are detected around 18 h post-fertilization (hpf) and a few hours later (24 hpf) these cells were associated with the developing gut (Bisgrove et al. 1997). At 48 hpf nerve cells are present within the gut and subsequently begin to grow axons, projecting to other cells both within the myenteric plexus and in other layers of the gut wall (Bisgrove et al. 1997; Holmberg et al. 2003; Wallace et al. 2005). At 3 dpf (days post-fertilization) the neurons have begun producing transmitters like acetylcholine and nitric oxide (Holmberg et al., 2004, 2006). For information on other species, see Olsson and Holmgren (2009).

Many other transmitters, including various neuropeptides, are expressed early in development, in both zebrafish and other species (Tagliafierro et al. 1988a; Reinecke et al. 1997; Holmberg et al. 2004; Pederzoli et al. 2004; Olsson et al. 2008). The exact order and timing of transmitters may vary among species and gut regions. Whether or when the enteric nervous system is functional also depends on the presence of the correct receptors and second messenger systems in the effector cells. In zebrafish, motility studies have shown effects of NKA and PACAP as early as 5 dpf (Holmberg et al. 2004).

The fate of a precursor cell depends on several chemical guidance cues, few of which have been established for fish. In mammals, two of the most important signaling pathways for a normal development are the glial-derived neurotrophic factor (GDNF)/Ret-Gfrα1 and endothelin-3 (ET-3)/endothelin B receptor systems. *c-ret* is transient expressed in enteric nerves in zebrafish during development. If either GDNF or the receptor components Ret or Gfrα1 is absent, enteric neurons do not develop appropriately (Shepherd et al. 2001, 2004).

For more detailed overviews on the molecular background of the development of the enteric nervous system as well as on functional aspects of the development in zebrafish, see Holmberg et al. (2008) and Wallace (2010).

6. NEURONAL CONTROL OF EFFECTOR FUNCTIONS

6.1. Effector Systems

Enteric nerves control a number of different gastrointestinal functions. Consequently the nerves make synaptic contacts with a variety of effector systems and cell types (see Furness 2006). Looking at a cross-section of the gut wall, it is apparent that nerve fibers reach all layers (Fig. 8.1). Most enteric fibers are varicose, making *en passant* contact with the target cells.

Generally, both muscle layers are densely innervated, indicating the important role of the enteric nervous system in control of gut motility. This will be discussed in more detail below. In mammals, ICCs may be involved in the signal transmission from nerves to muscle cells (Sanders et al. 2006) but, as mentioned above, information is lacking on the role of ICCs in the fish gut.

In the mucosal layer, enteric and/or extrinsic nerves innervate various glands and epithelial cells, thereby controlling secretion of acid, pepsinogen, mucus, hormones, etc. Acid and pepsinogen are secreted from oxyntico-peptic cells in the stomach and the secretion is tightly controlled by interacting hormonal and nervous signals. As in mammals, acetylcholine stimulates acid secretion in elasmobranchs and teleosts (Smit 1968; Holstein 1976, 1977; Holstein and Cederberg 1980). Since vagotomy strongly reduces acid secretion, extrinsic (vagal) innervation is believed to play an important role. Cholinergic nerves may act directly on oxynticopeptic cells and also via release of histamine from nearby cells (Holstein 1976). In contrast, acetylcholine has only weak effect on pepsinogen secretion (Holstein and Cederberg 1984). Different neuropeptides like VIP, somatostatin and tachykinins as well as serotonin also affect secretion; this effect may be either neuronal or hormonal (e.g. Holstein and Cederberg 1984, 1986).

The enteric innervation of the mucosa may also take part in the control of barrier functions, water and ion transport and nutrient uptake (Olsson and Holmgren 2009). For example, cholinergic and adrenergic nerves reduce ion permeability in the European eel intestine (Trischitta et al. 1999). In mammals, there is cross-talk between the immune and nervous system (for recent reviews, see Van Nassauw et al. 2007; Lomax et al. 2009). This area needs to be further investigated in fish.

Small blood vessels (arteries and arterioles) both within the gut wall and reaching it from outside, show a dense innervation in many fish species (Fig. 8.1). The nerves generally make contact with the smooth muscle layer. As with other functions, vascular control depends on both neuronal and hormonal signals. Adrenaline and/or noradrenaline are major vasoconstric-tors in the gut, as elsewhere in the body (Holmgren and Nilsson 1974, 1983; Nilsson et al. 1975; Anderson 1983). Although the relative contribution from nerves and circulating catecholamines in fish is not clear, the latter may presumably dominate. A population of the adrenergic (extrinsic) nerves innervating gut blood vessels in the Atlantic cod co-express NPY (Karila et al. 1997). A perivascular innervation has also been reported in several elasmobranch species (Bjenning et al. 1989, 1991; Preston et al. 1998). NPY increases blood flow in cod and spiny dogfish while in several other elasmobranchs it decreases blood flow (Holmgren et al. 1992b; Preston et al. 1998; Shahbazi et al. 2002).

Tachykinins also have variable effects on blood flow. For example, substance P reduces gut blood flow in Australian lungfish while increasing it in spiny dogfish (Holmgren et al. 1992a, 1994). The response in Atlantic cod includes both an increase and a decrease (Jensen et al. 1991). The presence of perivascular nerves and pharmacological results also indicates that nitric oxide, CGRP and VIP are involved in the control of gut blood flow.

6.2. Nerves and Reflex Pathways

Nerve signaling in the gut is initiated by stimulation of sensory nerve endings; these nerves in turn make synaptic contact with other neurons. The simplest enteric multicellular reflexes include sensory neurons, interneurons and motorneurons. The interneurons may either stimulate or inhibit excitatory or inhibitory motorneurons. In, e.g., mammals, most reflexes are more complex, often including several interneurons, both inhibitory and excitatory (Furness 2006). While very little is known about synaptic coupling and the electrical properties of enteric neurons in fish, it may be assumed that many features are similar to other vertebrates (Fig. 8.4).

The extrinsic reflexes may include varying numbers of neurons. Extrinsic afferent nerves often synapse with preganglionic autonomic nerves, either in the spinal cord/brain stem or directly in sympathetic ganglia (Fig. 8.4). Efferent nerves mainly innervate enteric neurons, but may also control smooth muscle in the wall, blood vessels or secretory cells. Extrinsic sensory nerves also convey information to the brain about hunger and pain.

In mammals, extrinsic and intrinsic sensory neurons express various receptors responding to mechanical and/or chemical stimuli. These include distension (mechanical) of the gut wall, pH, composition of food, temperature, etc. The stimuli may be noxious as well as physiological. Mammalian vagal afferent nerve endings have morphologically specialized characteristics like the intraganglionic laminar endings (IGLEs) that respond to mechanical stimulation (Berthoud and Powley 1992; Zagorodnyuk et al. 2001). Their possible presence in fish has not received a lot of attention and little is also known about the types of receptors present on fish sensory nerves. Many of the mammalian receptors are ion channels, like the TRP (transient receptor potential) family (Holzer 2004). The corresponding genes have been isolated from several fish species (see NCBI; www.ncbi.nlm. nih.gov) but again, almost nothing has been published about their distribution in the fish gut.

Mammalian studies have demonstrated that intrinsic reflexes controlling propagating gut motility are polarized, i.e. one stimulus activates ascending

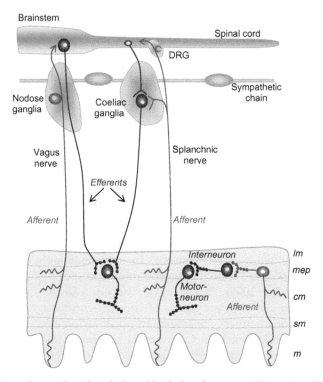

Fig. 8.4. Schematic overview of extrinsic and intrinsic reflexes controlling gut motility, based on mammalian models. Vagal afferents have their nerve cell bodies in the nodose ganglia while spinal afferents have theirs in the dorsal root ganglia (DRG). Spinal afferents may synapse with efferent nerves in the celiac ganglion in the sympathetic chain or in the spinal cord. Vagal and spinal efferents innervate the enteric nervous system but also, e.g., blood vessels in the gut wall. Local enteric reflexes include sensory (afferent), inter- and motor neurons. cm, circular muscle; lm, longitudinal muscle; m, mucosa; mep, myenteric plexus; sm, submucosa. See color plate section.

(excitatory) and descending (inhibitory) pathways simultaneously (Furness and Costa 1987). In Atlantic cod intestine, electrical stimulation in one region similarly caused contractions anally and relaxation orally (Karila and Holmgren 1995).

In vertebrates in general, the extrinsic reflexes coordinate events in different regions of the gut, e.g. a stimulus sensed in the intestine will affect motility or secretion in the stomach. The reflexes can also be more local. Accommodation or receptive relaxation of the stomach, i.e. relaxation in response to distension of either the stomach or esophagus, involves vagal pathways in mammals (see Furness 2006). In Atlantic cod and rainbow trout, accommodation is not affected by cutting vagal and splanchnic input, suggesting that local control mechanisms dominate (Grove and Holmgren 1992a,b). Vagal reflexes stimulate acid secretion in Atlantic cod (Holstein and Cederberg 1980).

The relative importance of different stimuli in triggering the reflexes in fish is so far not understood and further studies are needed. Mechanical stimulation can be mimicked by inflating a balloon placed in the gut (e.g. Grove and Holmgren 1992a,b), while the impact of food composition may be studied after local injections of different breakdown products (e.g. fatty acids or peptides) or by lowering the pH in the gut.

6.3. Neuronal Control of Motility

Rhythmic contractions of the smooth muscle, without any apparent stimulation, may be recorded both *in vivo* and from isolated gut segments. Neuronal control is demonstrated by blocking the nerves. Application of the sodium channel blocker TTX or the nicotinic receptor antagonist (ganglionic blocker) hexamethonium generally reduces and/or changes the contraction patterns. In zebrafish larvae (7 dpf), the frequency but not speed of the contractions was reduced *in vivo* by TTX (Holmberg et al. 2007). In Atlantic cod, rhythmic contractions of isolated intestinal strip preparations are reduced by TTX (Jensen and Holmgren 1985). While *in vivo* recordings reflect the integrated effects of both intrinsic and extrinsic innervation, the inhibitory effects seen *in vitro* are most likely by the enteric innervation. Another way to study the importance of enteric innervation comes from mutant or gene knockdown experiments in zebrafish where the severity of neuronal loss has been correlated to motility dysfunctions (Field et al. 2009).

A number of studies have examined the effects of different neurotransmitters on gut motility in different fish species. The two best-studied species are rainbow trout and Atlantic cod, with zebrafish as a recent addition. However, there are few consistent series of investigations trying to reveal the relative importance of each transmitter and how the signaling pathways interact. When it comes to integrated responses in whole animals, even less is known. Furthermore, exogenous application of transmitters does not necessarily discriminate between neuronal and hormonal responses.

The effect of a transmitter may differ among species and gut regions. Commonly, the response varies even within the same region and species, depending on, e.g., feeding status of the individual fish. Some transmitters also give rise to biphasic responses (e.g. relaxation followed by contraction) or reduce frequency while increasing amplitude, or vice versa. The following attempts to pinpoint the best-studied responses (see also Table 8.2 and Gräns and Olsson 2010).

Table 8.2

The effect of some common transmitters in the gastrointestinal tract of selected fish species. If the response differs between stomach and intestine both effects are noted. For references, see Gräns and Olsson, 2010

	Ach	Adr	5-HT	NO	BM/GRP	CGRP	Gal	NPY	PACAP	Som	Tach	VIP
Cephalaspidomorphi												
European river lamprey *Lampetra fluviatilis*	(−)		+								no	
Chondrichthyes												
Spin dogfish *Squalus acanthias*	+	+/−			+					+/−	+	R−
Actinopterygii												
Zebrafish *Danio rerio*	+			−					−			
Rainbow trout *Oncorhynchus mykiss*	+	+/−	+	−	+				−	+/−	+	?/−
Atlantic cod *Gadus morhua*	+	+/−[1]	+	−	+/−	−	+	+	−	+/−	+	−/no[2]
Sarcopterygii												
South American lungfish *Lepidosiren paradoxa*											+	

+ contraction; − inhibition/relaxation.
[1] Inhibition followed by contraction in some intestinal prep.
[2] Results on stomach differs between studies.

6.3.1. ACETYLCHOLINE

Application of acetylcholine (or the more stable analogue carbachol) on any gut region of most vertebrates, including fish, causes contractions. Depending on the initial state of the gut, this may be manifested as an increase in contraction amplitude, contraction frequency or simply basal tone, or a combination of these effects (von Euler and Östlund 1957; Holmgren and Fänge 1981; Nilsson and Holmgren 1983; Jensen and Holmgren 1985; Burka et al. 1989; see also Nilsson 1983; Jensen and Holmgren 1994).

In Atlantic cod intestine, application of specific receptor antagonists and nerve blockers (TTX) in combination with nerve tracings have led to the suggestion that the cholinergic response is mediated mainly by ascending enteric motorneurons acting via muscarinic receptors on the muscle (Jensen and Holmgren 1985; Karila and Holmgren 1995). It is most likely, however, that the cholinergic innervation also includes other cell types. In mammals, both ascending and descending interneurons are cholinergic (Steele et al. 1991).

Cholinergic nerves may also dominate the vagal innervation of the gut. The muscarinic cholinergic antagonist atropine reduces/abolishes contractions evoked by electrical stimulation of the vagus in European plaice (*Pleuronectes platessa*) (Edwards 1972; Stevenson and Grove 1977; see also Nilsson 1983). Stimulation of the vagus nerve gives weak and inconclusive responses in hagfish (Patterson and Fair 1933) and, in elasmobranchs, the contractile activity in the stomach was inhibited (Campbell 1975; Young 1983).

6.3.2. TACHYKININS

Tachykinins contract the gut in most fish except hagfish and lamprey (Jensen and Holmgren 1985, 1991; Jensen et al. 1987, 1993a,b; Karila et al. 1998). In Atlantic cod stomach, as well as in South American lungfish (*Lepidosiren paradoxa*) and some elasmobranch species, TTX does not affect the response to substance P, indicating that the effect is directly on the muscle (Holmgren 1985; Andrews and Young 1988; Jensen and Holmgren 1991; Jensen et al. 1993b). In contrast, in Atlantic cod intestine, TTX reduces the effect of substance P (Jensen 1997). Similarly, there is an indirect effect of substance P in rainbow trout stomach and intestine and in the intestinal bulb of common carp (*Cyprinus carpio*) (Holmgren et al. 1985; Kitazawa et al. 1988a,b; Jensen and Holmgren 1991). The indirect effect can be mediated by serotonin and/or acetylcholine since the response to substance P is reduced by atropine and methysergide.

6.3.3. SEROTONIN

The effect of serotonin depends on the type of receptor present but in most fish species, including cyclostomes, elasmobranchs and teleosts, it has a contractile effect on the gut (Johnels and Östlund 1958; Young 1980a,b, 1983;

Nilsson and Holmgren 1983; Jensen and Holmgren 1994). In several species of rays and skates, however, serotonin caused an increase of frequency while contraction amplitude in the stomach was decreased (Young 1983).

There seem to be species differences in the serotonergic signaling pathways. TTX reduces the response in Atlantic cod intestine. In some other species, neither atropine nor TTX affects the contractile effect of serotonin on rainbow trout gut (Grove and Campbell 1979; Holmgren et al. 1985; Jensen and Holmgren 1985; Burka et al. 1989; Kiliaan et al. 1989; Grove and Holmgren 1992a). The indirect pathway in Atlantic cod may involve release of acetylcholine and/or substance P (Holmgren et al. 1985; Jensen and Holmgren 1985, 1991; Jensen et al. 1987). Serotonin is also suggested to be released from excitatory descending interneurons in the Atlantic cod (Karila 1997). Whether serotonin plays a role in the sensory pathways, similar to the situation in mammals where it is released from endocrine cells and stimulates intrinsic sensory neurons, needs to be evaluated.

6.3.4. Nitric Oxide

Nitric oxide relaxes gut smooth muscles in fish (Green and Campbell 1994; Karila and Holmgren 1995; Olsson and Holmgren 2000; Holmberg et al. 2006). This can be seen as a reduction in basal tension and/or an inhibition of rhythmic contractions. Blocking the nitric oxide synthesis, both *in vivo* and *in vitro*, often reveals a nitrergic inhibitory tone (Karila and Holmgren 1995; Olsson and Holmgren 2000; Holmberg et al. 2006). In rainbow trout stomach, however, basal activity was not altered by NOS inhibition (Olsson et al. 1999). Judging from immunohistochemical and tracing studies, nitrergic innervation mainly involves descending motor (and possibly inter-) neurons (Karila 1997).

6.3.5. VIP and PACAP

VIP and PACAP are believed to act via the same receptors; however, there are clear regional and species differences in the effect of the two peptides. Although VIP is the better studied of the two, PACAP seems to be more potent. For example, PACAP reduces spontaneous contractions in cod intestinal strip preparations while neither mammalian nor cod VIP had any effect (Jensen and Holmgren 1985; Olsson and Holmgren 2000). In contrast, mammalian VIP reduces rhythmic contractions in the stomach *in vivo* (Jensen et al. 1991).

PACAP also inhibits activity in rainbow trout and zebrafish intestine and Japanese stargazer (*Uranoscopus japonicus*) rectum (Matsuda et al. 2000; Holmberg et al. 2004; C. Olsson unpublished). The effect of VIP on rainbow trout generally includes inhibition of the intestine, while the effect on the stomach could be either inhibitory or excitatory, or no effect at all depending on the preparation (Holmgren 1983). Similarly, in elasmobranchs, the response differs between species. While VIP inhibits activity in spiny dogfish rectum no effect was seen in either part of the gut in skates (Lundin et al. 1984; Andrews and Young 1988).

6.3.6. CATECHOLAMINES

The effect of adrenaline and/or noradrenaline is mainly mediated via the extrinsic sympathetic innervation and their effects vary among gut regions and species (Nilsson 1983; Jensen and Holmgren 1994). In teleosts, the predominating response to stimulation of the splanchnic nerves or application of adrenaline/noradrenaline is contraction of the stomach and relaxation of the intestine. This has been observed in several species including rainbow trout, brown trout (*Salmo trutta*) and European eel (von Euler and Östlund 1957; Burnstock 1958; Nilsson and Fänge 1967; Goddard 1973; Kitazawa et al. 1986a). However, in angler (*Lophius piscatorius*), the stomach relaxes in response to adrenaline (von Euler and Östlund 1957; Young 1980b), while Atlantic cod intestine shows a mixed response (von Euler and Östlund 1957; Nilsson and Fänge 1969; Jensen and Holmgren 1985).

Studies on common carp suggest that relaxation in response to adrenaline depends on inhibition of acetylcholine release (Kitazawa et al. 1986b). The effect of adrenaline in elasmobranchs is usually excitatory on the stomach and intestine and inhibitory on the rectum (Young 1980a; Nilsson and Holmgren 1983; Young 1983, 1988). The effect on isolated Atlantic hagfish intestine is weak and inconsistent (Holmgren and Fänge 1981). Also dopamine may have an excitatory effect on the stomach (Groisman and Shparkovskii 1989).

7. CONCLUDING REMARKS

The enteric innervation is a key regulator of gut functions. While much is known about the distribution of nerves and their transmitters in species representing the major fish groups, surprisingly little is still known about how the nerve cells function and interact in the control of gut functions. The innervation and the overall control of gut motility in fish follows a pattern similar to most other vertebrates groups, indicating the important roles of these control systems.

REFERENCES

Anderson, C. (1983). Evidence for 5-HT-containing intrinsic neurons in the teleost intestine. *Cell Tissue Res.* **230**, 377–386.
Anderson, C., and Campbell, G. (1988). Immunohistochemical study of 5-HT-containing neurons in the teleost intestine: relationship to the presence of enterochromaffin cells. *Cell Tissue Res.* **254**, 553–559.

Andrews, P. L., and Young, J. Z. (1988). The effect of peptides on the motility of the stomach, intestine and rectum in the skate (*Raja*). *Comp. Biochem. Physiol.* C **89**, 343–348.

Arvidsson, U., Riedl, M., Elde, R., and Meister, B. (1997). Vesicular acetylcholine transporter (VAChT) protein: a novel and unique marker for cholinergic neurons in the central and peripheral nervous systems. *J. Comp. Neurol.* **378**, 454–467.

Baumgarten, H. G., Björklund, A., Lachenmayer, L., Nobin, A., and Rosengren, E. (1973). Evidence for the existence of serotonin-, dopamine-, and noradrenaline-containing neurons in the gut of *Lampetra fluviatilis. Z. Zellforsch. Mikrosk. Anat.* **141**, 33–54.

Berthoud, H. R., and Powley, T. L. (1992). Vagal afferent innervation of the rat fundic stomach: morphological characterization of the gastric tension receptor. *J. Comp. Neurol.* **319**, 261–276.

Bisgrove, B. W., Raible, D. W., Walter, V., Eisen, J. S., and Grunwald, D. J. (1997). Expression of c-ret in the zebrafish embryo: potential roles in motoneuronal development. *J. Neurobiol.* **33**, 749–768.

Bjenning, C., and Holmgren, S. (1988). Neuropeptides in the fish gut. An immunohistochemical study of evolutionary patterns. *Histochemistry* **88**, 155–163.

Bjenning, C., Driedzic, W. R., and Holmgren, S. (1989). Neuropeptide Y-like immunoreactivity in the cardiovascular nerve plexus of the elasmobranchs *Raja erinacea* and *Raja radiata. Cell Tissue Res.* **255**, 481–486.

Bjenning, C., Farrell, A. P., and Holmgren, S. (1991). Bombesin-like immunoreactivity in skates and the in vitro effect of bombesin on coronary vessels from the longnose skate, *Raja rhina. Regul. Pept.* **35**, 207–219.

Bosi, G., Bermudez, R., and Domeneghini, C. (2007). The galaninergic enteric nervous system of pleuronectiformes (Pisces, Osteichthyes): an immunohistochemical and confocal laser scanning immunofluorescence study. *Gen. Comp. Endocrinol.* **152**, 22–29.

Brüning, G., Hattwig, K., and Mayer, B. (1996). Nitric oxide synthase in the peripheral nervous system of the goldfish, *Carassius auratus. Cell Tissue Res.* **284**, 87–98.

Burka, J. F., Blair, R. M., and Hogan, J. E. (1989). Characterization of the muscarinic and serotoninergic receptors of the intestine of the rainbow trout (*Salmo gairdneri*). *Can. J. Physiol. Pharmacol.* **67**, 477–482.

Burkhardt-Holm, P., and Holmgren, S. (1989). A comparative study of neuropeptides in the intestine of two stomachless teleosts (*Poecilia reticulata, Leuciscus idus melanotus*) under conditions of feeding and starvation. *Cell Tissue Res.* **255**, 245–254.

Burnstock, G. (1958). The effect of drugs on spontaneous motility and on response to stimulation of the extrinsic nerves of the gut of a teleostean fish. *Br. J. Pharmacol. Chemother.* **13**, 216–226.

Burnstock, G. (1959). The innervation of the gut of the brown trout *Salmo trutta. Quart. J. Microsc. Sci.* **100**, 199–220.

Burnstock, G. (1969). Evolution of the autonomic innervation of visceral and cardiovascular systems in vertebrates. *Pharmacol. Rev.* **21**, 247–324.

Campbell, G. (1975). Inhibitory vagal innervation of the stomach in fish. *Comp. Biochem. Physiol.* **50C**, 169–170.

Campbell, R. M., and Scanes, C. G. (1992). Minireview—evolution of the growth hormone-releasing factor (GRF) family of peptides. *Growth Regul.* **2**, 175–191.

Cimini, V., Van Noorden, S., Giordano-Lanza, G., Nardini, V., McGregor, G. P., Bloom, S. R., and Polak, J. M. (1985). Neuropeptides and 5-HT immunoreactivity in the gastric nerves of the dogfish (*Scyliorhinus stellaris*). *Peptides* **6**(Suppl. 3), 373–377.

Conlon, J. M., Deacon, C. F., O'Toole, L., and Thim, L. (1986). Scyliorhinin I and II: two novel tachykinins from dogfish gut. *FEBS Lett.* **200**, 111–116.

Costa, M., Brookes, S. J., Steele, P. A., Gibbins, I., Burcher, E., and Kandiah, C. J. (1996). Neurochemical classification of myenteric neurons in the guinea-pig ileum. *Neuroscience* **75**, 949–967.

Domeneghini, C., Radaelli, G., Arrighi, S., Mascarello, F., and Veggetti, A. (2000). Neurotransmitters and putative neuromodulators in the gut of *Anguilla anguilla* (L.). Localizations in the enteric nervous and endocrine systems. *Eur. J. Histochem.* **44**, 295–306.

Edwards., D. J. (1972). Reactions of the isolated plaice stomach to applied drugs. *Comp. Gen. Pharmacol.* **3**, 235–242.

Falk, B., and Owman, C. (1965). A detailed methodological description of the fluorescence method for the cellular demonstrations of biogenic monoamines. *Acta Univ. Lundensis II* **7**, 1–23.

Field, H. A., Kelley, K. A., Martell, L., Goldstein, A. M., and Serluca, F. C. (2009). Analysis of gastrointestinal physiology using a novel intestinal transit assay in zebrafish. *Neurogastroenterol. Motil.* **21**, 304–312.

Furness, J. B. (2006). *The Enteric Nervous System.* Blackwell Publishing, Oxford.

Furness, J. B., and Costa, M. (1987). *The Enteric Nervous System.* Churchill Livingstone, Edingburgh.

Gabella, G. (1981). Ultrastructure of the nerve plexuses of the mammalian intestine: the enteric glial cells. *Neuroscience* **6**, 425–436.

Gábriel, R., Halasy, K., Fekete, É., Eckert, M., and Benedeczki, I. (1990). Distribution of GABA-like immunoreactivity in myenteric plexus of carp, frog and chicken. *Histochemistry* **94**, 323–328.

Goddard, J. S. (1973). The effects of cholinergic drugs on the motility of the alimentary canal of *Blennius pholis* L. *Cell. Mol. Life Sci.* **29**, 974–975.

Goodrich, J. T., Bernd, P., Sherman, D., and Gershon, M. D. (1980). Phylogeny of enteric serotonergic neurons. *J. Comp. Neurol.* **190**, 15–28.

Gourine, A. V., Wood, J. D., and Burnstock, G. (2009). Purinergic signalling in autonomic control. *Trends Neurosci.* **32**, 241–248.

Gräns, A., and Olsson, C. (2010). Gut motility. In: *Encyclopedia of Fish Physiology: From Genome to Environment* (C. Olsson, S. Holmgren, E. D. Stevens and A. P. Farrell, eds). Elsevier, New York.

Green, K., and Campbell, G. (1994). Nitric oxide formation is involved in vagal inhibition of the stomach of the trout (*Salmo gairdneri*). *J. Auton. Nerv. Syst.* **50**, 221–229.

Green, T., and Dockray, G. J. (1988). Characterization of the peptidergic afferent innervation of the stomach in the rat, mouse and guinea-pig. *Neuroscience* **25**, 181–193.

Groisman, S. D., and Shparkovskii, I. A. (1989). Effect of dopamine and DOPA on electrical activity of stomach muscles in the skate *Raja radiata* and cod *Gadus morhua*. *Zh. Evolyut. Biokhim. Fiziol.* **25**, 505–511.

Grove, D. J., and Campbell, G. (1979). The role of extrinsic and intrinsic nerves in the coordination of gut motility in the stomachless flatfish *Rhombosolea tapirina* and *Ammotretis rostrata* Guenther. *Comp. Biochem. Physiol.* **63C**, 143–159.

Grove, D. J., and Holmgren, S. (1992a). Intrinsic mechanisms controlling cardiac stomach volume of the rainbow trout (*Oncorhynchus mykiss*) following gastric distension. *J. Exp. Biol.* **163**, 33–48.

Grove, D. J., and Holmgren, S. (1992b). Mechanisms controlling stomach volume of the Atlantic cod *Gadus morhua* following gastric distension. *J. Exp. Biol.* **163**, 49–63.

Grundy, D., and Schemann, M. (2002). Motor control of the stomach. In: *Innervation of the Gastrointestinal Tract* (S. Brookes and M. Costa, eds), pp. 57–102. Taylor and Francis, London, New York.

Hagström, C., and Olsson, C. (2010). Glial cells revealed by GFAP immunoreactivity in fish gut. *Cell Tissue Res.* DOI: 10.1007/500441-010-0979-3.

Holmgren, S. (1983). The effects of putative non-adrenergic, non-cholinergic autonomic transmitters on isolated strips from the stomach of the rainbow trout, *Salmo gairdneri*. *Comp. Biochem. Physiol. C* **74**, 229–238.

Holmgren, S. (1985). Substance P in the gastrointestinal tract of *Squalus acanthias*. *Molecular Physiol.* **8**, 119–130.

Holmgren, S., and Fänge, R. (1981). Effects of cholinergic drugs on the intestine and gallbladder of the hagfish, *Myxine glutinosa* L., with a report on the inconsistent effects of catecholamines. *Mar. Biol. Lett.* **2**, 265–277.

Holmgren, S., and Jensen, J. (2001). Evolution of vertebrate neuropeptides. *Brain Res. Bull.* **55**, 723–735.

Holmgren, S., and Nilsson, S. (1974). Drug effects on isolated artery strips from two teleosts, *Gadus morhua* and *Salmo gairdneri*. *Acta Physiol. Scand.* **90**, 431–437.

Holmgren, S., and Nilsson, S. (1983). Bombesin-, gastrin/CCK-, 5-hydroxytryptamine-, neurotensin-, somatostatin-, and VIP-like immunoreactivity and catecholamine fluorescence in the gut of the elasmobranch, *Squalus acanthias*. *Cell Tissue Res* **234**, 595–618.

Holmgren, S., and Olsson, C. (2010). Nervous system of the gut. In: *Encyclopedia of Fish Physiology: From Genome to Environment* (C. Olsson, S. Holmgren, E. D. Stevens and A. P. Farrell, eds), Elsevier, New York.

Holmgren, S., Grove, D. J., and Nilsson, S. (1985). Substance P acts by releasing 5-hydroxytryptamine from enteric neurons in the stomach of the rainbow trout, *Salmo gairdneri*. *Neuroscience* **14**, 683–693.

Holmgren, S., Axelsson, M., and Farrell, A. P. (1992a). The effect of catecholamines, substance P and vasoactive intestinal polypeptide on blood flow to the gut in the dogfish *Squalus acanthias*. *J. Exp. Biol.* **168**, 161–175.

Holmgren, S., Axelsson, M., and Farrell, A. P. (1992b). The effects of neuropeptide Y and bombesin on blood flow to the gut in dogfish *Squalus acanthias*. *Regul. Pept.* **40**, 169.

Holmgren, S., Fritsche, R., Karila, P., Gibbins, I., Axelsson, M., Franklin, C., Grigg, G., and Nilsson, S. (1994). Neuropeptides in the Australian lungfish *Neoceratodus forsteri*: effects in vivo and presence in autonomic nerves. *Am. J. Physiol.* **266**, R1568–R1577.

Holmberg, A., Schwerte, T., Fritsche, R., Pelster, B., and Holmgren, S. (2003). Ontogeny of intestinal motility in correlation to neuronal development in zebrafish embryos and larvae. *J. Fish Biol.* **63**, 318–331.

Holmberg, A., Schwerte, T., Pelster, B., and Holmgren, S. (2004). Ontogeny of the gut motility control system in zebrafish *Danio rerio* embryos and larvae. *J. Exp. Biol.* **207**, 4085–4094.

Holmberg, A., Olsson, C., and Holmgren, S. (2006). The effects of endogenous and exogenous nitric oxide on gut motility in zebrafish *Danio rerio* embryos and larvae. *J. Exp. Biol.* **209**, 2472–2479.

Holmberg, A., Olsson, C., and Hennig, G. W. (2007). TTX-sensitive and TTX-insensitive control of spontaneous gut motility in the developing zebrafish (*Danio rerio*) larvae. *J. Exp. Biol.* **210**, 1084–1091.

Holmberg, A., Holmgren, S., and Olsson, C. (2008). Enteric control. In: *Fish Larval Physiology* (R. Finn, ed.), pp. 553–572. Science Publishers, Enfield, NH.

Holstein, B. (1976). Effect of the H2-receptor antagonist metiamide on carbachol- and histamine-induced gastric acid secretion in the Atlantic cod, *Gadus morhua*. *Acta Physiol. Scand.* **97**, 189–195.

Holstein, B. (1977). Effect of atropine and SC-15396 on stimulated gastric acid secretion in the Atlantic cod, *Gadus morhua*. *Acta Physiol. Scand.* **101**, 185–193.

Holstein, B., and Cederberg, C. (1980). Effect of vagotomy and glucose administration on gastric acid secretion in the Atlantic cod, *Gadus morhua*. *Acta Physiol. Scand.* **109**, 37–44.

Holstein, B., and Cederberg, C. (1984). Effect of 5-HT on basal and stimulated secretions of acid and pepsin and on gastric volume outflow in the in vivo gastrically and intestinally perfused cod, *Gadus morhua. Agents Actions* **15**, 291–305.

Holstein, B., and Cederberg, C. (1986). Effects of tachykinins on gastric acid and pepsin secretion and on gastric outflow in the Atlantic cod, *Gadus morhua. Am. J. Physiol.* **250**, G309–G315.

Holzer, P. (2004). TRPV1 and the gut: from a tasty receptor for a painful vanilloid to a key player in hyperalgesia. *Eur. J. Pharmacol.* **500**, 231–241.

Hoyle, C. H. (1998). Neuropeptide families: evolutionary perspectives. *Regul. Pept.* **73**, 1–33.

Jensen, J. (1997). Co-release of substance P and neurokinin A from the Atlantic cod stomach. *Peptides* **18**, 717–722.

Jensen, J., and Holmgren, S. (1985). Neurotransmitters in the intestine of the Atlantic cod, *Gadus morhua. Comp. Biochem. Physiol.* **82C**, 81–89.

Jensen, J., and Holmgren, S. (1991). Tachykinins and intestinal motility in different fish groups. *Gen. Comp. Endocrinol.* **83**, 388–396.

Jensen, J., and Holmgren, S. (1994). The gastrointestinal canal. In: *Comparative Physiology and Evolution of the Autonomic Nervous System* (S. Nilsson and S. Holmgren, eds), pp. 119–167. Harwood Academic Publisher, Chur, Switzerland.

Jensen, J., Holmgren, S., and Jonsson, A. C. (1987). Substance P-like immunoreactivity and the effects of tachykinins in the intestine of the Atlantic cod, *Gadus morhua. J. Auton. Nerv. Syst.* **20**, 25–33.

Jensen, J., Axelsson, M., and Holmgren, S. (1991). Effects of substance P and VIP on the gastrointestinal blood flow in the Atlantic cod, *Gadus morhua. J. Exp. Biol.* **156**, 361–373.

Jensen, J., Olson, K. R., and Conlon, J. M. (1993a). Primary structures and effects on gastrointestinal motility of tachykinins from the rainbow trout. *Am. J. Physiol.* **265**, R804–R810.

Jensen, J., Karila, P., Jönsson, A.-C., Aldman, G., and Holmgren, S. (1993b). Effects of substance P and distribution of substance P-like immunoreactivity in nerves supplying the stomach of the cod *Gadus morhua. Fish Physiol. Biochem.* **12**, 237–247.

Johnels, A. G., and Östlund, E. (1958). Anatomical and physiological studies on the enteron of *Lampetra fluviatilis* (L.). *Acta Zool. (Stockh.)* **39**, 9–12.

Kågström, J., and Holmgren, S. (1998). Calcitonin gene-related peptide (CGRP), but not tachykinins, cause relaxation of small arteries from the rainbow trout gut. *Peptides* **19**, 577–584.

Kågström, J., Axelsson, M., Jensen, J., Farrell, A. P., and Holmgren, S. (1996). Vasoactivity and immunoreactivity of fish tachykinins in the vascular system of the spiny dogfish. *Am. J. Physiol. Regul. Integr. Comp. Physiol.* **270**, R585–R593.

Karila, P. (1997). Nervous control of gastrointestinal motility in the Atlantic cod, *Gadus morhua*. University of Gothenburg Göteborg, Sweden.

Karila, P., and Holmgren, S. (1995). Enteric reflexes and nitric oxide in the fish intestine. *J. Exp. Biol.* **198**, 2405–2411.

Karila, P., Jönsson, A.-C., Jensen, J., and Holmgren, S. (1993). Galanin-like immunoreactivity in extrinsic and intrinsic nerves to the gut of the Atlantic cod, *Gadus morhua*, and the effect of galanin on the smooth muscle of the gut. *Cell Tissue Res.* **271**, 537–544.

Karila, P., Messenger, J., and Holmgren, S. (1997). Nitric oxide synthase- and neuropeptide Y-containing subpopulations of sympathetic neurons in the coeliac ganglion of the Atlantic cod, *Gadus morhua*, revealed by immunohistochemistry and retrograde tracing from the stomach. *J. Auton. Nerv. Syst.* **66**, 35–45.

Karila, P., Shahbazi, F., Jensen, J., and Holmgren, S. (1998). Projections and actions of tachykininergic, cholinergic, and serotonergic neurones in the intestine of the Atlantic cod. *Cell Tissue Res.* **291**, 403–413.

Kelsh, R. N., and Eisen, J. S. (2000). The zebrafish colourless gene regulates development of non-ectomesenchymal neural crest derivatives. *Development* **127**, 515–525.

Kiliaan, A. J., Joosten, H. W., Bakker, R., Dekker, K., and Groot, J. A. (1989). Serotonergic neurons in the intestine of two teleosts, *Carassius auratus* and *Oreochromis mossambicus*, and the effect of serotonin on transepithelial ion-selectivity and muscle tension. *Neuroscience* **31**, 817–824.

Kirtisinghe, P. (1940). The myenteric nerve plexus in some lower chordates. *Quart. J. Microscop. Sci.* **81**, 521–539.

Kitazawa, T., Kondo, H., and Temma, K. (1986a). Alpha 2-adrenoceptor-mediated contractile response to catecholamines in smooth muscle strips isolated from rainbow trout stomach (*Salmo gairdneri*). *Br. J. Pharmacol.* **89**, 259–266.

Kitazawa, T., Temma, K., and Kondo, H. (1986b). Presynaptic alpha-adrenoceptor mediated inhibition of the neurogenic cholinergic contraction of the isolated intestinal bulb of the carp (*Cyprinus carpio*). *Comp. Biochem. Physiol.* **83C**, 271–277.

Kitazawa, T., Kimura, A., Furahashi, H., Temma, K., and Kondo, H. (1988a). Contractile responses to substance P in isolated smooth muscle strips from the intestinal bulb of the carp (*Cyprinus carpio*). *Comp. Biochem. Physiol.* **89C**, 277–285.

Kitazawa, T., Kudo, K., Ishigami, M., Furuhashi, H., Temma, K., and Kondo, H. (1988b). Evidence that a substance P-like peptide mediates the non-cholinergic excitatory response of the carp intestinal bulb (*Cyprinus carpio*). *Naunyn Schmiedebergs Arch. Pharmacol.* **338**, 68–73.

Lamers, C. H., Rombout, J. W., and Timmermans, L. P. (1981). An experimental study on neural crest migration in *Barbus conchonius* (Cyprinidae, Teleostei), with special reference to the origin of the enteroendocrine cells. *J. Embryol. Exp. Morphol.* **62**, 309–323.

Langer, M., Van Noorden, S., Polak, J. M., and Pearse, A. G. (1979). Peptide hormone-like immunoreactivity in the gastrointestinal tract and endocrine pancreas of eleven teleost species. *Cell Tissue Res.* **199**, 493–508.

Li, Z. S., and Furness, J. B. (1993). Nitric oxide synthase in the enteric nervous system of the rainbow trout, *Salmo gairdneri*. *Arch. Histol. Cytol.* **56**, 185–193.

Lincoln, J., Hoyle, C. H. V., and Burnstock, G. (1997). Nitric oxide in health and disease. In: *Biomedical Research Topics* (J. A. Lucy, ed.). Cambridge University Press, Cambridge.

Lomax, A. E., Sharkey, K. A. and Furness, J. B. (2010). The participation of the sympathetic innervation of the gastrointestinal tract in disease states. *Neurogastroenterol. Motil.* **22**, 7–18.

Lundin, K., Holmgren, S., and Nilsson, S. (1984). Peptidergic functions in the dogfish rectum. *Acta Physiol. Scand.* **121**, 46A.

Matsuda, K., Kashimoto, K., Higuchi, T., Yoshida, T., Uchiyama, M., Shioda, S., Arimura, A., and Okamura, T. (2000). Presence of pituitary adenylate cyclase-activating polypeptide (PACAP) and its relaxant activity in the rectum of a teleost, the stargazer, *Uranoscopus japonicus*. *Peptides* **21**, 821–827.

Mellgren, E. M., and Johnson, S. L. (2005). kitb, a second zebrafish ortholog of mouse Kit. *Dev. Genes Evol.* **215**, 470–477.

Monti, R. (1895). Contribution à la connaissance des nerfs du tube digestif des poissons. *Arch. Ital. de Biol.* **24**, 188–195.

Nicol, J. A. C. (1952). Autonomic nervous systems in lower chordates. *Biol. Rev. Cambridge Philos. Soc.* **27**, 1–49.

Nilsson, S. (1983). *Autonomic Nerve Function in the Vertebrates*. Springer Verlag, Berlin, Heidelberg, New York.

Nilsson, S., and Fänge, R. (1967). Adrenergic receptors in the swimbladder and gut of a teleost (*Anguilla anguilla*). *Comp. Biochem. Physiol.* **C 23**, 661–664.

Nilsson, S., and Fänge, R. (1969). Adrenergic and cholinergic vagal effects on the stomach of a teleost (*Gadus morhua*). *Comp. Biochem. Physiol.* **30**, 691–694.

Nilsson, S., and Holmgren, S. (1983). Splanchnic nervous control of the stomach of the spiny dogfish, *Squalus acanthias*. *Comp. Biochem. Physiol.* **C76**, 271–276.

Nilsson, S., Holmgren, S., and Grove, D. J. (1975). Effects of drugs and nerve stimulation on the spleen and arteries of two species of dogfish, *Scyliorhinus canicula* and *Squalus acanthias*. *Acta Physiol. Scand.* **95**, 219–230.

Olsson, C. (2009). Autonomic innervation of the fish gut. *Acta Histochem.* **111**, 185–195.

Olsson, C. (2010). Calbindin-immunoreactive cells in the fish enteric nervous system. *Autonomic Neuroscience: Basic and Clinical*, in press.

Olsson, C., and Holmgren, S. (1994). Distribution of PACAP (pituitary adenylate cyclase-activating polypeptide)-like and helospectin-like peptides in the teleost gut. *Cell Tissue Res.* **277**, 539–547.

Olsson, C., and Holmgren, S. (2000). PACAP and nitric oxide inhibit contractions in proximal intestine of the Atlantic cod, *Gadus morhua*. *J. Exp. Biol.* **203**, 575–583.

Olsson, C., and Holmgren, S. (2009). The neuroendocrine regulation of gut function. In: *Fish Physiology, Vol. 28: Fish Neuroendocrinology* (N. J. Bernier, G. Van Der Kraak, A. P. Farrell and C. J. Brauner, eds), Vol. 28, pp. 467–512. New York, Academic Press.

Olsson, C., and Karila, P. (1995). Coexistence of NADPH-diaphorase and vasoactive intestinal polypeptide in the enteric nervous system of the Atlantic cod (*Gadus morhua*) and the spiny dogfish (*Squalus acanthias*). *Cell Tissue Res.* **280**, 297–305.

Olsson, C., Aldman, G., Larsson, A., and Holmgren, S. (1999). Cholecystokinin affects gastric emptying and stomach motility in the rainbow trout *Oncorhynchus mykiss*. *J. Exp. Biol.* **202**, 161–170.

Olsson, C., Holmberg, A., and Holmgren, S. (2008). Development of enteric and vagal innervation of the zebrafish (*Danio rerio*) gut. *J. Comp. Neurol.* **508**, 756–770.

Parichy, D. M., Rawls, J. F., Pratt, S. J., Whitfield, T. T., and Johnson, S. L. (1999). Zebrafish sparse corresponds to an orthologue of c-kit and is required for the morphogenesis of a subpopulation of melanocytes, but is not essential for hematopoiesis or primordial germ cell development. *Development* **126**, 3425–3436.

Patterson, T. L., and Fair, E. (1933). The action of the vagus on the stomach-intestine of the hagfish. Comparative studies. VIII. *J. Cell. Comp. Physiol.* **3**, 113–199.

Pederzoli, A., Trevisan, P., and Bolognani Fantin, A. M. (1996). Immunocytochemical study of endocrine cells in the gut of goldfish *Carassius carassius* (L.) *var. auratus* submitted to experimental lead intoxication. *Eur. J. Histochem.* **40**, 305–314.

Pederzoli, A., Bertacchi, I., Gambarelli, A., and Mola, L. (2004). Immunolocalisation of vasoactive intestinal peptide and substance P in the developing gut of *Dicentrarchus labrax* (L.). *Eur. J. Histochem.* **48**, 179–184.

Pederzoli, A., Conte, A., Tagliazucchi, D., Gambarelli, A., and Mola, L. (2007). Occurrence of two NOS isoforms in the developing gut of sea bass *Dicentrarchus labrax* (L.). *Histol. Histopathol.* **22**, 1057–1064.

Preston, E., Mcmanus, C. D., Jonsson, A. C., and Courtice, G. P. (1995). Vasoconstrictor effects of galanin and distribution of galanin containing fibres in three species of elasmobranch fish. *Regul. Pept.* **58**, 123–134.

Preston, E., Jonsson, A. C., McManus, C. D., Conlon, J. M., and Courtice, G. P. (1998). Comparative vascular responses in elasmobranchs to different structures of neuropeptide Y and peptide YY. *Regul. Pept.* **78**, 57–67.

Raible, D. W., Wood, A., Hodsdon, W., Henion, P. D., Weston, J. A., and Eisen, J. S. (1992). Segregation and early dispersal of neural crest cells in the embryonic zebrafish. *Dev. Dyn.* **195**, 29–42.

Rajjo, I. M., Vigna, S. R., and Crim, J. W. (1989). Immunocytochemical localization of bombesin-like peptides in the digestive tract of the bowfin, *Amia calva*. *Comp. Biochem. Physiol.* **94C**, 405–409.

Reinecke, M., Schluter, P., Yanaihara, N., and Forssmann, W. G. (1981). VIP immunoreactivity in enteric nerves and endocrine cells of the vertebrate gut. *Peptides* **2**, 149–156.

Reinecke, M., Muller, C., and Segner, H. (1997). An immunohistochemical analysis of the ontogeny, distribution and coexistence of 12 regulatory peptides and serotonin in endocrine cells and nerve fibers of the digestive tract of the turbot, *Scophthalmus maximus* (Teleostei). *Anat. Embryol. (Berl.)* **195**, 87–101.

Rich, A., Leddon, S. A., Hess, S. L., Gibbons, S. J., Miller, S., Xu, X., and Farrugia, G. (2007). Kit-like immunoreactivity in the zebrafish gastrointestinal tract reveals putative ICC. *Dev. Dyn.* **236**, 903–911.

Rombout, J. H., and Reinecke, M. (1984). Immunohistochemical localization of (neuro)peptide hormones in endocrine cells and nerves of the gut of a stomachless teleost fish, *Barbus conchonius* (*Cyprinidae*). *Cell Tissue Res* **237**, 57–65.

Ruhl, A., Nasser, Y., and Sharkey, K. A. (2004). Enteric glia. *Neurogastroenterol. Motil.* **16** (Suppl. 1), 44–49.

Sadaghiani, B., and Vielkind, J. R. (1990). Distribution and migration pathways of HNK-1-immunoreactive neural crest cells in teleost fish embryos. *Development* **110**, 197–209.

Sakharov, D. A., and Salimova, N. B. (1980). Serotonin neurons in the peripheral nervous system of the larval lamprey *Lampetra planeri*; a histochemical, microspectrofluorimetric and ultrastructural study. *Zool. Jb. Physiol.* **84**, 231–239.

Salimova, N., and Fehér, E. (1982). Innervation of the alimentary tract in Chondrostean fish (*Acipenseridae*). A histochemical, microspectro-fluorimetric and ultrastructural study. *Acta Morphol. Acad. Sci. Hung* **30**, 213–222.

Sanders, K. M., Koh, S. D., and Ward, S. M. (2006). Interstitial cells of Cajal as pacemakers in the gastrointestinal tract. *Annu. Rev. Physiol.* **68**, 307–343.

Schemann, M., Sann, H., Schaaf, C., and Mader, M. (1993). Identification of cholinergic neurons in enteric nervous system by antibodies against choline acetyltransferase. *Am. J. Physiol.* **265**, G1005–G1009.

Severini, C., Improta, G., Falconieri-Erspamer, G., Salvadori, S., and Erspamer, V. (2002). The tachykinin peptide family. *Pharmacol. Rev.* **54**, 285–322.

Shahbazi, F., Karila, P., Olsson, C., Holmgren, S., and Jensen, J. (1998). Primary structure, distribution, and effects on motility of CGRP in the intestine of the cod *Gadus morhua*. *Am. J. Physiol.* **275**, R19–R28.

Shahbazi, F., Holmgren, S., Larhammar, D., and Jensen, J. (2002). Neuropeptide Y effects on vasorelaxation and intestinal contraction in the Atlantic cod, *Gadus morhua*. *Am. J. Physiol. Regul. Integr. Comp. Physiol.* **282**, R1414–R1421.

Shepherd, I. T., Beattie, C. E., and Raible, D. W. (2001). Functional analysis of zebrafish GDNF. *Dev. Biol.* **231**, 420–435.

Shepherd, I. T., Pietsch, J., Elworthy, S., Kelsh, R. N., and Raible, D. W. (2004). Roles for GFRα1 receptors in zebrafish enteric nervous system development. *Development* **131**, 241–249.

Smit, H. (1968). Gastric secretion in the lower vertebrates and birds. In: *Handbook of Physiology, Section 6: Alimentary Canal, Volume V: Bile, Digestion, Ruminal Physiology* (C. F. Code. ed.), pp. 2791–2805. American Physiological Society, Washington, DC, USA.

Steele, P. A., Brookes, S. J., and Costa, M. (1991). Immunohistochemical identification of cholinergic neurons in the myenteric plexus of guinea-pig small intestine. *Neuroscience* **45**, 227–239.

Stevenson, S. V., and Grove, D. J. (1977). The extrinsic innervation of the stomach of the plaice, *Pleuronectes platessa* L. I. The vagal nerve supply. *Comp. Biochem. Physiol.* **58C**, 143–151.

Sundström, G., Larsson, T. A., Brenner, S., Venkatesh, B., and Larhammar, D. (2008). Evolution of the neuropeptide Y family: new genes by chromosome duplications in early vertebrates and in teleost fishes. *Gen. Comp. Endocrinol.* **155**, 705–716.

Tagliafierro, G., Bonini, E., Faraldi, G., Farina, L., and Rossi, G. G. (1988a). Distribution and ontogeny of VIP-like immunoreactivity in the gastro-entero-pancreatic system of a cartilaginous fish, *Scyliorhinus stellaris. Cell Tissue Res.* **253**, 23–28.

Tagliafierro, G., Zaccone, G., Bonini, E., Faraldi, G., Farina, L., Fasulo, S., and Rossi, G. G. (1988b). Bombesin-like immunoreactivity in the gastrointestinal tract of some lower vertebrates. *Ann. NY Acad. Sci.* **547**, 458–460.

Tobin, G., Giglio, D., and Lundgren, O. (2009). Muscarinic receptor subtypes in the alimentary tract. *J. Physiol. Pharmacol.* **60**, 3–21.

Trischitta, F., Denaro, M. G., and Faggio, C. (1999). Effects of acetylcholine, serotonin and noradrenalin on ion transport in the middle and posterior part of *Anguilla anguilla* intestine. *J. Comp. Physiol. B* **169**, 370–376.

Van Nassauw, L., Adriaensen, D., and Timmermans, J. P. (2007). The bidirectional communication between neurons and mast cells within the gastrointestinal tract. *Auton. Neurosci.* **133**, 91–103.

Van Noorden, S. (1990). Gut hormones in cyclostomes. *Fish Physiol. Biochem.* **8**, 399–408.

von Euler, U. S., and Östlund, E. (1957). Effects of certain biologically occurring substances on the isolated intestine of fish. *Acta Physiol. Scand.* **38**, 364–372.

Wallace, K. N. (2010). Development of the gut. In: *Encyclopedia of Fish Physiology: From Genome to Environment* (C. Olsson, S. Holmgren, E. D. Stevens and A. P. Farrell, eds.). Elsevier, New York.

Wallace, K. N., Akhter, S., Smith, E. M., Lorent, K., and Pack, M. (2005). Intestinal growth and differentiation in zebrafish. *Mech. Dev.* **122**, 157–173.

Young, J. Z. (1931). On the autonomic nervous system of the teleostean fish. *Uranoscopus scaber. Q. J. Microsc. Sci.* **74**, 491–535.

Young, J. Z. (1980a). Nervous control of stomach movements in dogfishes and rays. *J. Mar. Biol. Assoc. UK* **60**, 1–17.

Young, J. Z. (1980b). Nervous control of gut movements in *Lophius. J. Mar. Biol. Assoc. UK* **60**, 19–30.

Young, J. Z. (1983). Control of movements of the stomach and spiral intestine of *Raja* and *Scyliorinus. J. Mar. Biol. Assoc. UK* **63**, 557–574.

Young, J. Z. (1988). Sympathetic innervation of the rectum and bladder of the skate and parallel effects of ATP and adrenalin. *Comp. Biochem. Physiol. C* **89**, 101–107.

Yui, R., Nagata, Y., and Fujita, T. (1988). Immunocytochemical studies on the islet and the gut of the arctic lamprey, *Lampetra japonica. Arch. Histol. Cytol.* **51**, 109–119.

Zagorodnyuk, V. P., Chen, B. N., and Brookes, S. J. (2001). Intraganglionic laminar endings are mechano-transduction sites of vagal tension receptors in the guinea-pig stomach. *J. Physiol.* **534**, 255–268.

9

THE CIRCULATION AND METABOLISM OF THE GASTROINTESTINAL TRACT

HENRIK SETH

MICHAEL AXELSSON

ANTHONY P. FARRELL

1. INTRODUCTION

Along with the evolution of metazoans came the specialization of different cells into discrete tissues and organs. Organ systems, such as the cardiovascular system and the gastrointestinal tract, represent a higher level of organization that enabled organisms to become more efficient and adapt to different environments. Today, we see the result of this evolutionary trajectory, with animals tailored to lifestyles in environments ranging from the dry deserts to the hydrothermal vents of the deep oceans.

351

The evolution of a gastrointestinal (GI) tract enabled efficient digestive and absorptive processes for meals varying widely in their origin and in their nutritional quality, as well as composition. Correspondingly, animals show an enormous diversity when it comes to the morphology and physiology of the gastrointestinal tract. Fish are no different in this regard. Thus, within the paraphyletic group of piscine species, there are a wide variety of specializations or adaptations that enable the efficient digestion, absorption and internal distribution of nutrients.

This chapter summarizes the current knowledge of the circulation and metabolism of the GI tract of fish. It uses a comparative approach among fish with some comparisons to mammals where this is of importance for basic understanding. The intent is to identify evolutionary diversity beyond the obvious similarities that exist between fish and mammalian species for many basic concepts of gastrointestinal blood flow regulation.

Gastrointestinal blood flow (GBF) is of critical importance to the digestion, absorption and distribution of nutrients. Therefore, it should come as no surprise that GBF is tightly regulated in unfed animals, as well as after feeding (postprandial) when nutrient composition and amount of food ingested become of great importance. Regulation of postprandial GBF is important for two reasons. First, like any tissue, the GI tract requires an adequate blood flow to supply its tissue with the required oxygen and nutrients, and to remove carbon dioxide and other waste products. With feeding, the blood flow requirement necessarily increases both to aid the processing of the ingested food and to support the activity of the gastrointestinal uptake mechanisms. Second, once absorbed, the metabolized/hydrolyzed nutrients must be transported from the GI mucosa to the liver, as well as to other parts of the intestine and beyond.

Remarkably, it is still unclear as to how GBF is regulated in the unfed state, postprandially and under adverse environmental or pathological conditions. Furthermore, there is limited knowledge concerning the link between the postprandial hyperemia and the postprandial increase in total body oxygen consumption, which is most often referred to as specific dynamic action (SDA) or the heat increment (HI) of feeding (see below). Thus, the following represents the sum state of our current knowledge rather than a complete account.

2. VASCULAR ANATOMY OF THE GASTROINTESTINAL TRACT: DIVERSITY AMONG FISHES

2.1. Gross Vascular Anatomy and Vascular Patterns

The vascular anatomy of teleosts and elasmobranchs is best visualized in three dimensions using a corrosion cast technique (Murakami 1971). In this technique, the blood vessels of a euthanized animal are cleared of blood cells by perfusing the vascular bed of interest with saline containing a vasodilator compound such as sodium nitroprusside. Then, a two-component epoxy-plastic is carefully injected and allowed to solidify. The organic tissue is then digested away with potassium hydroxide, in combination with acid to remove calcified tissue and reveal an epoxy cast of the vasculature (Fig. 9.1A).

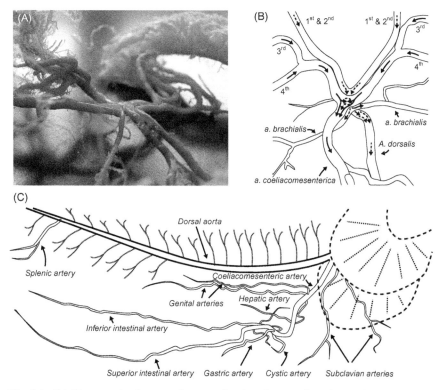

Fig. 9.1. (A) Photograph of a corrosion cast showing the complicated vascular arrangement between the efferent branchial arteries, the coeliacomesenteric artery and the dorsal aorta in the shorthorn sculpin. (B) Schematic drawing of the isolated part of the vasculature of the shorthorn sculpin that is based on several corrosion casts and which shows the connections between the efferent branchial arteries (1st, 2nd, 3rd and 4th) coming from the gills, the dorsal aorta and the coeliacomesenteric artery. The vessels from the 1st and 2nd branchial arteries are connected with the 3rd and 4th through a thin connecting vessel/anastomosis (hatched circle). Solid arrows show blood coming from the 3rd and 4th branchial arteries and hatched arrows show blood coming from the 1st and 2nd branchial arteries. (C) The entire arterial vasculature of a rainbow trout with focus on the GI tract blood supply, smaller vessels have been removed for clarity.

The primary blood vessels supplying the GI tract differ between elasmobranchs and teleosts. In teleosts, the primary blood supply to the GI tract is via a single major vessel, the *coeliacomesenteric artery* (*CMA*), which is the first caudal branch of the dorsal aorta. The CMA then divides progressively to supply the stomach, intestine and liver, as well as the gonads. Consequently, the terms CMA blood flow and GBF are often used interchangeably. Here, we use GBF unless a distinction is needed.

Elasmobranchs (sharks) show a more typical vertebrate pattern (Matheson et al. 2000) of blood supply to the GI tract. Three major vessels branch off the dorsal aorta: the coeliac artery supplying the liver, spleen, stomach and part of the intestine; the anterior mesenteric artery, supplying most of the small intestine; and the posterior mesenteric artery supplying the large intestine (Fig. 9.2D). As a curiosity, in lampreys (family, *Petromyzonidae*) part of the posterior digestive tract is supplied with blood via an artery that runs on the inside of a vein. Furthermore, the venous return from the intestine in this group does not enter the hepatic portal system, but returns via a suprarenal venous system to the cardinal vein (Farrell 2007). Since there is very little information about the regulation of blood flow to the GI tract of elasmobranchs (and none for cyclostomes), the remainder of the chapter focuses on the teleost system.

In teleosts, the coeliac and mesenteric arteries are the two major branches of the CMA. There are only minor anatomical differences among teleosts that possibly indicate specialized circulatory adaptations to specific habitats and feeding regimes. The bottom-living shorthorn sculpin (*Myoxocephalus scorpius*) has a highly vascularized GI tract that receives a relatively large proportion of cardiac output. A peculiar anastomosis exists at the origin of the CMA near the efferent branchial arteries, one that possibly enables oxygenated blood from the gills to be sent directly to the gastrointestinal tract (Figs 9.1A, B and 9.2B). However, at present, information about the functional significance of this vascular arrangement is lacking. In contrast, in rainbow trout (*Oncorhynchus mykiss*) and other salmonids, as well as in many other studied teleosts, a very distinctive branch forms a single CMA that clearly originates well downstream of the efferent branchial arteries (Figs 9.1C and 9.2A, C).

Although the major portion of the GI blood supply is via the CMA, teleosts also have an additional blood supply to the hindgut via two unpaired arteries originating from the posterior dorsal aorta (Thorarensen et al. 1991). These vessels are much smaller in diameter than the CMA, but so far nobody has tried to quantify blood flow in them. Therefore, the reported measurements of CMA blood flow underestimate total GBF, but we do not know by how much.

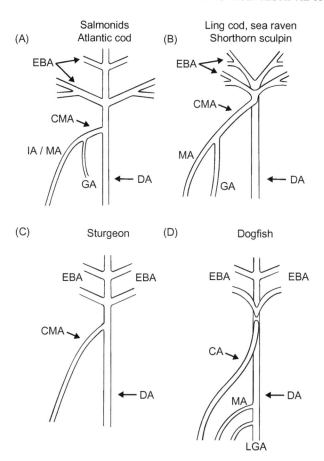

Fig. 9.2. Schematic drawings of the post-branchial vascular tree in different fish species showing where the gastrointestinal blood supply originates. *Abbreviations:* EBA, efferent branchial arteries; CMA, coeliacomesenteric artery; DA, dorsal aorta; GA, gastric artery; IA, intestinal artery; MA, mesenteric artery; LGA, lienogastric artery. Redrawn form Farrell et al. (2001).

2.2. Microvasculature

Cyclostomes have a poorly developed microcirculation. Elasmobranchs and teleosts are the first chordates that display an abundance of capillaries and small diameter arterioles (Soldatov 2006). Most of what is known about the vascular arrangement, at the level of an organ, in fish is from studies of the gills (Dunel-Erb and Laurent 1980; Laurent and Dunel 1980; Sundin and

Nilsson 1997, 1992; Olson 2002; Wilson and Laurent 2002), skeletal muscle (Mosse 1978, 1979; Egginton and Johnston 1982; Davison 1987) and coronary circulation (Axelsson and Farrell 1993).

In contrast the gastrointestinal microvasculature has been extensively reviewed in mammals from both anatomical (Rhodin 1967; Gore and Bohlen 1977) and physiological perspectives. In mammals, the microvascular branching pattern is very conserved, and the three major layers of the gut wall (the mucosa, the submucosa and muscular tissue) all receive blood from the same mesenteric arteries and arterioles. Arterioles in mammals, which closely resemble those found in fish, are smaller-diameter vessels (10–50 μm) that reside mainly within the submucosa and form a meshwork of interconnected second- and third-order vessels. These vessels then supply the distinct arterioles that supply the capillary system of one or several villi, as well as the surrounding glandular tissue.

3. CONTROL OF GASTROINTESTINAL BLOOD FLOW UNDER ROUTINE CONDITIONS

Under routine conditions, i.e. before the ingestion of a meal in a stress-free environment, GBF is regulated to a level sufficient for the housekeeping needs of the GI tract. In fish, routine GBF is typically 10–40% of cardiac output, which is in the same range reported for mammals. The lowest percentages in fish are reported for sea raven (*Hemitripterus americanus*) and sea bass (*Dicentrarchus labrax*), and the highest for the Atlantic cod (*Gadus morhua*) (Table 9.1).

Unfed sea raven and red Irish lord (*Hemilepidotus hemilepidotus*) maintain routine GBF by actively vasoconstricting the entire GI tract with an α-adrenergic tonus. This was discovered by injecting an α-adrenergic antagonist (phentolamine). Phentolamine lowered the vascular resistance of the coeliac and mesenteric circuits of these two species, but only the mesenteric circulation in Atlantic cod. Following a meal this vascular tone decreases, allowing blood flow to increase. In contrast, during stress, the vascular tone can increase to such an extent that blood flow completely ceases in the GI tract (Axelsson et al. 2000). Thus, an α-adrenergic vascular tonus is a primary means of regulating (both increasing and decreasing) GBF.

The α-adrenergic tonus is generally believed to be of nervous origin. However, this suggestion has not been experimentally verified, primarily because of the difficulty in separating adrenergic effects that are due to either circulating catecholamines (CA) or neural tone. The routine

Table 9.1

Postprandial cardiovascular responses in selected fish species, measured as cardiac output and gut blood flow plus the percentage of cardiac output that passes through the coeliacomesenteric artery, in a selection of several different species. The meal sizes used are also given as percentage of body weight

Species	°C	Meal size (% of b.w.)	Cardiac output (ml min⁻¹ kg⁻¹)		Gut blood flow (ml min⁻¹ kg⁻¹)		Gut blood flow/ Cardiac output		Source
			Routine	Feeding (%)	Routine	Feeding (%)	Routine (%)	Feeding (%)	
Sea bass (*Dicentrarchus labrax*)	22–23	2.7	51.4	22	13.8	82	27	40	Altimiras et al. 2008
Red Irish lord (*Hemilepidotus hemilepidotus*)	7–9	10–15	24	90	4.1/4.9	112/94	34	40	Axelsson et al. 2000
Sea raven (*Hemitripterus americanus*)	10–12	10–20	18.8	15.4	2.9	100	15	27–30	Axelsson et al. 1989
Atlantic cod (*Gadus morhua*)	10–11	2.2–3.5	19	23	4.1/3.5	72/42	40	52	Axelsson and Fritsche 1991
Sea bass (*Dicentrarchus labrax*)	16	2.9	40	13.5	9.6	71	24	34	Axelsson et al. 2002
Sea bass (*Dicentrarchus labrax*)	19	3	43.4	27	4.3	160	10	20	Dupont-Prinet et al. 2009
Rainbow trout (*Oncorhynchus mykiss*)	11–16	2	N/A	N/A	4–6	136	N/A	N/A	Eliasson et al. 2008
Rainbow trout (*Oncorhynchus mykiss*)	9–10	2	N/A	23–42	N/A	156	N/A	N/A	Gräns et al. 2009
Short-horn sculpin (*Myoxocephalus scorpius*)	10	8–10	N/A	50	N/A	93	N/A	N/A	Seth et al. 2008
Chinook salmon (*Oncorhynchus tshawytscha*)	9–10	2	N/A	N/A	N/A	81	N/A	N/A	Thorarensen and Farrell 2006
Chinook salmon (*Oncorhynchus tshawytscha*)	8–11	N/A	33	N/A	12–18	N/A	36	N/A	Thorarensen and Farrell 1993

concentration of plasma CA in Atlantic cod is likely too low (2.5/3.5 nM; adrenaline/noradrenaline) to contribute significantly to an adrenergic tonus on the heart and circulation, even though the concentration may be sufficient to have effects on branchial vascular tone. Plasma CA concentrations change with different activities and external stimuli. Low-intensity exercise does not alter plasma CA levels, but when a fish approaches its maximal swimming speeds plasma CA levels can reach 30/19 nM (adrenaline/noradrenaline). These plasma CA levels are likely high enough to directly affect the systemic circulation. In the rainbow trout, especially at cold temperatures, even the low concentration of circulating catecholamines (5 nM) may exert systemic vascular and cardiac effects, indicating both inter- and intra-species differences in the sensitivity to plasma CA (Axelsson and Nilsson 1986; Axelsson et al. 1987; Axelsson 1988; Graham and Farrell 1989). Nothing is known about autoregulation of GBF in fishes.

A recent study in rainbow trout has addressed the question to what extent a reflex-mediated α-adrenergic tonus influences the gastrointestinal vasculature both under routine and postprandial conditions (Seth and Axelsson 2010). Sectioning sympathetic innervation to the GI tract (the splanchnic nerve) markedly decreased (~40%) GI vascular resistance and increased blood flow almost proportionally (~50%). Quantitatively, these changes are comparable to the decrease in GI tract vascular resistance (55%/40%; coeliac/mesenteric) in the Irish lord after the administration of phentolamine (Axelsson et al. 2000). The α-adrenergic tonus is quantitatively similar in the sea raven (Axelsson et al. 1989).

3.1. Control at the Level of the Microvasculature

There is little knowledge concerning microvasculature regulation of the GBF in fish. Infusing adrenaline in the sea raven substantially increased the dorsal aortic pressure and systemic vascular resistance, with little difference in blood pressure between the dorsal aorta and the CMA (Axelsson et al. 1989). This result indicates that the major control of the vasculature resides at the level of the arterioles, but the relative importance of nervous or humoral mechanisms, including local or paracrine factors, is unknown.

Aside from hypoxia and pH that affect the local circulation of the GI and the gill vasculature (Smith 1999; Smith et al. 2001), several substances have been proposed to affect the vasculature within the GI tract. These vasoactive substances include substance P, neurokinin A, vasoactive intestinal polypeptide (VIP) and scyliorhinins (Jensen et al. 1991; Holmgren et al.

1992; Kågström et al. 1996a,b; Kågström and Holmgren 1997; Haverinen et al. 2007). Many of these substances influence both cardiac output and GBF, but responses vary among species, showing a bi-phasic pattern in one species and a tri-phasic response in another. For example, VIP increases GBF in Atlantic cod, whereas VIP decreases GFB in the spiny dogfish (*Squalus acanthias*). Many of these substances might also modulate the postprandial vascular response as discussed below.

In mammals, most of the control of GBF resides within microcirculation (i.e. arterioles of < 50 µm in diameter) of the GI tract. Capillary blood flow is controlled by general changes in arteriolar diameter and pre-capillary sphincters. Arteriolar vasoconstriction in the GI tract is mainly under metabolic/myogenic control, whereas autonomic innervation predominates in the control of the pre-capillary sphincters (Shepherd 1982). Arteriolar sphincters, resembling the pre-capillary sphincters found in mammals, have been found in the spleen of rainbow trout (Kita and Itazawa 1990). Atypical pericyte-like arteriolar/capillary structures, possibly with a similar function, have also been found in the vasculature of sheepshead minnow (*Cyprinodon variegatus*) (Couch 1990). How vascular smooth muscle is regulated, and especially whether or not these structures can be regulated independently (as in mammals), is presently unknown.

Arteriolar diameter in general regulates vascular resistance and flow rate, thereby adjusting the overall convective flux of oxygen to tissues. Vascular muscle sphincters in pre-capillary arterioles regulate blood flow at a local level and adjust the capillary perfusion pattern, and hence the diffusive distance of oxygen between capillaries and tissues. Consequently, it appears that general arteriolar diameter regulates and maintains a high capillary arterial PO_2, while pre-capillary sphincters regulate diffusion distance.

In theory, arteriolar diameter and pre-capillary sphincters can be controlled independently, although a change in one will affect the other and both affect the rate of oxygen delivery to tissues. During partial arterial occlusion there is improved tissue oxygen extraction via increased capillary recruitment that is mediated by the opening of pre-capillary sphincters. Nevertheless, sympathetic nerve stimulation can abolish this effect by closing pre-capillary sphincters in mammals. Thus, local metabolic control of the pre-capillary sphincters can be superseded by nervous input (Shepherd 1979), something that is specific for the GI tract and is not seen in mammalian skeletal microvasculature. Even so, distal GI vessels (i.e. in the colon) are poorly autoregulated, that is, there is a limited intrinsic ability of these vessels to maintain a relatively steady blood flow during changes in perfusion pressure (Kvietys et al. 1980b; Granger et al. 1982). Instead, its blood flow varies directly with arterial blood pressure. How blood flow in this part of the GI tract is regulated is still not fully understood (Kvietys et al. 1980b).

4. CONTROL OF GASTROINTESTINAL BLOOD FLOW UNDER VARIOUS CONDITIONS

4.1. Feeding

Tight regulation of GBF is critical following feeding and during the subsequent processing of food. In fish, information about postprandial changes in GBF is restricted to a few species and due care should be exercised when drawing general conclusions from the species-specific information presented below. Moreover, with regards to the control mechanisms, most studies have focused on vasoactive substances rather than neural controls. A further complicating factor is that most studies have examined force-fed fish, with the response to feeding being followed over several hours and up to several days. Feeding by gavage has the advantage that the amount of food given and timing of feeding are controlled with some precision. The disadvantage, of course, is that gavage induces some stress, which affects GBF (see below); fish are normally lightly anesthetized, something that in itself affects gastric emptying time, and therefore the entire dynamics of GBF and SDA development. In addition, the instrumentation of the fish to measure GBF affects gastric emptying time (or at least the dynamics), and thus the entire dynamics of the GBF response (Axelsson et al. 2002) unless there is a sufficient recovery from surgery.

4.1.1. MAGNITUDE OF THE POSTPRANDIAL RESPONSE

The increase in GBF after the ingestion of a normal-sized meal (Table 9.1) ranges from around 70% in the sea bass (Axelsson et al. 2002) to over 150% in the rainbow trout (Gräns et al. 2009). Due to the substantial variation in the measured values of GBF, even within a single species, it is hard, if not impossible, to detect any trends within the available data, but there appears to be no difference between sedentary, ambush predators such as shorthorn sculpin (Seth and Axelsson 2009) or red Irish lord (Axelsson et al. 2000) and agile swimmers such as Chinook salmon (*Oncorhynchus tshawytscha*) (Thorarensen and Farrell 2006), rainbow trout (Eliason et al. 2008) and sea bass (Axelsson et al. 2002; Altimiras et al. 2008; Dupont-Prinet et al. 2009). Even so, such species comparisons must be made with caution because of important methodological differences, including meal size, temperature, the physical status of the animal and the surgical protocol involved (types of flow probes, number of cannula and other invasive measuring devices). In sea raven, for example, GBF was measured before the bifurcation of the CMA, whereas mesenteric and coeliac artery blood flows were measured separately in red Irish lord. In the latter species, the coeliac artery is the dominant vessel and blood flow increased by 115%; the increase in the somewhat smaller mesenteric artery was 94%.

The postprandial increase in GBF in fish is quantitatively similar to that reported for mammalian species (Matheson et al. 2000) (Table 9.2). In mongrel dogs, the increase in intestinal blood flow ranged from 70% (Hopkinson and Schenk 1968; Burns and Schenk 1969) to 130% (Vatner et al. 1970a). In beagle dogs, the increase in GBF was 85% (Takagi et al. 1988) and in baboons it was approximately 75% (Vatner et al. 1974). Some studies, however, show increases in GBF of 270% (Kato et al. 1989) to 400% (Gallavan et al. 1980), while smaller responses can range from 22 to 33% (Fronek and Fronek 1970) for dogs (Pawlik et al. 1980). This variability likely reflects interspecies differences, measurement methods, diet and physiological status (stress level, anesthesia, etc.).

4.1.2. TEMPORAL NATURE OF THE POSTPRANDIAL RESPONSE

The timing and duration of the postprandial response in GBF can vary significantly among animals. This is perhaps even more pronounced in fish given the potentially large differences in water temperature and relative meal (ration) size. For example, salmonids typically consume 1–2% of body mass daily to reach satiation, whereas the red Irish lord can consume more than 50% of their body mass in a single meal and take many days to complete digestion of this meal. The roles of temperature and ration size in these temporal differences have not been systematically studied among diverse species, but are likely to be large given their known effects on SDA amplitude and duration within the catfish family (see Section 5; Fu and Xie 2004; Fu et al. 2005, 2006, 2009).

Many fish species slowly increase GBF, with a peak occurring around 24 h post-feeding. Elevated GBF continues for several days in red Irish lord (Axelsson et al. 2000), shorthorn sculpin (Seth and Axelsson 2009), Atlantic cod (Axelsson and Fritsche 1991) and sea raven (Axelsson et al. 1989), followed by a slow return to routine GBF. Sea bass, however, increased GBF within 1 h, reaching a maximum of 82% above pre-feeding levels after just 6 h (Altimiras et al. 2008). GBF is rapidly increased also in rainbow trout, with the maximal increase occurring after about 12 h (Eliason et al. 2008; Gräns et al. 2009) (Fig. 9.3).

The reason for the delayed increase in GBF in some species following feeding appears to be correlated with postprandial delay before the gastric emptying starts (termed the gastric lag phase) as well as the total gastric

Table 9.2

Postprandial cardiovascular responses in mammals, measured at the three major vessels of the gastrointestinal tract. Lines indicate a combined total mesenteric blood flow. CA: coeliac artery; SMA: superior mesenteric artery; IMA: inferior mesenteric artery

Species	Meal	CA Blood flow (%)	SMA Blood flow (%)	IMA Blood flow (%)	Source
Mongrel dog (anaesthetized)	5% glucose solution	22.2%			Pawlik et al. 1980
Mongrel dog (conscious)	According to taste and appetite			——— 132% ———	Vatner et al. 1970
Mongrel dogs (conscious)	Horsemeat (15 ounces)	100% (duodenum)		——— 71% ———	Burns and Schenk 1969
Mongrel dogs (conscious)	Dog food (1 can)	145% (proximal jejunum) 75% (distal jejunum)	83% (Ileum)	2% (proximal colon) −10% (distal colon)	Gallavan et al. 1980
Mongrel dogs (conscious)	Dog food (450 g)		$14.6\ \text{ml min}^{-1}\ \text{kg}^{-1}$ to $19.0\ \text{ml min}^{-1}\ \text{kg}^{-1}$ (30%)		Gallavan et al. 1980
Mongrel dogs (conscious)	Dog food (450 g)		$13.97\ \text{ml min}^{-1}\ \text{kg}^{-1}$ to $17.9\ \text{ml min}^{-1}\ \text{kg}^{-1}$ (29%)		Fronek et al. 1970
Sprague-Dawley Rat (conscious)	Mixed meal (1.5 ml)	57% (duodenum) 80% (proximal jejunum)	40% (Ileum)		Hernandez et al. 1986
Beagle dogs (conscious)	Dog food (1 can)		230%		Takagi et al. 1988
Baboons (conscious)	Mixed fruits		$212\ \text{ml min}^{-1}$ to $372\ \text{ml min}^{-1}$ (75%)		Vatner et al. 1974

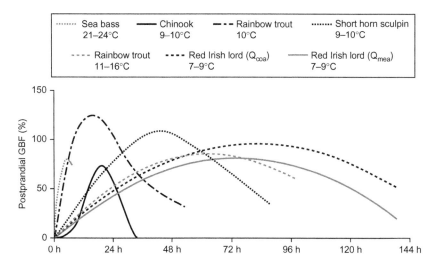

Fig. 9.3. Temporal pattern of the postprandial changes in gastrointestinal blood flow in fish. Feeding induces a substantial increase in the blood flow reaching the gastrointestinal tract, and the timing and duration of the response varies between species and the ingested diet, but usually develops over 10–20 h and may persist for over 70 h (Axelsson et al. 2000; Thorarensen and Farrell 2006; Altimiras et al. 2008; Eliason et al. 2008; Gräns et al. 2009; Seth and Axelsson 2009). From Seth (2010).

emptying time. It is commonly assumed that the postprandial increase in GBF primarily occurs as food enters the intestine and is hydrolyzed to release the composite macromolecules (as discussed below). Gastric emptying time depends on both the texture of the meal (solid versus liquid) (Ruohonen et al. 1997; Bucking and Wood 2006) and temperature. In general, there is a lag phase of at least 2–3 h (Windell et al. 1969; Ruohonen et al. 1997; Olsson et al. 1999; Bucking and Wood 2006), although virtually no lag phase is reported for the shorthorn sculpin fed a wet, partially pre-digested diet (Seth and Axelsson 2009).

Gastric emptying in fish also varies with temperature (Jobling and Davies 1979), meal size (Jobling et al. 1977) and type of food (Windell et al. 1976) as well as energy content (Jobling and Davies 1980), but in general lasts 15–96 h (total GI tract evacuation is around 15–150 h), with GBF showing a similar profile, returning to a routine level after 48–72 h.

Not surprising, given the high body temperature, gastric emptying is more rapid and can, depending on the texture of the diet, sometimes start almost instantaneously in mammals. Correspondingly, there is a more rapid increase in GBF (within 5–10 min) compared with fish. The maximal GBF in mammals occurs during the next 6 h, after which GBF returns to baseline/routine over the next 12–24 h (Fronek and Stahlgren 1968; Hopkinson and

Schenk 1968; Fronek and Fronek 1970; Vatner et al. 1970a, 1974; Takagi et al. 1988).

Maybe more surprising is the rapid increase in GBF seen in fish like the shorthorn sculpin and sea bass. This occurs within 1 h after feeding and presumably precedes appreciable hydrolysis of the ingested meal in the proximal intestine. This response suggests the potential involvement of gastric chemoreceptors that sense the presence of food in the stomach (Dupont-Prinet et al. 2009; Seth and Axelsson 2009). However, the presence of gastric chemoreceptors has yet to be established in fish, although such a mechanism, considering its benefits, probably evolved early in the vertebrate lineage.

In mammals, gastric chemoreceptors have been described that sense specific components of the diet composition, such as glutamate (Nakamura et al. 2008; Tsurugizawa et al. 2009), as well as and importantly the presence of noxious agents (Rozengurt 2006). There could, potentially, also be an additional effect from the actual ingestion or swallowing of the meal that would induce a very rapid increase in GBF. In reptiles, at least, there is a substantial increase in the oxygen consumption (SDA; discussed below) even before the meal enters the stomach (Secor 2003), but whether or not this increase in oxygen consumption is coupled to an increase in GBF remains to be determined.

In both fish (Axelsson et al. 2000) and mammals (Takagi et al. 1988), the increase in GBF that occurs following feeding is due to a local decrease in GI vascular resistance. Below, we use the term hyperemia to collectively describe the lowering of the GI vascular resistance as well as the accompanying increase in blood flow.

4.1.3. POSTPRANDIAL CARDIOVASCULAR ADJUSTMENTS TO MAINTAIN AN INCREASED GBF

Across all vertebrates, GBF can be regulated via a change in cardiac output, a change in systemic vascular resistance and a change in GI vascular resistance, or some combination. Fishes and mammals both change GI vascular resistance, as noted above. However, the relative contributions of these three mechanisms to the postprandial increase in GBF in fish and mammals differ. Increased GBF in fish mostly results from an increase in cardiac output. In contrast, mammals mostly redistribute blood flow to the GI tract from other vascular beds, which occurs through either an increase in the systemic vascular resistance and/or a substantial decrease in the GI vascular resistance decreases and little change in cardiac output (Hopkinson and Schenk 1968; Burns and Schenk 1969; Vatner et al. 1970b, 1974; Gallavan et al. 1980). Even so, the relative contribution of each mechanism can be strongly modulated by the animal's physical status (i.e. exercise or other stressors), as well as surrounding environmental factors such as oxygen availability.

In the shorthorn sculpin, the postprandial increase in GBF involves a 35% lowering of GI vascular resistance as food enters the intestine (Seth and Axelsson 2009). Decreasing GI vascular resistance without any other cardiovascular change (either a systemic vasoconstriction or a compensatory increase in cardiac output) would decrease arterial blood pressure. However, in most fish species studied to date (e.g. shorthorn sculpin, rainbow trout, Atlantic cod, sea raven, red Irish lord and sea bass), dorsal aortic pressure is unchanged or increases slightly because the increase in GBF is compensated for by an equivalent or larger increase in cardiac output (Axelsson et al. 1989, 2000; Axelsson and Fritsche 1991). The feedback mechanism behind the postprandial regulation of cardiac output is not known, but it possibly involves a barostatic reflex mediating cardiac and vascular changes.

The amount of cardiac output reaching the GI tract can shift even without a redistribution of blood from the remaining systemic circulation. As discussed above, about 10–40% of cardiac output reaches the GI tract under pre-feeding or post-absorptive routine conditions. After feeding, as much as 52% of the cardiac output reaches the GI tract of Atlantic cod (Axelsson and Fritsche 1991). In sea raven and sea bass, the proportion of postprandial cardiac output reaching the GI tract increases from 25 to 35–40% (Axelsson et al. 1989; Dupont-Prinet et al. 2009). A shift in the amount of cardiac output reaching the GI vasculature is mediated via a decrease in the resistance of the GI vasculature and either a maintained or increased resistance of other systemic vascular beds.

4.1.4. TRIGGERS FOR POSTPRANDIAL INCREASE IN GASTROINTESTINAL BLOOD FLOW

Ingestion of food can trigger the postprandial cardiovascular changes in several possible ways. The first phase, the cephalic phase, involves increased salivation and gastric acid secretion, which is unimportant in mammals (dogs; Takagi et al. 1988). It has been shown that fish can react to a number of odors in the water and some species have acute sight, but whether this triggers any changes in GBF before the food actually is ingested is not known at present.

4.1.4.1. Mechanical Stimuli. Mechanical stimuli during stomach distension are sensed by gastric stretch receptors, which are important in relaxing the stomach when ingesting food. This process is commonly referred to as accommodation (Grove and Holmgren 1992a,b). Mechanical stimuli obviously vary with the ration size (the level of stretch) and stomach compliance (ability to stretch). Rainbow trout, which often feed on small

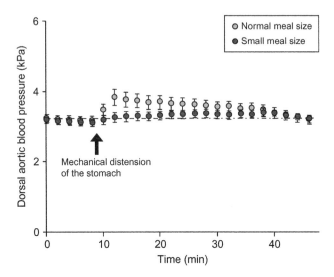

Fig. 9.4. Distension of the stomach produces a size-dependent increase in dorsal aortic pressure via an α-adrenoceptor-mediated reflex.

insects and larvae, stretch their stomach less than an ambush predator such as shorthorn sculpin that ingests up to 50% of body mass. Even so, stomach distension *per se* does not increase GBF in either species (Seth et al. 2008; Seth and Axelsson 2009). Instead, stomach distension elicits a rapid increase in dorsal aortic blood pressure via an increase in systemic vascular resistance (Fig. 9.4). This pressor response is mediated via an α-adrenergic mechanism and likely involves a generalized increase in sympathetic activity, similar to that reported for mammals (Fronek and Stahlgren 1968; Vatner et al. 1970b, 1974). The purpose of this pressor response is unclear. In mammals, several studies (Longhurst et al. 1980, 1981; Longhurst and Ibarra 1982, 1984) concluded that mechanical distension of the stomach accounts for most if not all of the increase in sympathetic activity that occurs soon after feeding.

While mechanical stimuli in fish do not elicit GI hyperemia, they may be important to increasing the dorsal aortic blood pressure, which is the driving force for GI perfusion (Seth and Axelsson 2009).

4.1.4.2. Chemical Stimuli.   In some fish, such as the shorthorn sculpin and sea bass, there is a very rapid increase in GBF that occurs within hours after feeding and when most of the food still remains in the stomach (Altimiras et al. 2008; Seth and Axelsson 2009). However, whether or not these processes involve central or local chemoreception remains undetermined.

This initial response is smaller than the subsequent response after the entry of hydrolyzed food into the intestine.

When food enters the intestine, the meal is broken down by several different enzyme systems. In fish, these enzymes include carbohydrases, esterases and proteases (Kitamikado and Tachino 1960a,b,c). Food breakdown and nutrient uptake are coordinated in ways similar to those common for mammalian species. These hydrolyzed products are most likely the main trigger for GI hyperemia since undigested food entering the intestine elicits no or only a small hyperemia.

In the red Irish lord and perhaps the Atlantic cod (Axelsson and Fritsche 1991; Axelsson et al. 2000), blood flow distribution shifts over time from the coeliac to the mesenteric circulation, which supplies most of the proximal and distal intestine, as well as other tissues within the GI tract excluding the stomach. This local increase in blood flow coincides with the presence of food and its breakdown products, as seen in most mammals (Gallavan and Chou 1985; Matheson et al. 2000), and is best explained by a chemically induced hyperemia. The fact that the hydrolyzed components of ingested food are the most important inducer of the postprandial hyperemia is well known for mammals (Chou et al. 1978; Chou and Coatney 1994). However, the importance of hydrolyzed components has not been established in fish until recently. There was significant hyperemia with a bolus injection of a predigested diet into the intestine of rainbow trout without a response to saline injection (Seth et al. 2009) and mechanical stimuli noted above.

It is far from clear which nutrients influence this hyperemic response in fishes. In rainbow trout, while pre-digested food injected into the proximal intestine of free-swimming fish rapidly increases GBF, the amplitude depends on the nutrient composition (Seth et al. 2009). The diet composition resembling the normal diet gave the largest hyperemic and metabolic response (as discussed below). The temporal pattern of the response also differed, perhaps reflecting differences in the way nutrients are digested, absorbed and metabolized.

The postprandial increase in total body oxygen consumption (SDA) is in most cases larger for a high-protein diet compared with a high-fat diet (discussed below). However, it is unclear whether or not the caloric content of the food is important in the hyperemic response. In fact, how caloric content of the diet might be "sensed" by the GI tract is unclear. Recent results using isocaloric diets of different composition show similar effects on GBF in rainbow trout (Eliason et al. 2008) and isocaloric diets of either a high protein or high lipid content on the metabolic response of Atlantic cod (Juan et al. 2009).

One contributing factor is most likely the fact that the GI tract is adapted to the "normal" diet composition (Buddington et al. 1987, 1997;

Buddington and Hilton 1987). For example, a carnivorous fish, like the rainbow trout, normally ingests limited carbohydrates and relies mostly on proteins and lipids/fats and will likely respond weakly to changes in carbohydrate content. This is most likely in contrast to the response in omnivorous or herbivorous fish like, for example, the carps (*Carassius* sp.).

Another factor, which in combination with nutrient composition could affect the gastrointestinal blood flow, is the amount of bile secreted, which at least in mammals has been shown to potentiate the effects of certain nutrients on gut blood flow (Kvietys et al. 1980a, 1981).

4.1.5. REGULATION OF POSTPRANDIAL BLOOD FLOW

4.1.5.1. Central Nervous Control.   The coordination of blood flow within an animal in order to supply different tissues with oxygen, based on the oxygen demand of the tissues, must involve some sort of central control mechanism. Central control of the cardiovascular system is mediated via the autonomic nervous system, which is divided into the sympathetic, parasympathetic and enteric nervous systems.

The GI tract of fish (e.g. brown trout, *Salmo trutta*) is innervated via the splanchnic as well as the vagus nerve (Fig. 9.5), with the former innervating most of the intestine and hind gut, and the latter innervating the major portion of the stomach (Burnstock 1959). In fish, the vascular system is under the control of sympathetic innervation, as well as plasma CA (Axelsson et al. 1989, 2000; Axelsson and Fritsche 1991). The branchial

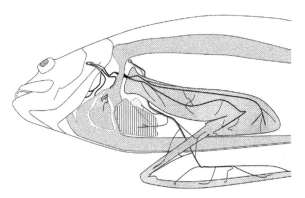

Fig. 9.5. Illustration of the gastrointestinal tract and its innervation via the paired vagus (only one side is shown) and the splanchnic nerve that runs along the coeliacomesenteric artery and its branches. The vagus nerve, originating from the nodose ganglia, supplies the larger portion of the stomach (as well as the heart, etc.) with few or no fibers reaching the proximal intestine. The coeliacomesenteric artery innervates most of the intestines, as well as several other organs including the stomach. From Seth and Axelsson (2010).

vasculature also has a parasympathetic innervation (Sundin and Nilsson 1992, 1997; Nilsson and Sundin 1998), but whether or not this part of the autonomic nervous system innervates the GI tract remains to be established.

In rainbow trout, a substantial portion of the routine GI vasoconstrictory tonus of the GI tract is mediated via the splanchnic nerve, but additional mechanisms are involved in the control of postprandial GBF. Local innervation via the enteric nervous system likely becomes more important during the postprandial state (Seth and Axelsson 2010). A predominance of local control mechanisms is also seen in the sea bass (Axelsson et al. 2002). During hypoxia there is a coordinated decrease in the perfusion of the gastrointestinal tract, which in fed sea bass is counteracted by a simultaneous vasodilation that depends on local control within the GI tract. Local control mechanisms also predominate in the red Irish lord (Axelsson et al. 2000). The same is true in mammals, with the sympathetic and parasympathetic nervous systems being of little importance for the postprandial hyperemia. Sectioning the nerves to the GI tract or blocking the effects of these nerves with antagonistic drugs such as atropine (a muscarinic antagonist) have little effect on GBF (Nyhof and Chou 1981, 1983; Nyhof et al. 1985).

Central neural system (CNS) control of GI vasculature, however, could be important, during stomach distension following feeding. As discussed above, there is a rapid vasoconstriction (within seconds) following mechanical stimuli of the stomach (Seth et al. 2008; Seth and Axelsson 2009). Stretch is most likely recognized via gastric mechanoceptors and an efferent adrenergic pathway regulates vascular resistance in the peripheral systemic vasculature via α-adrenoceptors. In mammals, these mechano-/stretch-receptors are located inside the circular muscular layer and relay signals via the vagus to the CNS for processing (Ozaki et al. 1999; Ozaki and Gebhart 2001; Carmagnola et al. 2005).

Mammals also have mucosal chemoceptors in the stomach that send afferent information to the CNS via the vagus (Tsurugizawa et al. 2009). A few studies in fish indicated gastric chemoceptors could be involved in the regulation of GBF. For example, the shorthorn sculpin rapidly increases GBF after feeding, a response that is independent of mechanical stimuli and food entering the intestine (Seth and Axelsson 2009). A similar rapid increase in GBF is seen with sea bass, but the independence of mechanical stimulus is untested (Altimiras et al. 2008).

4.1.5.2. Hormonal Control. Hormones such as cholecystokinin (CCK) and gastrin are probably important in regulating and coordinating GBF with other GI functions, especially given its wide range of functions. Several isoforms of CCK have been described in rainbow trout (Jensen et al. 2001).

CCK controls gastric emptying (Olsson et al. 1999), as well as bile production and secretion in rainbow trout (Aldman et al. 1992). A possible involvement of CCK in the regulation of GBF was shown for rainbow trout, with GBF increasing with administration of physiological concentrations of CCK (Seth et al. 2010). Therefore, because plasma levels of CCK are influenced by diet composition (Jönsson et al. 2006), CCK could modulate GBF in relation to diet composition. A possible mechanism for the effect of CCK and other hormones is illustrated in Fig. 9.6.

In mammals, several studies have shown a CCK involvement in the modulation of the postprandial GI hyperemia (Granger et al. 1980) via a mechanism probably involving chemical stimulation of gastric mucosal cells, which triggers CCK release from neuroendocrine type 1 cells (Verberne et al. 2003). Even so, postprandial plasma CCK levels in most studies are below those needed to induce an increase in GBF (Premen et al. 1985; Chou and

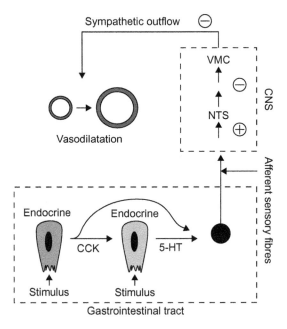

Fig. 9.6. A schematic representation of a reflex pathway mediating a decrease in the tonus of the gastrointestinal vasculature leading to an increase in blood flow. Cholecystokinin (CCK) and/or 5-HT (serotonin) is released upon a chemical stimulus and this triggers a sensory pathway that projects to the NTS region in the brain. This illustration is based on the results from several mammalian studies, but a similarly intricate series of events might very well occur in teleost fish in order to modulate the postprandial gastrointestinal blood flow. NTS, nucleus tractus solitarius; VMC, vasomotor center; CNS, central nervous system. + and − denote an activation or inhibition, respectively. From Seth (2010).

Coatney 1994). In this context, CCK probably works in a paracrine fashion, reaching much higher concentrations locally around type 1 cells; plasma levels probably reflect only the "spillover" of CCK into the circulation. The effects of CCK could also be secondary to an increased pancreatic metabolism (Nakajima et al. 2001).

Whether or not the vascular effects in fish are mediated in a similar way to those reported in mammals remains to be determined, but there seems to be limited effect on the vasculature itself (Seth et al. 2010). Exposing isolated vascular preparations revealed no response and the response is probably reflex mediated, as reported for mammals.

4.1.5.3. Enteric Nervous System/Paracrine Signaling. The enteric nervous system is an independent part of the autonomic nervous system (reviewed for fish by Olsson and Holmgren 2009). Even though the enteric nervous system has the capacity to function on its own, it receives sympathetic and parasympathetic nerves to coordinate the activities within an animal. It also sends sensory information from the GI tract to the CNS. Since the enteric nervous system is involved in most aspects of the GI function, such as motility and secretions, it should also have the capacity to modulate the vasculature.

However, difficulties in discriminating between intrinsic and extrinsic nerves make it difficult to determine to what extent enteric nerves regulate GBF. Substances could be released from local nerves, influencing GBF in a paracrine fashion, but these effects could also be endocrine as discussed above.

Despite such difficulties, a crucial involvement of the enteric nervous system in the induction of a postprandial hyperemia was confirmed for rainbow trout (Seth and Axelsson 2010). The enteric nervous system, in addition to the extrinsic innervation via sympathetic and parasympathetic nerves, was blocked using the voltage-gated sodium channel antagonist, tetrodotoxin (TTX). TTX completely repressed postprandial hyperemia, with only limited effects on other cardiovascular parameters. At present, it is unknown which parts of the enteric nervous system are of such crucial importance. One potential explanation for the result is that TTX prevented release of one or several of the vasoactive substances known to be present in the GI tract of fish. Tachykinins, including substance P and neurokinin A, have been identified in the GI tract of spiny dogfish (*Squalus acanthias*) (Kågström et al. 1996a) and Atlantic cod (Jensen et al. 1993), as well as rainbow trout (Kågström et al. 1996b). Bradykinin could exert effects on Atlantic cod blood vessels (Shahbazi et al. 2001).

Tachykinins stimulate gut motility and have vasoactive effects on gut vessels (Table 9.3). Substance P increases dorsal aortic pressure through

Table 9.3

In vivo cardiovascular responses to selected gastrointestinal-related substances. $\uparrow\downarrow$ and $\uparrow\downarrow\uparrow$ denote a bi-phasic and a tri-phasic response, respectively. GBF, gut blood flow; CO, cardiac output; R_{coel}, vascular resistance of gastrointestinal (coeliac) artery; HR, heart rate

Species	Temp °C	Substance P	Neurokinin A	VIP	Scyliorhinin I	Scyliorhinin II	Source
Spiny dogfish (*Squalus acanthias*)	10	GBF $\uparrow$; CO $\uparrow$; HR $\uparrow$	N/A	GBF $\downarrow$; R_{coel} $\uparrow$; HR $\uparrow$	N/A	N/A	Holmgren et al. 1992
Atlantic cod (*Gadus morhua*)	10–11	GBF $\uparrow\downarrow\uparrow$; CO $\uparrow$	N/A	GBF $\uparrow$; CO $\uparrow$	N/A	N/A	Jensen et al. 1991
Rainbow trout (*Oncorhynchus mykiss*)	10	N/A	N/A	N/A	R_{coel} $\downarrow\uparrow$; CO $\downarrow\uparrow$	R_{coel} $\downarrow\uparrow$; CO $\downarrow\uparrow$	Kågström et al. 1994
Rainbow trout (*Oncorhynchus mykiss*)	10	R_{coel} $\uparrow$; CO $\downarrow$	R_{coel} $\uparrow$; CO $\downarrow$	N/A	N/A	N/A	Kågström et al. 1996b
Spiny dogfish (*Squalus acanthias*)	10	GBF $\uparrow$; R_{coel} $\downarrow$	GBF $\uparrow\downarrow$; R_{coel} $\downarrow\uparrow$	N/A	GBF $\uparrow$; R_{coel} $\downarrow$	GBF $\uparrow$; R_{coel} $\uparrow\downarrow$	Kågström et al. 1996a

both a splanchnic and systemic vasoconstriction in rainbow trout (Kågström et al. 1996b). Substance P has a complex triphasic effect on GBF, an initial increase, followed by a decrease to baseline and a secondary increase. A similar effect of substance P is seen for spiny dogfish (Holmgren et al. 1992).

In contrast to substance P, the tachykinin VIP is a potent systemic vasorelaxant in teleost fish, acting through an endothelium/NO-independent mechanism in rainbow trout (Kågström and Holmgren 1997). Because VIP decreases gut motility, again unlike substance P, increased GBF is not necessarily coupled with increased gut motility. In contrast to the effects on GBF in rainbow trout and Atlantic cod, VIP decreases GBF in the spiny dogfish (Holmgren et al. 1992). The reason for this species difference is unclear, but the presence of two different scyliorhinins in elasmobranchs could be important. These scyliorhinins have responses in elasmobranchs similar to those of VIP in teleosts and are even vasoactive in the rainbow trout (Kågström and Holmgren 1997). Whether these vasoactive tachykinins are important in controlling postprandial GBF or even post-absorptive but this remains to be determined.

In mammals, enteric neuron effects are variable. In some cases, significant effects on GBF are evident (Chou et al. 1972; Neild et al. 1990; Surprenant 1994), but not in others (Nyhof and Chou 1983). Cholinergic control of postprandial GBF is also observed (Nyhof et al. 1985).

4.1.5.4. Metabolite-Induced Control.   Changes in the oxygen tension and tissue osmolarity are the most probable causes of a chemically induced hyperemia in mammals (Bohlen 1998a; Matheson et al. 2000). Tissue hypoxia could act directly on blood vessels or could induce surrounding tissues to release vasoactive metabolites such as lactate, hydrogen ions, ADP and NO from (Bohlen 1980a,b, 1998a,b). A lowering of the oxygen tension or content could also be sensed by red blood cells, leading to the release of ATP (Gonzalez-Alonso et al. 2002), thus further contributing to the hyperemia. Little is known for fishes in this regard.

The emergence of hydrogen sulfide (H_2S) as a possible oxygen sensor in fish provides a possible mechanism for a metabolite-induced control of the GI vasculature (Olson 2008, 2009). H_2S is synthesized within the vasculature and depending on the species as well as the tissue under study; H_2S might act as either a vasoconstrictor or a vasodilator. The response depends on both endothelium-dependent mechanisms as well as direct effects on the vascular smooth muscle. In lampreys (*Petromyzon marinus*), H_2S produces a vasoconstriction similar to that occurring with hypoxia; inhibitors of H_2S production remove the hypoxic vasoconstriction (Olson et al. 2006, 2008).

The mechanism for H_2S production is for the most part unknown, but it could be that mitochondria release biologically active H_2S during hypoxia.

During digestion and absorption, there could be an increase in the mucosal H_2S that is sensed by chemosensitive enteric sensory nerves. The signal would be relayed via interneurons to vasomotor neurons innervating the submucosal arterioles, where a substantial portion of the vascular resistance resides. There could also be additional control by oxygen sensors such as hemoglobin (Wood et al. 1975; Dietrich et al. 2000; Gonzalez-Alonso et al. 2002). This would explain the requirement of local enteric nerves in controlling the postprandial hyperemia as discussed above.

4.1.5.5. Indirect Effects of Gut Motility on Gastrointestinal Blood Flow. Gut motility and peristalsis are probably of limited importance in the postprandial hyperemia, but this remains to be tested in fish. Non-absorbable ballotini beads (Bucking and Wood 2006), or other inert materials that are not broken down or absorbed by the gut, such as kaolin, could be used to determine the relative contribution of gut motility in fish. In mammals, increased gut motility leads to a redistribution of blood within the gut wall providing the muscular tissue layers with more blood, with no increase in overall blood flow (Chou 1982). Gut motility, however, could be important by means of mixing the chyme and enable an efficient absorption of the hydrolyzed food particles, and therefore a secondary increase in GBF.

4.2. Hypoxia and Hypercapnia

Most animals studied so far cannot maximally and simultaneously perfuse all vascular beds, the integration of the different demands of different tissues is vital and a continuously ongoing process. Some organs, like the heart and brain, are highly prioritized while other tissue types or organs can withstand periods of lower blood supply without any long-lasting effect. In spite of the high routine GBF, GBF varies greatly due to prioritization of other vascular beds. Hypoxia, hypercapnia, exercise and general stress are all examples of situations where GBF has a low priority and is reduced to allow more blood to be directed for sustaining other tissues.

Environmental hypoxia may or may not be accompanied by increased environmental carbon dioxide (hypercarbia). Physically dissolved oxygen in water is relatively small. For example, freshwater at $0°C$ has only 10.2 ml oxygen per liter water. Increasing temperature and salinity decreases soluble oxygen, such that seawater at $35°C$ and 35 ppt has just 4.0 ml per liter water.

During environmental hypoxia, fish make several cardiorespiratory adjustments with the overall aim of maintaining the oxygen transport. Most studies show a rapid and marked reduction in GBF upon exposure to hypoxic water which serves to allocate more blood to other, more oxygen-sensitive tissues (Axelsson and Fritsche 1991). The postprandial situation is more complicated, however. Axelsson et al. (2002) reported a decline in routine GBF in both fed and unfed sea bass during hypoxia, but the ratio of cardiac output directed to the GI was unchanged between the fed and unfed animals, likely because the metabolic hyperemia out-competed the hypoxic reflex down-regulation of GBF.

Exercise in combination with hypoxia was tested in fed and unfed sea bass. GBF was reduced to unfed values at U_{crit} during both normoxia and in hypoxia. An interesting difference was that during the combined challenge the animals also reduced the SDA completely, showing that sea bass have an ability to prioritize aerobic exercise (Dupont-Prinet et al. 2009).

The effects of hypercapnia on GBF have only been studied in one species so far. White sturgeon (*Acipenser transmontanus*) exposed to hypercapnia similar to concentrations reported from an aquaculture setting maintained GBF while overall systemic vascular resistance was lowered and cardiac output was increased (Crocker et al. 2000). However, the decrease in GBF in response to struggling was accentuated.

4.3. Exercise

Most studies with fish show a decrease in routine GBF during prolonged, steady exercise. However, there is a large inter-species variation with this response (Farrell et al. 2001). In Atlantic cod, celiac and mesenteric artery blood flow decreased by 29% and 36%, respectively, during exercise as a result of a 75% increase in GI vascular resistance (Axelsson and Fritsche 1991). Sea bass decrease routine GBF with steady swimming and the postprandial increase in GBF becomes attenuated during swimming (Altimiras et al. 2008). Dupont-Prinet et al. (2009) also showed that sea bass have little capacity to support postprandial GBF during exercise. Even so, the SDA response was maintained throughout the exercise regime and maximal swimming speed was unaffected by feeding.

In Chinook salmon and rainbow trout, however, while routine GBF also decreased during steady swimming, U_{crit} was reduced in fed fish presumably to maintain some GBF. Presumably, maximum cardiac output cannot supply the maximum blood flow needs of sustained exercise and digestion (Stevens 1968; Thorarensen and Farrell 2006). Similarly unrestrained dogs and primates (Vatner 1978) do not decrease either routine or postprandial GBF, but increase cardiac output instead to sustain the increased demand for blood flow of exercise and digestion (Hopkinson and Schenk 1968; Burns and Schenk 1969; Fronek and Fronek 1970). Dogs actually show a small decrease in peripheral/muscular blood flow after feeding, which is restored once the animal becomes ambulatory (Gallavan et al. 1980). This is in sharp contrast to what happens in humans during exercise where superior mesenteric blood flow is reduced during (measured just post-exercise) exercise, both in the fasting and fed state (Qamar and Read 1987).

4.4. Temperature

The fish cardiovascular system is influenced considerably by environmental temperature (Farrell 1996). Yet little is known about how GBF, particularly after feeding, is affected by temperature. In the green sturgeon (*Acipenser medirostris*), absolute GBF increased with temperature (19°C to >26°C) (Gräns et al. 2009a). On the other hand, rainbow trout acclimated to 16°C have a lower absolute GBF and a smaller postprandial increase when compared with individuals acclimated to 10°C (Gräns et al. 2009). The preferred temperature for a particular species will affect the response to temperature (McCauley et al. 1977), and so species differences in preferred temperatures may explain observed differences among studies and species. Wurtsbaugh and Neverman (1988) showed that the Bear Lake sculpin (*Cottus extensus*) has a daily migration to feed in colder, deeper water during daytime and back to shallower, 10°C warmer water at night. This allowed an increase in the night-time rate of digestion, thereby increasing overall feeding rate and growth. Unfortunately there are no data on GBF from this study. Conversely, Brett has argued that sockeye salmon (*Oncorhynchus nerka*) surface at dusk and dawn to feed, and in between migrate to colder, deeper water to minimize routine metabolic rate while digesting their meal and thereby maximizing growth rate. Measurements of GBF during diurnal feeding and temperature migration would be informative but are technically challenging.

5. GASTROINTESTINAL METABOLISM

Despite the fact that the postprandial metabolic response is one of the most studied physiological phenomena in fish and more than 250 vertebrate and invertebrate species (Secor 2009), virtually nothing is known about the routine metabolism of the GI tract itself. The metabolic processes involved in digestion lead to a rapid and sustained increase in postprandial oxygen consumption. This increase in metabolic rate is followed by a gradual return to the fasted metabolic rate, and is commonly known as the specific dynamic action (SDA). For economically important species (commercial or recreational) such as the salmonids, cod, carps, catfish and tunas, it has been important to optimize meal size and composition, as well as environmental conditions in terms of the SDA response so that the maximum amount of energy can be allocated to growth (Medland and Beamish 1985; LeGrow and Beamish 1986; Fu et al. 2005; Couto et al. 2008; Enes et al. 2009). For comparative purposes there are several available reviews concerning the postprandial metabolic response in mammals (McCue 2006), in fish (Fu et al. 2009), as well as a very comprehensive review covering the most studied vertebrates (Secor 2009).

5.1. Routine Gastrointestinal Metabolism

Given that the GI tract receives a substantial portion of cardiac output and assuming that the GI tract has a similar oxygen extraction to other tissues, oxygen consumption by the GI tract could constitute a large proportion of total metabolic rate. However, there is a very limited amount of data available on the oxygen consumption by the GI tract during routine post-absorptive condition, especially *in vivo*. Isolated GI tissue segments from the gulf toadfish (*Opsanus beta*) have an oxygen consumption of approximately $0.076 \, \text{ml} \, \text{min}^{-1}$ in a 20 g fish or 5.6% of the total oxygen consumption (Taylor and Grosell 2009).

A few mammalian studies have measured metabolism of the GI tract (Yen et al. 1989; Vaugelade et al. 1994). These studies demonstrate that the gastrointestinal tissue consumes approximately $1.5 \, \text{ml} \, \text{min}^{-1} \, \text{kg}^{-1}$ (body weight) of oxygen or around 25% of total oxygen consumption. So far no *in vivo* studies of the metabolic cost of the GI tract have been published even if the same methodology as has been used in the few mammalian studies could be applied.

5.2. Postprandial Gastrointestinal Metabolism

Only part of SDA directly involves the GI tract (as discussed below), but few studies have quantified this. In isolated gulf toadfish gastrointestinal tissues, an 85% increase in oxygen consumption accounted for less than 5% of SDA (Taylor and Grosell 2009). This modest contribution supports the notion that most of the postprandial metabolic response resides outside of the GI tract. The few mammalian reports provide a similar result. For example, a 46% increase in postprandial GI metabolism of pigs represents around 25% of SDA (Yen et al. 1989). An increased knowledge about the isolated metabolic response of the GI tract following feeding would aid in deciphering the link between GI metabolism and GBF. Changes in GI metabolism are not necessarily correlated with GBF and a possible explanation for this is that there can be a change in the tissue oxygen extraction. Such changes would be detected when isolating the GI metabolism *in vivo*, as described above.

In the absence of additional data, the remaining part of this section will focus on SDA data, which are more abundant. Besides, without feeding and absorption across the gut, there would be no SDA response.

5.2.1. SPECIFIC DYNAMIC ACTION (SDA) OR HEAT INCREMENT (HI) OF FEEDING

Originally referred to as "spezifisch dynamische wirkung" (Rubner 1902) and later specific dynamic action or effect, this term describes the postprandial increase in heat production due to an increase in metabolism. Initially, direct calorimetry was used to estimate heat production (hence the term heat increment (HI) of feeding), but this method has been replaced by measuring the increase in oxygen consumption or CO_2 production. In water-breathing animals such as fish, measuring oxygen consumption using an intermittent flow respirometer, as described by Steffensen (Steffensen 1989), has proven to be an efficient and reliable method of estimating SDA. Some scientists still use HI or apparent HI. This acknowledges that the absolute increase in heat production and hence oxygen consumption is the result of several metabolic processes that accumulate to give a certain outcome (Secor 2009), rather than an "action." Other terms such as thermogenic effects of feeding and postprandial thermogenesis have been used. For traditional reasons and clarity, we use the term SDA.

The SDA can also be represented in terms of the SDA coefficient, which is SDA expressed as a proportion of the dietary energy content, i.e. the cost of digestion and assimilation. Because the SDA coefficient is independent of meal size, as shown for Atlantic cod under normoxic conditions (Jordan and Steffensen 2007), it can be used to make reliable comparisons among studies and species. It might even be more useful to express SDA as a proportion of the digested and absorbed energy (Beamish 1974), since most of the SDA is supposed to reside within post-absorptive processes in fish (Jobling and Davies 1980; Brown and Cameron 1991a,b), reptiles (Coulson et al. 1978) and mammals (Wilhelmj and Bollman 1928; Ashworth 1969).

5.2.2. EFFECTS OF FEEDING

It is generally assumed that the SDA is largely a post-absorptive phenomenon, yet the majority of studies concerning postprandial metabolism in fish have focused on the GI portion. Evidence for the importance of post-absorptive processes stems from the fact that there is little difference in the SDA response whether nutrients are ingested or infused intravenously in fish (Jobling and Davies 1980; Brown and Cameron 1991a). Therefore, SDA probably involves intestinal catabolic processes (deamination and formation of excretory products) and anabolic processes both within the intestine and outside of the intestine, including the liver. Brown and Cameron (1991b) showed that by injecting cyclohexamide, a potent inhibitor of protein synthesis, into channel catfish they were able to almost completely abolish the SDA response. The same has also been shown in reptiles, where cyclohexamide decreased SDA by about 70% (McCue et al. 2005).

The ingestion or swallowing of food could contribute to a mechanically dependent SDA that is part of the GI tract metabolism itself via the increase in the smooth muscle of activity of the gastrointestinal tract. Added onto this is the cost for the production and secretion of gastric proteolytic enzymes. In mammals these contributions are small but in reptiles, the gastric phase can constitute a major portion of the SDA (Secor 2003). Ultimately, stomach residence time influences its contribution. The motility of the intestine probably contributes little to the SDA in fish, as has been shown in plaice (*Pleuronectes platessa*) fed an indigestible diet (Jobling and Davies 1980). To what extent absorption (as well as secretion) contributes to the SDA remains to be determined in fish, but in alligators there seems to be a limited contribution from the absorption and transport of amino acids (Coulson and Hernandez 1979). In contrast, processes such as secretion, within the rat intestine, such as secretion, constitute a substantial portion of the initial SDA. After the resection of 70% of the intestine, there was a substantial decrease in oxygen consumption, even before the food entered the intestine (Luz et al. 2000).

Values of SDA amplitude in fishes range from an impressive 1,000% increase in metabolic rate for the European eel (*Anguilla anguilla*) (Owen 2001) to a minor increase of 29% for the bluegill (*Lepomis macrochirus*) (Pierce and Wissing 1974). The duration of SDA ranges from 3 to 6 h for rainbow trout and Atlantic cod (Smith et al. 1978; Peck et al. 2003) to 390 h for the Antarctic spiny plunderfish (*Harpagifer antarcticus*) (Boyce and Clarke 1997). Longer durations are almost always associated with low-temperature conditions (Table 9.4). The magnitude of the SDA response is similar in many mammalian species, but the duration is a lot shorter compared with most fishes (McCue 2006; Secor 2009), presumably in relation to their warmer temperature. The SDA coefficient ranges from 1.6% in rainbow trout (Smith et al. 1978) to 58% in the cyprinid topmouth gudgeon (*Pseudorasbora parva*) (Cui and Liu 1990). Larger values indicate a higher cost of digestion and absorption, with less absorbed energy available for growth.

Table 9.4

Postprandial changes in oxygen consumption in selected species of fish, represented as specific dynamic action (SDA) and the corresponding coefficient. The duration of the response is also given along with size and type of the ingested meal as well as the water temperature

Species	Temp °C	Body size (g)	Meal (% b.w.)	SDA (kJ)	SDA coefficient (%)	Duration (h)	Source
Southern catfish (*Silurus meridionalis*)	27–28	40	6.0% mixed formula	2.64	15.6	56	Fu et al. 2005
European plaice (*Pleuronectes platessa*)	10	32	3.8% fish paste	1.12	16.5	70	Jobling and Davies 1980
Shorthorn sculpin (*Myoxocephalus scorpius*)	15	74.5	12.7% crustaceans	7.64	16.0	162	Johnston and Battram 1993
Atlantic cod (*Gadus morhua*)	10	147	5.0% fish	7.10	9.7	95	Jordan and Steffensen 2007
Atlantic cod (*Gadus morhua*)	10–11	180	6.5% fish	3.61	4.4	48	Lyndon et al. 1992
Rainbow trout (*Oncorhynchus mykiss*)	15	15.0	2.0% mixed formula	0.38	7.6	N/A	Medland and Beamish 1985
Sea bass (*Dicentrarchus labrax*)	25	42	2.0% mixed formula	2.49	14.9	N/A	Peres and Oliva-Teles 2001
Atlantic salmon (*Salmo salar*)	–	3.7	2.7% mixed formula	0.041	2.5	3.5	Smith et al. 1978
Chinook salmon (*Oncorhynchus tshawytscha*)	10	520	2.4% mixed formula	0.68	16.6	30	Thorarensen and Farrell 2006

5.2.2.1. Effects of Chemical Composition. In most animals, including fish, there is a larger SDA with a protein-rich meal. In both largemouth bass (*Micropterus salmoides*) (Tandler and Beamish 1980) and rainbow trout (LeGrow and Beamish 1986), there was an increase in SDA with increased dietary protein. However, there are exceptions. There was no additional increase in SDA with increasing protein content for sea bass (Peres and Oliva-Teles 2001). The reason for these very different results is not obvious. It could be that the caloric content is a key factor in determining the SDA among diets. For example, Tandler and Beamish (1981) noted an increase in SDA amplitude as well as duration with an increase in caloric content (Tandler and Beamish 1981). Also, when iosocaloric diets were compared, increased dietary protein or lipid had no effect on SDA in rainbow trout, Atlantic cod and haddock (*Melanogrammus aeglefinus*) (Eliason et al. 2007; Juan et al. 2009). The response to a change in dietary lipid content also varies among species (Medland and Beamish 1985; Ross et al. 1992).

5.2.2.2. Effects of Ration Size. A strong correlation exists between ration or meal size and the corresponding peak SDA and duration of SDA in all animal groups: fishes (Fu et al. 2005), reptiles (Secor and Diamond 1997), amphibians (Secor and Faulkner 2002) and mammals (LeBlanc and Diamond 1986). An increase in magnitude and duration of the SDA response with an increase in ration size has been reported for the Atlantic cod (Jordan and Steffensen 2007) and plaice (Jobling and Davies 1979). In the Chinese (*Silurus asotus*) and southern catfish (*Silurus meridionalis*) (Fu et al. 2005, 2006), the magnitude of the SDA response increased with both meal size and temperature.

5.2.3. EFFECTS OF EXERCISE

Salmonids maintain SDA during prolonged swimming and this is presumed to limit the allocation of blood flow to the skeletal muscles (Thorarensen and Farrell 2006). The consequence of this is a 15% decrease in maximal swimming performance after feeding. At a lower swimming speed, the postprandial metabolism can be added to that of the exercise metabolism until maximal metabolic rate is reached. Gastrointestinal metabolism also limits maximal swimming performance in juvenile rainbow trout (Alsop and Wood 1997). In Atlantic cod, the efficiency of digestion was reduced as the swimming speed was increased, and this led to an increase in the duration of the SDA (Blaikie and Kerr 1996). However, in the largemouth bass sustained swimming at sub-maximal speeds had little effect on SDA (Beamish 1974).

Three studies in sea bass provide additional evidence that SDA must reflect also metabolic processes outside of the GI tract. Sea bass increase metabolic rate when swimming and maintain SDA (Altimiras et al. 2008; Dupont-Prinet et al., 2009; Jourdana-Pineau et al., 2010) despite a decrease in GBF (Altimiras et al. 2008). This argues against a sparing of GBF as shown for salmonids. The extent to which SDA is influenced by swimming or activity level probably reflects behavioral foraging strategy. Sedentary bottom feeders such as the southern catfish (*Silurus meridionalis*) have a low activity label during routine activities and display a substantial SDA after the ingestion of a large meal. Consequently, southern catfish show a larger decrease in swimming performance compared to actively foraging crucian carp (*Carassius auratus*) (Fu et al. 2009).

5.2.4. EFFECTS OF OTHER EXTERNAL STIMULI

Temperature is a fundamental determinant of routine metabolic rate in fish (Clarke and Johnston 1999). Both SDA amplitude and routine metabolic rate increased with temperature in mulloway (*Argyrosomus japonicus*) (Pirozzi and Booth 2009). The increased SDA amplitude reflects a more rapid processing, which is to be expected considering that the total amount of consumed oxygen should be constant, assuming no change in meal composition and size. A similar dependence on temperature is seen in southern catfish (Luo and Xie 2008). On the other hand, increased temperature did not affect the maximal amplitude of the SDA, but reduced the duration of the SDA response in the plaice (Jobling and Davies 1980).

In Atlantic cod (Jordan and Steffensen 2007) and mulloway (Fitzgibbon et al. 2007), environmental hypoxia decreased SDA amplitude and increased SDA duration to maintain the total amount of consumed oxygen during SDA. In sea bass, hypoxia decreased metabolic rate and swimming performance (Dupont-Prinet et al. 2009; Jourdana-Pineau et al. 2010). Furthermore, at low swimming speeds sea bass maintained SDA during hypoxia, but as swimming speed increased, SDA decreased until at high speed there was no SDA at all.

5.2.5. INTEGRATIVE RESPONSES OF SPECIFIC DYNAMIC ACTION

Few studies have considered the affects of how a combination of several factors might influence or modify the integrative response of the gastrointestinal postprandial metabolism such as during simultaneous hypoxia and exercise. Fitzgibbon et al. (2007) investigated the effects of hypoxia and exercise on the overall metabolism of the mulloway and found that although metabolic scope or metabolic capacity was reduced during mild hypoxia, swimming performance was maintained. This compensatory mechanism may reflect their estuarian lifestyle. Moreover, since combined hypoxia and exercise severely limited SDA in sea bass, sea bass presumably change the allocation of oxygen to maintain swimming capacity at the cost of digestive capacity/efficacy (Dupont-Prinet et al. 2009; Jourdana-Pineau et al. 2010). Similarly, swimming capacity is spared at the cost of digestion in more agile swimmers, compared to benthic ambush predators, as discussed above (Fu et al. 2009). Despite this limited information, future studies are clearly needed that integrate the effects of several stimuli to broaden our overall knowledge in this field.

6. CONCLUDING REMARKS

Given the multiple functions served by the GI tract it is perhaps not strange that the GI vasculature is under the influence of a wide array of control mechanisms (Fig. 9.7). The regulation of blood supply is of fundamental importance during digestion for an adequate digestion, absorption and redistribution of nutrients following feeding. The cardio-vascular system must respond rapidly to the presence of food and adjust its response depending on the composition and quality of food. The

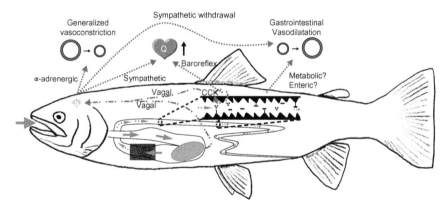

Fig. 9.7. An overview of the most important regulatory mechanisms involved in the regulation of gastrointestinal blood flow after feeding. As food is passed along the gastrointestinal tract there is first an initial stretching of the stomach, which leads to an α-adrenergic vasoconstriction of the general vasculature and possibly also an increase in the cardiac output. This reflex is most likely mediated via vagal mechanoreceptors. Then, as food enters the intestine and is hydrolyzed to release amino acids/peptides, lipids/triglycerides as well as carbohydrates/glucose, there is a subsequent chemically induced hyperemia, i.e. a local vasodilation of the gastrointestinal vasculature. This leads to a possible redistribution of blood to the gastrointestinal tract as well as a baroreflex-mediated (and other mechanisms) increase in cardiac output. Several mechanisms modulate this response and substances such as cholecystokinin (CCK) and serotonin (5-HT) probably influence the response via both a vagal pathway leading to a reduced sympathetic outflow (Fig. 9.4) as well as local reflex, intrinsic to the gastrointestinal tract. From Seth (2010).

cardiovascular system serves two main purposes in this regard; first to supply the metabolically active tissue with oxygenated blood and second to enable efficient transport of nutrients.

The postprandial hyperemia in fish, which leads to an increased GBF, is most likely triggered by the increase in oxygen consumption by the GI tract tissue, initially leading to an intestinal oxygen deficit, and increased tissue osmolarity due to the absorptive processes that follow after feeding. Other factors such as external and internal innervation, as well as hormonal control mechanisms, are also involved in the fine tuning of the GI tract blood supply and metabolism.

Although little is known about the routine metabolism of the gastro-intestinal tract itself the increase in total oxygen consumption associated with feeding, known as the SDA, continues to be one of the more studied phenomena in fish. The SDA depends on nutrient uptake and the secretion of hydrolytic compounds, but predominantly post-absorptive catabolic and anabolic processes including the biosynthesis of macromolecules. However, the connection between GBF and SDA remains to be resolved in more detail. A more integrative approach is most likely needed in order to determine several of the aspects of GBF regulation and the origin of the SDA. Looking at single components provides valuable information but does not necessarily provide the entire integrative picture.

Increased knowledge will not only benefit the scientific community, but it could also prove useful in increasing the producer efficiency within the aquaculture industry. By, for instance, developing fish feeds that can be utilized more efficiently by the cultured fish for growth, with less energy being devoted to the cost of digestion, the biomass obtained per weight unit of fed could hopefully increase.

REFERENCES

Aldman, G., Grove, D., and Holmgren, S. (1992). Duodenal acidification and intra-arterial injection of CCK8 increase gallbladder motility in the rainbow trout, *Oncorhynchus mykiss*. *Gen. Comp. Endocrinol.* **86**, 20–25.

Alsop, D., and Wood, C. (1997). The interactive effects of feeding and exercise on oxygen consumption, swimming performance and protein usage in juvenile rainbow trout (*Oncorhynchus mykiss*). *J. Exp. Biol.* **200**, 2337–2346.

Altimiras, J., Claireaux, G., Sandblom, E., Farrell, A. P., McKenzie, D. J., and Axelsson, M. (2008). Gastrointestinal blood flow and postprandial metabolism in swimming sea bass, *Dicentrarchus labrax*. *Physiol. Biochem. Zool.* **81**, 663–672.

Ashworth, A. (1969). Metabolic rates during recovery from protein-calorie malnutrition: the need for a new concept of specific dynamic action. *Nature* **223**, 407–409.

Axelsson, M. (1988). The importance of nervous and humoral mechanisms in the control of cardiac performance in the Atlantic cod (*Gadus morhua*) at rest and during non-exhaustive exercise. *J. Exp. Biol.* **137**, 287–301.

Axelsson, M., and Fritsche, R. (1991). Effects of exercise, hypoxia and feeding on the gastrointestinal blood flow in the Atlantic cod, *Gadus morhua. J. Exp. Biol.* **158**, 181–198.

Axelsson, M., and Nilsson, S. (1986). Blood pressure control during exercise in the Atlantic cod, *Gadus morhua. J. Exp. Biol.* **126**, 225–236.

Axelsson, M., Ehrenstrom, F., and Nilsson, S. (1987). Cholinergic and adrenergic influence on the teleost heart *in vivo. J. Exp. Biol.* **46**, 179–186.

Axelsson, M., Driedzic, W. R., Farrell, A. P., and Nilsson, S. (1989). Regulation of cardiac output and gut blood flow in the searaven, *Hemitripterus americanus. Fish Physiol. Biochem.* **6**, 315–326.

Axelsson, M., Thorarensen, H., Nilsson, S., and Farrell, A. P. (2000). Gastrointestinal blood flow in the red Irish lord, *Hemilepidotus hemilepidotus*: long-term effects of feeding and adrenergic control. *J. Comp. Physiol. [B]* **170**, 145–152.

Axelsson, M., Altimiras, J., and Claireaux, G. (2002). Postprandial blood flow to the gastrointestinal tract is not compromised during hypoxia in the sea bass, *Dicentrarchus labrax. J. Exp. Biol.* **205**, 2891–2896.

Beamish, F. W. H. (1974). Apparent specific dynamic action of largemouth bass, *Micropterus salmoides. J. Fish Res. B Can.* **31**, 1763–1769.

Blaikie, H. B., and Kerr, S. R. (1996). Effect of activity level on apparent heat increment in Atlantic cod, *Gadus morhua. Can. J. Fish Aquat. Sci.* **53**, 2093–2099.

Bohlen, H. G. (1980a). Intestinal mucosal oxygenation influences absorptive hyperemia. *Am. J. Physiol.* **239**, H489–H493.

Bohlen, H. G. (1980b). Intestinal tissue PO2 and microvascular responses during glucose exposure. *Am. J. Physiol.* **238**, H164–H171.

Bohlen, H. G. (1998a). Integration of intestinal structure, function, and microvascular regulation. *Microcirculation* **5**, 27–37.

Bohlen, H. G. (1998b). Mechanism of increased vessel wall nitric oxide concentrations during intestinal absorption. *Am. J. Physiol.* **275**, 542–550.

Boyce, S. J., and Clarke, A. (1997). Effect of body size and ration on specific dynamic action in the Antarctic plunderfish, *Harpagifer antarcticus* (Nybelin—1947). *Physiol. Zool.* **70**, 679–690.

Brett, J. R. (1971). Energetic responses of salmon to temperature. A study of some thermal relations in the physiology and freshwater ecology of sockage salmon (*Oncorhynchus nerka*). *Am. Zool.* **81**, 663–672.

Brown, C. R., and Cameron, J. N. (1991a). The induction of specific dynamic action in channel catfish by infusion of essential amino acids. *Physiol. Zool.* **64**, 276–297.

Brown, C. R., and Cameron, J. N. (1991b). The relationship between specific dynamic action (SDA) and protein synthesis rates in the channel catfish. *Physiol. Zool.* **64**, 298–309.

Bucking, C., and Wood, C. M. (2006). Water dynamics in the digestive tract of the freshwater rainbow trout during the processing of a single meal. *J. Exp. Biol.* **209**, 1883–1893.

Buddington, R. K., and Hilton, J. W. (1987). Intestinal adaptations of rainbow trout to changes in dietary carbohydrate. *Am. J. Physiol.* **253**, G489–G496.

Buddington, R. K., Chen, J. W., and Diamond, J. (1987). Genetic and phenotypic adaptation of intestinal nutrient transport to diet in fish. *J. Physiol.* **393**, 261–281.

Buddington, R. K., Krogdahl, A., and Bakke-Mckellep, A. M. (1997). The intestines of carnivorous fish: structure and functions and the relations with diet. *Acta Physiol. Scand. Suppl.* **638**, 67–80.

Burns, G. P., and Schenk, W. G., Jr. (1969). Effect of digestion and exercise on intestinal blood flow and cardiac output. An experimental study in the conscious dog. *Arch. Surg.* **98**, 790–794.

Burnstock, G. (1959). The innervation of the gut of the brown trout (*Salmo trutta*). *Quart. J. Microsc. Sci.* **100**, 199–219.

Carmagnola, S., Cantu, P., and Penagini, R. (2005). Mechanoreceptors of the proximal stomach and perception of gastric distension. *Am. J. Gastroenterol.* **100**, 1704–1710.

Chou, C. C. (1982). Relationship between intestinal blood flow and motility. *Annu. Rev. Physiol.* **44**, 29–42.

Chou, C. C., and Coatney, R. W. (1994). Nutrient-induced changes in intestinal blood flow in the dog. *Br. Vet. J.* **150**, 423–437.

Chou, C. C., Burns, T. D., Hsieh, C. P., and Dabney, J. M. (1972). Mechanisms of local vasodilation with hypertonic glucose in the jejunum. *Surgery* **71**, 380–387.

Chou, C. C., Kvietys, P., Post, J., and Sit, S. P. (1978). Constituents of chyme responsible for postprandial intestinal hyperemia. *Am. J. Physiol.* **235**, H677–H682.

Clarke, A., and Johnston, N. M. (1999). Scaling of metabolic rate with body mass and temperature in teleost fish. *J. Animal Ecol.* **68**, 893–905.

Couch, J. A. (1990). Pericyte of a teleost fish: ultrastructure, position, and role in neoplasia as revealed by a fish model. *Anatomical Record* **228**, 7–14.

Coulson, R. A., and Hernandez, T. (1979). Increase in metabolic rate of the alligator fed proteins or amino acids. *J. Nut.* **109**, 538–550.

Coulson, R. A., Herbert, J. D., and Hernandez, T. (1978). Energy for amino acid absorption, transport and protein synthesis *in vivo. Comp. Biochem. Physiol.* **60**, 13–20.

Couto, A., Enes, P., Peres, H., and Oliva-Teles, A. (2008). Effect of water temperature and dietary starch on growth and metabolic utilization of diets in gilthead sea bream (*Sparus aurata*) juveniles. *Comp. Biochem. Physiol. A* **151**, 45–50.

Crocker, C. E., Farrell, A. P., Gamperl, A. K., and Cech, J. J., Jr. (2000). Cardiorespiratory responses of white sturgeon to environmental hypercapnia. *Am. J. Physiol.* **279**, R617–R628.

Cui, Y., and Liu, J. (1990). Comparison of energy budget among six teleosts—II. Metabolic rates. *Comp. Biochem. Physiol. A* **97**, 169–174.

Davison, W. (1987). Arterioles in the swimming muscles of the leatherjacket Parika scaber (Pisces: Balistidae). *Cell Tissue Res.* **248**, 703–708.

Dietrich, H. H., Ellsworth, M. L., Sprague, R. S., and Dacey, R. G., Jr. (2000). Red blood cell regulation of microvascular tone through adenosine triphosphate. *Am. J. Physiol. Heart Circ. Physiol.* **278**, H1294–H1298.

Dunel-Erb, S., and Laurent, P. (1980). Ultrastructure of marine teleost gill epithelia: SEM and TEM study of the chloride cell apical membrane. *J. Morphol.* **165**, 175–186.

Dupont-Prinet, A., Claireaux, G., and McKenzie, D. J. (2009). Effects of feeding and hypoxia on cardiac performance and gastrointestinal blood flow during critical speed swimming in the sea bass *Dicentrarchus labrax. Comp. Biochem. Physiol. A* **154**, 233–240.

Egginton, S., and Johnston, I. A. (1982). Muscle fibre differentiation and vascularisation in the juvenile European eel (*Anguilla anguilla* L.). *Cell Tissue Res.* **222**, 563–577.

Eliason, E. J., Higgs, D. A., and Farrell, A. P. (2007). Effect of isoenergetic diets with different protein and lipid content on the growth performance and heat increment of rainbow trout. *Aquaculture* **272**, 723–736.

Eliason, E. J., Higgs, D. A., and Farrell, A. P. (2008). Postprandial gastrointestinal blood flow, oxygen consumption and heart rate in rainbow trout (*Oncorhynchus mykiss*). *Comp. Biochem. Physiol. A* **149**, 380–388.

Enes, P., Peres, H., Couto, A. and Oliva-Teles, A. (2009). Growth performance and metabolic utilization of diets including starch, dextrin, maltose or glucose as carbohydrate source by gilthead sea bream (*Sparus aurata*) juveniles. *Fish Physiol. Biochem.* DOI 10.1007/s10695-009-9366-y.

Farrell, A. P. (1996). Effects of temperature on cardiovascular performance. *Soc. Exp. Biol. Sem. Ser.* 135–158.

Farrell, A. P. (2007). Tribute to P. L. Lutz: a message from the heart—why hypoxic bradycardia in fishes? *J. Exp. Biol.* **210**, 1715–1725.

Farrell, A. P., Thorarensen, H., Axelsson, M., Crocker, C. E., Gamperl, A. K., and Cech, J. J., Jr. (2001). Gut blood flow in fish during exercise and severe hypercapnia. *Comp. Biochem. Physiol. A* **128**, 551–563.

Fitzgibbon, Q. P., Strawbridge, A., and Seymour, R. S. (2007). Metabolic scope, swimming performance and the effects of hypoxia in the mulloway, *Argyrosomus Japonicus* (Pisces: Sciaenidae). *Aquaculture* **270**, 358–368.

Fronek, K., and Fronek, A. (1970). Combined effect of exercise and digestion on hemodynamics in conscious dogs. *Am. J. Physiol.* **218**, 555–559.

Fronek, K., and Stahlgren, L. H. (1968). Systemic and regional hemodynamic changes during food intake and digestion in nonanesthetized dogs. *Circ. Res.* **23**, 687–692.

Fu, S. J., and Xie, X. J. (2004). Nutritional homeostasis in carnivorous southern catfish (*Silurus meridionalis*): is there a mechanism for increased energy expenditure during carbohydrate overfeeding? *Comp. Biochem. Physiol. A Mol. Integr. Physiol.* **139**, 359–363.

Fu, S. J., Xie, X. J., and Cao, Z. D. (2005). Effect of meal size on postprandial metabolic response in southern catfish (*Silurus meridionalis*). *Comp. Biochem. Physiol. A* **140**, 445–451.

Fu, S. J., Cao, Z. D., and Peng, J. L. (2006). Effect of meal size on postprandial metabolic response in Chinese catfish (*Silurus asotus*) (Linnaeus). *J. Comp. Physiol. B* **176**, 489–495.

Fu, S. J., Zeng, L. Q., Li, X. M., Pang, X., Cao, Z. D., Peng, J. L., and Wang, Y. X. (2009). The behavioural, digestive and metabolic characteristics of fishes with different foraging strategies. *J. Exp. Biol.* **212**, 2296–2302.

Gallavan, R. H., Jr., and Chou, C. C. (1985). Possible mechanisms for the initiation and maintenance of postprandial intestinal hyperemia. *Am. J. Physiol.* **249**, G301–G308.

Gallavan, R. H., Jr., Chou, C. C., Kvietys, P. R., and Sit, S. P. (1980). Regional blood flow during digestion in the conscious dog. *Am. J. Physiol.* **238**, H220–H225.

Gonzalez-Alonso, J., Olsen, D. B., and Saltin, B. (2002). Erythrocyte and the regulation of human skeletal muscle blood flow and oxygen delivery: role of circulating ATP. *Circ. Res.* **91**, 1046–1055.

Gore, R. W., and Bohlen, H. G. (1977). Microvascular pressures in rat intestinal muscle and mucosal villi. *Am. J. Physiol.* **233**, H685–H693.

Graham, M. S., and Farrell, A. P. (1989). The effect of temperature acclimation and adrenergic stimulation on performance of a perfused trout heart. *Physiol. Zool.* **62**, 38–61.

Granger, D. N., Richardson, P. D., Kvietys, P. R., and Mortillaro, N. A. (1980). Intestinal blood flow. *Gastroenterology* **78**, 837–863.

Granger, D. N., Mortillaro, N. A., Perry, M. A., and Kvietys, P. R. (1982). Autoregulation of intestinal capillary filtration rate. *Am. J. Physiol.* **243**, G475–G483.

Grove, D. J., and Holmgren, S. (1992a). Intrinsic mechanisms controlling cardiac stomach volume of the rainbow trout (*Oncorhynchus mykiss*) following gastric distension. *J. Exp. Biol.* **163**, 33–48.

Grove, D. J., and Holmgren, S. (1992b). Mechanisms controlling stomach volume of the Atlantic cod (*Gadus morhua*) following gastric distension. *J. Exp. Biol.* **163**, 49–63.

Haverinen, J., Hassinen, M., and Vornanen, M. (2007). Fish cardiac sodium channels are tetrodotoxin sensitive. *Acta Physiol. (Oxf.)* **191**, 197–204.

Holmgren, S., Axelsson, M., and Farrell, A. P. (1992). The effect of catecholamines, substance P and vasoactive intestinal polypeptide (VIP) on blood flow to the gut in the dogfish *Squalus acanthias*. *J. Exp. Biol.* **168**, 161–175.

Hopkinson, B. R., and Schenk, W. G., Jr. (1968). The electromagnetic measurement of liver blood flow and cardiac output in conscious dogs during feeding and exercise. *Surgery* **63**, 970–975.

Jensen, J., Axelsson, M., and Holmgren, S. (1991). Effects of substance P and vasoactive intestinal polypeptide on the gastrointestinal blood flow in the Atlantic cod, *Gadus morhua*. *J. Exp. Biol.* **156**, 361–373.

Jensen, J., Olson, K. R., and Conlon, J. M. (1993). Primary structures and effects on gastrointestinal motility of tachykinins from the rainbow trout. *Am. J. Physiol.* **265**, R804–R810.

Jensen, H., Rourke, I. J., Moller, M., Jonson, L., and Johnsen, A. H. (2001). Identification and distribution of CCK-related peptides and mRNAs in the rainbow trout, *Oncorhynchus mykiss*. *Biochem. Biophys. Acta* **1517**, 190–201.

Jobling, M., and Davies, P. S. (1979). Gastric evacuation in plaice (*Pleuronectes platessa*): effects of temperature and meal size. *J. Fish Biol.* **14**, 539–546.

Jobling, M., and Davies, P. S. (1980). Effects of feeding on metabolic rate, and the specific dynamic action in plaice, *Pleuronectes platessa*. *J. Fish Biol.* **16**, 629–638.

Jobling, M., Gwyther, D., and Grove, D. J. (1977). Some effects of temperature, meal size and body weight on gastric evacuation time in the dab *Limanda limanda* (L). *J. Fish Biol.* **10**, 291–298.

Jönsson, E., Forsman, A., Einarsdottir, I. E., Egner, B., Ruohonen, K., and Björnsson, B. T. (2006). Circulating levels of cholecystokinin and gastrin-releasing peptide in rainbow trout fed different diets. *Gen. Comp. Endocrinol.* **148**, 187–194.

Jordan, A., and Steffensen, J. (2007). Effects of ration size and hypoxia on specific dynamic action in the cod. *Physiol. Biochem. Zool.* **80**, 178–185.

Jourdana-Pineau, H., Dupont-Prinet, A., Claireaux, G., and McKenzie, D. J. (2010). An investigation of metabolic prioritization in the European sea bass, *Dicentrarchus labrax*. *Physiol. Biochem. Zool.* **83**, 68–77.

Juan, C. P.-C., Santosh, P. L., and Gamperl, A. K. (2009). Effects of dietary protein and lipid level, and water temperature, on the post-feeding oxygen consumption of Atlantic cod and haddock. *Aquaculture* **294**, 228–235.

Kågström, J., and Holmgren, S. (1997). VIP-induced relaxation of small arteries of the rainbow trout, *Oncorhynchus mykiss*, involves prostaglandin synthesis but not nitric oxide. *J. Auton. Nerv. Syst.* **63**, 68–76.

Kågström, J., Axelsson, M., Jensen, J., Farrell, A. P., and Holmgren, S. (1996a). Vasoactivity and immunoreactivity of fish tachykinins in the vascular system of the spiny dogfish. *Am. J. Physiol.* **270**, R585–R593.

Kågström, J., Holmgren, S., Olson, K. R., Conlon, J. M., and Jensen, J. (1996b). Vasoconstrictive effects of native tachykinins in the rainbow trout, *Oncorhynchus mykiss*. *Peptides* **17**, 39–45.

Kato, M., Naruse, S., Takagi, T., and Shionoya, S. (1989). Postprandial gastric blood flow in conscious dogs. *Am. J. Physiol.* **257**, G111–G117.

Kita, J., and Itazawa, Y. (1990). Microcirculatory pathways in the spleen of the rainbow trout *Oncorhynchus*. *Jap. J. Ichthyol.* **37**, 265–272.

Kitamikado, M., and Tachino, S. (1960a). Studies on the digestive enzymes of rainbow trout— I, Carbohydrates. *Bull. Jap. Soc. Sci. Fish* **26**, 679–684.

Kitamikado, M., and Tachino, S. (1960b). Studies on the digestive enzymes of rainbow trout—II, Proteases. *Bull. Jap. Soc. Sci. Fish* **26**, 685–690.

Kitamikado, M., and Tachino, S. (1960c). Studies on the digestive enzymes of rainbow trout— III, Esterases. *Bull. Jap. Soc. Sci. Fish* **26**, 691–694.

Kvietys, P. R., Gallavan, R. H., and Chou, C. C. (1980a). Contribution of bile to postprandial intestinal hyperemia. *Am. J. Physiol.* **238**, G284–G288.

Kvietys, P. R., Miller, T., and Granger, D. N. (1980b). Intrinsic control of colonic blood flow and oxygenation. *Am. J. Physiol.* **238**, G478–G484.

Kvietys, P. R., McLendon, J. M., and Granger, D. N. (1981). Postprandial intestinal hyperemia: role of bile salts in the ileum. *Am. J. Physiol.* **241**, G469–G477.

Laurent, P., and Dunel, S. (1980). Morphology of gill epithelia in fish. *Am. J. Physiol.* **238**, R147–R159.

LeBlanc, J., and Diamond, P. (1986). Effect of meal size and frequency on postprandial thermogenesis in dogs. *Am. J. Physiol.* **250**, E144–E147.

LeGrow, S. M., and Beamish, F. W. H. (1986). Influence of dietary protein and lipid on apparent heat increment of rainbow trout, *Salmo gairdneri. Can. J. Fish Aquat. Sci.* **43**, 19–25.

Longhurst, J. C., and Ibarra, J. (1982). Sympathoadrenal mechanisms in hemodynamic responses to gastric distension in cats. *Am. J. Physiol.* **243**, H748–H753.

Longhurst, J. C., and Ibarra, J. (1984). Reflex regional vascular responses during passive gastric distension in cats. *Am. J. Physiol.* **247**, R257–R265.

Longhurst, J. C., Ashton, J. H., and Iwamoto, G. A. (1980). Cardiovascular reflexes resulting from capsaicin-stimulated gastric receptors in anesthetized dogs. *Circ. Res.* **46**, 780–788.

Longhurst, J. C., Spilker, H. L., and Ordway, G. A. (1981). Cardiovascular reflexes elicited by passive gastric distension in anesthetized cats. *Am. J. Physiol.* **240**, H539–H545.

Luo, Y., and Xie, X. (2008). Effects of temperature on the specific dynamic action of the southern catfish, *Silurus meridionalis. Comp. Biochem. Physiol. A* **149**, 150–156.

Luz, J., Griggio, M. A., Fagundes, D. J., Araujo, R. M., and Marcondes, W. (2000). Oxygen consumption of rats with broad intestinal resection. *Braz. J. Med. Biol. Res.* **33**, 1497–1500.

Matheson, P. J., Wilson, M. A., and Garrison, R. N. (2000). Regulation of intestinal blood flow. *J. Surg. Res.* **93**, 182–196.

McCauley, R. W., Elliott, J. R., and Read, L. A. A. (1977). Influence of acclimation temperature on preferred temperature in the rainbow trout *Salmo gairdneri. Am. Fish Soc.* **106**, 362–365.

McCue, M. D. (2006). Specific dynamic action: a century of investigation. *Comp. Biochem. Physiol. A Mol. Integr. Physiol.* **144**, 381–394.

McCue, M. D., Bennett, A. F., and Hicks, J. W. (2005). The effect of meal composition on specific dynamic action in Burmese pythons (*Python molurus*). *Physiol. Biochem. Zool.* **78**, 182–192.

Medland, T. E., and Beamish, F. W. H. (1985). The influence of diet and fish density on apparent heat increment in rainbow trout, *Salmo gairdneri. Aquaculture* **47**, 1–10.

Mosse, P. R. (1979). Capillary distribution and metabolic histochemistry of the lateral propulsive musculature of pelagic teleost fish. *Cell Tissue Res.* **203**, 141–160.

Murakami, T. (1971). Application of the scanning electron microscope to the study of the fine distribution of the blood vessels. *Arch. Histol. Jpn* **32**, 445–454.

Nakajima, M., Naruse, S., Kitagawa, M., Ishiguro, H., Jin, C., Ito, O., and Hayakawa, T. (2001). Role of cholecystokinin in the intestinal phase of pancreatic circulation in dogs. *Am. J. Physiol. Gastrointest. Liver Physiol.* **280**, G614–G620.

Nakamura, E., Torii, K., and Uneyama, H. (2008). Physiological roles of dietary free glutamate in gastrointestinal functions. *Biol. Pharm. Bull.* **31**, 1841–1843.

Neild, T. O., Shen, K. Z., and Surprenant, A. (1990). Vasodilatation of arterioles by acetylcholine released from single neurones in the guinea-pig submucosal plexus. *J. Physiol.* **420**, 247–265.

Nilsson, S., and Sundin, L. (1998). Gill blood flow control. *Comp. Biochem. Physiol. A Mol. Integr. Physiol.* **119**, 137–147.

Nyhof, R., and Chou, C. C. (1981). Absence of cholinergic or serotonergic mediation in food induced intestinal hyperemia (Abstract). *Federation Proc.* **40**, 491.

Nyhof, R. A., and Chou, C. C. (1983). Evidence against local neural mechanism for intestinal postprandial hyperemia. *Am. J. Physiol.* **245**, H437–H446.

Nyhof, R. A., Ingold-Wilcox, D., and Chou, C. C. (1985). Effect of atropine on digested food-induced intestinal hyperemia. *Am. J. Physiol.* **249**, G685–G690.

Olson, K. R. (2002). Vascular anatomy of the fish gill. *J. Exp. Zool.* **293**, 214–231.

Olson, K. R. (2008). Hydrogen sulfide and oxygen sensing: implications in cardiorespiratory control. *J. Exp. Biol.* **211**, 2727–2734.

Olson, K. R. (2009). Is hydrogen sulfide a circulating "gasotransmitter" in vertebrate blood? *Biochim. Biophys. Acta* **1787**, 856–863.

Olson, K. R., Dombkowski, R. A., Russell, M. J., Doellman, M. M., Head, S. K., Whitfield, N. L., and Madden, J. A. (2006). Hydrogen sulfide as an oxygen sensor/transducer in vertebrate hypoxic vasoconstriction and hypoxic vasodilation. *J. Exp. Biol.* **209**, 4011–4023.

Olson, K. R., Forgan, L. G., Dombkowski, R. A., and Forster, M. E. (2008). Oxygen dependency of hydrogen sulfide-mediated vasoconstriction in cyclostome aortas. *J. Exp. Biol.* **211**, 2205–2213.

Olsson, C., and Holmgren, S. (2009). The neuronal and endocrine regulation of gut function. In: *Fish Neuroendocrinology* (N.J. Bernier, G. Van Der Kraak, A.P. Farrell and C.J Brauner, eds), vol. 28, pp. 467–538. Academic Press, London.

Olsson, C., Aldman, G., Larsson, A., and Holmgren, S. (1999). Cholecystokinin affects gastric emptying and stomach motility in the rainbow trout *Oncorhynchus mykiss*. *J. Exp. Biol.* **202**, 161–170.

Owen, S. F. (2001). Meeting energy budgets by modulation of behaviour and physiology in the eel (*Anguilla anguilla* L.). *Comp. Biochem. Physiol. A* **128**, 631–644.

Ozaki, N., and Gebhart, G. F. (2001). Characterization of mechanosensitive splanchnic nerve afferent fibers innervating the rat stomach. *Am. J. Physiol. Gastrointest. Liver Physiol.* **281**, G1449–G1459.

Ozaki, N., Sengupta, J. N., and Gebhart, G. F. (1999). Mechanosensitive properties of gastric vagal afferent fibers in the rat. *J. Neurophysiol.* **82**, 2210–2220.

Pawlik, W. W., Fondacaro, J. D., and Jacobson, E. D. (1980). Metabolic hyperemia in canine gut. *Am. J. Physiol.* **239**, G12–G17.

Peck, M. A., Buckley, L. J., and Bengtsson, D. A. (2003). Energy losses due to routine and feeding metabolism in young-of-the-year juvenile Atlantic cod (*Gadus morhua*). *Can. J. Fish Aquat. Sci.* **60**, 927–937.

Peres, H., and Oliva-Teles, A. (2001). Effect of dietary protein and lipid level on metabolic utilization of diets by European sea bass (*Dicentrarchus labrax*) juveniles. *Fish Physiol. Biochem.* **25**, 269–275.

Pierce, R. J., and Wissing, T. E. (1974). Energy cost of food utilization in the bluegill (*Lepomis macrochirus*). *Am. Fish Soc.* **103**, 38–45.

Pirozzi, I., and Booth, M. A. (2009). The effect of temperature and body weight on the routine metabolic rate and postprandial metabolic response in mulloway, *Argyrosomus japonicus*. *Comp. Biochem. Physiol. A* **154**, 110–118.

Premen, A. J., Kvietys, P. R., and Granger, D. N. (1985). Postprandial regulation of intestinal blood flow: role of gastrointestinal hormones. *Am. J. Physiol.* **249**, G250–G255.

Qamar, M. I., and Read, A. E. (1987). Effects of exercise on mesenteric blood flow in man. *Gut* **28**, 583–587.

Rhodin, J. A. (1967). The ultrastructure of mammalian arterioles and precapillary sphincters. *J. Ultrastruct. Res.* **18**, 181–223.

Ross, L. G., Mckinney, R. W., Cardwell, S. K., Fullarton, J. G., Roberts, S. E. J., and Ross, B. (1992). The effects of dietary protein content, lipid content and ration level on oxygen consumption and specific dynamic action in *Oreochromis niloticus. Comp. Biochem. Physiol. A* **103**, 573–578.

Rozengurt, E. (2006). Taste receptors in the gastrointestinal tract. I. Bitter taste receptors and alpha-gustducin in the mammalian gut. *Am. J. Physiol. Gastrointest. Liver Physiol.* **291**, G171–G177.

Rubner, M. (1902). *Die Gesetze des Energiever Brauchs bei der Ernahrung.* Leipzig/Wien, Franz Dauticke.

Ruohonen, K., Grove, D. J., and McIlroy, J. T. (1997). The amount of food ingested in a single meal by rainbow trout offered chopped herring, dry and wet diets. *J. Fish Biol.* **51**, 93–105.

Secor, S. M. (2003). Gastric function and its contribution to the postprandial metabolic response of the Burmese python *Python molurus. J. Exp. Biol.* **206**, 1621–1630.

Secor, S. M. (2009). Specific dynamic action: a review of the postprandial metabolic response. *J. Comp. Physiol. B* **179**, 1–56.

Secor, S. M., and Diamond, J. (1997). Effects of meal size on postprandial responses in juvenile Burmese pythons (*Python molurus*). *Am. J. Physiol.* **272**, R902–R912.

Secor, S. M., and Faulkner, A. C. (2002). Effects of meal size, meal type, body temperature, and body size on the specific dynamic action of the marine toad, *Bufo marinus. Physiol. Biochem. Zool.* **75**, 557–571.

Seth, H. (2010). On the regulation of gastrointestinal blood flow in teleost fish. PhD Thesis. University of Gotheburg. ISBN 978-91-628-8058-3.

Seth, H., and Axelsson, M. (2009). Effects of gastric distension and feeding on cardiovascular variables in the shorthorn sculpin (*Myoxocephalus scorpius*). *Am. J. Physiol.* **296**, R171–R177.

Seth, H. and Axelsson , M. (2010). Sympathetic, parasympathetic and enteric regulation of the gastrointestinal vasculature in rainbow trout (*Oncorhynchus mykiss*) during routine and postprandial conditions. *J. Exp. Biol.* In press.

Seth, H., Sandblom, E., Holmgren, S., and Axelsson, M. (2008). Effects of gastric distension on the cardiovascular system in rainbow trout (*Oncorhynchus mykiss*). *Am. J. Physiol.* **294**, R1648–R1656.

Seth, H., Sandblom, E., and Axelsson, M. (2009). Nutrient-induced gastrointestinal hyperemia and specific dynamic action in rainbow trout (*Oncorhynchus mykiss*)—importance of proteins and lipids. *Am. J. Physiol.* **296**, R345–R352.

Seth, H., Gräns, A. and Axelsson , M. (2010). Cholecystokinin (CCK) as a regulator of cardiac function and postprandial gastrointestinal blood flow in rainbow trout (*Oncorhynchus mykiss*). *Am. J. Physiol. Gastrointest. Liver Physiol.* **298**, 1240–1248.

Shahbazi, F., Conlon, J. M., Holmgren, S., and Jensen, J. (2001). Effects of cod bradykinin and its analogs on vascular and intestinal smooth muscle of the Atlantic cod, *Gadus morhua. Peptides* **22**, 1023–1029.

Shepherd, A. P. (1979). Intestinal O2 uptake during sympathetic stimulation and partial arterial occlusion. *Am. J. Physiol.* **236**, H731–H735.

Shepherd, A. P. (1982). Role of capillary recruitment in the regulation of intestinal oxygenation. *Am. J. Physiol.* **242**, G435–G441.

Smith, M. P. (1999). Local control of vascular smooth muscle tone in the rainbow trout. *Diss. Abst. Int. Pt B Sci. Eng.* **59**, 3817.

Smith, M. P., Russell, M. J., Wincko, J. T., and Olson, K. R. (2001). Effects of hypoxia on isolated vessels and perfused gills of rainbow trout. *Comp. Biochem. Physiol.—Pt A: Molec. Integr. Physiol.* **130**, 171–181.

Smith, R. R., Rumsey, G. L., and Scott, M. L. (1978). Heat increment associated with dietary protein, fat, carbohydrate and complete diets in salmonids. *Comp. Energ. Effic.* **108**, 1025–1032.

Soldatov, A. (2006). Organ blood flow and vessels of microcirculatory bed in fish. *J. Evol. Biochem. Physiol.* **42**, 243–252.

Steffensen, J. (1989). Some errors in respirometry of aquatic breathers: how to avoid and correct for them. *Fish Physiol. Biochem.* **6**, 49–59.

Stevens, E. D. (1968). The effect of exercise on the distribution of blood to various organs in rainbow trout. *Comp. Biochem. Physiol.* **25**, 615–625.

Sundin, L., and Nilsson, S. (1992). Arterio-venous branchial blood flow in the Atlantic cod *Gadus morhua. J. Exp. Biol.* **165**, 73–84.

Sundin, L., and Nilsson, G. E. (1997). Neurochemical mechanisms behind gill microcirculatory responses to hypoxia in trout: *in vivo* microscopy study. *Am. J. Physiol.* **272**, R576–R585.

Surprenant, A. (1994). Control of the gastrointestinal tract by enteric neurons. *Annu. Rev. Physiol.* **56**, 117–140.

Takagi, T., Naruse, S., and Shionoya, S. (1988). Postprandial celiac and superior mesenteric blood flows in conscious dogs. *Am. J. Physiol.* **255**, 522–528.

Tandler, A., and Beamish, F. W. H. (1980). Specific dynamic action and diet in largemouth bass, *Micropterus salmoides* (Lacepede). *J. Nutr.* **110**, 750–764.

Tandler, A., and Beamish, F. W. H. (1981). Apparent specific dynamic action (SDA), fish weight and level of caloric intake in largemouth bass, *Micropterus salmoides* Lacepede. *Aquaculture* **23**, 231–242.

Taylor, J. R., and Grosell, M. (2009). The intestinal response to feeding in seawater gulf toadfish, *Opsanus beta*, includes elevated base secretion and increased epithelial oxygen consumption. *J. Exp. Biol.* **212**, 3873–3881.

Thorarensen, H., and Farrell, A. P. (2006). Postprandial intestinal blood flow, metabolic rates, and exercise in Chinook salmon (*Oncorhynchus tshawytscha*). *Physiol. Biochem. Zool.* **79**, 688–694.

Thorarensen, H., McLean, E., Donaldson, E. M., and Farrell, A. P. (1991). The blood vasculature of the gastrointestinal tract in Chinook, *Oncorhynchus tshawytscha* (Walbaum) and coho, *O. kisutch* (Walbaum), salmon. *J. Fish Biol.* **38**, 525–532.

Tsurugizawa, T., Uematsu, A., Nakamura, E., Hasumura, M., Hirota, M., Kondoh, T., Uneyama, H., and Torii, K. (2009). Mechanisms of neural response to gastrointestinal nutritive stimuli: the gut–brain axis. *Gastroenterology* **137**, 262–273.

Vatner, S. F. (1978). Effects of exercise and excitement on mesenteric and renal dynamics in conscious, unrestrained baboons. *Am. J. Physiol.* **234**, H210–H214.

Vatner, S. F., Franklin, D., and Van Citters, R. L. (1970a). Coronary and visceral vasoactivity associated with eating and digestion in the conscious dog. *Am. J. Physiol.* **219**, 1380–1385.

Vatner, S. F., Franklin, D., and Van Citters, R. L. (1970b). Mesenteric vasoactivity associated with eating and digestion in the conscious dog. *Am. J. Physiol.* **219**, 170–174.

Vatner, S. F., Patrick, T. A., Higgins, C. B., and Franklin, D. (1974). Regional circulatory adjustments to eating and digestion in conscious unrestrained primates. *J. Appl. Physiol.* **36**, 524–529.

Vaugelade, P., Posho, L., Darcy-Vrillon, B., Bernard, F., Morel, M. T., and Duee, P. H. (1994). Intestinal oxygen uptake and glucose metabolism during nutrient absorption in the pig. *Proc. Soc. Exp. Biol. Med.* **207**, 309–316.

Verberne, A. J., Saita, M., and Sartor, D. M. (2003). Chemical stimulation of vagal afferent neurons and sympathetic vasomotor tone. *Brain Res. Brain Res. Rev.* **41**, 288–305.

Wilhelmj, C. M., and Bollman, J. L. (1928). The specific dynamic action and nitrogen elimination following intravenous administration of various amino acids. *J. Biol. Chem.* **77**, 127–149.

Wilson, J. M., and Laurent, P. (2002). Fish gill morphology: inside out. *J. Exp. Zool.* **293**, 192–213.

Windell, J. T., Norris, D. O., Kitchell, J. F., and Norris, J. S. (1969). Digestive response of rainbow trout, *Salmo geirderneri*, to pellet diets. *J. Fish Res. Board Can.* **26**, 1801–1812.

Windell, J. T., Kitchell, J. F., Norris, D. O., Norris, J. S., and Foltz, J. W. (1976). Temperature and rate of gastric evacuation by rainbow trout, *Salmo gairdneri*. *Trans. Am. Fish Soc.* **105**, 712–717.

Wood, S. C., Johansen, K., and Weber, R. E. (1975). Effects of ambient PO2 on hemoglobin-oxygen affinity and red cell ATP concentrations in a benthic fish, *Pleuronectes platessa*. *Respir. Physiol.* **25**, 259–267.

Wurtsbaugh, W. A., and Neverman, D. (1988). Post-feeding thermotaxis and daily vertical migration in a larval fish. *Nature* **333**, 846–848.

Yen, J. T., Nienaber, J. A., Hill, D. A., and Pond, W. G. (1989). Oxygen consumption by portal vein-drained organs and by whole animal in conscious growing swine. *Proc. Soc. Exp. Biol. Med.* **190**, 393–398.

ADDITIONAL READING

Altimiras, J., Claireaux, G., Sandblom, E., Farrell, A. P., McKenzie, D. J., Alsop, D., and Wood, C. (1997). The interactive effects of feeding and exercise on oxygen consumption, swimming performance and protein usage in juvenile rainbow trout (*Oncorhynchus mykiss*). *J. Exp. Biol.* **200**, 2337–2346.

Axelsson, M. (2008). Gastrointestinal blood flow and postprandial metabolism in swimming sea bass, *Dicentrarchus labrax*. *Physiol. Biochem. Zool.* **81**, 663–672.

Gräns, A., Axelsson, M., Pitsillides, K., Olsson, C., Höjesjö, J., Kaufman, R., and Cech, J. (2009a). A fully implantable multi-channel biotelemetry system for measurement of blood flow and temperature: a first evaluation in the green sturgeon. *Hydrobiologia* **619**, 11–25.

Gräns, A., Albertsson, F., Axelsson, M., and Olsson, C. (2009b). Postprandial changes in enteric electrical activity and gut blood flow in rainbow trout (*Oncorhynchus mykiss*) acclimated to different temperatures. *J. Exp. Biol.* **212**, 2550–2557.

Mosse, P. R. (1978). The distribution of capillaries in the somatic musculature of two vertebrate types with particular reference to teleost fish. *Cell Tissue Res.* **187**, 281–303.

10

THE GI TRACT IN AIR BREATHING

JAY A. NELSON

A. MICKEY DEHN

1. INTRODUCTION

The difficulty in obtaining oxygen from many aquatic environments has led to the frequent evolution of air breathing among fishes. Among several groups of fish, elements of the gastrointestinal (GI) tract have been exploited to extract oxygen from the air and have become air-breathing organs (ABOs). Despite the perceived difficulties in balancing digestive and respiratory function, gut air breathing (GAB) in fishes has evolved multiple times and GAB fishes have become very successful. The modification of esophagus, stomach, or intestine into ABOs always involved increased

The Multifunctional Gut of Fish: Volume 30
FISH PHYSIOLOGY

vascularization with capillaries embedded in the epithelium close enough to inspired air for significant O_2 diffusion to occur. The gut wall in ABOs has generally undergone substantial reduction, is separated from digestive portions of the GI tract with sphincters and is capable of producing surfactant. GAB fishes tend to be facultative air breathers that use air breathing to supplement aquatic respiration in hypoxic waters, but some hind-gut breathers also appear to be continuous, but not obligate, air breathers. Gut ABOs are generally used for oxygen uptake; CO_2 elimination occurs primarily via the gills and/or skin in tested species. Aerial ventilation in GAB fishes is driven primarily by the partial pressure of oxygen in the water (P_wO_2) and possibly by metabolic demand. Hypoxic water elicits a bradycardia from GAB fishes that is often, but not always, ameliorated following an air breath. Blood from GAB fish generally has a low hemoglobin-oxygen half-saturation pressure (P_{50}) with a very low erythrocytic (nucleotide tri-phosphate (NTP)). Gut air-breathing behavior in nature depends upon ecological as well as physiological factors. Evidence for the role of the gut of GAB fish in nitrogen excretion is limited.

Fish have followed an evolutionary path that requires the additional energy extraction from foodstuff molecules that exploitation of oxygen as a terminal electron acceptor allows. Thermodynamics dictates that the primary goal of fish existence is to obtain sufficient energy from food to offset randomization, grow and reproduce, but a close secondary goal is to position itself in an environment where oxygen can be exploited to transform this energy efficiently. Formation of ATP chemical bond energy from food anaerobically extracts about 8.1% of the energy from food that can be extracted aerobically (Gnaiger 1993). Adopting the conventions of Fry (1971), oxygen can then be considered a limiting factor for fish; i.e. exploitable niche space will be a complex function of oxygen availability, extraction ability and utilization rate. Some fish can survive extended periods in anoxic water (Nilsson 2001), but all fish are obligate aerobes and require at least some minimal amount of oxygen to complete their lifecycle. Because oxygen availability in water is very dependent upon environmental conditions (Diaz and Breitburg 2009), but always far less available than in air, the evolution of air breathing has occurred on the order of 50 times in the vertebrate lineage (Graham 1997).

Gas solubility in water obeys Henry's law:

$$[gas]_{H_2O} = P_{gas} * \alpha_{gas} * V_{H_2O}$$

which states that the concentration of a gas in water will be a simple product of the partial pressure of the gas (P) in air above that water, its unique solubility coefficient (α) and the volume of water (V) being considered. The solubility coefficient for oxygen in water is fairly low and variably dependent upon water temperature and salinity. Oxygen solubility in pure liquid water is maximal at 0°C and falls to $\sim \frac{1}{2}$ that value over the 40°C temperature range experienced by fish on earth (Withers 1992). Oxygen solubility in liquid water at a given temperature is also maximal in pure water and is reduced by up to 20% across the 40‰ salinity gradient that fish experience on earth (Withers 1992). Thus the prediction from the solubility coefficient and Henry's law is that warm, salty waters such as tropical oceans would have the least amount of oxygen to support fish life and thus foster the evolution of air breathing. However, biotic and climatic factors generally overwhelm the physical constraints of Henry's law in aquatic ecosystems so that tropical oceans turn out to be relatively well oxygenated compared to other aquatic habitats, most of which occur in fresh water and, thus, most air-breathing fishes are also freshwater fishes (Graham 1997).

Diaz and Breitburg (2009) cover the development of aquatic hypoxia in a previous edition of this series, so this topic will just be briefly summarized here. Aquatic hypoxia/anoxia develops due to various combinations of: (1) aquatic respiratory rates exceeding photosynthetic rates, (2) poor or no mixing at the aerial/aquatic interface, (3) poor light penetration due to shading or turbidity, (4) isolation of water bodies, and (5) various combinations of the above (Junk 1984; Wetzel 2001; Diaz and Breitburg 2009). In temperate to polar regions, the most common scenarios resulting in hypoxia/anoxia involve isolation of the water body from the oxygen-rich atmosphere. Other temperate fish habitats subject to hypoxia include soft sediments that do not mix well with the water column and many swamp and marsh habitats such as sphagnum bogs.

Hypoxic and anoxic waters form in tropical waters through a diversity of processes that generally deplete oxygen faster and are harder to characterize than temperate systems (Junk 1984). In addition to hypolimnetic isolation due to thermal stratification, there are large annual rainfall and water level fluctuations that create flooded forests, seasonal lakes, remnant river channels and bring nutrient pulses to existing water bodies. The combinations of: (1) high temperature; (2) waters that can be nutrient rich; (3) dense terrestrial vegetation that can block sunlight and wind; (4) dense surface vegetation that can also block sunlight and wind but can also contribute photosynthetic oxygen to the water; and (5) a dense biota that can influence the water chemistry as much as physical factors, all create a mosaic of hypoxic and anoxic aquatic habitats across the tropics. It is common for tropical waters of the Amazon flooded forest (varzea) to have their top 10 cm be the only predictable aquatic oxygen source for months at a time (Val and Almeida-Val 1995). It is also not uncommon to find waters that cycle between near oxygen saturation conditions during the day and complete anoxia at night (Junk 1984).

2. EVOLUTION OF AIR BREATHING IN FISH

Against this backdrop of worldwide hypoxic habitats a number of fish lineages have taken advantage of the much richer and more predictable source of oxygen in the atmosphere by evolving the ability to breathe air. Extant air-breathing fishes include several marine species, several groups of temperate freshwater air breathers and a large number of small intertidal species that will obtain oxygen from the air when air exposed, but are not particularly specialized for air breathing (Graham 1997; Chapman and McKenzie 2009). However, the majority of specialized air-breathing fishes are found in tropical freshwaters (Graham 1997). Graham (1997) points out that since most of these fishes still breathe water, "air-breathing fishes" is not the most accurate term, but appears to be a term in use by the majority of workers in this field and will be adopted here. There is some speculation that air breathing in fish originated in salt water and that even tetrapod air breathing had a marine origin (Shultze 1999), yet the most parsimonious origin for lungfish and tetrapod air breathing is tropical swamp-like habitats where most specialized air-breathing taxa are found today (Graham and Lee 2004). Graham and Lee (2004) count more than 370 air-breathing fish species distributed among 49 fish families, but these numbers are certain underestimates due to poorly described families like the Loricariidae (Graham 1999). The Neotropical catfish family Loricariidae is the most diverse siluriform family and occupies most hypoxia-prone habitats in the Neotropics. Loricariids comprise 83 genera with over 825 nominal species, 709 of which are considered valid (Armbruster 2006). Although air-breathing is not synapomorphic in this family (Armbruster 1998), most loricariids examined to date will facultatively breath air upon exposure to hypoxia using their gut as an ABO (Gee 1976; Graham 1997), or show morphological evidence of air-breathing capabilities (Armbruster 1998). As the loricariids and other speciose air-breathing families like the Callichthyidae, Trichomycteridae, Gobidae, among others, are examined in more detail, it is likely that the number of extant air-breathing fish species will exceed 1,000. Despite extensive study of a few air-breathing fish species such as the lungfishes (Dipnoi), the morphology and physiology of gas exchange has not been studied in the vast majority of these air-breathing fish.

Of the more than 370 confirmed air-breathing species, members of the families listed in Table 10.1 have been confirmed to use elements of their alimentary tract proper (esophagus, stomach or intestine) as a specialized ABO. Since the Siluriform families listed in Table 10.1 are among the poorest described taxonomically and are already listed at 1,026 species by the All Catfish Species Inventory (2009) with an estimated 305 newly

Table 10.1

Diversity and number of species that potentially use a region of the gastrointestinal tract for air breathing. Species number data for Siluriforms come from the All Catfish Species Inventory (2009) and Froese and Pauly (2010) for the other families. Estimates of the number of undescribed species were not available for all families

GI region used for air breathing	Family	No. of species in that family	Estimated # of undescribed species in that family
Esophagus	Umbridae	5	0
	Blenniidae	420	?
	Synbrachidae	15	?
Stomach	Loricariidae	673	205
	Trichomycteridae	176	55
Intestine	Cobitidae	110	?
	Callicthyidae	177	45

discovered species yet to be described, fishes that use their alimentary tract to breathe air could eventually account for the majority of all air-breathing fish species. Lungs were the presumed original ABO in fishes and date to the earliest gnathostomes of the Devonian period (360–387 MYA; Perry 2007). Once the gas bladder appeared as an organ of buoyancy in fish evolution, it appears to have canalized ABO development and was frequently exploited for a respiratory function (Graham 1997). This was facilitated by a pneumatic duct connecting the gas bladder to the esophagus in the ancestral physostomous condition as well as the requisite musculature and valves for filling the bladder. Since a respiratory gas bladder is the only ABO other than lungs found in extant representatives from this early phase of fish evolution, we can assume that whenever selection for exploiting abundant atmospheric oxygen arose, a gas-filled organ with a direct connection to the atmosphere was a convenient target for selection. Once the physoclistous condition evolved and the pneumatic duct lost, or the swim bladder became encased in bone (as in the loricarioids; Schaefer and Lauder 1986), it appears that the gas bladder was less available as an ABO (a constraint was lifted) and natural selection could act on other body parts. Thus in teleosts, a diversity of ABOs has evolved, including the alimentary tract that is the subject of this chapter (Graham 1997). GAB fishes fall into two general categories: (1) facultative air breathing (FAB), which refers to those animals that only breath air when oxygen in the aquatic medium does not meet biological requirements, and (2) continuous air breathing (CAB), which refers to those animals that do not suffocate when denied access to air but still breathe air continuously when allowed access to the surface, even in normoxic water (Graham 1997; Chapman and McKenzie 2009). A GAB fish that is an obligate air breather has yet to be described.

2.1. Evolution of the Gut as an ABO

Considering the phylogenetic location of the families in Table 10.1, each evolution of the gut proper (esophagus, stomach or intestine) as an ABO (hereafter gut air-breathing, GAB) appears to be a unique evolutionary event (Fig. 10.1). Armbruster (1998) suggests, based on morphological evidence mapped onto the phylogenies of Schaefer and Lauder (1986) and de Pinna (1993), that just within the loricariids and scoloplacids, there may have been as many as five separate origins of GAB. The respiratory diverticula emanating from the esophagus in another loricariid (*Loricarichthys*) (Silva et al. 1997) may be yet another unique evolutionary origin of GAB. This proclivity for GAB to evolve suggests some distinct advantages to using the gut for this purpose that overwhelm any perceived disadvantages. One might argue that a more parsimonious interpretation is that a common loricarioid ancestor was a GAB fish and that the frequent reappearance of GAB in the descendent trichomycterids, loricariids (scoloplacids) and callichthyids (Fig. 10.1) are atavisms, since these are

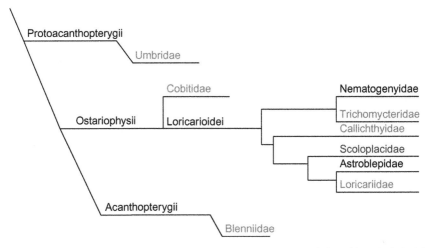

Fig. 10.1. Phylogenetic distribution of known or suspected gut air-breathing (GAB) fish families. Families in red have at least one species with published documentation of gastrointestinal tract oxygen extraction from the air. Families in blue are suspected of gut air breathing based on histological evidence. Families in black have not been demonstrated to breathe air. Ostariophysan relationships were drawn from the All Catfish Species Inventory (2009). See color plate section.

thought to be more likely to occur than *de novo* evolutionary innovations (Riegler 2008). This latter interpretation appears unlikely considering that the callichthyids use their hindgut for respiration whereas the other two families (and presumably the scoloplacids) use their foregut and there is no evidence for homology between the structures or even within many of the foregut breathers (Armbruster 1998). Certainly the evolution of esophageal ABOs in *Dallia* (Umbridae) and *Lipophrys* (Blennidae) and the evolution of an intestinal ABO in the loaches (Cobitidae) were independent evolutionary events. Interestingly, a synapomorphy for the loricarioidea that encompasses all of the GAB catfish families as well as the scoloplacids (Fig. 10.1) is a bony encapsulation of the swim bladder that may constrain its use as an ABO or organ of buoyancy (Schaefer and Lauder 1986).

Aside from obtaining access to the large amount of oxygen in the atmosphere, which is true of any air-breathing structure, the question then arises as to why the gut evolved into an ABO so frequently. One argument could be that portions of the gut already function as a transport epithelium and therefore must satisfy the constraints of the diffusion equation for some substances:

$$D = \Delta C^* S / d$$

where D is the rate of diffusion, ΔC the difference in concentration between the gut lumen and the adjacent epithelial cell, S is the surface area of diffusional contact and d is the distance separating the two compartments. Since small fatty acids and some carbohydrates are generally absorbed in vertebrate guts by diffusion, the argument would be that the criteria for effective oxygen uptake by diffusion was already present and would involve relatively minimal evolutionary change for use as an ABO. This hypothesis would carry more weight if all gut air breathers used their intestines, the usual site of diffusional nutrient uptake, or if GAB fishes did not require substantial morphological remodeling to their gut (Sections 3 and 4) to breath air. However, the fact that taxa exploit regions of the gut for air breathing that were not originally specialized for foodstuff absorption (e.g. esophagus and stomach; Table 10.1) suggests that other criteria were important in driving the frequent evolution of the gut as an ABO.

A second hypothesis for the frequent evolution of GAB is that since the musculature, skeletal structures and cavities for ingesting and directing food from the mouth through the gastrointestinal (GI) tract were already present and had already been used to ingest air in the filling of physostomous swim bladders, this facilitated the evolution of gut air breathing. However, this same morphology and physiology can be used to fill the buccal cavity and pharyngeal region with air and thus does not explain why some fish evolved transport of the air to the GI tract as opposed to exploiting the head region as an ABO which many other fish taxa have done (Graham 1997).

Since many of the GAB families are benthically oriented fishes, the answer to the frequent evolution of GAB may lie in buoyancy. Air in the head region would make the head more buoyant than the center of mass and tend to lift it off the substrate away from sources of benthic food and possibly increase visibility to predators or potential prey. As Gee (1976) discussed for buoyancy in general, asymmetric head buoyancy could also make it difficult to hold position in lotic waters. Displacing air to the GI tract would bring it closer to the center of mass and minimize buoyancy asymmetries. GAB fish generally hold on to air until taking a subsequent breath even after most of the oxygen is depleted (e.g. Gee and Graham 1978; McMahon and Burggren 1987), and Gee (1976) found that in a number of air-breathing loricarioids, the buoyancy attributable to air in the gut far surpassed buoyancy attributable to air in the swim bladder. Thus, buoyancy attributable to respiration represents a significant component of the animal's "buoyancy budget," especially for CABs, and could be envisioned as a target for natural selection. Certainly for the loricariids, where many species use an oral sucker to remain inverted under logs or rocks in the water column, the possession of a gut ABO would seem advantageous to a head region ABO. A head region ABO would presumably require extensive vascularization of both the dorsal and ventral buccal surfaces to accommodate extraction of oxygen in both normal and inverted positions, respectively. Thus, buoyancy issues and problems with maneuvering/ positioning in the benthic environment may have been key factors favoring the frequent evolution of gut ABOs.

A final potential explanation for the frequent evolution of gut ABOs relates to diet. Many of the species in these gut air-breathing families are herbivorous or detritivorous (e.g. the majority of the loricarioids), so it is possible that ancestral forms were as well. Thus, one scenario is that dietary-driven changes to the GI tracts of ancestors facilitated (pre-adapted) evolution of gut ABOs. Lobel (1981) outlined four basic types of herbivory in fishes: (1) acid lysis in a thin-walled stomach; (2) trituration in a gizzard-like stomach; (3) trituration with pharyngeal jaws; and (4) microbial fermentation in the hindgut. It is entirely possible that morphological changes to the gut that facilitated digestion, for example thinning of GI epithelia to facilitate absorption of end-products of symbiont fermentations (Kihara and Sakata 1997), could have helped promulgate use of the gut for air breathing. Little is known about the trophic ecology of most of the fishes in these groups or the evolution of their digestive tract morphology, but one factor may be the sheer size of the GI tract in herbivorous fishes. For example, Nelson et al. (2007) report an astounding 4-meter gut length (relative intestinal length of 20) in *Hypostomus regani*, a herbivorous loricariid. Whether any morphological changes driven by diet predisposed the evolution of gut air breathing in these groups remains a matter of speculation. Interestingly, Carter and Beadle (1931) and Persaud et al. (2006a) show that the respiratory intestine of *Hoplosternum* (Callichthyidae) develops from a fully functional digestive gut, well after normal maturation of the gut.

2.2. Challenges in Using the Gut as an ABO

The use of the gut as an ABO carries with it a number of perceived disadvantages. Gut air breathing potentially compromises digestive function in fishes. Vertebrate digestive processes generally take place under anaerobic conditions (van Soest 1994) and many of the GABs are herbivorous or detritivorous and presumably require an anaerobic gut to facilitate energy extraction from fermentative processes, as seen in marine herbivorous fishes (Clements et al. 2009), although the extent of fermentative processes in these freshwater herbivores is largely unknown. Thus, the use of the gut as an ABO could oxygenate portions of the gut and compromise digestive performance. In addition, the specialized respiratory portions of the gut are not thought to be involved in either the secretory or absorptive components of the digestive process (Persaud et al. 2006b). Thus turning over some of the gut to respiration may require additional structures, or modifications to existing structures, to accommodate digestive requirements. Based on histological evidence, Podkowa and Goniakowska-Witalińska (2002) suggest, however, that some absorption may be occurring in the respiratory segment of the *Corydoras* (Callichthyidae) intestine. This idea was confirmed by Gonçalves et al. (2007) in *Misgurnus* (Cobitidae) who found evidence that transport proteins normally involved in the digestive absorption of glucose are expressed in the respiratory portion of the intestine.

Conversely, food in the GI tract could potentially compromise the respiratory function of the ABO. The thin epithelium necessary for efficient diffusive gas exchange (Section 3) would seemingly be vulnerable to damage by the acid, alkali, digestive enzymes and dietary items that are normal constituents of the vertebrate GI tract. In addition, the physical presence of food and digestive juices would tend to increase the diffusive distance for oxygen, potentially limiting oxygen uptake in a gut ABO. Finally, use of the gut as an ABO seemingly places digestion and respiration in competition for some of the interfacing morphology and physiology. For example, it is hard to imagine that transport of substances with large differences in viscosity such as air and digesta is best accomplished by the same arrangement of visceral smooth muscle. In addition, both respiration and digestion are characterized by blood flows regulated in accordance with activity; simultaneous digestion and air breathing would seemingly place different regions of the GI tract in competition for the available blood (Fig. 10.2). Full perfusion of the fish GI tract generally occurs postprandially (Thorarensen and Farrell 2006). Although ventilation–perfusion matching has not been studied in gut air-breathing fishes, presumably perfusion of gut ABOs is maximized when fresh air is present as in other air-breathing fishes (Burggren and Johansen 1986), possibly diverting blood flow from digestion during air breathing. Thus, GAB fishes had a number of interesting morphological and physiological challenges to overcome to successfully evolve the multi-functional gut into an ABO. The number of times this has occurred and the evolutionary success of the groups (Table 10.1) would attest to the viability of the strategy.

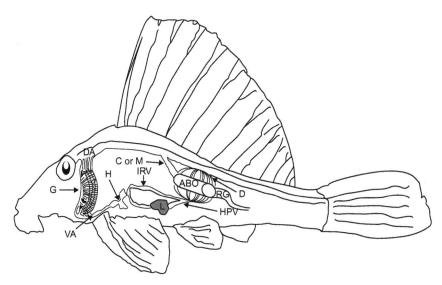

Fig. 10.2. General circulatory design among GAB fishes. Composite drawing incorporating the major features of all the different GAB fishes. Blood supply to the more anterior gut air-breathing organs (ABOs) (e.g. *Dallia* and *Ancistrus*) is via the coeliac artery whereas the more posterior intestinal ABOs tend to be supplied by the anterior mesenteric artery (e.g. *Misgurnus*) or directly from the dorsal aorta (e.g. *Callichthys* and *Hoplosternum*). ABO effluent blood travels back to the heart via the hepatic portal vein (Cobitidae) or the interrenal vein (all others). ABO = air-breathing organ; C or M = coeliac or anterior mesenteric artery; D = direct connection to the dorsal aorta through multiple vessels; DA = dorsal aorta; G = gills; H = heart; HPV = hepatic portal vein; IRV = interrenal vein; L = liver; RG = remaining gut; VA = ventral aorta.

3. MORPHOLOGY OF THE GUT AS AN AIR-BREATHING ORGAN

Across the families of GAB fish, common modifications of the GI tract consist of a reduction of digestive function and modification of features to facilitate respiration. The latter can be manifest as: (1) a high degree of vascularization; (2) capillaries embedded in an epithelium that is in direct contact with air; (3) a blood–air distance narrow enough for significant O_2 diffusion to occur; (4) increased surface area for gas diffusion to occur; (5) presence of surfactant producing lamellar bodies in epithelial cells; and (6) presence of muscular sphincters isolating the respiratory region from other regions of the GI tract. For the purposes of this review, we consider the esophagus, stomach, and intestine to be distinct regions of the gut proper.

3.1. Esophagus

Accounts of the esophagus serving as a respiratory region of the GI tract are limited to one species of umbrid (Crawford 1971, 1974), one species of blenniid (Laming et al. 1982; Pelster et al. 1988), although many blenniids have not been examined (Table 10.1), and a synbranchid (Liem 1967).

Though the majority of GAB fish are freshwater tropical species, the Alaska blackfish, *Dallia pecotralis* (Umbridae) inhabits the muskeg swamps of Alaska and Northern Siberia (McPhail and Lindsey 1970), and uses its esophagus as an ABO (Crawford 1974). *Dallia* possesses a highly vascularized swim bladder, but capillaries do not penetrate the inner epithelial lining of the swim bladder, and are separated from the epithelium by a layer of loose connective tissue and smooth muscle (Crawford 1974). This makes the swim bladder a good candidate for gas secretion (and certainly it may function in buoyancy regulation), but a poor candidate for gas absorption (Crawford 1974). However, the esophagus of *Dallia*, connected to the swim bladder at the pneumatic duct, is also highly vascularized but with extensive capillary penetration into the epithelial lining. Crawford's (1974) histological analysis revealed a blood–air barrier of < 1 µm which is similar to that reported in the ABO gas bladders of the bowfin, *Amia calva*, and the longnose gar, *Lepisosteus osseus* (Crawford 1971) (Table 10.2). Co-occurring with the extensive vascularization of the respiratory region of the esophagus was extensive folding of the mucosa, presumably for increasing surface area for gas absorption (Crawford 1974). In *Dallia*, gas absorption in the GIT is likely limited to the esophagus as there exists a distinct stricture between the near equally sized esophagus and stomach, and distal to the esophagus intraepithelial capillaries disappear and gastric glands appear (Crawford 1974).

Table 10.2

Air-blood diffusion distances and capillary density for the gut respiratory mucosal epithelium of gut air-breathing fishes

Respiratory region of the GI	Family	Species	Air-blood diffusion distance	Capillary density	Source
Esophagus	Umbridae	*Dallia pectoralis*	<1 μm		Crawford 1971
Stomach	Loricariidae	*Ancistrus multispinnus*	0.6 μm	7/100 μm	Satora 1998
		Hypostomus plecostomus	0.25-2.02 μm; arithmetic mean = 0.86 +/-0.0046 μm (SE)	3-4/100 μm	Podkowa and Goniakowska-Witalińska 2003
		Pterygoplichthys anisitsi	Harmonic mean = 0.4-0.74 μm; arithmetic mean = 1.52+/-0.07 μm (SE)		Cruz et al. 2009
Posterior intestine	Callichthyidae	*Hoplosternum thoracatum*	1-2 μm		Huebner and Chee 1978
		Corydoras aeneus	0.24-3.00 μm		Podkowa and Goniakowska-Witalińska 2002
	Cobitidae	*Lepidocephalichthys guntea*	0.86-1.08 μm	$0.15/\mu m^2$	Yadav and Singh 1980
		Lepidocephalichthys guntea	2.6 μm		Moitra et al. 1989
		Misgurnus anguillicaudatus	11.9+/-0.4 μm	6.8/100 μm	McMahon and Burggren 1987
		Misgurnus mizolepis	0.7+/-0.11 μm		Park and Kim 2001

Lipophrys pholis (Blenniidae), an intertidal species of northern Europe, also uses its esophagus as an ABO. Gross observations of the esophagus consisted of a rich blood supply, many longitudinal folds, and visible separation from the buccopharynx and stomach by muscular sphincters (Laming et al. 1982). Following 3 h of forced air exposure, X-rays confirmed the presence of air bubbles in the esophagus, which was inflated to three times its pre-emersion size (Laming et al. 1982). Though Laming et al. (1982) did not report a blood–air diffusion distance, histological observation revealed that capillaries were embedded in the esophageal folds and were described as "superficial."

Distinguishable from other regions of the GIT by the presence of many longitudinal folds and intraepithelial capillaries, the esophagus of *Monopterus albus* (Synbranchidae) may serve as an ABO (Liem 1967). Though blood–air diffusion distance was not measured, the capillaries are known to be embedded within the mucosal layer of the esophageal epithelium and are found only in the peaks of the papillary folds (Liem 1967).

3.2. Stomach

Reports of the stomach being used as an ABO are common for the Neotropical catfish families Loricariidae and Trichomycteridae. Attenuation of gastric activity in favor of oxygen uptake may have been feasible in these groups because of their exceptionally long, coiled intestines (Armbruster 1998; Nelson et al. 1999; Delariva and Agostinho 2001; Podkowa and Goniakowska-Witalińska 2003; Nelson et al. 2007) that could compensate for loss of stomach digestive function (Section 2.2).

3.2.1. GROSS MORPHOLOGY

Since the earliest observations in *Ancistrus* of Carter and Beadle (1931), it has been apparent that a thin-walled, translucent, highly vascularized stomach can function as an ABO in loricariids. This trend has been documented in many genera of loricariids (e.g. *Ancistrus*, *Hypostomus*, *Liposarcus*, *Pterygoplichthys*) (Carter and Beadle 1931; Carter 1935; Satora 1998; Souza and Intelizano 2000; Oliviera et al. 2001; Podkowa and Goniakowska-Witalińska 2003; Cruz et al. 2009), and is suggested in several closely related families (Gee 1976; Cala 1987; Armbruster 1998). Armbruster (1998) conducted the most extensive survey of GI gross morphology in loricarioids in which he scored modifications of the stomach for air breathing ranging from mere enlargement of the stomach to increasing degrees of stomach vascularization to the presence of one or two air-filled diverticula branching off the stomach. There are indications, however, that not all regions of a respiratory stomach may play an equal role in O_2 uptake.

Carter and Beadle (1931) described an elongated posterior region of the stomach in *Liposarcus* that was more vascularized than other regions.

3.2.2. Reduced Gastric Function

Indications of respiratory function restricted to specific regions of the stomach can also be seen when examining the distribution of digestive glands. While the stomachs of both *Ancistrus* (Satora 1998) and *Pterygoplichthys* (Cruz et al. 2009) contain gastric glands, there is a reduced number in the corpus region of the stomach compared to the cardia and pylorus regions. In *Ancistrus*, capillaries were only embedded in the mucosal epithelia in the corpus region (Satora 1998), further suggesting that the corpus is the site of O_2 uptake. Cruz et al. (2009) reported a reduction in gastric glands coincident with a high degree of longitudinal folding of the mucosa in the corpus region. Furthermore, no regions stained positive for Alcian blue or for PAS indicating a lack of acid and neutral mucopolysaccharide production, respectively, in all regions of the stomach. This suggests that digestion occurs elsewhere in the GI tract (Cruz et al. 2009). Similarly, only a weak positive Alcian blue reaction was seen in the entire stomach mucosal surface of *Hypostomus* and the PAS reaction was negative (Podkowa and Goniakowska-Witalińska 2003). These authors also noted that food was never observed in the stomach of *Hypostomus*, also suggestive of the hypothesis that most digestion is occurring elsewhere in the GI tract.

Carter and Beadle (1931) were the first to report a reduction in digestive glands in *Ancistrus*. This observation was corroborated in *Liposarcus* by Oliveira et al. (2001) who described the stomach mucosa as non-glandular simple cuboidal epithelia. No food was observed in the stomach of any specimens despite regular feeding throughout captivity in the Oliveira et al. (2001) study. The epithelial mucosa develops into many folds and projections distal to the stomach, providing more surface area in the intestine for nutrient absorption (Oliveira et al. 2001).

3.2.3. Vascularization

A short blood–air barrier is essential for the diffusion of O_2 from the stomach lumen into surrounding capillaries. Reports of capillaries embedded within the stomach mucosa include *Ancistrus* (Satora 1998; Satora and Winnicki 2000), *Liposarcus* (Oliviera et al. 2001), *Hypostomus* (Podkowa and Goniakowska-Witalińska 2003), *Pterygoplichthys* (Cruz et al. 2009), and *Ancistrus, Hypostomus, Peckoltia, Pterygoplichthys*, and *Megaloancistrus* (Souza and Intelizano 2000). The shortness of the blood–air barrier is illustrated in Fig. 10.3 where it is evident that, in *Ancistrus*, the gut lumen is separated from a capillary erythrocyte by only

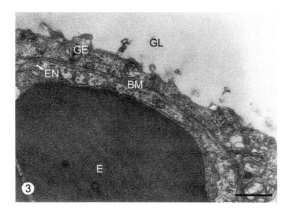

Fig. 10.3. Transmission electron micrograph of the corpus region of the stomach of *Ancistrus Multispinnis* (Loricariidae). The air space is top/right. BM = basement membrane; E = erythrocyte; EN = capillary endothelium; GE = gastric epithelium; GL = gastric lumen. The scale bar = 1 μm. Taken from Satora (1998).

three layers: (1) the gastric epithelium, (2) the basement membrane, and (3) the capillary endothelium. Published diffusion distances and capillary densities are summarized in Table 10.2.

Though all studies of stomach ABOs describe a dense capillary network for O_2 uptake, few have investigated the type of capillaries embedded within the mucosa. Visceral capillaries, normally associated with the GI tract, are characterized by an endothelium varying in thickness and containing many pores, while alveolar capillaries, such as those found in mammalian lungs, are characterized by a flattened endothelium with a continuous basement membrane lacking pores (Jasinski 1973). Satora and Winnicki (2000) liken the capillaries in the stomach mucosa of *Ancistrus* to the alveolar type in that the endothelium is flattened and the basement membrane is nearly continuous except for the existence of sparse pores, leading the authors to speculate that the stomach mucosa capillaries evolved from the visceral type to accommodate a respiratory function. In contrast, the capillary endothelium within the *Hypostomus* stomach mucosa contains "relatively numerous" pores and more closely resembles the visceral capillary type (Podkowa and Goniakowska-Witalińska 2003).

3.2.4. SURFACTANT PRODUCTION

Epithelial cells resembling mammalian type I and II pneumocytes are found in all known fish lungs and gas bladders (Graham 1997). These cells contain surfactant-producing lamellar bodies thought to aid in gas absorption by decreasing surface tension among other suggested functions

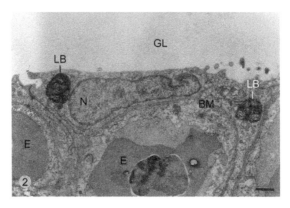

Fig. 10.4. Transmission electron micrograph from the corpus region of the stomach of *Ancistrus multispinnis*. Lamellar bodies (LB) are visible beneath the epithelial cell surface. BM = basement membrane; E = erythrocyte; GL = gastric lumen; N = nucleus. Scale bar = 1 μm. Taken from Satora (1998).

(Daniels and Orgeig 2003). Satora (1998) was the first to document the presence of lamellar bodies in the cytoplasm of stomach epithelial cells, less than 0.5 μm from the gut lumen in *Ancistrus* (Fig. 10.4). The presence of type I pneumocyte-like cells, specific to the corpus of the stomach mucosa, was later confirmed (Satora and Winnicki 2000). In the latter study, however, the authors reject the possibility that the abundant amount of surfactant produced in the stomach of *Ancistrus* reduces surface tension since the stomach is not vesicular like the alveoli of mammalian lungs. They instead favor the idea that the surfactant may serve to protect against desiccation and oxidative stress. Since then, lamellar bodies have also been documented in the stomach epithelial cells and gastric glands of *Hypostomus* (Podkowa and Goniakowska-Witalińska 2003) and *Pterygoplichthys* (Cruz et al. 2009). These authors favor the idea that surfactant aids in O_2 uptake.

3.3. Posterior Intestine

Gut air breathers in the families Cobitidae and Callichthyidae have been shown to use the intestine as an ABO (Table 10.1).

3.3.1. Gross Morphology

Similar to the stomach ABOs of loricariids and trichomycterids, the respiratory intestine of several cobitids and callicthyids has been described as thin-walled, translucent, and highly vascularized. This is seen in the posterior two-thirds of the intestine in *Lepidocephalichthys* (Cobitidae)

(Yadav and Singh 1980; Moitra et al. 1989). Additionally in this species, collagen fibers are present in lieu of the tunica propria in the respiratory segment of the intestine, allowing greater expandability to hold more air (Yadav and Singh 1980). In *Misgurnus* (Cobitidae), the distal 60% of the GI tract is estimated to be involved in O_2 uptake as the intestinal epithelia become thinner and more heavily vascularized in this region (McMahon and Burggren 1987).

The stomach and anterior intestine of *Corydoras* (Callichthyidae) are muscular, opaque and easily discernible from the thin-walled, translucent, air-filled posterior intestine, which occupies 30–40% of the entire intestine length (Kramer and McClure 1980; Podkowa and Goniakowska-Witalińska 2002). Kramer and McClure (1980) measured gas bubbles seen in the posterior intestine that occupied an average of 48% of the length of the entire intestine (range 21–69%) in experimentally induced air-breathing *Corydoras* (Callichthyidae). Huebner and Chee (1978) estimated that 50% of the GI tract of *Hoplosternum* (Callichthyidae) is dedicated to respiration, as it appears smooth, thin-walled, highly vascularized, and contained gas bubbles.

3.3.2. MUCUS PRODUCTION AND THE PRESENCE OF FOOD IN THE RESPIRATORY INTESTINE

There is a disparity between cobitids and callichthyids regarding the abundance of mucus-producing goblet cells in the respiratory intestinal epithelia. Sparse goblet cells are seen in the respiratory intestine of *Misgurnus* (Cobitidae) (Jasinski 1973) and *Lepidocephalichthys* (Cobitidae) (Yadav and Singh 1980; Moitra et al. 1989). Park and Kim (2001) reported a reduced density of goblet cells in the posterior intestine of *Misgurnus* (Cobitidae) relative to that seen in the esophagus and stomach. Numerous goblet cells are reported in the respiratory intestine of *Hoplosternum* (Callichthyidae) (Huebner and Chee 1978) and *Corydoras* (Callichthyidae) (Podkowa and Goniakowska-Witalińska 2002). The advantages of mucus may include lubrication for the passage of digesta and protection of the thin-walled, highly vascularized posterior intestine from desiccation and mechanical damage, but excessive mucus may increase gas diffusional distance.

Though numerous goblet cells in the respiratory intestine may be advantageous, it may not be a necessity, as seen in *Misgurnus,* where fecal matter emerges from the anterior spiral intestine already enclosed in a mucus pouch (McMahon and Burggren 1987). The respiratory intestines of the cobitids *Misgurnus* (Wu and Chang 1945) and *Lepidocephalichthys* (Yadav and Singh 1980) are usually filled with air and digesta are rarely seen, although Huebner and Chee (1978) observed fecal pellets between gas bubbles in the respiratory intestine of the callichthyid *Hoplosternum.*

3.3.3. Vascularization

Capillary penetration into the intestinal mucosa creating a short blood–air barrier for O_2 absorption has been recorded for the respiratory intestine of several species (Table 10.2). McMahon and Burggren (1987) examined the stomach, anterior, mid, and posterior intestine of *Misgurnus* and found a significant decrease in the blood–air barrier distance and a significant increase in capillary density progressing posteriorly along the GI tract. Histological preparations confirmed infiltration of capillaries into the posterior intestinal mucosal epithelium of *Misgurnus* (Jasinski 1973; Park and Kim 2001), *Hoplosternum* (Huebner and Chee 1978), and *Corydoras* (Podkowa and Goniakowska-Witalińska 2002).

Jasinski (1973) classified infiltrating capillaries seen in *Misgurnus* as the alveolar type based on the observations of flattened capillary endothelium, a continuous basement membrane, and only occasional pores. Though Podkowa and Goniakowska-Witalińska's (2002) description of *Corydoras* capillaries is similar to Jasinski's (1973) description of *Misgurnus*, they classify the infiltrating capillaries in the respiratory intestine of *Corydoras* as the visceral type, suggesting that some nutrient absorption does take place here. In order to better classify capillary type in the future, pore numbers should be quantified.

3.3.4. Surfactant Production

Goblet cells and their secretions into the intestinal lumen may protect against desiccation and oxidative stress, but the existence of surfactant producing lamellar bodies in the respiratory intestinal epithelia has only been confirmed in *Misgurnus* (Jasinski 1973) and *Corydoras* (Podkowa and Goniakowska-Witalińska 2002).

4. CIRCULATORY MODIFICATIONS ASSOCIATED WITH USE OF GUT ABOs

The potential for air-breathing fishes to lose oxygen to hypoxic waters across their gills has been extensively considered (e.g. Graham 1997) and will not be heavily reiterated here. Figure 10.2 shows the general circulatory plan of fishes with GABOs. Because oxygenated venous blood emanating from the ABO next encounters capillaries in the gills (or the liver and then the gills (Cobitidae)), oxygen will be lost to the water across the gill secondary lamellae if the water PO_2 is lower than that of the blood. Thus, the ability to shunt oxygenated blood away from the gill secondary lamellae during simultaneous aquatic hypoxia and air breathing could be favored by natural selection. Fernandes and Perna (1995) and Crawford (1971) present histological evidence for the possibility of shunting in *Hypostomus* and *Dallia*, respectively, and blood chemistry studies showing CO_2 retention in *Hypostomus* during air breathing (Wood et al. 1979) also imply shunting, but to our knowledge there has yet to be direct physiological demonstration of shunting oxygenated blood away from the gill gas exchange surface of GAB fish while breathing air. Interestingly, some GAB fish do not alter their rate of branchial ventilation upon breathing air in hypoxic water and some appear to actually increase gill ventilation under air-breathing conditions (Section 5.1.2).

The major variation in circulatory design among GAB fishes is the ABO drainage returning to the heart either via the hepatic portal circulation (Cobitidae) or the systemic circulation (all others) (Carter and Beadle 1931; Graham 1997; Fig. 10.2). Venous return through the systemic circulation is perceived as advantageous because bypassing the capillary beds of the liver will facilitate a more even corporal distribution of oxygen and avoid a pressure drop between the ABO and the heart (Fig. 10.2). Systemic venous return also exposes the ABO effluent to more direct suction from the heart during cardiac relaxation than would the hepatic portal return pathway (Olson 1994). In addition to the differential return paths to the heart, there is some variation in the path arterial blood takes to the ABO among the different groups, supporting contentions of independent evolutionary origins for GAB among the different taxa (Section 2.1). Although all GABOs are supplied with blood from branches of the dorsal aorta (Fig. 10.2), the more anterior gut ABOs tend to be supplied by branches of the coeliac artery (e.g. *Dallia* (Crawford 1971) and *Ancistrus* (Carter and Beadle 1931)) whereas the more posterior intestinal ABOs tend to be supplied by the anterior mesenteric artery (e.g. *Misgurnus* (McMahon and Burggren 1987)) or directly from the dorsal aorta (e.g. *Callichthys* and *Hoplosternum* (Callichthyidae); Carter and Beadle 1931). Yadav and Singh (1980) report the presence of a possible shunt vessel between the ventral aorta and the intestinal ABO in *Lepidocephalus* (Cobitidae), but this vessel has not been reported by other workers nor has its physiological function been investigated.

An additional major circulatory adjustment found in GAB fishes is the intrusion of the capillaries of the ABO into the gut luminal epithelial layer (Section 3). The normal position for capillaries of the GI tract would be in the *lamina propria*, basal to the gut epithelium. Although this capillary position functions perfectly well for the diffusive uptake of some foodstuffs such as small fatty acids (Fänge and Grove 1979), apparently, rates of diffusive oxygen uptake with such an arrangement were insufficient as the ABOs evolved, so a general histological finding is that capillaries have migrated into the epithelial layer, but do not directly abut the luminal surface of the gut (Fig. 10.3). Table 10.2 presents estimated diffusional distances for gut ABOs from a number of histological studies. The ranges reported fall within those of other fish ABOs and are in the range of mammalian lung diffusive distances (Weibel 1984). Capillary densities of gut ABOs are typical for fish ABOs in general (Podkowa and Goniakowska-Witalińska 2002, 2003; Cruz et al. 2009). The Cruz et al. (2009) study of *Pterygoplichthys* (Loricariidae) is the only known estimate of the diffusing capacity of a gut ABO. Cruz et al. (2009) found the diffusing capacity of this stomach ABO to be higher than for most other fish ABOs, but lower than published values for the lungs of lungfish (*Lepidosiren* and *Protopterus*). This estimated diffusing capacity of a gut ABO (Cruz et al. 2009) is also substantially below estimates for lungs of similar-sized mammals (Weibel 1984).

5. PHYSIOLOGY OF GUT AIR BREATHING

5.1. Ventilation

5.1.1. Ventilation of Gut Air-breathing Organs (GABOs)

Surprisingly, there are no studies on the mechanics of GABO inflation. Most authors describe inspiration as "gulping," presumably a negative buccal cavity pressure derived from buccal and perhaps opercular cavity expansion as described for other air-breathing fishes (Liem 1989). A further assumption is that positive pressure is then generated by buccal cavity contraction with both oral and opercular valves closed, driving the air posteriorly to the ABO as detailed for lungs and swim bladder ABOs (Liem 1989). Expiration and air transport through the GI tract are equally as mysterious in GAB fishes. Crawford (1971) comments on the presence of a well-developed esophageal skeletal musculature that may be involved in expelling gas from the *Dallia* (Umbridae) esophagus; however, there is no confirmation of this. Likewise, Gradwell (1971) discusses three ways that *Plecostomus* (Loricariidae) could exhale while submerged: (1) hydrostatic pressure alone in conjunction with elastic recoil and appropriate valving; (2) contraction of visceral smooth muscle of the respiratory stomach; and (3) positive abdominal cavity pressure developed through contraction of the *rectus abdominis* skeletal muscle, but there has been no experimental differentiation among these alternatives. The two families with intestinal ABOs (Callichthyidae and Cobitidae) have a unique transport problem. Since they exhale through their vent (Gee and Graham 1978; McMahon and Burggren 1987), they must transport air the entire length of the GI tract.

While this unidirectional transport of air potentially minimizes respiratory dead space, it raises a new problem of coordinating transport of air with transport of digesta. Both of these families have a transitional zone between the digestive and respiratory portions of their intestine (McMahon and Burggren 1987; Persaud et al. 2006b) that is richly endowed with smooth muscle and appears to function in compacting the digesta and possibly encasing it in mucus. This compacted digesta would then minimally interfere with gas exchange in the respiratory (posterior) segment of the intestine. Persaud et al. (2006b) found that depriving two callicthyiids of access to the surface in normoxic water caused them to curtail feeding activity and diminish the transport of digesta posteriorly. Since Persaud et al. (2006b) and other authors (e.g. Gee and Graham 1978) have noted the synchronous release of air from the vent immediately following inspiration, Persaud et al. (2006b) propose that the column of air present in the guts of these continuous air-breathing fishes is necessary for normal digesta transport. Furthermore, Persaud et al. (2006b) claim that the amount of visceral smooth muscle in the respiratory portion of the intestine is insufficient for peristaltic transport of air. Thus, the pressure generated by the buccal/opercular pump speculatively provides the force for inspiration, air transport, digesta transport across the respiratory zone of the intestine and expiration in intestinal GAB fishes.

5.1.2. GILL VENTILATION DURING AIR BREATHING

The general pattern of branchial ventilation in fish exposed to progressive hypoxia is to defend arterial PO_2 (oxygen regulation) by increasing total gill ventilation (Holeton 1980; Fig. 10.5). This is manifested by either increases in ventilatory frequency f_v (Affonso and Rantin 2005) or ventilation volume V_t (Nelson et al. 2007) with the latter considered more common, possibly as an energy-saving strategy (Mattias et al. 1998; Fig. 10.5). As environmental oxygen falls to levels where this strategy waxes futile, the animal ceases oxygen regulation at a point called the critical oxygen tension (P_cO_2) and metabolic rate begins to drop in concert with environmental $[O_2]$ (oxygen conforming) or the animal switches to air-breathing to maintain oxygen regulation if it has evolved that capacity (Fig. 10.5). As an animal starts using atmospheric oxygen, the prediction is that gill ventilation should diminish if not stop. Fish expend substantial energy, around 10% of their resting metabolic rate, in ventilating their gills (Holeton 1980; Glass and Rantin 2009), and although air-breathing fish generally still use their gills or skin to eliminate CO_2 (Johansen 1970), the potential loss of O_2 to the water across the secondary lamellae would seem to make ventilating the gills in hypoxic water generally counterproductive. Indeed, fish from one of the GAB taxa that completely emerses (*Lipophrys*)

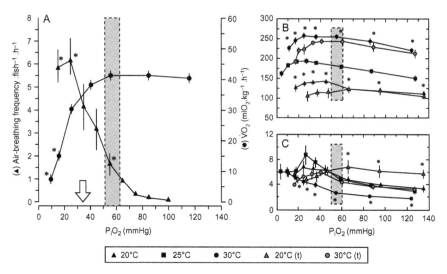

Fig. 10.5. (A) Air-breathing frequency (▲) of 50 *Hypostomus regani* exposed to progressive hypoxia and allowed free access to the surface and aquatic mass-specific oxygen consumption (●) of 17 *Hypostomus regani* exposed to progressive hypoxia but denied access to the surface. The stippled bar designates the range of PO₂ over which air breathing normally commences (Mattias et al. 1998). **(B)** Respiratory frequency (breaths · min⁻¹) (f_R; top panel) and (C) tidal volume (ml · kg⁻¹ · min⁻¹) (V_T; bottom panel) of *Hypostomus regani* exposed to progressive hypoxia over a period of 9 h and denied access to the surface. Closed symbols represent animals chronically acclimated to a temperature whereas the stippled symbols represent animals acutely transferred to the experimental temperature. Each symbol represents the mean for that group ± 1 SE. The stippled bar designates the range of PO₂ over which air breathing would normally commence (modified from Nelson et al. 2007).

has been reported to cease branchial ventilation upon leaving the water (Laming et al. 1982), although Pelster et al. (1988) claim that ventilation continues during emersion in a tidal fashion with the opercular valve closed. The situation is also unclear and certainly not generalizeable to those GAB fishes that remain submersed. Some investigators have reported cessation of branchial ventilation in loricariids (Gradwell 1971; Wood et al. 1979) but these were visual observations and appear to have been erroneous. Graham (1983) reports a 20% drop in gill f_v in air-breathing *Ancistrus* when acclimated to hypoxic water, but Gee and Graham (1978) report a 33% increase in *Hoplosternum* gill f_v when induced to increase its frequency of air breathing through aquatic hypoxia and Affonso and Rantin (2005) report a 31% increase in hypoxic *Hoplosternum* gill f_v when not allowed to breathe air, although *Hoplosternum* has also been reported to have poorly developed gill secondary lamellae (Carter and Beadle 1931). Finally, no change in the rate of gill ventilation with air breathing has been reported for both foregut breathers *Hypostomus* (Nelson et al. 2007) and hindgut breathers *Misgurnus*

(Cobitidae) (McMahon and Burggren 1987). Although V_t has not been measured in GAB fish while air breathing, we have to conclude at this juncture, based upon f_v measurements alone, that reduced gill ventilation while air breathing is not a general strategy GAB fish have employed to either save energy or reduce O_2 loss to hypoxic water.

5.2. Gas Exchange and Metabolic Rate in GAB Fishes Breathing Air

Although the attempt to characterize the gas composition of GABOs dates to the nineteenth century (Jobert 1877), there is a surprisingly small amount of information on gas exchange in GAB fishes. Despite this, a general consensus has emerged that GABOs are only minimally involved in CO_2 excretion. Regardless of whether the fish is a foregut breather (*Ancistrus* Loricariidae) (Carter and Beadle 1931; Graham 1983) or hindgut breather *Misgurnus* (Cobitidae) (McMahon and Burggren 1987) or *Hoplosternum* (Callichthyidae) (Carter and Beadle 1931), the general finding is for very minimal amounts or no CO_2 to be released via the ABO. The high capacitance of water for CO_2 (Schmidt-Nielsen 1997) allows for efficient CO_2 excretion by the gills and skin. Blennies, represented by *Lipophrys pholis* are the only GAB fish that are classified as amphibious and have had their plasma CO_2 measured under emersion conditions (Pelster et al. 1988). (Note: *Hoplosternum* have been reported to migrate over land between ponds (Carter and Beadle 1931), *Misgurnus* have been reported to endure droughts in dry burrows (Ip et al. 2004) and some Loricariids will self-emerse in the laboratory (J. A. Nelson personal observation).) Pelster et al. (1988) report a 53% rise (from 2.43 to 3.71 mmHg) in plasma PCO_2 in emersed *Lipophrys pholis*, consistent with the general vertebrate emersion expectation (Schmidt-Nielsen 1997), but do not report ABO [CO_2]. Based on this limited physiology, GABOs do not appear to be significant organs of CO_2 excretion.

The frequent evolution of GAB (Section 2.1) most likely transpired to provide supplementary oxygen to animals when dissolved oxygen became limiting. Other proposed functions such as buoyancy (Gee 1976) and propelling of digesta (Persaud et al. 2006b) appear to be secondary functions (see below) and possible exaptations (Gould and Vrba 1982). Almost without exception, GAB fish exposed to hypoxia do not go hypometabolic when allowed to breathe air and instead retain oxygen regulation and normal levels of activity (Fig. 10.5). Rates of air breathing will thus be complex functions of metabolic rate and all of its inherent variance and problems in accurately assessing. Metabolic rate in fishes depends on size, species and multiple additional factors including experimental and environmental variables such as stress level, temperature and water chemistry

(Nelson and Chabot 2010). Thus, attempts to characterize ABO ventilation rates across GAB fishes would be premature and disingenuous based upon the limited information available. The same can be said for respiratory partitioning (the fraction of oxygen contributed by aerial versus aquatic respiration) under various conditions and rates of oxygen uptake from the ABO; the number of studies are so few, and the probability of experimental and environmental effects overwhelming the results so real, that there can be little heuristic value in creating generalizations here. Graham (1983) for *Ancistrus* (Loricariidae) at 25°C and McMahon and Burggren (1987) for *Misgurnus* (Cobitidae) at 20°C give the most detailed accountings of changes in ABO gas composition over time, and the reader is referred to these if more detailed information is desired. The following consideration of gas exchange in GAB fishes follows the organization of Table 10.1.

5.2.1. ESOPHAGEAL GAS EXCHANGE

For esophageal breathers (Table 10.1), Crawford (1971) showed that three individual *Dallia* (Umbridae) maintained normal metabolic rates below an [O_2] of 2 mg/l by supplementing aquatic respiration with air breathing. Likewise, Laming et al. (1982) show that metabolic rates of *Lipophrys* (Blenniidae) stay relatively constant throughout cycles of immersion/emersion and that skin oxygen consumption is only a minor contributor to oxygen uptake, suggesting that the GABO is able to maintain normal resting rates of oxygen consumption in emergent members of this species.

5.2.2. FOREGUT (STOMACH) GAS EXCHANGE

Of the two stomach-breathing families (Table 10.1), gas exchange has only been studied in the loricariids. Cala et al. (1990) report that the hypoxia-exposed trichomycterid *Eremophilus* must increase its air-breathing frequency to survive, but there have been no studies of actual gas exchange in this family. The loricariids that have been studied conform to the general facultative air-breathing (FAB) pattern (Section 5.1.2; Fig. 10.5). As environmental [O_2] drops, metabolic rate is regulated until a P_wO_2 between 25 and 60 mmHg is reached (Graham and Baird 1982; Graham 1983; Nelson et al. 2007), at which time, many loricariids begin breathing air with their stomachs (Gee 1978; Graham and Baird 1982; Graham 1983; Mattias et al. 1998; Nelson et al. 2007). Graham (1983) demonstrated utilization of the oxygen component of air in the loricariid ABO by showing a progressive decline in [O_2] of ABO gas with breathhold length in *Ancistrus*. Oxygen uptake from the ABO can also be inferred from the sub-atmospheric gut oxygen levels reported by Carter and Beadle (1931) for *Ancistrus* and Nelson et al. (2007) for *Hypostomus*. Graham (1983) also showed that increased oxygen demand in *Ancistrus* was met entirely through increases in air-breathing frequency (f_v); ventilation volume

(V_t) of air breaths remained constant throughout changes in air-breathing demand in this species. Interestingly, *Ancistrus* acclimates to 2–3 weeks of hypoxia by expanding the size of its ABO 25% and increasing its ability to extract oxygen from the ABO (Graham 1983). Although most authors report no air breathing from loricariids in normoxic water, MacCormack et al. (2006) report surfacing (presumably air-breathing) behavior that was independent of environmental [O_2] in a telemetered loricariid (*Glyptoperichthyes*) held in cages in a natural environment. This observation, coupled with observations that another loricariid genus (*Panaque*) will voluntarily emerse in normoxic water (Nelson personal observation), suggests that our understanding of air-breathing behavior and ventilation dynamics in GAB loricariids will benefit from more studies on additional species under field-relevant conditions.

5.2.3. HINDGUT (STOMACH) GAS EXCHANGE

Both families of intestinal air breathers (Table 10.1) appear to be continuous, but not obligate air breathers (Gee and Graham 1978; Kramer and McClure 1980; McMahon and Burggren 1987). Wu and Chang (1945) claim that *Misgurnus* (Cobitidae) abandons continuous air breathing at low temperature, and it is possible that McMahon and Burggren (1987) did not lower temperature enough (10°C) to observe this effect, but based on the currently available information, it seems safest to classify both the callichthyids and cobitids as CAB fishes. This implies that some other physiological function such as buoyancy (Gee 1976), digesta transport (Persaud et al. 2006b) or enhanced scope for activity (Gee and Graham 1978; Almeida-Val and Farias 1996) is served by breathing air in these fishes. However, since representatives from both families conform to the FAB pattern of responding to aquatic hypoxia with increased ventilation of their ABO to regulate metabolic rate (Gee and Graham 1978; McMahon and Burggren 1987), we conjecture that these factors were secondary to aquatic hypoxia in driving the original evolution of intestinal respiration. McMahon and Burggren (1987) found that when *Misgurnus* was exposed to aquatic hypoxia, it increased the rate of intestinal ventilation entirely through increases in f_v. Changes in aerial V_t were not involved, similar to what Graham (1983) reported for stomach-breathing *Ancistrus* (Section 5.2.2). Gee and Graham (1978), Kramer and McClure (1980), and Affonso and Rantin (2005) also report increases in f_v with progressive aquatic hypoxia in three species of intestinal-breathing callichthyiids. McMahon and Burggren (1987) estimated an ~70% turnover of ABO air (30% deadspace) with each subsequent breath in *Misgurnus* intestines. Interestingly, they report no correlation between gas composition of the ABO and breath interval durations ranging from 5 min to 1 h, suggesting that the available oxygen is extracted from the air relatively rapidly and very little gas exchange occurs subsequently. Jucá-Chagas (2004) reports that the intestinal breathing *Hoplosternum* (Callichthyidae) can extract more oxygen per unit body mass from an air breath than either a gas-bladder-breathing erythrinid (*Hoplerythrinus*) or a lungfish (*Lepidosiren*).

5.3. Gut Air-breathing Ventilatory Drive

Manipulations of the gas composition of the aquatic and aerial medium by Gee and Graham (1978) and McMahon and Burggren (1987) conclusively show that representatives from neither intestinal air-breathing family (Table 10.1) are sensing the chemical composition of the gas in the ABO to set ABO ventilation rate, although ABO volume may play a role (Gee and Graham 1978). The main factors that appear to set the rate of aerial ventilation in GAB fish are P_wO_2 and metabolic rate (Kramer and McClure 1980; Graham and Baird 1982; McMahon and Burggren 1987). McMahon and Burggren (1987) report a modest sensitivity of *Misgurnus* (Cobitidae) f_v to water PCO_2 as do Graham and Baird (1982) for stomach-breathing *Ancistrus* and *Hypostomus*, but the primary drive for ventilation appears to be chemosensation of P_wO_2 as evidenced by the inverse relation of f_v with P_wO_2 in all species measured. This would accord well with Oliviero et al. (2004) who found ventilation of a gas-bladder ABO in the erythrinid *Hoplerythrinus* to be driven by O_2 chemoreceptors on the gills. Graham and Baird (1982) report no change in the threshold P_wO_2 for air breathing to commence even after hypoxia acclimation had produced improvements in the size and extraction efficiency of the ABO, also strongly suggestive of environmental chemosensation of P_wO_2 setting this parameter. In contrast, Brauner et al. (1995) show a leveling off of the f_v/P_wO_2 relationship in *Hoplosternum* (Callichthyidae) such that reductions in P_wO_2 below 100 mmHg did not elicit further increases in f_v. Brauner et al. (1995) also demonstrated an inverse relation between f_v and pH and a direct relationship between f_v and water hydrosulfide (HS^-) concentration in this species and suggest that these latter two variables (indicative of water parcels that have gone anaerobic) may be as important as P_wO_2 in setting rates of ABO ventilation in nature.

Rates of aerial ventilation in GAB fishes are also sensitive to metabolic demand influenced by changes in temperature, size or activity level. McMahon and Burggren (1987) report a linear increase in ABO ventilation between 10° and 30°C ($Q_{10} \sim 2$ for 10°–20°C and $Q_{10} \sim 1.5$ for 20°–30°C) for *Misgurnus* and Graham and Baird (1982) report a steady increase in f_v between 20° and 30°C ($Q_{10} \sim 1.4$) for *Ancistrus*. Sloman et al. (2009) found that smaller, and therefore metabolically more active, *Hoplosternum* (Callichthyidae) had a higher P_wO_2 air-breathing threshold during exposure to progressive hypoxia than larger animals when held in isolation. This relationship disappeared when animals were held in groups, casting doubt upon the field relevance of the former result, but still suggestive of the idea that metabolic demand may partially drive ABO ventilation rate. Conversely, Mattias et al. (1998) found no relationship between body mass (M_b) and air-breathing threshold in 50 *Hypostomus regani* over a 600 g size range. Perna and Fernandes (1996) also found no effect of size on air-breathing threshold for *Hypostomus plecostomus* over an 83 g size range. Activity appears to correlate with air-breathing frequency in CAB callichthyiids, but has not been studied in other GAB fish taxa. Gee and Graham (1978) found a significant correlation between air-breathing frequency and activity in two species of CAB callichthyiids (*Hoplosternum* and *Brochis*). Kramer and McClure (1980) also found significant relationships between activity and f_v in a third species of callichthyiid (*Corydoras*), but only at three of the five depths they tested. Boujard et al. (1990) report a robust circadian cycle in *Hoplosternum* wherein feeding, activity and air-breathing activity all peaked during the hours of darkness in these nocturnal fish, and Sloman et al. (2009) report a significant relationship between activity and air-breathing in *Hoplosternum,* but only when in groups. Almeida-Val and Farias (1996) found one species of CAB (*Hoplosternum*) to have an exceptionally high rate of tissue oxygen consumption suggesting that CAB may have evolved to support high tissue rates of ATP turnover, but since an FAB foregut-breathing loricariid (*Liposarcus*) had the lowest rate of tissue ATP turnover among analyzed species this is not a general property of GAB fishes.

5.4. Cardiovascular Response to Gut Air Breathing

Although there is a fairly rich literature on cardiovascular responses to air breathing in fish, most of this literature relates to lungfishes and several other large (mostly swim bladder ABO) species (Graham 1997). There is very little cardiac, and no vascular, information on GAB fishes while breathing air. GAB fishes can exhibit an almost immediate reflex bradycardia when exposed to hypoxia as shown in the loricariid *Hypostomus* by Nelson et al. (2007) (Fig. 10.6). This is presumably the same generalized, although not universal, vagally mediated reflex bradycardia response to hypoxia seen in water-breathing fishes (Taylor 1992). This hypoxic bradycardia appears to have a metabolic component, as Nelson et al. (2007) report a more established hypoxic bradycardia at 30°C than at either 20°C or 25°C. The development of a significant bradycardia in loricariids during progressive hypoxia, when metabolic rate remains unchanged (Nelson et al. 2007), suggests that the cardiac response to hypoxia may be somewhat

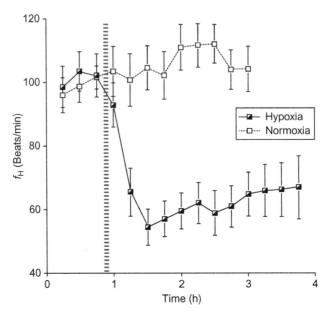

Fig. 10.6. Heart rate (f_H) in *Hypostomus regani* exposed to 20 mmHg PO_2 for 3 h (half-closed symbols) or normoxic conditions for 3 h (open symbols). Each symbol represents the mean for that group ± 1 SEM. The striped bar designates the onset of hypoxia. Hypoxia exposure immediately initiated an approximate 50% reduction in heart rate that gradually recovered over time. Mean heart rate was significantly lower in the hypoxia-exposed animals throughout the exposure period (MANOVA $P < 0.001$), although individual animals would briefly elevate their heart rate back to control levels or even higher when surfacing to breath air (modified from Nelson et al. 2007).

similar to that for carp (*Cyprinus carpio*) at 15°C (Stecyk and Farrell 2002), where a reduction in f_H is somewhat compensated for by increases in stroke volume. Reimmersion of emergent, air-breathing *Lipophrys* could be considered a descent into hypoxic conditions and does result in a transient 37% bradycardia, but this is followed almost immediately by a 15% tachycardia that is sustained for at least 10 min (Laming et al. 1982).

The precepts of ventilation–perfusion matching dictate that inflation of the ABO by air-breathing fish in hypoxic water should produce an increase in cardiac output coincident with diversion of blood to the ABO (Johansen 1970). This is often, but not universally, manifest by an immediate post-breath tachycardia in air-breathing fishes (Table 6.5 in Graham 1997). The only pre- and post-breath heart rates published for hypoxic GAB fishes are for loricariids. Nelson et al. (2007) report an average 34% increase in heart rate immediately post-breath in *Hypostomus*, similar to the 32% air-breath tachycardia reported by Graham (1983) for the co-familiar *Ancistrus*. Nelson et al. (2007) also report a large individual variance in post-air-breath tachycardia that was somewhat related to an animal's predilection to breath air. Interestingly, emersion from normoxic water produces a transient bradycardia in the GAB blenny, *Lipophrys* (Laming et al. 1982; Pelster et al. 1988).

5.5. Blood Chemistry of Gut Air-breathing Fishes

Perhaps the most heavily studied aspect of air-breathing fish physiology is the properties of their hemoglobins. Thus, the literature is replete with blood chemistry information, often collected at one time of the year at a single location (e.g. Powers et al. 1979). These types of measurements are often the basis for speculation about the action of natural selection on hemoglobin, but as Graham (1997) points out, without standardized collection conditions or methodology for analyzing hemoglobin, the foundations on which these adaptive scenarios are built perhaps are weak or non-existent. Wells (1990) also provides cautionary evidence against painting adaptive scenarios based upon differences in hemoglobin amount or properties. Graham (1997) provides an extensive review of the blood chemistry of all air-breathing fishes, for which, in this case, much of the information comes from GAB fishes.

From an oxygen transport perspective, there are two major issues for the blood of a GAB fish. First, there is the problem of venous admixture with the (presumably saturated) blood leaving the ABO (Fig. 10.2) and, second, there is the potential loss of O_2 to hypoxic water at the gills (Section 4). The amphibious blenny *Lipophrys* (Blenniidae) or any other GAB fish that emerses has the additional challenge of CO_2 accumulation. The blood of GAB can be generally characterized as having a moderate to low P_{50} with coincident low erythrocyte [NTP] (Powers et al. 1979; Graham 1997; Marcon et al. 1999). However, since many of the GAB fishes acclimatize/ acclimate to hypoxia by increasing blood oxygen affinity at least partially through reductions in erythrocyte [NTP] (Graham 1983, 1985; Wilhelm and Weber 1983; Val et al. 1990), reported values will be heavily dependent on the environmental conditions at the time of collection. The high-oxygen-affinity blood may be associated with the general occupancy of hypoxic habitats by these fishes, but is in contrast to the hypothesis that air-breathing fish should have reduced blood oxygen affinity. Riggs (1979) summarized the extensive data set from the second 1976 *Alpha Helix* expedition and concluded that there was no evidence for a right shift in the blood oxygen equilibrium curves of air-breathing fishes, although Morris and Bridges (1994) expanded that data set and provide some modest evidence for a right shift across all measured air-breathing fish.

The blood of GAB fishes can be further generalized as having a moderate to strong Bohr effect and no Root effect (Focesi et al. 1979; Powers et al. 1979; Bridges et al. 1984). The Bohr effect would seem maladaptive, at least in the case of one GAB loricariid *Liposarcus*, which appears not to tightly regulate their plasma pH in the face of hypercarbic acidosis (Brauner et al., 2004); in this case, the Bohr effect would exacerbate both the desaturation of hemoglobin in ABO effluent and the potential loss of O_2 to hypoxic water during gill transit by blood under acidotic conditions. The general absence of a Root effect in the GAB fishes would seemingly support the swim bladder inflation function of this property as GAB fishes generally have reduced swim bladders and swim bladder function (Gee 1976; Schaefer and Lauder 1986).

The blood of GAB fishes can also be characterized, with the exception of the amphibious *Lipophrys* (Bridges et al. 1984), as having a high volume of erythrocytes (hematocrit; Hct) with a coincident high blood hemoglobin concentration ([Hb]) (Graham 1997). These values are also subject to hypoxia acclimation/acclimatization in some loricariids (Graham 1983, 1985; Val et al. 1990) although Graham (1985) also demonstrated a lack of [Hb]/Hct acclimation in one callichthyid (*Hoplosternum*) and one loricariid (*Loricaria*). Interestingly, Sloman et al. (2009) show that there was no relationship between blood [Hb] or Hct and surfacing behavior in a species of *Hoplosternum*. Fernandes et al. (1999) and Nelson et al. (2007) show that changes in blood erythrocytic content begin almost immediately upon hypoxia exposure in *Hypostomus*. *Hypostomus regani* exposed to three hours of hypoxia or eight hours of graded hypoxia were characterized by having significantly higher [Hb] and smaller erythrocytes that contained more hemoglobin per erythrocyte than normoxic animals (Fernandes et al. 1999; Nelson et al. 2007). This result is most likely due to the hypoxic animals releasing a store of erythrocytes to enhance oxygen transport because the time is too short for new erythrocyte production. Val et al. (1990) also reported a higher cell hemoglobin concentration in another loricariid exposed to hypoxia for 30 days or captured from hypoxia-prone habitats. However, Weber et al. (1979) reported cell swelling and decreased cell hemoglobin concentrations in loricariids exposed to hypoxia for 4–7 days, suggesting that there may not be a generalized loricariid or GAB blood chemistry response to hypoxia or that stress interacts to differential degrees with hypoxia in laboratory settings. Although most of the GAB fishes carry multiple forms of hemoglobin, there is presently no evidence that air-breathing fishes generally adjust the relative expression of various hemoglobin isoforms in response to hypoxia or air breathing (Almeida-Val et al. 1999).

6. BEHAVIORAL ECOLOGY OF GUT AIR BREATHING

Considering the stochasticity of oxygen availability in most of the environments occupied by GAB fishes (Section 1), it is not unreasonable to assume that most of the ecology and behavior of these organisms will be influenced by environmental oxygen availability (Kramer 1987), yet experimental verification is mostly lacking. Predator–prey dynamics are undoubtedly a function of environmental [O_2] whether a fish is a water, air or bi-modal breather (Domenici et al. 2007). The increased aquatic ventilation required in hypoxic habitats (Section 5.1.2) will increase an animal's energetic requirements and potentially its detectability by both potential predators and prey. Certainly, the diminished scope for activity under hypoxia (Chabot and Claireaux 2008) could compromise predator–prey performance; however, the diminished feeding activity and scope for growth in hypoxic water (Chabot and Claireaux 2008) may be just as important on a different time scale.

The evolution of air breathing solved some of these problems for GAB fishes, but opened up several new ones (Kramer 1987). Air breathing could expose animals to a new class of aerial predators. In a direct test of this hypothesis, Kramer et al. (1983) showed that air-breathing fishes (one GAB fish) were more vulnerable to predation from a green heron (*Buterides striatus*) when forced to breath air than were water-breathing fishes under similar conditions. Kramer and McClure (1980) offer some evidence that an hypoxic GAB fish (*Corydoras*) is less likely to surface the deeper they reside and Power (1984) provides evidence from the field that GAB loricariids will avoid shallow waters where they are vulnerable to avian predation despite an abundance of food. These results, coupled with observations of synchronized air-breathing behavior in several GAB fishes (e.g. Gee and Graham (1978), Kramer and Graham (1976) and Sloman et al. (2009) for *Hoplosternum*; Kramer and Graham (1976) for *Ancistrus*) suggest that the evolution of gut air breathing may be associated with anti-predator behaviors to compensate for the increased visibility to surface predators or to aquatic predators that are potentially more adept at detecting and capturing an animal while surfacing, or both. Indeed, many GAB fishes have adopted a nocturnal lifestyle where they are less likely to be visible to visual predators when most active. Boujard et al. (1990) demonstrate nocturnal maxima of activity, feeding and air breathing in *Hoplosternum* that very quickly tracked experimental changes to the timing of the daylight cycle. Likewise, MacCormack et al. (2006) report only nocturnal surfacing (presumably air-breathing) behavior in the loricariid (*Glyptoperichthyes*) telemetered and held in cages in a natural environment.

Considering the potential interactions between digestive and respiratory function when the gut is used for both (Section 2.2), there is surprisingly little information on the partitioning between feeding activity and aerial respiration in GAB fishes. Persaud et al. (2006b) found that two callichthyids would stop eating when denied access to air and propose that air breathing is essential to move digesta through the poorly muscled hindgut. Presumably air pressure is also used to propel sperm caudally in the *Corydoras* (Callichthyidae) that fertilize their eggs by ingesting sperm (Kohda et al. 1995). Nelson et al. (2007) found no predilection between fed or starved *Hypostomus* to breath air or any differences between fed and unfed animals in other physiological parameters upon hypoxia exposure. Kramer and Braun (1983) report that air-breathing frequency after feeding in *Corydoras* (Callichthyidae) is variably dependent on P_wO_2. Above 50% saturation there was a decrease in air-breathing frequency after feeding whereas at a P_wO_2 of 44 mmHg there was no change and at a P_wO_2 of 24 mmHg there was an increase in air-breathing frequency (Kramer and Braun 1983). Certainly our understanding of how GAB fishes balance their digestive and respiratory gut functions is in its infancy.

7. AMMONIA VOLATILIZATION BY THE GI TRACT

Most fish are ammonotelic, that is, they excrete waste nitrogen as molecular ammonia or ammonium ion (NH_4^+). This is the least energetically costly method of nitrogen excretion and in circumneutral waters of low [NH_3] is thought to occur largely by passive diffusion of NH_3 to the water aided by diffusional trapping of NH_3 as NH_4^+ through acidification of the surface boundary layer (Moreira-Silva et al. 2010). This mechanism of nitrogen excretion is no longer available to animals that emerse or inhabit waters of high [NH_3] where the diffusive gradient would be lowered or reversed. Thus, amphibious fishes have adopted a variety of strategies to tolerate higher body [NH_3], reduce endogenous NH_3 production, develop alternative modes of NH_3 excretion or employ various combinations of these strategies (Ip et al. 2004). One alternative mechanism of NH_3 excretion that has been demonstrated in emersed fish is the release of NH_3 gas to the atmosphere, or ammonia volatilization (Frick and Wright 2002; Tsui et al. 2002). Tsui et al. (2002) show a progressive increase in NH_3 volatilization with time of emersion in *Misgurnus*. They conclude that volatilization is occurring in the gut because NH_3 volatilization ceased in animals that were not allowed access to the surface and because emersed and NH_3-exposed fish had a significantly more alkaline anterior intestine than controls (Tsui et al. 2002). This would be an exciting new function for the gut of air-breathing fishes and might help explain the frequent evolution of GABOs (Section 2.1). Unfortunately, this finding has yet to be confirmed by other investigators and since Moreira-Silva et al. (2010) also demonstrate a significant increase in the membrane fluidity of gills in emersed *Misgurnus* that would aid in NH_3 loss to the atmosphere, the role of the gut in NH_3 excretion clearly represents an interesting area for further studies.

REFERENCES

Affonso, E. G., and Rantin, F. T. (2005). Respiratory responses of the air-breathing fish *Hoplosternum littorale* to hypoxia and hydrogen sulfide. *Comp. Biochem. Physiol.* **141C**, 275–280.

All Catfish Species Inventory (2009). Sabaj, M. H., Armbruster, J. W., Ferraris, C. J., Friel, J. P., Lundberg, J. G. and Page, L. M. (eds.). Internet address: http://silurus.acnatsci.org/

Almeida-Val, V. M. F., and Farias, I. P. (1996). Respiration in fish of the Amazon: metabolic adjustments to chronic hypoxia. In: *Physiology and Biochemistry of the Fishes of the Amazon* (A.L. Val, V.M.F. Almeida-Val and D.J. Randall, eds), pp. 257–271. INPA, Manaus.

Almeida-Val, V. M. F., Val, A. L., and Walker, I. (1999). Long- and short-term adaptation of Amazon fishes to varying O_2 levels: intraspecific phenotypic plasticity and interspecific variation. In: *The Biology of Tropical Fishes* (A.L. Val and V.M.F. Almeida-Val, eds), pp. 185–206. INPA, Manaus.

Armbruster, J. W. (1998). Modifications of the digestive tract for holding air in loricariid and scoloplacid catfishes. *Copeia* **3**, 663–675.

Armbruster, J. W. (2006). The Loricariidae. Internet address: http://www.auburn.edu/academic/science_math/res_area/loricariid/fish_key/about.html

Boujard, T., Keith, P., and Luquet, P. (1990). Diel cycle in *Hoplosternum littorale* (Teleostei): evidence for synchronization of locomotor, air breathing and feeding activity by circadian alternation of light and dark. *J. Fish Biol.* **36**, 133–140.

Brauner, C. J., Ballantyne, C. L., Randall, D. J., and Val, A. L. (1995). Air breathing in the armoured catfish (*Hoplosternum littorale*) as an adaptation to hypoxic, acidic, and hydrogen sulphide rich waters. *Can. J. Zool.* **73**, 739–744.

Brauner, C. J., Wang, T., Wang, Y., Richards, J. G., Gonzalez, R. J., Bernier, N. J., Xi, W., Patrick, M., and Val, A. L. (2004). Limited extracellular but complete intracellular acid–base regulation during short-term environmental hypercapnia in the armoured catfish, *Liposarcus pardalis*. *J. Exp. Biol.* **207**, 3381–3390.

Bridges, C. R., Taylor, A. C., Morris, S. J., and Grieshaber, M. K. (1984). Ecophysiological adaptations in *Blennius pholis* (L.) blood to intertidal rockpool environments. *J. Exp. Mar. Biol. Ecol.* **77**, 151–167.

Burggren, W. W., and Johansen, K. (1986). Circulation and respiration in lungfishes (dipnoi). *J. Morphol. Suppl.* **1**, 217–236.

Cala, P. (1987). Aerial respiration in the catfish, *Eremophilus mutisii* (Trichomycteridae, Siluriformes), in the Rio Bogota Basin, Colombia. *J. Fish Biol.* **31**, 301–303.

Cala, P., Castillo, B. D., and Garzon, B. (1990). Air-breathing behaviour of the Colombian catfish *Eremophilus mutisii* (Trichomycteridae, Siluriformes). *J. Exp. Biol.* **48**, 357–360.

Carter, G. S. (1935). Reports of the Cambridge expedition to British Guiana, 1933. Respiratory adaptations of the fishes of the forest waters, with descriptions of the accessory respiratory organs of *Electrophorus electricus* (Linn.) (=*Gymnotus electricus* auctt.) and *Plecostomus plecostomus* (Linn). *J. Linn. Soc. London (Zool.)* **39**, 219–233.

Carter, G. S., and Beadle, L. C. (1931). The fauna of the swamps of the Paraguayana Chaco in relation to its environment. II. Respiratory adaptations in the fishes. *J. Linn. Soc. London (Zool.)* **37**, 327–368.

Chabot, D., and Claireaux, G. (2008). Environmental hypoxia as a metabolic constraint on fish: the case of Atlantic cod, *Gadus morhua. Mar. Pollut. Bull.* **57**, 287–294.

Chapman, L. J., and McKenzie, D. J. (2009). Behavioral responses and ecological consequences. In: *Fish Physiology* (J.G Richards, A.P Farrell and C.J Brauner, eds), Vol. 27, pp. 25–77. Academic Press, New York.

Clements, K. D., Raubenheimer, D., and Choat, J. H. (2009). Nutritional ecology of marine herbivorous fishes: ten years on. *Functional Ecology* **23**, 79–92.

Crawford, R. H. (1971). Aquatic and aerial respiration in the bowfin, longnose gar and Alaska blackfish. Ph.D. Thesis, Univ. of Toronto, Toronto, Canada, 202 pp.

Crawford, R. H. (1974). Structure of an air-breathing organ and the swim bladder in the Alaska blackfish, *Dallia pectoralis* Bean. *Can. J. Zool.* **52**, 1221–1225.

Cruz, A. L. d., Pedretti, A. C. E., and Fernandes, M. N. (2009). Stereological estimation of the surface area and oxygen diffusing capacity of the respiratory stomach of the air-breathing armored catfish *Pterygoplichthys anisitsi* (Teleostei: Loricariidae). *J. Morphol.* **270**, 601–614.

Daniels, C. B., and Orgeig, S. (2003). Pulmonary surfactant: the key to the evolution of air breathing. *News Physiol. Sci.* **18**, 151–157.

Delariva, R. L., and Agostinho, A. A. (2001). Relationship between morphology and diets of six neotropical loricariids. *J. Fish Biol.* **58**, 832–847.

de Pinna, M. C. (1993). Higher-level phylogeny of Siluriformes, with a new classification of the order (Teleostei, Ostariophysi). PhD thesis, City University of New York, New York.

Diaz, R. J., and Breitburg, D. L. (2009). The hypoxic environment. In: *Fish Physiology* (J.G Richards, A.P Farrell and C.J Brauner, eds), Vol. 27, pp. 1–23. Academic Press, New York.

Domenici, P., Lefrançois, C., and Shingles, A. (2007). Hypoxia and the antipredator behaviours of fishes. *Phil. Trans. R. Soc. B* **362**, 2105–2121.

Fänge, R., and Grove, D. (1979). Digestion. In: *Fish Physiology* (W.S Hoar and D.J Randall, eds), Vol. VIII, pp. 161–260. Academic Press, New York.

Fernandes, M. N., and Perna, S. A. (1995). Internal morphology of the gill of a loricariid fish, *Hypostomus plecostomus*: arterio-arterial vasculature and muscle organization. *Can. J. Zool.* **73**, 2259–2265.

Fernandes, M. N., Sanches, J. R., Matsuzaki, M., Panepucci, L., and Rantin, F. T. (1999). Aquatic respiration in facultative air-breathing fish: effects of temperature and hypoxia. In: *Biology of Tropical Fishes* (A.L. Val and V.M. Almeida-Val, eds), pp. 341–352. INPA, Manaus.

Focesi, A., Brunori, M., Bonaventura, J., Wilson, M. T., and Galdames-Portus, M. I. (1979). Effect of pH on the kinetics of oxygen and carbon dioxide reactions with hemoglobin from the air-breathing fish *Loricariichthys. Comp. Biochem. Physiol.* **62A**, 169–171.

Frick, N. T., and Wright, P. A. (2002). Nitrogen metabolism and excretion in the mangrove Killifish *Rivulus marmoratus* II: Significant ammonia volatilisation in a teleost during air-exposure. *J. Exp. Biol.* **205**, 91–100.

Froese, R., and D. Pauly (eds). (2010) *FishBase*. Online: www.fishbase.org [03/2010].

Fry, F. E. J. (1971). The effect of environmental factors on the physiology of fish. In: *Fish Physiology* (W.S. Hoar and D.J. Randall, eds), Vol. VI, pp. 1–98. Academic Press, New York.

Gee, J. H. (1976). Buoyancy and aerial respiration: factors influencing the evolution of reduced swim-bladder volume of some Central American catfishes (Trichomycteridae, Callichthyidae, Loricariidae, Astroblepidae). *Can. J. Zool.* **57**, 1030–1037.

Gee, J. H., and Graham, J. B. (1978). Respiratory and hydrostatic functions of the intestine of the catfishes *Hoplosternum thoracatum* and *Brochis splendens* (Callichthyidae). *J. Exp. Biol.* **74**, 1–16.

Glass, M. L., and Rantin, F. T. (2009). Gas exchange and control of respiration in air-breathing teleost fish. In: *Cardio-respiratory Control in Vertebrates* (M.L. Glass and S.C. Wood, eds), pp. 99–119. Springer-Verlag, Berlin, Heidelberg.

Gnaiger, E. (1993). Efficiency and power strategies under hypoxia. Is low efficiency at high glycolytic ATP production a paradox? In: *Surviving Hypoxia: Mechanisms of Control and Adaptation* (P.W. Hochachka, P.L. Lutz, T.J. Sick and M. Rosenthal, eds), pp. 77–110. Academic Press, New York.

Gonçalves, A., Castro, L., Pereira-Wilson, C., Coimbra, J., and Wilson, J. (2007). Is there a compromise between nutrient uptake and gas exchange in the gut of *Misgurnus anguillicaudatus*, an intestinal air-breathing fish?. *Comparative Biochemistry and Physiology Part D: Genomics and Proteomics* **2**(4), 345–355.

Gould, S. J., and Vrba, E. S. (1982). Exaptation—a missing term in the science of form. *Paleobiology* **8**, 4–15.

Gradwell, N. (1971). A photographic analysis of the air breathing behavior of the catfish, *Plecostomus punctatus*. *Can. J. Zool.* **49**, 1089–1094.

Graham, J. B. (1983). The transition to air breathing in fishes: II. Effects of hypoxia acclimation on the bimodal gas exchange of *Ancistrus chagresi* (Loricariidae). *J. Exp. Biol.* **102**, 157–173.

Graham, J. B. (1985). Seasonal and environmental effects on the blood hemoglobin concentrations of some Panamanian air-breathing fishes. *Env. Biol. Fishes* **12**, 291–301.

Graham, J. B. (1997). *Air-Breathing Fishes: Evolution, Diversity, and Adaptation*. Academic Press, San Diego.

Graham, J. B. (1999). Comparative aspects of air-breathing fish biology: an agenda for some Neotropical species. In: *The Biology of Tropical Fishes* (A.L. Val and V.M.F. Almeida-Val, eds), pp. 317–332. INPA, Manaus.

Graham, J. B., and Baird, T. A. (1982). The transition to air breathing in fishes: I. Environmental effects on the facultative air breathing of *Ancistrus chagresi* and *Hypostomus plecostomus* (Loricariidae). *J. Exp. Biol.* **96**, 53–67.

Graham, J. B., and Lee, H. J. (2004). Breathing air in air: in what ways might extant amphibious fish biology relate to prevailing concepts about early tetrapods, the evolution of vertebrate air breathing and the vertebrate land transition? *Physiol. Biochem. Zool.* **77**, 720–731.

Holeton, G. F. (1980). Oxygen as an environmental factor of fishes. In: *Environmental Physiology of Fishes* (M.A. Ali, ed.), pp. 7–32. Plenum Publishing Corporation, New York.

Huebner, E., and Chee, G. (1978). Histological and ultrastructural specialization of the digestive tract of the intestinal air breather *Hoplosternum thoracatum* (Teleost). *J. Morphol.* **157**, 301–328.

Ip, Y. K., Chew, S. F., and Randall, D. J. (2004). Five tropical air breathing fishes, six different strategies to defend against ammonia toxicity on land. *Physiol. Biochem. Zool.* **77**, 768–782.

Jasinski, A. (1973). Air–blood barrier in the respiratory intestine of the pond-loach, *Migurnus fossilis. L. Acta. Anat.* **86**, 376–393.

Jobert, M. (1877). Recerces pour servir a l'histoire de la respiration chez les poissons. *Annales des sciences naturelles (Zoologie et paleontologie)* **5**, 1–4.

Johansen, K. (1970). Air breathing in fishes. In: *Fish Physiology Series* (W.S. Hoar and D.J. Randall, eds), Vol. IV, pp. 361–411. Academic Press, New York.

Jucá-Chagas, R. (2004). Air breathing of the neotropical fishes *Lepidosiren paradoxa*, *Hoplerythrinus unitaeniatus* and *Hoplosternum littorale* during aquatic hypoxia. *Comp. Biochem. Physiol. A: Molecular & Integrative Physiology* **139**, 49–53.

Junk, W. J. (1984). Ecology of the varzea, floodplain of Amazonian whitewater rivers. In: *The Amazon. Limnology and Landscape Ecology of a Mighty Tropical River and its Basin* (H. Sioli, ed.), pp. 215–244. Dr. W. Junk Publishers, Dordrecht.

Kihara, M., and Sakata, T. (1997). Fermentation of dietary carbohydrates to short-chain fatty acids by gut microbes and its influence on intestinal morphology of a detritivorous teleost tilapia (*Oreochromis niloticus*). *Comp. Biochem. Physiol.* **118A**, 1201–1207.

Kohda, M., Tanimura, M., Nakamura, M. K., and Yamagishi, S. (1995). Sperm drinking by female catfishes: a novel mode of insemination. *Environ. Biol. Fishes* **42**, 1–6.

Kramer, D. L. (1987). Dissolved oxygen and fish behavior. *Environ. Biol. Fishes* **18**, 81–92.

Kramer, D. L., and Braun, E. A. (1983). Short-term effects of food availability on air-breathing frequency in the fish *Corydoras aeneus* (Callichthyidae). *Can. J. Zool.* **6**, 1964–1967.

Kramer, D. L., and Graham, J. B. (1976). Synchronous air breathing, a social component of respiration in fishes. *Copeia* **4**, 689–697.

Kramer, D. L., and McClure, M. (1980). Aerial respiration in the catfish, *Corydoras aeneus* (Callichthyidae). *Can. J. Zool.* **58**, 1984–1991.

Kramer, D. L., Manley, D., and Bourgeois, R. (1983). The effect of respiratory mode and oxygen concentration on the risk of aerial predation in fishes. *Can. J. Zool.* **61**, 653–665.

Laming, P. R., Funston, C. W., and Armstrong, M. J. (1982). Behavioural, physiological, and morphological adaptations of the shanny (*Blennius pholis*) to the intertidal habitat. *J. Mar. Biol. Assn. U.K.* **62**, 329–338.

Liem, K. F. (1967). Functional morphology of the integumentary, respiratory and digestive systems of the synbranchoid fish, *Monopterus albus. Copeia* **2**, 375–388.

Liem, K. F. (1989). Respiratory gas bladders in teleosts: functional conservatism and morphological diversity. *Am. Zool.* **29**, 333–352.

Lobel, P. S. (1981). Trophic biology of herbivorous reef fish: alimentary pH and digestive capabilities. *J. Fish Biol.* **19**, 365–397.

MacCormack, T. J., McKinley, R. S., Roubach, R., Almeida-Val, V. M. F., Val, A. L., and Driedzic, W. R. (2006). Changes in ventilation, metabolism, and behaviour, but not bradycardia, contribute to hypoxia survival in two species of Amazonian armoured catfish. *Can. J. Zool.* **81**, 272–280.

Marcon, J. L., Chagas, E. C., Kavassaki, J. M., and Val, A. L. (1999). Intraerythrocytic phosphates in 25 fish species of the Amazon: GTP as a key factor in the regulation of Hb-O_2 affinity. In: *The Biology of Tropical Fishes* (A.L. Val and V.M.F. Almeida-Val, eds), pp. 229–240. INPA, Manaus.

Mattias, A. T., Rantin, F. T., and Fernandes, M. N. (1998). Gill respirometry parameters during progressive hypoxia in the facultative air-breathing fish, *Hypostomus regani* (Loricariidae). *Comp. Biochem. Physiol.* **120A**, 311–315.

McMahon, B. R., and Burggren, W. W. (1987). Respiratory physiology of intestinal air breathing in the teleost fish *Misgurnus anguillicaudatus. J. Exp. Biol.* **133**, 371–393.

McPhail, J. D. and Lindsey, C. C. (1970). Freshwater fishes of northwestern Canada and Alaska. *Bull. Fish. Res. Board Can.* No. 173.

Moitra, A., Singh, O., and Munshi, J. (1989). Microanatomy and cytochemistry of the gastro-respiratory tract of an air-breathing cobitidid fish, *Lepidocephalichthys guntea. Japanese Journal of Ichthyology* **36**, 227–231.

Moreira-Silva, J., Tsui, T. K. N., Coimbra, J., Vijayan, M. M., Ip, Y. K., and Wilson, J. M. (2010). Branchial ammonia excretion in the Asian weatherloach *Misgurnus anguillicaudatus*. *Comp. Biochem. Physiol. C* **151**, 40–50.

Morris, S., and Bridges, C. R. (1994). Properties of respiratory pigments in bimodal breathing animals: air and water breathing by fish and crustaceans. *Amer. Zool.* **34**, 216–228.

Nelson J. A. and Chabot, D. 2010. Energy consumption: metabolism (general). In: *Encyclopedia of Fish Physiology: From Genome to Environment*. Chapter 148. In press.

Nelson, J. A., Wubah, D. A., Whitmer, M. E., Johnson, E. A., and Stewart, D. J. (1999). Wood-eating catfishes of the genus *Panaque*: gut microflora and cellulolytic enzyme activities. *J. Fish Biol.* **54**, 1069–1082.

Nelson, J. A., Rios, F. S. A., Sanches, J. R., Fernandes, M. N., and Rantin, F. T. (2007). Environmental influences on the respiratory physiology and gut chemistry of a facultatively air-breathing, tropical herbivorous fish *Hypostomus regani* (Ihering, 1905). In: *Fish Respiration and the Environment* (M.N. Fernandes, M.L. Glass and B.G. Kapoor, eds), pp. 191–217. Science Publisher Inc., Enfield.

Nilsson, G. E. (2001). Surviving anoxia with the brain turned on. *News Physiol. Sci.* **16**, 217–221.

Oliveira, C. A. D., Taboga, S. R., Smarra, A. L., and Bonilla-Rodriguez, G. O. (2001). Microscopical aspects of accessory air breathing through a modified stomach in the armoured catfish *Liposarcus anisitsi* (Siluriformes, Loricariidae). *Cytobios* **105**, 153–162.

Oliveira, R. B., Lopes, J. M., Sanchez, J. R., Kalinin, A. L., Glass, M. L., and Rantin, F. T. (2004). Cardio-respiratory responses of the facultative air-breathing fish jeju, *Hoplerythrinus unitaeniatus* (Teleostei, Erythrinidae), exposed to graded ambient hypoxia. *Comp. Biochem. Physiol. A* **139**, 479–485.

Olson, K. R. (1994). Circulatory anatomy in bimodally breathing fish. *Amer. Zool.* **34**, 280–288.

Park, J. Y., and Kim, I. S. (2001). Histology and mucin histochemistry of the gastrointestinal tract of the mud loach, in relation to respiration. *J. Fish Biol.* **58**, 861–872.

Pelster, B., Bridges, C. R., and Grieshaber, M. K. (1988). Physiological adaptations of the intertidal rockpool teleost *Lipophrys pholis* L. to aerial exposure. *Resp. Physiol.* **71**, 355–374.

Perna, S. A., and Fernandes, M. N. (1996). Gill morphometry of the facultative air-breathing loricariid fish, *Hypostomus plecostomus* (Walbaum) with special emphasis on aquatic respiration. *Fish Physiol. Biochem.* **15**, 213–220.

Perry, S. F. (2007). Swimbladder-lung homology in basal Osteichthyes revisited. In: *Fish Respiration and the Environment* (M.N. Fernandes and F.T. Rantin, eds), pp. 41–54. Science Publisher Inc., Enfield.

Persaud, D. I., Ramnarine, I. W., and Agard, J. B. R. (2006a). Ontogeny of the alimentary canal and respiratory physiology of larval *Hoplosternum littorale* (Hancock, 1828): an intestinal air-breathing teleost. *Environ. Biol. Fishes* **76**, 37–45.

Persaud, D. I., Ramnarine, I. W., and Agard, J. B. R. (2006b). Trade-off between digestion and respiration in two airbreathing callichthyid catfishes *Holposternum littorale* (Hancock) and *Corydoras aeneus* (Gill). *Environ. Biol. Fishes* **76**, 159–165.

Podkowa, D., and Goniakowska-Witalińska, L. (2002). Adaptations to the air breathing in the posterior intestine of the catfish (*Corydoras aeneus*, Callichthyidae). A histological and ultrastructural study. *Folia Biol.* **50**, 69–82.

Podkowa, D., and Goniakowska-Witalińska, L. (2003). Morphology of the air-breathing stomach of the catfish *Hypostomus plecostomus*. *J. Morphol.* **257**, 147–163.

Power, M. E. (1984). Depth distributions of armored catfish: predator-induced resource avoidance? *Ecology* **65**, 523–528.

Powers, D. A., Fyhn, H. J., Fyhn, U. E. H., Martin, J. P., Garlick, R. L., and Wood, S. C. (1979). A comparative study of the oxygen equilibria of blood from 40 genera of Amazonian fishes. *Comp. Biochem. Physiol.* **62A**, 67–85.

Riegler, A. (2008). Natural or internal selection? The case of canalization in complex evolutionary systems. *Artificial Life* **14**, 345–362.

Riggs, A. (1979). Studies of the hemoglobins of Amazonian fishes: an overview. *Comp. Biochem. Physiol.* **62A**, 257–272.

Satora, L. (1998). Histological and ultrastructural study of the stomach of the air-breathing *Ancistrus multispinnis* (Siluriformes, Teleostei). *Can. J. Zool.* **76**, 83–86.

Satora, L., and Winnicki, A. (2000). Stomach as an additional respiratory organ, as exemplified by *Ancistrus multispinnis* (Couvier et Valenciennes, 1937), Siluriformes, Teleostei. *Acta Ichthyol. Piscatoria* **30**, 73–79.

Schaefer, S. A., and Lauder, G. V. (1986). Historical transformation of functional design: evolutionary morphology of feeding mechanisms in loricarioid catfishes. *Syst. Biol.* **35**, 489–508.

Schmidt-Nielsen, K. (1997). *Animal Physiology: Adaptation and Environment* (5th edition). Cambridge University Press, Cambridge.

Schultze, H. P. (1999). The fossil record of the intertidal zone. In: *Intertidal Fishes: Life in Two Worlds* (M.H. Horn, K.L.M. Martin and M.A. Chotkowski, eds), pp. 373–392. Academic Press, San Diego.

Silva, J. M., Hernandez-Blazquez, F. J., and Julio, H. F. (1997). A new accessory respiratory organ in fishes: morphology of the respiratory purses of *Loricariichthys platymetopon* (Pisces, Loricariidae). *Annales des sciences naturelles, Zoologie, Paris* **18**, 93–103.

Souza, A. M., and Intelizano, W. (2000). Anatomy and histology of the stomach in some species of Loricariidae (Siluriformes, Ostariophysi). *Publs. Avulsas do Instituto Pau Brasil* **10**, dez. 2.

Sloman, K. A., Sloman, R. D., De Boeck, G., Scott, G. R., Iftikar, F. I., Wood, C. M., Almeida-Val, V. M. F., and Val, A. L. (2009). The role of size in synchronous air breathing of *Hoplosternum littorale*. *Physiol. Biochem. Zool.* **82**, 625–634.

Stecyk, J. A. W., and Farrell, A. P. (2002). Cardiorespiratory responses of the common carp (*Cyprinus carpio*) to severe hypoxia at three acclimation temperatures. *J. Exp. Biol.* **205**, 759–768.

Taylor, E. W. (1992). Nervous control of the heart and cardiorespiratory interactions. In: *Fish Physiology* (W.S. Hoar, D.J. Randall and A.P. Farrell, eds), Vol. XIIB, pp. 343–389. Academic Press, San Diego.

Thorarensen, H., and Farrell, A. P. (2006). Postprandial intestinal blood flow, metabolic rates, and exercise in chinook salmon (*Oncorhynchus tshawytscha*). *Physiol. Biochem. Zool.* **79**, 688–694.

Tsui, T. K. N., Randall, D. J., Chew, S. F., Jin, Y., Wilson, J. M., and Ip, Y. K. (2002). Accumulation of ammonia in the body and NH_3 volatilization from alkaline regions of the body surface during ammonia loading and exposure to air in the weather loach *Misgurnus anguillicaudatus*. *J. Exp. Biol.* **205**, 651–659.

Val, A. L., and Almeida-Val, V. M. F. (1995). *Fishes of the Amazon and their Environments. Physiological and Biochemical Features*. Springer, Heidelberg.

Val, A. L., Almeida-Val, V. M. F., and Affonso, E. G. (1990). Adaptive features of Amazonian fishes: hemoglobins, hematology, intraerythrocytic phosphates and whole blood Bohr effect of *Pterygoplichthyes multiradiatus* (Siluriformes). *Comp. Biochem. Physiol.* **97B**, 435–440.

van Soest, P. J. (1994). *Nutritional Ecology of the Ruminant* (2nd edition). Cornell Univ. Press, Ithaca.

Weber, R. E., Wood, S. C., and Davis, B. J. (1979). Acclimation to hypoxic water in facultative air-breathing fish: blood oxygen affinity and allosteric effectors. *Comp. Biochem. Physiol.* **62A**, 125–129.

Weibel, E. W. (1984). *The Pathway for Oxygen.* Harvard Univ. Press, Cambridge.

Wells, R. M. G. (1990). Hemoglobin physiology in vertebrate animals: a cautionary approach to adaptationist thinking. In: *Advances in Comparative and Environmental Physiology* (R.G. Boutilier, ed.), pp. 143–161. Springer-Verlag, Berlin.

Wetzel, R. G. (2001). *Limnology. Lake and River Ecosystems* (3rd edition). Academic Press, London.

Wilhelm, D., and Weber, R. E. (1983). Functional characterization of hemoglobins from south Brazilian freshwater teleosts I. Multiple hemoglobins from the gut/gill breather, *Callichthys callichthys. Comp. Biochem. Physiol.* **75A**, 475–482.

Withers, P. C. (1992). *Comparative Animal Physiology* (1st edition). Saunders College, Fort Worth, TX.

Wood, S. C., Weber, R. E., and Davis, B. J. (1979). Effects of air-breathing on acid–base balance in the catfish, *Hypostomus* sp. *Comp. Biochem.Physiol.* **62A**, 185–187.

Wu, H. W., and Chang, H. (1945). On the structures of the intestine of the Chinese pond loach with special reference to its adaptation for aerial respiration. *Sinensia* **16**, 1–8.

Yadav, A. N., and Singh, B. R. (1980). The gut of an intestinal air-breathing fish, *Lepidocephalus guntea* (Ham). *Arch. Biol. (Bruxelles)* **91**, 413–422.

INDEX

OTHER VOLUMES IN THE
FISH PHYSIOLOGY SERIES

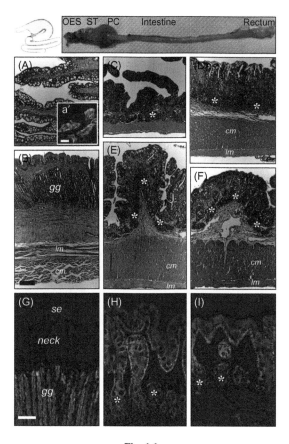

Fig. 1.1.

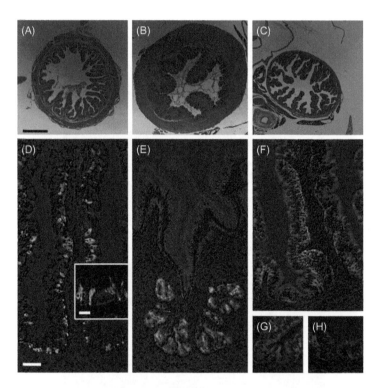

Fig. 1.2.

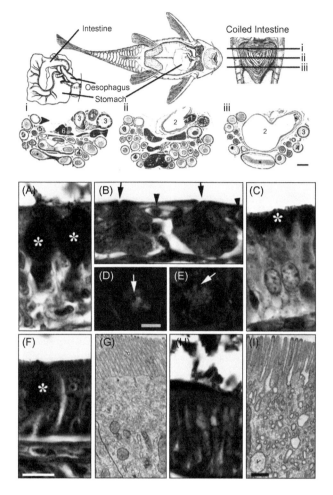

Fig. 1.3.

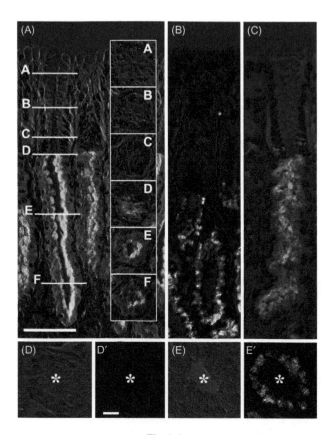

Fig. 1.4.

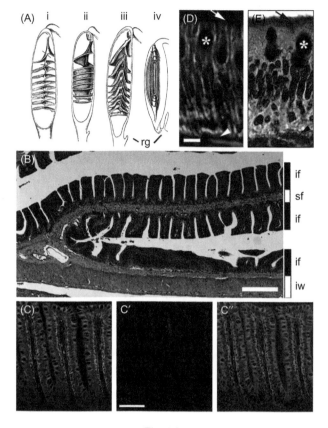

Fig. 1.5.

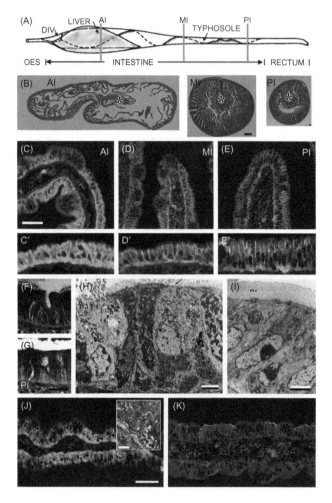

Fig. 1.6.

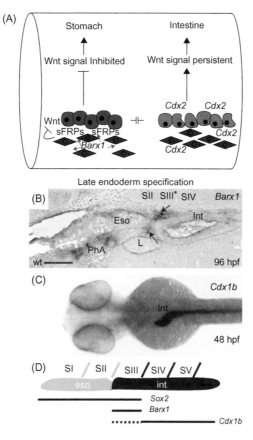

Fig. 1.8.

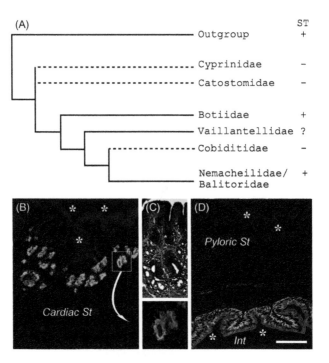

Fig. 1.9.

	Ingestion	Branchial			Rectal	Renal
		Efflux	Influx	Net flux		
Volume	2.6 ml/kg/h			−1.8 ml/kg/h	−0.7 ml/kg/h	−0.1 ml/kg/h
Na^+	1092	−7364	6296	−1068	−19	−5
Cl^-	1274	−21597	20394	−1203	−47	−24
Mg^{2+}	130			5	−120	−15
SO_4^{2-}	78			−9	−63	−6
Ca^{2+}	26			−21.5	−4	−0.5
K^+	26			−24.5	−1	−0.5
HCO_3^- eqv	6			230	−68	−0.3

Fig. 4.2.

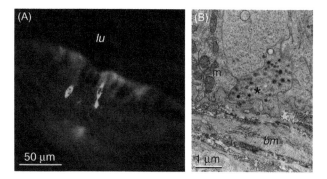

Fig. 7.2.

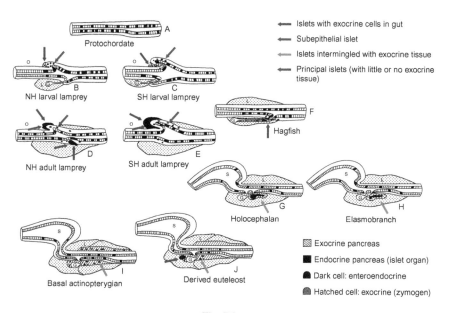

Fig. 7.3.

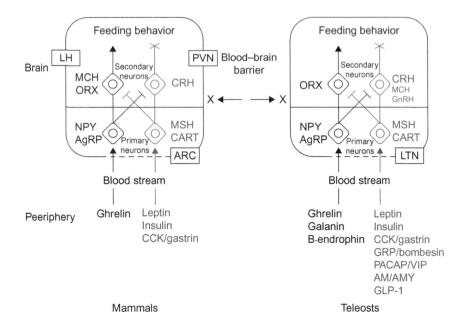

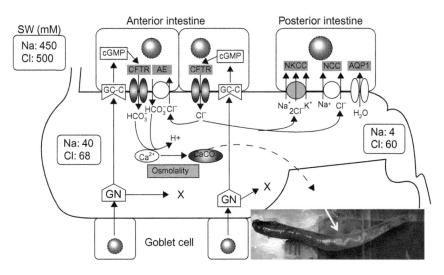

Fig. 7.5.

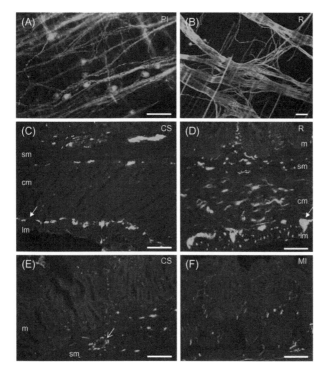

Fig. 8.1.

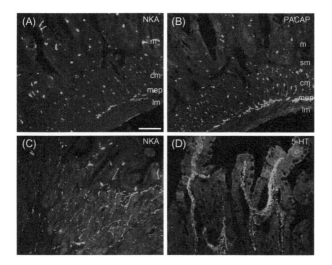

Fig. 8.3.

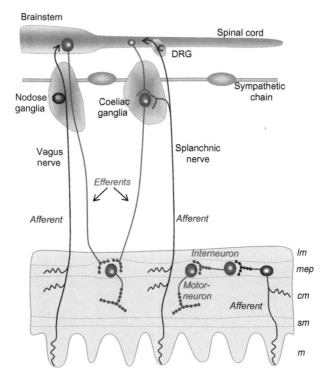

Fig. 8.4.

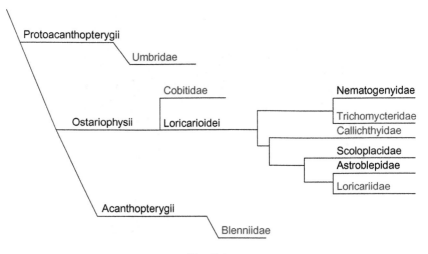

Fig. 10.1.

Printed and bound by CPI Group (UK) Ltd, Croydon, CR0 4YY

03/10/2024

01040415-0014